Plasma Kinetic Theory

Series in Plasma Physics

Series Editor:
Steve Cowley, Imperial College, UK and UCLA, USA

Other recent books in the series:

High Power Microwaves, *Second Edition*
J Benford, J A Swegle and E Schamiloglu

Controlled Fusion and Plasma Physics
K Miyamoto

Plasma Electronics: Applications in Microelectronic Device Fabrication
T Makabe and Z Petrović

An Introduction to Inertial Confinement Fusion
S Pfalzner

Aspects of Anomalous Transport in Plasmas
R Balescu

Non-Equilibrium Air Plasmas at Atmospheric Pressure
K H Becker, R J Barker and K H Schoenbach (Eds)

Magnetohydrodynamic Waves in Geospace: The Theory of ULF Waves and their Interaction with Energetic Particles in the Solar-Terrestrial Environment
A D M Walker

Plasma Physics via Computer Simulation (paperback edition)
C K Birdsall and A B Langdon

Plasma Waves, Second Edition
D G Swanson

Microscopic Dynamics of Plasmas and Chaos
Y Elskens and D Escande

Plasma and Fluid Turbulence: Theory and Modelling
A Yoshizawa, S-I Itoh and K Itoh

The Interaction of High-Power Lasers with Plasmas
S Eliezer

Introduction to Dusty Plasma Physics
P K Shukla and A A Mamun

The Theory of Photon Acceleration
J T Mendonça

Laser Aided Diagnostics of Plasmas and Gases
K Muraoka and M Maeda

Series in Plasma Physics

Plasma Kinetic Theory

D. G. Swanson
Auburn University
Alabama, USA

CRC Press
Taylor & Francis Group
Boca Raton London New York

CRC Press is an imprint of the
Taylor & Francis Group, an **informa** business
A CHAPMAN & HALL BOOK

Chapman & Hall/CRC
Taylor & Francis Group
6000 Broken Sound Parkway NW, Suite 300
Boca Raton, FL 33487-2742

ISBN-13: 978-1-4200-7580-9 (hbk)

Library of Congress Cataloging-in-Publication Data

Swanson, D. G. (Donald Gary), 1935-
Plasma kinetic theory / Donald Gary Swanson.
p. cm. -- (Series in plasma physics ; 23)
Includes bibliographical references and index.
ISBN 978-1-4200-7580-9 (hardback : alk. paper)
1. Plasma dynamics. 2. Kinetic theory of gases. I. Title. II. Series.

QC718.5.T5S93 2008
530.4'4--dc22 2008009175

Visit the Taylor & Francis Web site at
http://www.taylorandfrancis.com

and the CRC Press Web site at
http://www.crcpress.com

Contents

Preface

The first part of this book grew out of the lectures from the principal graduate plasma physics course at Auburn University in 1989-1990. Significant additions were included in 1996. The principal text for this first part was the assigned text by Davidson, **Introduction to Plasma Theory**, and the early chapters generally follow that text. The portion relating to the kinetic theory of plasma waves and Appendix A is largely drawn from my text, **Plasma Waves, 2nd Edition**. The last three chapters and Appendix B is largely from W. Marshall's **Kinetic Theory of an Ionized Gas**, parts I, II, and III (unpublished, issued as A.E.R.E. T/R 2247, 2352, and 2419). I have edited and corrected some parts of these reports and expanded his Appendix. These chapters were included because there is not an easily accessible source for the transport coefficients or their derivation.

The first four chapters and perhaps chapter 6 are suitable for an introductory course in plasma physics, but the other chapters are more appropriate for an intermediate or advanced course in kinetic theory. Chapters 4, 5, and 6 could be used for a course in plasma waves except that cold plasma waves would need to come from another source (check my webpage at electro.physics.auburn.edu/~swanson/ for download). My text, **Plasma Waves, 2nd Edition**, provides a complete source for such a course.

If typographical or other errors are discovered, please report them to swanson@physics.auburn.edu. Errata will be posted on my webpage.

SI units are used throughout, although sometimes the temperature or energy may be given in electron volts. In such cases, a subscript is added, such as T_{eV} or W_{eV}.

D. Gary Swanson
Physics Department
Auburn University, AL

Author

D. Gary Swanson is the principal developer (along with students and coworkers) of mode conversion theory that describes the absorption and emission of wave energy at plasma wave resonances. Swanson has been active in plasma wave research, both experimentally and theoretically, for nearly 50 years.

Born in Los Angeles, he received both a bachelor of theology from Northwest Christian College and a bachelor of science in physics from the University of Oregon in 1958. He continued at the California Institute of Technology, receiving his M.S.(1961) and his Ph.D.(1963) there. He taught for 10 years at the University of Texas at Austin, 6 years at the University of Southern California, and was on the faculty at Auburn University from 1980 until his retirement in 2006, where he is now a Professor Emeritus.

He has written over 50 articles in refereed journals, a chapter in the **Handbook of Plasma Physics** and **Basic Plasma Physics**, articles in two encyclopedias, two textbooks on **Plasma Waves** (original and expanded **2nd Edition**), written a monograph on mode conversion, and an undergraduate text, **Quantum Mechanics – Foundations and Applications**.

A fellow of the American Physical Society, he has refereed for *Physics of Fluids*, *Physics of Plasmas*, the *Journal of Applied Physics*, *Nuclear Fusion*, *Physical Review Letters*, and several others, including being a reviewer for the Department of Energy. He has been a consultant with the Oak Ridge National Laboratory, Los Alamos National Laboratory, the Princeton Plasma Physics Laboratory, and McDonnell-Douglas.

1

INTRODUCTION

1.1 Overview

Before we examine in detail the behavior of plasmas in a wide variety of limits and configurations, we first need to establish what a plasma is and learn some of the basic parameters which characterize a plasma. While one may speak quite generally of a collection of charged particles as being a plasma, there are restrictions which require that the interactions of the charged particles be dominated by electrostatic forces (as opposed to collisions with neutral atoms or boundaries), and that the force on an individual charged particle should be dominated by weak interactions from a large number of particles rather than by stronger interactions with its nearest neighbors. For example, cosmic rays produce ionization in air, but the collection of charged particles so formed is dominated by interactions with neutral molecules so that the charges effectively fail to "see" one another, and hence fail to form a plasma. In another limit, an electron beam is a collection of charges, but their mutual interactions are typically weak compared to the external fields, so only in very dense electron beams do plasma effects begin to play a role. While most plasmas are "quasineutral," which means that the deviation of the charge density from neutrality is a small percentage of the total charge, one-component plasmas do exist, even though the self-consistent electric field is not small.

In order to characterize a plasma, we will nearly always begin without a magnetic field in order to simplify the analysis, but will subsequently include magnetic field effects because of the rich variety of effects and applications. We will begin with a brief examination of a characteristic time (or frequency), length, and velocity, although we will see that any pair determines the third of these plasma fundamental quantities. The addition of a magnetic field will introduce additional frequencies and velocities which will further characterize the plasma state.

1.2 Plasma frequency

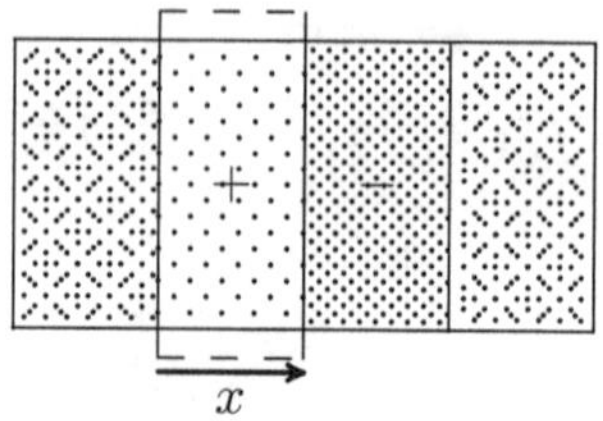

FIGURE 1.1
Electron slab displacement.

In order to estimate the characteristic time for an unmagnetized plasma, we shall displace some charge from equilibrium and estimate the time for the system to return to equilibrium. For this example, we shall imagine moving a slab of electrons to the right as in Figure 1.1, and assume the background ions are immobile. From Gauss' law, where we draw a Gaussian surface as indicated by the dashed lines, we obtain the electric field from the now unneutralized ions as

$$\oint \boldsymbol{D} \cdot \mathrm{d}\boldsymbol{S} = \epsilon_0 E_x A = Q = \rho A x = neAx\,,$$

where A is the area of the displaced charge, n is the particle density (#/m^3), and e is the electron charge ($1.6 \cdot 10^{-19}$ Coul.). From this, the electric field is $E_x = nex/\epsilon_0$. This electric field acts on the displaced electron charge, drawing it back towards equilibrium so that

$$F = QE_x = (-neAx)(nex/\epsilon_0) = nAxm_e \frac{\mathrm{d}^2 x}{\mathrm{d}t^2}\,,$$

or

$$\frac{\mathrm{d}^2 x}{\mathrm{d}t^2} = -\left(\frac{ne^2}{m_e \epsilon_0}\right) x\,. \tag{1.1}$$

This is the simple harmonic oscillator equation with frequency ω_{pe} given by

$$\omega_{pe}^2 = \frac{ne^2}{m_e \epsilon_0}\,, \tag{1.2}$$

and indicates that instead of simply relaxing to equilibrium, the disturbance will oscillate at the *plasma frequency.* The characteristic time is just the inverse of the electron plasma frequency.

Problem 1.1 *Plasma frequency.* Determine the accuracy of the simple formula $f_{pe} = \omega_{pe}/2\pi = 9\sqrt{n_e}$.

1.3 Debye length

There are several ways to infer a characteristic length in a plasma, depending upon which physical principles one wishes to emphasize. That each method of estimation gives essentially the same length is indicative of its fundamental nature.

The first and simplest way is to relate the characteristic time and velocity to find the length. Even though we frequently call a plasma "cold," it still has a mean thermal speed which characterizes the motions of the charges about equilibrium. Neglecting the subtleties between the most probable speed, rms speed, etc., we take the fundamental speed to be $v = \sqrt{k_B T/m}$, so that the length is given by

$$L = vt = \sqrt{k_B T/m}/\omega_{pe} = \sqrt{\frac{\epsilon_0 k_B T}{ne^2}} . \tag{1.3}$$

This length is called the *Debye length*, usually denoted by λ_D, and is fundamental to plasmas and even to the definition of what a plasma is.

The most definitive way to estimate this fundamental length is to investigate what is called "Debye shielding." In order to get an idea of what causes shielding, we imagine that in the presence of a positive/negative test charge q, ions will in general be deflected away/toward the test charge as they stream by, and electrons will be deflected toward/away. The result is that on the average, there will be less positive/negative charge inside any sphere than there would be if the streaming particles were undeflected. From Gauss' law, this means the net charge seen from outside the sphere will be reduced, or that some of the test charge will be "shielded," and the amount of shielding will increase with distance since the effect is cumulative. Debye originally estimated the shielding of ions in an electrolyte. In order to make this estimate, we use Poisson's equation with a test charge at the origin, so that

$$\nabla^2 \varphi = -\frac{\rho}{\epsilon_0} = -\frac{q}{\epsilon_0}\delta(\boldsymbol{r}) - \frac{e}{\epsilon_0}(n_i - n_e) ,$$

where n_i and n_e are the ion and electron densities. These densities will be influenced by the potential, so we estimate the effect by using a Boltzmann factor, such that

$$n_i = n_{i0} \mathrm{e}^{-e\varphi/k_B T_i} \simeq n_{i0}\left(1 - \frac{e\varphi}{k_B T_i}\right)$$
$$n_e = n_{e0} \mathrm{e}^{e\varphi/k_B T_e} \simeq n_{e0}\left(1 + \frac{e\varphi}{k_B T_e}\right) ,$$

where we have assumed that the electrostatic energy is small compared to the thermal energy. If the unperturbed plasma has no net charge, so that $n_{i0} = n_{e0} = n_0$, then Poisson's equation becomes

$$\nabla^2 \varphi - \frac{1}{\lambda_D^2}\varphi = -\frac{q}{\epsilon_0}\delta(\boldsymbol{r}) , \tag{1.4}$$

where

$$\frac{1}{\lambda_D^2} = \frac{n_0 e^2}{\epsilon_0}\left(\frac{1}{k_B T_i} + \frac{1}{k_B T_e}\right) . \tag{1.5}$$

The solution of this equation is

$$\varphi(r) = \frac{q}{4\pi\epsilon_0 r} e^{-r/\lambda_D} . \tag{1.6}$$

While equation (1.5) gives the definitive expression for the Debye length, it still has inadequacies when we begin to consider what a test charge means. The calculation is made for a test charge at rest only, but we may want to know what electron potentials are like, and while we may easily imagine electrons shielding other electrons, the thermal speed of electrons is so much higher than that of ions and the ion mass is so much larger, that it is difficult to imagine how ions could screen electrons. We also expect the faster electrons in the distribution to be less shielded than the slower electrons, so we have probably overestimated the shielding for a typical plasma particle, which is not at rest. One way to estimate this effect is to assume that we have a thermal distribution of test particles of specified mass and temperature, and have velocity $v_t = \sqrt{k_B T_t/m_t}$. This idea is developed in Section 4.3.2. Averaging the screening over the distribution, to lowest order, one finds from equation (4.65)

$$k_D^2 \equiv \frac{1}{\lambda_D^2} = \frac{\omega_{pe}^2}{v_e^2 + v_t^2} + \frac{\omega_{pi}^2}{v_i^2 + v_t^2} . \tag{1.7}$$

If the test particles are electrons, then $v_t^2 \gg v_i^2$ and the ion terms drop out (order m_e/m_i), while on the average, the electron shielding is only half as effective as for a test particle at rest. If the test particles are ions, then the electron contribution is undiminished (to order m_e/m_i) while the ion self-screening is halved. This calculation is not rigorous, but supports our intuitive expectations.

From this understanding of the fundamental length, it follows that a plasma must be many Debye lengths in overall dimensions. This is because we require external fields to be largely excluded from a plasma, and any external electric field will be shielded over this length scale. A collection of charged particles which is smaller in extent than this length will be dominated by the external fields, minimizing the mutual interactions, and while we may have a many-body problem, we do not have a plasma.

Problem 1.2 *Debye potential.* Show that the potential of equation (1.6) is a solution of Poisson's equation, equation (1.4), for $r \neq 0$ *and* $r = 0$.

1.4 Plasma parameter

Having estimated the characteristic length for a plasma, we are now in a position to define a plasma more definitively. If the electrostatic potential of an individual particle extends about a Debye length, and we require a particle to be influenced more by many distant particles than a few nearest neighbors, we must have many particles in a Debye sphere, or

$$N_D = nV_D = \frac{4\pi}{3} n\lambda_D^3 \gg 1\,. \tag{1.8}$$

N_D is called the *plasma parameter*, although sometimes the numerical factor is ignored. As will be clear in Section 1.7 when we study collisions more closely, the condition that the plasma parameter be large is a sufficient condition that the trajectory of a single particle will be relatively smooth, making a finite deflection via a succession of many small deflections rather than a few large ones as occurs in interactions among neutral particles.

In deriving the Debye shielding potential, we assumed that the electrostatic energy was small compared to the thermal energy. In order to estimate this ratio, we take the thermal energy per particle to be $\frac{3}{2}k_BT$ and the mean electrostatic energy per particle to be $e^2/4\pi\epsilon_0\lambda_D$, so the ratio is

$$\frac{e^2}{4\pi\epsilon_0\lambda_D} \Big/ \frac{3}{2}k_BT = \frac{2}{9N_D}\,.$$

This result shows that the largeness of the plasma parameter guarantees the validity of the Debye potential for a stationary test charge.

1.5 Distribution functions

Because of the enormous number of particles in a typical plasma, we never try to describe the motion of each of the particles (although certain plasma simulation computer codes do try to follow the motion of pseudo-particles which represent many real particles), but resort to describing them by a *distribution function*, $f(\boldsymbol{r}, \boldsymbol{v}, t)$, which gives some description of "typical" particles and allows the equilibrium to be characterized. Although rarely in true equilibrium, we most commonly describe the plasma by a Maxwellian distribution,

$$f(\boldsymbol{v}) = \left(\frac{m}{2\pi k_BT}\right)^{\frac{3}{2}} \mathrm{e}^{-mv^2/2k_BT}\,, \tag{1.9}$$

which is normalized to unity so that $f(v)\,dv$ represents the probability of finding a particle with velocity v between v and $v + dv$. This is frequently normalized to the number density of particles, so that one obtains the number of particles with velocity v between v and $v + dv$, but we shall try to be consistent with the definition given above.

We frequently will need to distinguish between various species, and shall use the notation

$$\Delta n_s = \bar{n}_s f_s(\boldsymbol{r}, \boldsymbol{v})$$

where $\bar{n}_s = N_s/V$ is the mean density per unit volume, with N_s being the number of particles of species s in volume V. In this expression, Δn_s is the number of particles of species s per unit volume between $\boldsymbol{r}$ and $\boldsymbol{r} + \mathrm{d}\boldsymbol{r}$ having velocity $\boldsymbol{v}$ between $\boldsymbol{v}$ and $\boldsymbol{v} + \mathrm{d}\boldsymbol{v}$, and

$$\int\int \bar{n}_s f(\boldsymbol{r}, \boldsymbol{v})\, \mathrm{d}\boldsymbol{r}\mathrm{d}\boldsymbol{v} = N_s\,. \tag{1.10}$$

1.5.1 Moments of the distribution function

Using the distribution function, we can evaluate the mean kinetic energy of a plasma, with the result

$$\int \tfrac{1}{2} m v^2 f\, \mathrm{d}\boldsymbol{v} = \frac{3}{2} k_B T\,. \tag{1.11}$$

This is exact for a Maxwellian distribution, and is effectively a definition of temperature. If the distribution is not Maxwellian, then the system is not in equilibrium, and there is no true temperature, but if the distribution is a *quasi-equilibrium*, or one which is near equilibrium and very slowly changing in time, then we may use the integral above to define an effective temperature. If the actual distribution is not close to a Maxwellian, then the temperature determined by this method will not be a good estimate of temperature, since none exists, but it is still useful to think of the mean kinetic energy.

Other important features of a plasma described by a distribution function are given by the various moments. In fact, a complete description may be given either by a distribution function or by *all* of the moments. Since all of the moments represent an infinitely complex system, we usually restrict ourselves to the first three or four, such as

$$n_s = \bar{n}_s \int f_s\, \mathrm{d}\boldsymbol{v}\,, \tag{1.12}$$

$$\langle \boldsymbol{v}_s \rangle = \frac{\int \boldsymbol{v} f_s\, \mathrm{d}\boldsymbol{v}}{\int f_s\, \mathrm{d}\boldsymbol{v}}\,, \tag{1.13}$$

$$\mathsf{P}_s = \frac{n_s m_s \int (\boldsymbol{v} - \langle \boldsymbol{v}_s \rangle)(\boldsymbol{v} - \langle \boldsymbol{v}_s \rangle) f_s\, \mathrm{d}\boldsymbol{v}}{\int f_s\, \mathrm{d}\boldsymbol{v}}\,, \tag{1.14}$$

$$\vdots$$

where the first gives the density, the second the mean velocity, and the third the pressure tensor. The next higher moment is the heat flux tensor, but it becomes harder and harder to give concrete physical interpretations to the higher moments. These parameters give a *macroscopic* description, whereas the distribution function gives a *microscopic* description of the plasma. We note that for an isotropic distribution at rest, the pressure tensor reduces to a scalar pressure $p_s = n_s k_B T_s$ times a unit tensor, so we can neglect the tensor character of the pressure.

1.5.2 Magnetic pressure

In many plasma examples, a magnetic field is either imbedded in the plasma or surrounds the plasma boundary. From the Maxwell stress tensor, a magnetic field may be characterized as having a pressure of $p_m = B^2/2\mu_0$ N/m^2 perpendicular to the local magnetic field. Since a plasma is characteristically diamagnetic, a sudden application of a magnetic field will subject the plasma to a pressure differential at the boundary which is balanced with the plasma pressure in equilibrium. When the magnetic field is imbedded in the plasma, it is useful to compare the ratio of these pressures, which is given by the parameter

$$\beta = \frac{n_i k_B T_i + n_e k_B T_e}{B^2/2\mu_0} \,. \tag{1.15}$$

This *plasma* β should not be confused with the relativistic $\beta = v/c$. For confinement of a plasma, a *low* β plasma means that the plasma pressure is only a perturbation (though not necessarily unimportant) to the equilibrium, while in a *high* β plasma, the plasma pressure plays a dominant role in the equilibrium and its stability.

1.6 Cyclotron frequencies

With the addition of a magnetic field a plasma adds two or more additional characteristic time scales or frequencies, and some additional characteristic lengths. These are the frequencies of the gyro motion of the charged particles as they rotate about a local field line, and the corresponding radii of these gyrations are the fundamental lengths. The cyclotron frequencies are given by $\omega_c = qB/m$ and the corresponding Larmor orbit radii are given by $\rho_L = v_t/\omega_c$, with $v_t \equiv \sqrt{2k_B T/m}$. Some authors let ω_c be a positive or negative quantity, depending on the sign of the charge, but we shall adopt the convention that $\omega_c = |qB/m|$ and $\epsilon\omega_c = qB/m$ or $\epsilon = q/|q| = \pm 1$.

1.7 Collisions

Although we usually like to describe plasmas where collisions play a minor role, we cannot shove collisions under a rug and ignore them because they are fundamental to all relaxation processes, and put limits on the validity of all analyses that ignore them. We have also required that individual particle motions be determined by many small deflections from many distant particles rather than a large angle binary collision which seems to be much simpler to comprehend. Thus collisions play a role in our very concept of what a plasma is, and hence we must carefully analyze the physics of the collision process in plasmas.

1.7.1 Binary Coulomb collisions

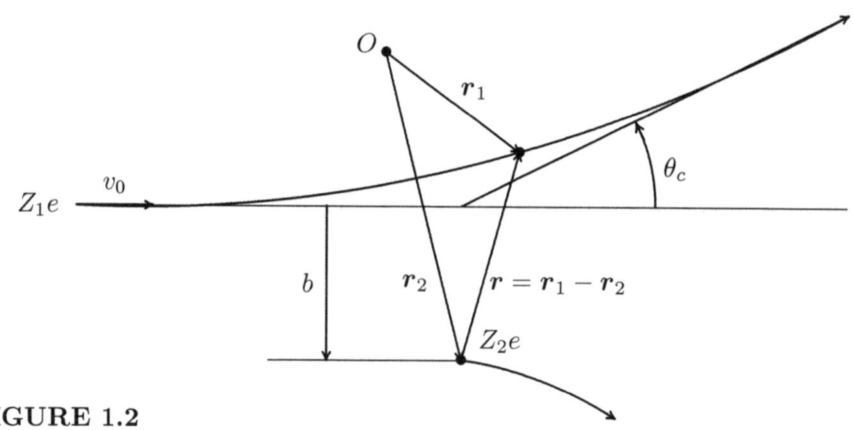

FIGURE 1.2
Sketch of scattering event between two ions with charges Z_1e and Z_2e.

We begin with simple binary Coulomb collisions which involve only two particles, a projectile with charge Z_1e and mass m_1 traveling with speed v_0 and an initially stationary target particle with charge Z_2e and mass m_2. The scattering encounter is illustrated in Figure 1.2. The equations of motion of these two particles are given by

$$m_1 \frac{\mathrm{d}^2 \boldsymbol{r}_1}{\mathrm{d}t^2} = \frac{Z_1 Z_2 e^2 (\boldsymbol{r}_1 - \boldsymbol{r}_2)}{4\pi\epsilon_0 |\boldsymbol{r}_1 - \boldsymbol{r}_2|^3} \tag{1.16}$$

$$m_2 \frac{\mathrm{d}^2 \boldsymbol{r}_2}{\mathrm{d}t^2} = \frac{Z_1 Z_2 e^2 (\boldsymbol{r}_2 - \boldsymbol{r}_1)}{4\pi\epsilon_0 |\boldsymbol{r}_1 - \boldsymbol{r}_2|^3} \,. \tag{1.17}$$

By adding and subtracting these, we find

$$m_1\frac{\mathrm{d}^2\boldsymbol{r}_1}{\mathrm{d}t^2}+m_2\frac{\mathrm{d}^2\boldsymbol{r}_2}{\mathrm{d}t^2}=(m_1+m_2)\frac{\mathrm{d}^2\boldsymbol{r}'}{\mathrm{d}t^2}=0\,, \tag{1.18}$$

$$\frac{\mathrm{d}^2\boldsymbol{r}_1}{\mathrm{d}t^2}-\frac{\mathrm{d}^2\boldsymbol{r}_2}{\mathrm{d}t^2}=\frac{\mathrm{d}^2\boldsymbol{r}}{\mathrm{d}t^2}=\left(\frac{m_1+m_2}{m_1m_2}\right)\frac{Z_1Z_2e^2\boldsymbol{r}}{4\pi\epsilon_0|\boldsymbol{r}|^3}\,, \tag{1.19}$$

where

$$\boldsymbol{r}'=\frac{m_1\boldsymbol{r}_1+m_2\boldsymbol{r}_2}{m_1+m_2}\qquad\text{and}\qquad\boldsymbol{r}=\boldsymbol{r}_1-\boldsymbol{r}_2\,.$$

Equation (1.18) describes the motion of the center of mass, and shows that the center of mass does not move. Equation (1.19) shows that the motion of the relative distance is the same as if a single particle of reduced mass $\mu=m_1m_2/(m_1+m_2)$ moved about a fixed center.

In polar coordinates, we must solve

$$\ddot{\boldsymbol{r}}=K\boldsymbol{r}/r^3\,, \tag{1.20}$$

with $K=Z_1Z_2e^2/4\pi\epsilon_0\mu$. Using

$$\begin{aligned}\boldsymbol{r}&=r\hat{e}_r\\ \dot{\boldsymbol{r}}&=\dot{r}\hat{e}_r+r\dot{\hat{e}}_r=\dot{r}\hat{e}_r+r\dot{\theta}\hat{e}_\theta\\ \ddot{\boldsymbol{r}}&=\ddot{r}\hat{e}_r+2\dot{r}\dot{\theta}\hat{e}_\theta+r\ddot{\theta}\hat{e}_\theta-r\dot{\theta}^2\hat{e}_r\,,\end{aligned}$$

we have from the $\boldsymbol{r}$ and θ components of equation (1.20)

$$\ddot{r}-r\dot{\theta}^2=K/r^2 \tag{1.21}$$

$$2\dot{r}\dot{\theta}+r\ddot{\theta}=0\,. \tag{1.22}$$

Equation (1.22) gives $r^2\dot{\theta}$ =const., and from the initial angular momentum, we have $r^2\dot{\theta}=v_0b$ where v_0 is the velocity of the projectile as $r\to\infty$, and b is the impact parameter, which is defined as the distance of closest approach if the trajectory were undeflected.

To solve equation (1.21), it is convenient to let $u=1/r$, and let θ be the independent variable, so that

$$\begin{aligned}r&=\frac{1}{u}\\ \dot{r}&=-\frac{\dot{u}}{u^2}=-\frac{\dot{\theta}}{u^2}\frac{\mathrm{d}u}{\mathrm{d}\theta}=v_0b\frac{\mathrm{d}u}{\mathrm{d}\theta}\\ \ddot{r}&=v_0b\dot{\theta}\frac{\mathrm{d}^2u}{\mathrm{d}\theta^2}=-v_0^2b^2u^2\frac{\mathrm{d}^2u}{\mathrm{d}\theta^2}\,,\end{aligned}$$

so that equation (1.21) becomes

$$-v_0^2b^2u^2\frac{\mathrm{d}^2u}{\mathrm{d}\theta^2}-v_0^2b^2u^3=Ku^2$$

or

$$\frac{\mathrm{d}^2 u}{\mathrm{d}\theta^2} + u = -\frac{K}{v_0^2 b^2}\,. \tag{1.23}$$

The solution of this equation is

$$u = \frac{1}{r} = A\cos(\theta + \delta) - \frac{K}{v_0^2 b^2}\,. \tag{1.24}$$

The initial conditions are $\theta_0 = \pi$, $r_0 = \infty$, $\dot{r}_0 = -v_0$, so

$$\tan\delta = \frac{v_0^2 b}{K} \tag{1.25}$$

$$A = -\frac{\sqrt{1 + K^2/v_0^4 b^2}}{b}\,. \tag{1.26}$$

From the solution of equation (1.24), the asymptotic deflection angle is given by

$$A\cos(\theta_c + \delta) = K/v_0^2 b^2\,. \tag{1.27}$$

One solution is of course $\theta_c = \pi$ which was an initial condition, and the other is $\theta_c = \pi - 2\delta$, since $\cos(\pi + \delta) = \cos(\pi - \delta)$. Using the expression for δ, this gives

$$\tan\frac{\theta_c}{2} = \frac{Z_1 Z_2 e^2}{4\pi\epsilon_0 \mu v_0^2 b} = \cot\delta\,. \tag{1.28}$$

We now deduce that all of the particles which are incident with impact parameters such that they pass through the annular ring between radius b and $b + \mathrm{d}b$ will be scattered into angles between θ_c and $\theta_c + \mathrm{d}\theta_c$. This fact, and the definition of differential cross section, gives the flux scattered into solid angle $\mathrm{d}\Omega$ as

$$\mathrm{d}\sigma(\theta_c) = \sigma(\theta_c)\,\mathrm{d}\Omega = \sigma(\theta_c) 2\pi \sin\theta_c\,\mathrm{d}\theta_c\,, \tag{1.29}$$

since $\Omega = 2\pi(1 - \cos\theta_c)$. Conservation of particles then requires the number entering the annulus to be the same as the number leaving through the solid angle, so that

$$2\pi b\,\mathrm{d}b = \sigma(\theta_c) 2\pi \sin\theta_c\,\mathrm{d}\theta_c\,, \tag{1.30}$$

from which we obtain

$$\sigma(\theta_c) = \frac{b}{\sin\theta_c}\frac{\mathrm{d}b}{\mathrm{d}\theta_c}\,. \tag{1.31}$$

This leads to the *Rutherford scattering cross section* using equation (1.28),

$$\sigma(\theta_c) = \left(\frac{K}{2v_0^2 \sin^2\frac{\theta_c}{2}}\right)^2\,. \tag{1.32}$$

(A negative sign is discarded because an increase in b leads to a decrease in θ_c.)

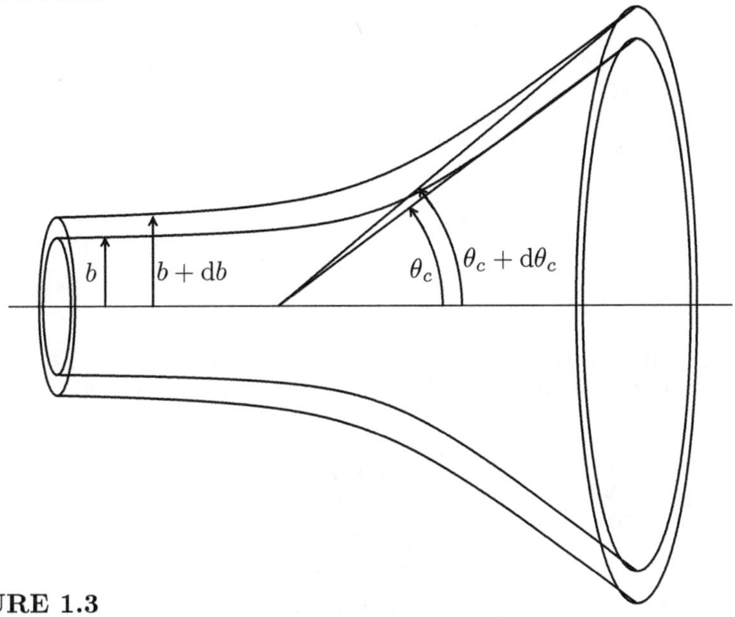

FIGURE 1.3
Sketch of scattering event between two ions with charges Z_1e and Z_2e showing the differential areas for the incident and scattered particles.

It is convenient to define an impact parameter which leads to 90° scattering for an electron from a singly charged ion (so $\mu \sim m_e$). From equation (1.28), this is

$$b_0 = \frac{e^2}{4\pi\epsilon_0 m_e v_0^2}, \qquad (1.33)$$

with the corresponding cross section

$$\sigma_{90^\circ} = \pi b_0^2 = \pi \left(\frac{e^2}{4\pi\epsilon_0 m_e v_0^2}\right)^2 \simeq \frac{1.63 \cdot 10^{-18}}{W_{\mathrm{eV}}^2} \quad \mathrm{m}^2 . \qquad (1.34)$$

Problem 1.3 *Unit vectors.* Show that $\dot{\hat{e}}_r = \dot{\theta}\hat{e}_\theta$ and $\dot{\hat{e}}_\theta = -\dot{\theta}\hat{e}_r$.

1.7.2 Deflection by multiple Coulomb collisions

For this calculation, we wish to find the multiple Coulomb collision cross section for electrons on massive ions of charge Z. Assuming it begins traveling in the z direction, it will acquire a transverse component of velocity as a result of these many collisions. In a single collision, where we assume the z-component of the velocity to be unchanged (weak collision), the perpendicular energy is changed by an amount

$$(\Delta v_\perp)^2 = v^2 \sin^2\theta_c = 4v^2 \sin^2\delta \cos^2\delta \qquad (1.35)$$

and since

$$\cos^2\delta = \frac{1}{1+\tan^2\delta}\,, \quad \text{and} \quad \sin^2\delta = \frac{\tan^2\delta}{1+\tan^2\delta}\,,$$

Equation (1.35) becomes

$$(\Delta v_\perp)^2 = \frac{4v^2\tan^2\delta}{(1+\tan^2\delta)^2} = \frac{4v^2(b/b_0)^2}{[1+(b/b_0)^2]^2} \tag{1.36}$$

where $\tan\delta \equiv b/b_0$ with $b_0 \equiv K/v_0^2$.

Now the number of encounters/second is $2\pi b\,db\,nv$, so averaging over b we find

$$\begin{aligned} <(\Delta v_\perp)^2> &= 4v^2(2\pi nv)b_0^2\int_0^{b_m/b_0}\frac{(b/b_0)^3 d(b/b_0)}{[1+(b/b_0)^2]^2} \\ &= 8\pi v^3 n b_0^2\int_0^{x_m}\frac{x^3 dx}{(1+x^2)^2} \\ &= 4\pi v^3 n b_0^2\left[\ln(1+x_m^2)+\frac{1}{1+x_m^2}-1\right] \\ &\simeq 8\pi v^3 n b_0^2\ln(x_m)\,, \end{aligned} \tag{1.37}$$

for $x_m = b_m/b_0 \gg 1$ where b_m is the maximum impact parameter which we take to be the Debye length so that $x_m = \Lambda \equiv 9N_D$ where $N_D = \frac{4\pi}{3}\lambda_D^3$ is the plasma parameter we introduced in Section 1.4. In Appendix B we will find this same value for the logarithmic term from a different analysis. If we now consider the number of collisions in a time corresponding to the distance L that the particle travels, so that $vt = L$, then we can divide by v^2 to estimate the mean angle over this distance, so that

$$<(\Delta\theta)^2> = \frac{<(\Delta v_\perp)^2>}{v^2} = 8\pi n L b_0^2\ln\Lambda\,. \tag{1.38}$$

At about 36 eV, some quantum mechanical diffraction effects modify the Λ term. As discussed by Spitzer[1], an electron wave passing through a circular aperture of radius p will be spread out by diffraction through an angle $\lambda/2\pi p$ where λ is the electron wavelength. If this deflection exceeds the classical value, then we must reduce Λ by the factor $2\alpha c/v_0$ where $\alpha = 1/137$ is the fine structure constant. This is equivalent to multiplying Λ by $\sqrt{36.2/T_{\mathrm{eV}}}$ for $T_{\mathrm{eV}} > 36.2$ for temperature measured in eV. Some values for $\ln\Lambda$ are given in Table 1.1.

If we now approximate that a large angle corresponds to $\langle(\Delta\theta)^2\rangle \sim 1$, which is of the order of 90°, then we have an immediate estimate for the distance the electron must travel to accumulate this deflection, so that

$$L_{90^\circ} \approx \left(\frac{nZ^2e^4}{2\pi\epsilon_0^2 m_e^2 v_0^4}\ln\Lambda\right)^{-1}\,. \tag{1.39}$$

TABLE 1.1
Values of $\ln \Lambda$ with $Z = 1$.

	electron density, n_e, m^{-3}							
k_BT, eV	10^3	10^6	10^9	10^{12}	10^{15}	10^{18}	10^{21}	10^{24}
10^{-1}	23.5	20.0	16.6	13.1	9.65	6.19		
10^0	26.9	23.5	20.0	16.6	13.1	9.65	6.19	
10^1	30.4	26.9	23.5	20.0	16.6	13.1	9.65	6.19
10^2	33.3	29.9	26.4	23.0	19.5	16.0	12.6	9.14
10^3	35.6	32.2	28.7	25.3	21.8	18.4	14.9	11.5
10^4	37.9	34.5	31.0	27.6	24.1	20.7	17.2	13.8

The corresponding collision frequency is given by $\nu_c L = v_0$ so that

$$\nu_c \approx \frac{nZ^2e^4}{2\pi\epsilon_0^2 m_e^2 v_0^3} \ln \Lambda \,. \tag{1.40}$$

We can finally relate the multiple collision deflection rate with the single collision rate by comparing the cross sections. For the multiple collision case, we use

$$\sigma_{90^\circ M} \equiv \frac{1}{nL_{90^\circ}} = \frac{1}{2\pi}\left(\frac{Ze^2}{\epsilon_0 m_e v_0^2}\right)^2 \ln \Lambda \,, \tag{1.41}$$

so comparing with equation (1.34), the ratio is

$$\frac{\sigma_{90^\circ M}}{\sigma_{90^\circ S}} = 8 \ln \Lambda \,. \tag{1.42}$$

This shows that our claim that the multiple scattering dominates is based on the largeness of the quantity Λ.

1.7.3 Charge-neutral collisions

When the plasma is not fully ionized, as is frequently the case in the ionosphere, the principal collisional mechanism may be collisions between charged particles and neutral atoms or molecules. Contrary to the conclusion for Coulomb collisions, this type of collision is more likely to lead to a single large angle deflection rather than a large number of small angle collisions. The effects of this type of collision are somewhat simpler to analyze, because a simpler model for their effect is possible. When we consider the various effects that cause the distribution function to evolve, the evolution equation takes the form

$$\frac{\mathrm{d}f}{\mathrm{d}t} = \left.\frac{\partial f}{\partial t}\right|_{\text{coll.}} ,$$

where the left-hand side is expanded, since $f = f(\boldsymbol{r}, \boldsymbol{v}, t)$, to

$$\frac{\mathrm{d}f}{\mathrm{d}t} = \frac{\partial f}{\partial t} + \boldsymbol{v}\cdot\nabla f + \boldsymbol{a}\cdot\nabla_{\boldsymbol{v}} f \,,$$

where $\boldsymbol{a}$ is the acceleration, typically $\boldsymbol{a} = (q/m)(\boldsymbol{E} + \boldsymbol{v} \times \boldsymbol{B})$. For the right-hand side, the expression depends on the type of collision. For Coulomb collisions, the expression on the right must include effects of the drag on the charged particles whereby particles traveling faster than the average are slowed down, and the effects of diffusion, which tends to bring the distribution back to equilibrium, which is a Maxwellian. These effects and the model for the right-hand side collisional term are described in Chapter 3. For charge-neutral collisions, however, we can use a much simpler model such that

$$\left.\frac{\partial f}{\partial t}\right|_{\text{coll.}} = -\nu(f - f_0)\,,$$

where ν is the charge-neutral collision frequency and f_0 is the equilibrium distribution function. This model is generally called the Krook model, due to Bhatnagar, Gross, and Krook[2] who showed that it satisfied the critical conservation laws.*

1.7.4 Charge-exchange collisions

Another type of collision in a partially ionized plasma is the charge-exchange collision, where an electron on the neutral particle "jumps" to the charged particle passing by. This is especially important for hydrogen where the cross section for such transfers is significant. In this "collision," it appears as a head-on collision for same-species transfers as the ionized neutral (the neutrals are presumably at a lower temperature so effectively at rest compared to the plasma) suddenly appears as a nearly stationary ion, and the higher energy ion suddenly disappears (and promptly leaves the plasma). This looks like a billiard ball head-on collision where the cue ball stops and the target moves ahead in the same direction and with the same velocity as the cue ball had originally. In a fluid model, this type of collision can be modeled as a "pseudo-ion mass," where the neutrals appear to be partially carried along with the charged ions, effectively reducing the charge-to-mass ratio since the ion charge remains the same, but the mass effectively increases [4, 5].

1.8 Particle drifts

While the majority of the study of plasma physics is the exploration of collective effects, many single particle effects are also important. The motions

*Although sometimes referred to as the BGK model, this must not be confused with the BGK *mode* that is a nonlinear plasma wave with a non-Maxwellian distribution due to Bernstein, Greene, and Kruskal[3].

of charged particles in electric, magnetic, and gravitational fields which may be homogeneous or inhomogeneous must first be understood and appreciated before many of the collective effects can be put in proper perspective. In a uniform magnetic field, without any collisions, a charged particle follows a helical trajectory following a single magnetic field line. The rotation frequency is the cyclotron frequency, ω_c, and the gyration is circular with radius $\rho_L = v_\perp/\omega_c$, which is the *Larmor radius.* The particle streams freely at uniform velocity along the magnetic field line.

When an electric field is added, an additional motion is added, called the $\boldsymbol{E} \times \boldsymbol{B}$ drift since the center of gyration drifts with velocity $\boldsymbol{v}_d = \boldsymbol{E} \times \boldsymbol{B}/B^2$. This drift is obtained by assuming there is no net force on the center of rotation, so the Lorentz force vanishes, or

$$0 = q(\boldsymbol{E} + \boldsymbol{v}_d \times \boldsymbol{B}),$$

whose only nontrivial solution is the $\boldsymbol{E} \times \boldsymbol{B}$ drift velocity.

More commonly, drifts occur because of some additional force, such as a gravitational force, or due to some inhomogeneity in the magnetic field. It is common to refer to this kind of drift motion as *guiding center* drift, since it is the center of gyration which is drifting. For a gravitational field, this drift is given by

$$\boldsymbol{v}_{dg} = \frac{m}{q}\frac{\boldsymbol{g} \times \boldsymbol{B}}{B^2}. \tag{1.43}$$

If we generalize the concept of the gravitational force to be any kind of average force on the particle, then we can see that curvature in the magnetic field will lead to centripetal acceleration, which induces *curvature drift.* This clearly depends on the velocity parallel to the magnetic field, and will be described in more detail in the next chapter. There is also a drift due to an inhomogeneity of the magnetic field transverse to the field. This is due to the fact that as the particle rotates with a finite Larmor radius, it samples a stronger field on one side and a weaker field on the other, so the trajectory is no longer a circle.

1.9 Waves

Although we will devote a large effort to the study of wave phenomena in plasmas in subsequent chapters, it is useful to introduce some of the fundamental wave types first, since they will be involved in many phenomena, and because we do not wish to lose sight of the fundamentals amidst the later complexity. We will also encounter both *linear* and *nonlinear* waves, where the distinction is based on wave amplitude. We will usually assume the amplitude is small, and *linearize* in order to simplify the analysis and extract the basic character of the wave, but we must always be aware that all waves in plasmas are fundamentally nonlinear, and in some cases we must focus attention on the effects of finite amplitudes. The amplitudes of linear waves may be written as $\boldsymbol{A}(\boldsymbol{r},t) = \boldsymbol{A}_0 \exp[\mathrm{i}(\boldsymbol{k}\cdot\boldsymbol{r} - \omega t)]$ where $\boldsymbol{A}_0$ is a complex constant. The characterization of waves is commonly indicated by the phase velocity,

$$\boldsymbol{v}_p = \frac{\omega}{k}\hat{\boldsymbol{e}}_k\,,$$

where $\boldsymbol{k}$ is the wave vector, and the group velocity,

$$\boldsymbol{v}_g = \nabla_k \omega\,,$$

which is often written as $\mathrm{d}\omega/\mathrm{d}k$ in an isotropic medium. The phase velocity indicates the speed and direction of a phase front of the wave, while the group velocity gives the speed and direction that energy is transported by the wave.

1.9.1 Sound waves

Sound waves are not unique to plasmas, but they do occur there. For wave frequencies well below the collision frequency, the characteristic speed is given by $v_s = \sqrt{\gamma p/\rho} = \sqrt{\gamma k_B T/m}$ where γ is the usual ratio of specific heats.

When the electron temperature is much greater than the ion temperature, there is a type of sound wave that is unique to a plasma called the *ion-acoustic wave*, which is characterized by the electron temperature (since the electron pressure dominates) and the ion mass (since the inertia is dominated by the ions). The waves propagate with speed

$$c_s = \sqrt{\frac{Z\gamma_e k_B T_e + \gamma_i k_B T_i}{m_i}}\,. \tag{1.44}$$

These waves are strongly damped unless the electron temperature is much greater than the ion temperature and the frequency is much higher than the collision frequency.

1.9.2 Electromagnetic waves

Just as in any homogeneous dielectric medium, electromagnetic waves propagate in a plasma with the wavenumber given by

$$k^2 = \frac{\omega^2}{c^2}K\,, \tag{1.45}$$

where K is the relative dielectric constant. In a cold plasma with no magnetic field, this is given by

$$K = 1 - \frac{\omega_p^2}{\omega^2}\,. \tag{1.46}$$

The corresponding dispersion relation ($\omega = \omega(\boldsymbol{k})$) for electromagnetic waves is

$$k^2 = \frac{\omega^2 - \omega_p^2}{c^2}\,. \tag{1.47}$$

From this relation, it is apparent that electromagnetic waves propagate only for $\omega > \omega_p$, and since k is imaginary for lower frequencies, a plasma will reflect electromagnetic radiation below the plasma frequency. These simple conclusions will be modified for inhomogeneous or magnetized plasmas or boundaries.

Problem 1.4 *Plasma dielectric constant.* Derive the plasma dielectric constant for a plasma of cold electrons and stationary ions by assuming a plane wave solution to the Maxwell equations using the equation of motion for the current. Also show that the product of the phase and group velocities is $v_p v_g = c^2$.

1.9.3 Alfvén waves

In his book, **Cosmical Dynamics**[6], Hannes Alfvén postulated the existence of waves whose speed was proportional to the magnetic field by considering an analogy between waves on a stretched string and the motion of a highly conducting fluid in a magnetic field. For waves on a stretched string, the speed is given by $v = \sqrt{T/\rho_m}$, where T is the tension in the string and ρ_m is the mass per unit length. By analogy, a magnetic field is under tension via the Maxwell stress tensor, with $T = B^2/\mu_0$ per unit area, and in a highly conducting fluid, the mass is tightly bound to the magnetic field lines (provided the frequency is well below the cyclotron frequency), so he proposed a wave should travel at a speed given by

$$V_A = \sqrt{B^2/\mu_0\rho_m} = B/\sqrt{\mu_0 n_i m_i}, \tag{1.48}$$

where in this case, $\rho_m = n_i m_i$ is the volume mass density. These Alfvén waves were subsequently found experimentally, and are a new fundamental velocity characterizing a magnetized plasma. As one would expect from the nature of the analogy, these waves are transverse waves and travel along magnetic field lines, and any low frequency perturbation in a magnetized plasma travels at this speed. Another analogy may be made with sound waves, since the Maxwell stress tensor also indicates the plasma exerts a pressure $p_m = B^2/2\mu_0$. Using the expression for sound speed, we find

$$v = \sqrt{\gamma p/\rho} = \sqrt{\frac{2B^2}{2\mu_0\rho_m}} = V_A\,, \tag{1.49}$$

if we take $\gamma = 2$. This is also a sound-like wave that propagates in all directions, and is called a compressional wave since the magnetic field lines are alternately compressed and expanded, while the original Alfvén wave is torsional, since the field lines are twisted. The compressional Alfvén wave continues to propagate above the ion cyclotron frequency, while the torsional Alfvén wave does not.

1.9.4 Electrostatic waves

Both the Alfvén waves and the electromagnetic waves described above have significant transverse components, but there are a number of waves in plasmas that are essentially longitudinal and have no appreciable wave magnetic field. These waves may be characterized by an electric potential, and hence are called electrostatic waves.

1.9.5 Plasma waves

Without any magnetic field, these electrostatic waves are called *plasma waves, space-charge waves*, or *Langmuir waves.* These are the finite temperature generalization of simple plasma oscillations, with dispersion relation

$$\omega^2 = \omega_p^2 + \frac{3k_BT_e}{m_e}k^2 , \tag{1.50}$$

and are exponentially damped even in the absence of collisions. It is obvious from the dispersion relation that the waves travel near the electron thermal speed for $\omega \gg \omega_p$.

Problem 1.5 *Plasma waves.* Find the phase and group velocities for plasma waves if $\omega \simeq \omega_p$.

1.9.6 Bernstein waves

When a magnetic field is added to the finite temperature plasma, an entirely new set of electrostatic waves appears near each harmonic of both the electron and ion cyclotron frequencies. These waves are called *Bernstein waves* and propagate without damping (in a collisionless plasma) across a magnetic field, and their group velocity is typically of the order of the ion thermal speed near the ion cyclotron harmonics and near the electron thermal speed near the electron cyclotron harmonics.

1.9.7 Wave damping

The exponential damping mentioned above in connection with plasma waves is a fundamental process in plasmas, and will need careful analysis, since it is a dissipationless damping mechanism. Without any collisions, there is no loss of information with such damping, even though the potential dies away. The damping is caused by the thermal spreading of the particles which initially formed the wave. If the distribution function is monotonically decreasing with velocity, the wave is always damped, but if the distribution function has a bump, then the damping may change to growth and with a finite amplitude, nonlinear effects modify the distribution to eliminate the bump. The damping is called *Landau damping*, and leads to very interesting nonlinear phenomena, such as plasma echoes where apparently lost information can be recovered. It may be shown that nonlinear phenomena eventually dominate linear processes without some collisions, so that very weak collisions are required for the validity of linear theory. In addition to damping plasma waves, this same kind of collisionless damping is involved in waves near the electron and ion cyclotron frequencies and their harmonics.

1.9.8 Large amplitude waves

As the amplitude of a wave grows in a plasma, a variety of nonlinear processes may dramatically change the propagation characteristics of the wave. The wave energy density may be large enough to actually expel some of the background plasma, creating a density cavity, or *caviton.* This can lead to collapse and filamentation of the wave. It is also possible that solitary waves may propagate where the nonlinear processes support a single, localized disturbance whose velocity is amplitude dependent. In some cases, these may be solitons, which are solitary waves which are very stable and survive collisions. It is also possible for a large amplitude wave to modify the distribution function to such an extent that Landau damping no longer occurs. Keeping in mind that all plasma waves are fundamentally nonlinear, it will be important to note at what levels these effects become noticeable.

1.10 Plasma radiation

Even as there is a variety of ways a plasma may absorb energy, there is a variety of ways plasmas may radiate energy. From the fluctuation-dissipation theorem, every absorber is an emitter, so we expect plasmas to radiate. The radiation may be grouped into two major categories: (1) radiation from atomic processes and (2) radiation from accelerated charges. Some of this radiation is of concern because it represents a plasma energy loss mechanism, and some is of interest as a diagnostic for the interior of a plasma, and some falls into both categories.

1.10.1 Bremsstrahlung

While in the strict sense, Bremsstrahlung refers to both free-bound and free-free transitions, which could include almost everything, we restrict our use of the word in plasma physics to the free-free radiation when particles make large angle collisions. The radiation intensity for a hydrogen plasma is

$$P_B = 5.35 \cdot 10^{-37} n^2 T_{\rm keV}^{1/2} \quad \text{Watt/m}^3\,. \tag{1.51}$$

Since this is calculated under the assumption that none of the emitted radiation is reabsorbed, this is the lowest radiation rate possible. For fusion plasmas, even for temperatures as high as 100 keV, this is an insignificant loss rate compared to the fusion production rate.

1.10.2 Blackbody radiation

Blackbody radiation assumes that all internal radiation from whatever source is absorbed, so that only surface radiation escapes the plasma. From the Stefan-Boltzmann law, the radiation is given by

$$P_{bb} = \sigma T^4 \quad \text{Watt/m}^2 ,$$

where $\sigma = 5.67 \cdot 10^{-8}$ W/m^2 K^4. In contrast to Bremsstrahlung radiation, this radiation would greatly *exceed* the fusion production of a fusion plasma, but this is not of great concern since fusion plasmas are optically thin over most of the radiation spectrum.

1.10.3 Impurity radiation

When high Z impurities enter high temperature plasmas and become highly stripped, they emit both line emission and Bremsstrahlung radiation at very high rates. By impurity control, this otherwise serious energy loss can be kept within acceptable limits, so the principal interest is in plasma diagnostics. Impurity atoms provide spectroscopic information which is invaluable in estimating the internal temperature, and even the density if it is high enough.

1.10.4 Cyclotron or synchrotron radiation

Substantial radiation is measured from high temperature plasmas in a magnetic field at harmonics of both the electron and ion cyclotron frequencies. In fusion plasmas, the electrons are even weakly relativistic, so that some synchrotron radiation is also emitted. If electron temperatures in fusion plasmas exceed the ion temperature by a large amount, synchrotron radiation could exceed the fusion production, so it is desirable to keep the electron temperature similar to the ion temperature.

1.10.5 Cherenkov radiation

Cherenkov radiation is emitted whenever a particle travels faster than the speed of light in the medium. In a plasma, energetic particles commonly exceed the speed of various waves, and hence emit some Cherenkov radiation. Though not important for energy balance, this radiation could provide some diagnostic information on velocity distributions.

2

BASIC KINETIC EQUATIONS

2.1 The Klimontovich equation

2.1.1 Introduction

It is possible to give either a formally exact *microscopic* description of plasma behavior which has too much information for use, or to give a *macroscopic* description which has less information, but is tractable for many problems of interest. We shall begin with the exact description, for it can give us some information, but then give the alternative description which we shall use subsequently. The exact microscopic theory is given by the *Klimontovich equation*, and its average, if taken with suitable precautions, will enable us to glean some macroscopic information, and also give us something of the ordering so that we can estimate what is left out in the zero-order approximations. Then we shall examine the *Liouville equation*, which again is exact, but is more readily adaptable to an ordering scheme so that we can examine the zero-order equation and get a more detailed form for the first and higher order corrections.

2.1.2 The density function

We first describe the plasma in terms of the individual particles by introducing a density function which describes the location and velocity of each particle,

$$N_s(\boldsymbol{r},\boldsymbol{v},t)=\sum_{i=1}^{N_0}\delta[\boldsymbol{r}-\boldsymbol{R}_i(t)]\delta[\boldsymbol{v}-\boldsymbol{V}_i(t)]\,, \tag{2.1}$$

where there are N_0 particles of species s. We note that $\boldsymbol{R}_i$, $\boldsymbol{V}_i$, are the Lagrangian coordinates of the particles themselves, while $\boldsymbol{r}$, $\boldsymbol{v}$, are the Eulerian coordinates of a 6-dimensional space we call *phase space*. Integrating over phase space, we obtain the total number of particles, so this is a density of particles in phase space. Integrating over a small volume $(\Delta\boldsymbol{r})(\Delta\boldsymbol{v})$ gives the number of particles in that particular volume of phase space, and because of the delta function character of the density function, this number is not smoothly varying. If one chooses the volume large enough to include many particles, but small enough to show variations in the density and electromagnetic fields, we may be able to approximate the function by a smooth function, but it is not smooth as it stands. In order to include a complete description of the plasma, we must include all species, so this we indicate by the sum over species

$$N(\boldsymbol{r},\boldsymbol{v},t) = \sum_{s=e,i} N_s(\boldsymbol{r},\boldsymbol{v},t)\,. \tag{2.2}$$

Knowing the exact location and velocity of every particle at one point in time, we in principle know the position and velocity at every time, since

$$\dot{\boldsymbol{R}}_i(t) = \boldsymbol{V}_i(t) \tag{2.3}$$

for all i. The velocity is related back to the coordinates through the Lorentz force law, since

$$m_s\dot{\boldsymbol{V}}_i(t) = q_s\{\boldsymbol{E}^m[\boldsymbol{R}_i(t),t] + \boldsymbol{V}_i(t)\times\boldsymbol{B}^m[\boldsymbol{R}_i(t),t]\}\,, \tag{2.4}$$

where the superscript m denotes that the fields are a combination of external electric and magnetic fields and the self-consistent fields from all the particles *except* particle i. These fields then satisfy the Maxwell equations:

$$\nabla\cdot\boldsymbol{E}^m(\boldsymbol{r},t) = \rho^m(\boldsymbol{r},t)/\epsilon_0\,, \tag{2.5}$$

$$\nabla\cdot\boldsymbol{B}^m(\boldsymbol{r},t) = 0\,, \tag{2.6}$$

$$\nabla\times\boldsymbol{E}^m(\boldsymbol{r},t) = -\frac{\partial\boldsymbol{B}^m(\boldsymbol{r},t)}{\partial t}\,, \tag{2.7}$$

$$\nabla\times\boldsymbol{B}^m(\boldsymbol{r},t) = \mu_0\boldsymbol{J}^m(\boldsymbol{r},t) + \frac{1}{c^2}\frac{\partial\boldsymbol{E}^m(\boldsymbol{r},t)}{\partial t}\,. \tag{2.8}$$

The charge and current densities are given by

$$\rho^m(\boldsymbol{r},t) = \sum_{e,i} q_s\int \mathrm{d}\boldsymbol{v}\,N_s(\boldsymbol{r},\boldsymbol{v},t) \tag{2.9}$$

$$\boldsymbol{J}^m(\boldsymbol{r},t) = \sum_{e,i} q_s\int \mathrm{d}\boldsymbol{v}\,\boldsymbol{v}N_s(\boldsymbol{r},\boldsymbol{v},t)\,. \tag{2.10}$$

These equations comprise a set of $6N_0$ differential equations which in classical mechanics completely determines the system. The difficulty of solving this large number of equations prevents us from learning anything significant. The number of equations must be reduced and in the process, we will lose some of the information. Before reducing the set of equations, we examine the full set in more detail.

2.1.3 The Klimontovich equation

To find the time evolution of the plasma, we use the time derivative of the density function, N_s. In order to see how the derivatives are taken we note

$$\frac{\partial}{\partial t}\delta[\boldsymbol{r}-\boldsymbol{R}_i(t)] = \frac{\partial\boldsymbol{R}_i}{\partial t}\cdot\frac{\partial}{\partial\boldsymbol{R}_i}\delta[\boldsymbol{r}-\boldsymbol{R}_i(t)] = \dot{\boldsymbol{R}}_i\cdot\left(-\frac{\partial}{\partial\boldsymbol{r}}\right)\delta[\boldsymbol{r}-\boldsymbol{R}_i(t)]$$

$$= -\dot{\boldsymbol{R}}_i \cdot \nabla\delta[\boldsymbol{r} - \boldsymbol{R}_i(t)] \,, \tag{2.11}$$

so that

$$\frac{\partial}{\partial t}\sum_i \delta[\boldsymbol{r}-\boldsymbol{R}_i(t)]\delta[\boldsymbol{v}-\boldsymbol{V}_i(t)] = -\sum_i \dot{\boldsymbol{R}}_i\cdot\nabla\delta[\boldsymbol{r}-\boldsymbol{R}_i(t)] - \sum_i \dot{\boldsymbol{V}}_i\cdot\nabla_{\boldsymbol{v}}\delta[\boldsymbol{v}-\boldsymbol{V}_i(t)] \,. \tag{2.12}$$

We know $\dot{\boldsymbol{V}}_i$ from the Lorentz force equation, however, so this becomes

$$\begin{aligned}\frac{\partial N_s}{\partial t} &= -\sum_i \boldsymbol{V}_i \cdot \nabla\delta[\boldsymbol{r} - \boldsymbol{R}_i(t)]\delta[\boldsymbol{v} - \boldsymbol{V}_i(t)] \\ &\quad -\sum_i \frac{q_s}{m_s}(\boldsymbol{E}^m + \boldsymbol{V}_i \times \boldsymbol{B}^m) \cdot \nabla_{\boldsymbol{v}}\delta[\boldsymbol{r} - \boldsymbol{R}_i(t)]\delta[\boldsymbol{v} - \boldsymbol{V}_i(t)] \,. \end{aligned} \tag{2.13}$$

But, since the delta functions guarantee that they will be evaluated only where there are particles, we may let $\boldsymbol{V}_i \to \boldsymbol{v}$, so that

$$\begin{aligned}\frac{\partial N_s}{\partial t} &= -\sum_i \boldsymbol{v} \cdot \nabla\delta[\boldsymbol{r} - \boldsymbol{R}_i(t)]\delta[\boldsymbol{v} - \boldsymbol{V}_i(t)] \\ &\quad -\sum_i \frac{q_s}{m_s}(\boldsymbol{E}^m + \boldsymbol{v} \times \boldsymbol{B}^m) \cdot \nabla_{\boldsymbol{v}}\delta[\boldsymbol{r} - \boldsymbol{R}_i(t)]\delta[\boldsymbol{v} - \boldsymbol{V}_i(t)] \\ &= -\boldsymbol{v} \cdot \nabla \sum_i \delta[\boldsymbol{r} - \boldsymbol{R}_i(t)]\delta[\boldsymbol{v} - \boldsymbol{V}_i(t)] \\ &\quad -\frac{q_s}{m_s}(\boldsymbol{E}^m + \boldsymbol{v} \times \boldsymbol{B}^m) \cdot \nabla_{\boldsymbol{v}} \sum_i \delta[\boldsymbol{r} - \boldsymbol{R}_i(t)]\delta[\boldsymbol{v} - \boldsymbol{V}_i(t)] \\ &= -\boldsymbol{v} \cdot \nabla N_s - \frac{q_s}{m_s}(\boldsymbol{E}^m + \boldsymbol{v} \times \boldsymbol{B}^m) \cdot \nabla_{\boldsymbol{v}} N_s \,. \end{aligned} \tag{2.14}$$

This may be written as a conservation law

$$\frac{\mathrm{D}}{\mathrm{D}t}N_s = \frac{\partial N_s}{\partial t} + \boldsymbol{v} \cdot \nabla N_s + \frac{q_s}{m_s}(\boldsymbol{E}^m + \boldsymbol{v} \times \boldsymbol{B}^m) \cdot \nabla_{\boldsymbol{v}} N_s = 0 \,, \tag{2.15}$$

which expresses the "incompressibility" of the plasma.

2.1.4 Kinetic equation

The complexity of the Klimontovich equation makes it of little direct use, but by averaging over a small volume, we may be able to resolve many features of plasma behavior. The size of the averaging volume must be chosen carefully, since if it is too large, we lose resolution in the variations of the plasma which we may wish to investigate, while if the volume is too small, the change in density from one volume to its neighboring volume may be too large for a description in terms of continuous functions, which is our goal. In order to guarantee "smoothness," we will require many particles (so changes of a few particles will represent small percentage changes), and this can be obtained by letting the volume in coordinate space be of the order of a Debye sphere and make the plasma parameter N_D large. We will then be able to resolve variations on the Debye length scale and larger, but variations smaller than a Debye length will not be well described. The volume is six dimensional, so we must also pick a velocity range large enough to include many particles.

One way to get many particles in a given volume is to consider not one group of N particles, but to pick an ensemble of groups of N particles, and take ensemble averages. This adds some complexity, but allows greater resolution.

For cases where the scale lengths of the electric and magnetic fields are much longer than the Debye length, we choose a distance large compared to the mean interparticle spacing, but less than the Debye length, and if $N_D \gg 1$, we know there exists such a distance. The smoothed function is then given by

$$f(\boldsymbol{r},\boldsymbol{v},t) = \frac{N(\boldsymbol{r} < \boldsymbol{r}_i < \boldsymbol{r} + \mathrm{d}\boldsymbol{r}, \boldsymbol{v} < \boldsymbol{v}_i < \boldsymbol{v} + \mathrm{d}\boldsymbol{v}, t)}{\Delta x \Delta y \Delta z \Delta v_x \Delta v_y \Delta v_z} \,. \tag{2.16}$$

If we now distinguish between the exact and averaged value of N_s, we may define the variations from average by

$$N_s = f_s + \delta N_s \,, \tag{2.17}$$
$$\boldsymbol{E}^m(\boldsymbol{r},t) = \boldsymbol{E}(\boldsymbol{r},t) + \delta \boldsymbol{E}^m(\boldsymbol{r},t) \,, \tag{2.18}$$
$$\boldsymbol{B}^m(\boldsymbol{r},t) = \boldsymbol{B}(\boldsymbol{r},t) + \delta \boldsymbol{B}^m(\boldsymbol{r},t) \,, \tag{2.19}$$

where $\boldsymbol{B} = \langle \boldsymbol{B}^m \rangle$ so that $\langle \delta \boldsymbol{B} \rangle = 0$, etc. Then averaging the various terms of equation (2.15) leads to

$$\left\langle \frac{\partial N_s}{\partial t} \right\rangle = \frac{\partial f_s}{\partial t} \,,$$
$$\langle \boldsymbol{v} \cdot \nabla N_s \rangle = \boldsymbol{v} \cdot \nabla f_s \,,$$

since $\langle \delta N_s \rangle = 0$, and

$$\begin{aligned} \frac{q_s}{m_s} \langle (\boldsymbol{E}^m + \boldsymbol{v} \times \boldsymbol{B}^m) \cdot \nabla_{\boldsymbol{v}} N_s \rangle = & \frac{q_s}{m_s} (\boldsymbol{E} + \boldsymbol{v} \times \boldsymbol{B}) \cdot \nabla_{\boldsymbol{v}} f_s \\ & + \frac{q_s}{m_s} \langle (\delta \boldsymbol{E} + \boldsymbol{v} \times \delta \boldsymbol{B}) \cdot \nabla_{\boldsymbol{v}} \delta N_s \rangle \end{aligned}$$

since $\langle \delta \boldsymbol{E}^m f_s \rangle = \langle \delta \boldsymbol{B}^m f_s \rangle = \langle \boldsymbol{E} \delta N_s \rangle = \langle \boldsymbol{B} \delta N_s \rangle = 0$. Combining these elements, we establish the *Kinetic equation*,

$$\frac{\partial f_s}{\partial t} + \boldsymbol{v} \cdot \nabla f_s + \frac{q_s}{m_s} (\boldsymbol{E} + \boldsymbol{v} \times \boldsymbol{B}) \cdot \nabla_{\boldsymbol{v}} f_s = -\frac{q_s}{m_s} \langle (\delta \boldsymbol{E}^m + \boldsymbol{v} \times \delta \boldsymbol{B}^m) \cdot \nabla_{\boldsymbol{v}} \delta N_s \rangle \,. \tag{2.20}$$

Since f_s is smooth by construction, all of the discrete, particle-like behavior is on the right hand side of equation (2.20), which is due to collisions and correlations. The average behavior is on the left, and describes the macroscopic behavior, while the microscopic variations from average are on the right. Estimating the importance of collisions, if we take $v_0 \sim \sqrt{k_B T_e/m_e}$ in equation (1.40), then the ratio of the collision frequency to the plasma frequency is

$$\frac{\nu_c}{\omega_{pe}} = \frac{\ln \Lambda}{2\pi n \lambda_D^3} = \frac{2 \ln \Lambda}{3 N_D}, \tag{2.21}$$

so collisional terms are small compared to the average plasma terms by the order of N_D^{-1}.

In order to estimate the importance of the discrete behavior more carefully, we imagine a scheme where we subdivide each of the particles, so that we will have in the limit an infinite number of particles, but each is too small to influence its neighbor. To this end, we let $e \to 0$, $m_e \to 0$, but hold e/m_e constant. We also let $n \to \infty$, holding ne constant. In order to hold the thermal speed, $v_e \sim \sqrt{k_B T_e/m_e}$, constant, we also must let $T_e \to 0$. Neglecting ion motions, we shall assume there is simply a uniform positive background charge density to neutralize the plasma on the average. We may now write the Debye length as

$$\lambda_D^2 = \frac{\epsilon_0 k_B T_e}{ne^2} = \left(\frac{k_B T_e}{m_e}\right) \frac{\epsilon_0}{(ne)} \left(\frac{m_e}{e}\right), \tag{2.22}$$

which is constant. This means that the plasma parameter, $N_D \sim n\lambda_D^3 \to \infty$, so the collisional terms will vanish as $\ln N_D/N_D$ from equation (2.21).

Another way of estimating the smallness of the collisional term is to use the law of large numbers which indicates that deviations from average scale as the square root of the total number, so that $\delta N_s \sim N_0^{1/2} \sim N_D^{1/2}$. Then, since $\delta \boldsymbol{E}$ and $\delta \boldsymbol{B}$ are produced by δN_s through the Maxwell equations, we can argue from Poisson's equation that $\delta \boldsymbol{E} \sim e\delta N_s \sim N_0^{-1} N_0^{1/2} \sim N_0^{-1/2} \sim N_D^{-1/2}$. Since the right hand side of equation (2.20) involves products of these kinds of terms, the right hand side will be approximately constant as $N_D \to \infty$. As the density tends toward infinity, however, the left hand side will also tend toward infinity, so in *relative* terms, the right hand side will again vanish as $1/N_D$. This leaves us with

$$\frac{\partial f_s}{\partial t} + \boldsymbol{v} \cdot \nabla f_s + \frac{q_s}{m_s}(\boldsymbol{E} + \boldsymbol{v} \times \boldsymbol{B}) \cdot \nabla_{\boldsymbol{v}} f_s = 0\,. \tag{2.23}$$

Along with the corresponding Maxwell equations

$$\nabla \cdot \boldsymbol{E}(\boldsymbol{r},t) = \rho(\boldsymbol{r},t)/\epsilon_0\,, \tag{2.24}$$

$$\nabla \cdot \boldsymbol{B}(\boldsymbol{r},t) = 0\,, \tag{2.25}$$

$$\nabla \times \boldsymbol{E}(\boldsymbol{r},t) = -\frac{\partial \boldsymbol{B}(\boldsymbol{r},t)}{\partial t}\,, \tag{2.26}$$

$$\nabla \times \boldsymbol{B}(\boldsymbol{r},t) = \mu_0 \boldsymbol{J}(\boldsymbol{r},t) + \frac{1}{c^2}\frac{\partial \boldsymbol{E}(\boldsymbol{r},t)}{\partial t} \tag{2.27}$$

with the charge and current densities given by

$$\rho(\boldsymbol{r},t) = \sum_{e,i} q_s \int \mathrm{d}\boldsymbol{v}\, f_s(\boldsymbol{r},\boldsymbol{v},t) \tag{2.28}$$

$$\boldsymbol{J}(\boldsymbol{r},t) = \sum_{e,i} q_s \int \mathrm{d}\boldsymbol{v}\, \boldsymbol{v} f_s(\boldsymbol{r},\boldsymbol{v},t)\,. \tag{2.29}$$

Equations (2.23) through (2.29) form the *Vlasov equations*, although equation (2.23) alone is frequently called the Vlasov equation. This set of equations forms the foundation for most of plasma physics when collisions are not a dominant effect.

2.2 The Liouville equation

2.2.1 Introduction

In deriving the Klimontovich equation, we used a six-dimensional phase space to describe N_0 particles, whereas in this section, we will describe *systems of particles*. We will use this description to derive the *Liouville equation*, which will also give an exact result which is also too difficult to use, but will lead to a systematic method for ordering our approximations. One result of this approximation method will be to give an explicit form for the right-hand side term which describes collisions and correlations, and gives us an ordering with which to estimate the error.

In using the description of systems, we first consider a system consisting of one particle, then two particles, and then generalize to N_0 systems of N_0 particles. For the one particle "system," we describe the location in terms of the coordinate $\boldsymbol{r}_1$, and the particle is located at $\boldsymbol{R}_1(t)$ at time t. The velocity of this particle is measured in the velocity coordinate system $\boldsymbol{v}_1$, where the particle velocity is given by $\boldsymbol{V}_1(t) = \dot{\boldsymbol{R}}_1(t)$. Thus the trajectory is described by $\boldsymbol{r}_1(t)$ and $\boldsymbol{v}_1(t)$ in a six-dimensional phase space $\boldsymbol{r}_1$, $\boldsymbol{v}_1$. In such a simple system, the subscript is of course unnecessary, but as we add more particles to our system, particle one will *always* be described in the same coordinate system. Although it is trivial for one particle, we can also define a *system density* in this space as

$$N(\boldsymbol{r}_1,\boldsymbol{v}_1,t) = \delta[\boldsymbol{r}_1 - \boldsymbol{R}_1(t)]\delta[\boldsymbol{v}_1 - \boldsymbol{V}_1(t)]\,. \tag{2.30}$$

When we consider two particles, we want a *system* of two particles, and in this system, we will use separate coordinates for the second particle. The second particle's trajectory is described by $\boldsymbol{R}_2(t)$ and $\boldsymbol{V}_2(t)$ in a six-dimensional phase space $\boldsymbol{r}_2$, $\boldsymbol{v}_2$, resulting in the system having a twelve-dimensional phase space. We can similarly define a system density in this twelve-dimensional space as

$$N(\boldsymbol{r}_1, \boldsymbol{v}_1, \boldsymbol{r}_2, \boldsymbol{v}_2, t) = \delta[\boldsymbol{r}_1 - \boldsymbol{R}_1(t)]\delta[\boldsymbol{v}_1 - \boldsymbol{V}_1(t)]\delta[\boldsymbol{r}_2 - \boldsymbol{R}_2(t)]\delta[\boldsymbol{v}_2 - \boldsymbol{V}_2(t)]\,. \tag{2.31}$$

In this space, there is one system which occupies the point $\boldsymbol{r}_1 = \boldsymbol{R}_1(t)$, $\boldsymbol{v}_1 = \boldsymbol{V}_1(t)$, $\boldsymbol{r}_2 = \boldsymbol{R}_2(t)$, and $\boldsymbol{v}_2 = \boldsymbol{V}_2(t)$ at time t.

Note that this density is completely different from N_s in the previous chapter, where there we had the density of *particles* in a six-dimensional phase space, whereas here we have the density of systems (*each* of which has two particles) in twelve-dimensional phase space. Generalizing to N_0 particles, we wish to describe N_0 systems, each comprising N_0 particles in a $6N_0$-dimensional phase space, using a separate coordinate system for each particle. By analogy from equation (2.31), we have the density

$$N(\boldsymbol{r}_1, \boldsymbol{v}_1, \boldsymbol{r}_2, \boldsymbol{v}_2, \ldots, \boldsymbol{r}_{N_0}, \boldsymbol{v}_{N_0}, t) = \prod_{i=1}^{N_0} \delta[\boldsymbol{r}_i - \boldsymbol{R}_i(t)]\delta[\boldsymbol{v}_i - \boldsymbol{V}_i(t)]\,, \tag{2.32}$$

where $\prod_{j=1}^{n} f_n \equiv f_1 f_2 \cdots f_n$.

2.2.2 Liouville equation — standard form

We now wish to describe the evolution of the system density as before. Since our system density now involves the product of $6N_0$ terms, its derivative will involve the sum of $6N_0$ terms. Using relations of the form

$$\frac{\partial}{\partial t}\delta[\boldsymbol{r}_i - \boldsymbol{R}_i(t)] = -\frac{\partial \boldsymbol{R}_i}{\partial t} \cdot \nabla_{\boldsymbol{r}_i}\delta[\boldsymbol{r}_i - \boldsymbol{R}_i(t)]\,, \tag{2.33}$$

we find

$$\begin{aligned}\frac{\partial N}{\partial t} &+ \sum_{i=1}^{N_0} \boldsymbol{V}_i \cdot \nabla_{\boldsymbol{r}_i} \prod_{j=1}^{N_0} \delta[\boldsymbol{r}_j - \boldsymbol{R}_j(t)]\delta[\boldsymbol{v}_j - \boldsymbol{V}_j(t)] \\ &+ \sum_{i=1}^{N_0} \dot{\boldsymbol{V}}_i \cdot \nabla_{\boldsymbol{v}_i} \prod_{j=1}^{N_0} \delta[\boldsymbol{r}_j - \boldsymbol{R}_j(t)]\delta[\boldsymbol{v}_j - \boldsymbol{V}_j(t)] = 0\,. \end{aligned}\tag{2.34}$$

Using the same delta function relations as in the previous chapter, we can write for $\dot{\boldsymbol{V}}_i$ the expression

$$\dot{\boldsymbol{V}}_i(t) = \frac{q_s}{m_s}(\boldsymbol{E}^m(\boldsymbol{r}_i, t) + \boldsymbol{v}_i \times \boldsymbol{B}^m(\boldsymbol{r}_i, t)\,. \tag{2.35}$$

Then if we note that the products are just the density, then we may write equation (2.34) as

$$\frac{\partial N}{\partial t} + \sum_{i=1}^{N_0} \boldsymbol{V}_i \cdot \nabla_{\boldsymbol{r}_i} N + \sum_{i=1}^{N_0} \dot{\boldsymbol{V}}_i \cdot \nabla_{\boldsymbol{v}_i} N = 0\,, \tag{2.36}$$

which is the *Liouville equation.* Along with the Maxwell equations and equation (2.35) for the acceleration, this is a complete and exact description of the evolution of the plasma, but is of little practical use, since we still cannot deal with a $6N_0$-dimensional phase space. Before we try to reduce the complexity of the system, we note that again the equation is of the form of what is called a *convective time derivative* except that in this case the dimensionality is higher, such that

$$\frac{\mathrm{D}}{\mathrm{D}t}N(\boldsymbol{r}_1,\boldsymbol{v}_1,\boldsymbol{r}_2,\boldsymbol{v}_2,\ldots,\boldsymbol{r}_{N_0},\boldsymbol{v}_{N_0},t)=0 \tag{2.37}$$

where

$$\frac{\mathrm{D}}{\mathrm{D}t}\equiv\frac{\partial}{\partial t}+\sum_{i=1}^{N_0}\boldsymbol{v}_i\cdot\nabla_{\boldsymbol{r}_i}+\sum_{i=1}^{N_0}\dot{\boldsymbol{V}}_i\cdot\nabla_{\boldsymbol{v}_i}\,. \tag{2.38}$$

This is again the expression of the fact that the density of systems is incompressible.

We can also put the Liouville equation in the form of a continuity equation by noting some vector identities. First, from $\nabla\cdot(a\mathbf{b})=\mathbf{b}\cdot\nabla a+a\nabla\cdot\mathbf{b}$, we have

$$\boldsymbol{v}_i\cdot\nabla_{\boldsymbol{r}_i}N=\nabla_{\boldsymbol{r}_i}\cdot(\boldsymbol{v}_iN) \tag{2.39}$$

since $\boldsymbol{r}_i$ and $\boldsymbol{v}_i$ are independent. Also, we find

$$\dot{\boldsymbol{V}}_i\cdot\nabla_{\boldsymbol{v}_i}N=\nabla_{\boldsymbol{v}_i}\cdot(\dot{\boldsymbol{V}}_iN) \tag{2.40}$$

since

$$\nabla_{\boldsymbol{v}_i}\cdot\dot{\boldsymbol{V}}_i=\nabla_{\boldsymbol{v}_i}\cdot\left\{\frac{q_{s_i}}{m_{s_i}}[\boldsymbol{E}^m(\boldsymbol{r}_i,t)+\boldsymbol{v}_i\times\boldsymbol{B}^m(\boldsymbol{r}_i,t)]\right\}=0\,,$$

since $\boldsymbol{E}^m$ and $\boldsymbol{B}^m$ are independent of velocity and $\nabla_{\boldsymbol{v}_i}\times\boldsymbol{v}_i=0$. Then we may write the Liouville equation as

$$\frac{\partial N}{\partial t}+\sum_{i=1}^{N_0}\nabla_{\boldsymbol{r}_i}\cdot(\boldsymbol{v}_iN)+\sum_{i=1}^{N_0}\nabla_{\boldsymbol{v}_i}\cdot(\dot{\boldsymbol{V}}_iN)=0\,. \tag{2.41}$$

In this form, the Liouville equation expresses the conservation of systems in our $6N_0$-dimensional phase space.

2.3 System ensembles

If now we consider not one system, but an ensemble of N_0 systems, each described by the same set of coordinates in $6N_0$-dimensional phase space, then we can introduce the probability of finding any one of these systems inside a particular volume in phase space. Each of these individual systems will be initialized at t_0 and evolve separately, each one following a *single trajectory* in this phase space. Denoting the probability by

$$f_{N_0}(\boldsymbol{r}_1, \boldsymbol{v}_1, \boldsymbol{r}_2, \boldsymbol{v}_2, \ldots, \boldsymbol{r}_{N_0}, \boldsymbol{v}_{N_0}, t)\mathrm{d}\boldsymbol{r}_1\mathrm{d}\boldsymbol{v}_1\mathrm{d}\boldsymbol{r}_2\mathrm{d}\boldsymbol{v}_2 \cdots \mathrm{d}\boldsymbol{r}_{N_0}\mathrm{d}\boldsymbol{v}_{N_0}\,, \tag{2.42}$$

we interpret this f_{N_0} as representing the probability that there exists a *system* at the point $(\boldsymbol{r}_1, \boldsymbol{v}_1, \boldsymbol{r}_2, \boldsymbol{v}_2, \ldots, \boldsymbol{r}_{N_0}, \boldsymbol{v}_{N_0}, t)$ in the $6N_0$-dimensional phase space, or that $\boldsymbol{R}_1(t)$ lies between $\boldsymbol{r}_1$ and $\boldsymbol{r}_1 + \mathrm{d}\boldsymbol{r}_1$, *and* $\boldsymbol{V}_1(t)$ lies between $\boldsymbol{v}_1$ and $\boldsymbol{v}_1 + \mathrm{d}\boldsymbol{v}_1$, *and* $\boldsymbol{R}_2(t)$ lies between $\boldsymbol{r}_2$ and $\boldsymbol{r}_2 + \mathrm{d}\boldsymbol{r}_2$, *and* $\boldsymbol{V}_2(t)$ lies between $\boldsymbol{v}_2$ and $\boldsymbol{v}_2 + \mathrm{d}\boldsymbol{v}_2$, *and* etc. If we integrate over all space, the probability is unity that a system will be found. In this picture, each *system* has a *single trajectory*, and since each trajectory is smooth, with no system being created or destroyed, then this probability density must likewise satisfy a conservation of probability density, or

$$\frac{\partial f_{N_0}}{\partial t} + \sum_{i=1}^{N_0} \boldsymbol{v}_i \cdot \nabla_{\boldsymbol{r}_i} f_{N_0} + \sum_{i=1}^{N_0} \dot{\boldsymbol{V}}_i \cdot \nabla_{\boldsymbol{v}_i} f_{N_0} = 0\,, \tag{2.43}$$

which may also be written, using equation (2.39) and equation (2.40), as

$$\frac{\mathrm{D} f_{N_0}}{\mathrm{D}t} = 0\,. \tag{2.44}$$

Again, this indicates that the probability density is incompressible.

2.3.1 BBGKY hierarchy

In addition to the probability of finding a particle at *some* point in phase space, we may consider reduced probability distributions, defined by

$$f_k(\boldsymbol{r}_1, \boldsymbol{v}_1, \boldsymbol{r}_2, \boldsymbol{v}_2, \ldots, \boldsymbol{r}_k, \boldsymbol{v}_k, t) \equiv V^{k-N_0} \int \mathrm{d}\boldsymbol{r}_{k+1}\mathrm{d}\boldsymbol{v}_{k+1} \cdots \mathrm{d}\boldsymbol{r}_{N_0}\mathrm{d}\boldsymbol{v}_{N_0} f_{N_0}\,, \tag{2.45}$$

which gives the joint probability of finding particles 1 through k in the volume between $(\boldsymbol{r}_1, \boldsymbol{v}_1)$ and $(\boldsymbol{r}_1 + \mathrm{d}\boldsymbol{r}_1, \boldsymbol{v}_1 + \mathrm{d}\boldsymbol{v}_1)$, *and... and* between $(\boldsymbol{r}_k, \boldsymbol{v}_k)$ and $(\boldsymbol{r}_k + \mathrm{d}\boldsymbol{r}_k, \boldsymbol{v}_k + \mathrm{d}\boldsymbol{v}_k)$, irrespective of the coordinates of the other particles $k+1$, $k+2$,..., N_0. In equation (2.45), V is a finite spatial volume where f_{N_0} is nonzero. In the following development, we shall assume that $f_{N_0} \to 0$ as $|\boldsymbol{v}_i| \to \infty$ in order that the energy be bounded, and either use periodic boundary conditions in space, or assume that $f_{N_0} \to 0$ as $|\boldsymbol{r}_i| \to \infty$. We shall assume the particles are indistinguishable, so that the distribution is symmetric with respect to the exchange of any pair of particles. This means we do not care which particle is labeled 1, or which is labeled 2, etc. If we were to reduce the function by integrating over all the sets of coordinates but one, so that only one set of coordinates were left, we would have $f_1(\boldsymbol{r}_1, \boldsymbol{v}_1, t)$

which now is the probability (within a normalization constant) of finding a particle between $(\boldsymbol{r}_1, \boldsymbol{v}_1)$ and $(\boldsymbol{r}_1 + \mathrm{d}\boldsymbol{r}_1, \boldsymbol{v}_1 + \mathrm{d}\boldsymbol{v}_1)$, and then this distribution function would be equivalent to the function $f_s(\boldsymbol{r}, \boldsymbol{v}, t)$ which appeared in the kinetic equation and the Vlasov equations.

We now wish to actually *do* these integrations and reduce the complexity of the equations. To simplify this calculation, enabling us to see the structure of the arguments without becoming too complex algebraically, we will

1. Consider only one species of particles (could be extended, but will consider ions to be a uniform positive background),
2. Exclude any external electric and magnetic fields,
3. Ignore the self-generated magnetic fields, and
4. Adopt the *Coulomb Model* where only mutual electrostatic forces are considered.

In this model,

$$\dot{\boldsymbol{V}}_i(t) = \sum_{j \neq i}^{N_0} \boldsymbol{a}_{ij} \,, \tag{2.46}$$

$$\boldsymbol{a}_{ij} = \frac{q_s^2}{4\pi\epsilon_0 m_s |\boldsymbol{r}_i - \boldsymbol{r}_j|^3} (\boldsymbol{r}_i - \boldsymbol{r}_j) \,, \tag{2.47}$$

where $\boldsymbol{a}_{ij}$ is the acceleration of particle i *due* to particle j, and the sum obviously excludes self forces ($j = i$). The Liouville equation may then be written in the form of equation (2.44) as

$$\frac{\partial f_{N_0}}{\partial t} + \sum_{i=1}^{N_0} \boldsymbol{v}_i \cdot \nabla_{\boldsymbol{r}_i} f_{N_0} + \sum_{i=1}^{N_0} \sum_{j \neq i}^{N_0} \boldsymbol{a}_{ij} \cdot \nabla_{\boldsymbol{v}_i} f_{N_0} = 0 \,. \tag{2.48}$$

Now we integrate over $\boldsymbol{r}_{N_0}$ and $\boldsymbol{v}_{N_0}$ to obtain f_{N_0}, so that

$$\int \mathrm{d}\boldsymbol{r}_{N_0} \mathrm{d}\boldsymbol{v}_{N_0} \frac{\partial f_{N_0}}{\partial t} + \int \mathrm{d}\boldsymbol{r}_{N_0} \mathrm{d}\boldsymbol{v}_{N_0} \sum_{i=1}^{N_0} \boldsymbol{v}_i \cdot \nabla_{\boldsymbol{r}_i} f_{N_0} + \int \mathrm{d}\boldsymbol{r}_{N_0} \mathrm{d}\boldsymbol{v}_{N_0} \sum_{i=1}^{N_0} \sum_{j \neq i}^{N_0} \boldsymbol{a}_{ij} \cdot \nabla_{\boldsymbol{v}_i} f_{N_0} = I_1 + I_2 + I_3 = 0. \tag{2.49}$$

Treating the first integral first, we have

$$I_1 = \int \mathrm{d}\boldsymbol{r}_{N_0} \mathrm{d}\boldsymbol{v}_{N_0} \frac{\partial f_{N_0}}{\partial t} = \frac{\partial}{\partial t} \int \mathrm{d}\boldsymbol{r}_{N_0} \mathrm{d}\boldsymbol{v}_{N_0} \, f_{N_0} = \frac{\partial}{\partial t} V f_{N_0 - 1} \,, \tag{2.50}$$

since the coordinates are independent of time, noting that

$$f_{N_0 - 1} = \frac{1}{V} \int \mathrm{d}\boldsymbol{r}_{N_0} d\boldsymbol{v}_{N_0} \, f_{N_0} \,. \tag{2.51}$$

For the second integral of equation (2.49), coordinates of the first $N_0 - 1$ terms are independent of the integration variables, so we can interchange the order of the integration with the $\boldsymbol{v}_i \cdot \nabla_{\boldsymbol{r}_i}$ operator, with the result

$$I_2 = \sum_{i=1}^{N_0-1} \boldsymbol{v}_i \cdot \nabla_{\boldsymbol{r}_i} \int \mathrm{d}\boldsymbol{r}_{N_0} \mathrm{d}\boldsymbol{v}_{N_0}\, f_{N_0} + \int \mathrm{d}\boldsymbol{r}_{N_0} \mathrm{d}\boldsymbol{v}_{N_0}\, \boldsymbol{v}_{N_0} \cdot \nabla_{\boldsymbol{r}_{N_0}} f_{N_0} \,. \tag{2.52}$$

Looking at the form of the second term, however, we can integrate by parts to obtain

$$\int \mathrm{d}\boldsymbol{r}\, \mathrm{d}\boldsymbol{v} \left(v_x \frac{\partial}{\partial x} + v_y \frac{\partial}{\partial y} + v_z \frac{\partial}{\partial z} \right) f = \int \mathrm{d}y\, \mathrm{d}z\, \mathrm{d}\boldsymbol{v}\, v_x f \Big|_{-\infty}^{\infty} + \text{cyclic perm.}$$
$$- \int \mathrm{d}\boldsymbol{r}\, \mathrm{d}\boldsymbol{v} f \nabla \cdot \boldsymbol{v} = 0 \tag{2.53}$$

using either periodic boundary conditions or $f \to 0$ as $|\boldsymbol{r}| \to \infty$. The second integral thus becomes

$$I_2 = V \sum_{i=1}^{N_0-1} \boldsymbol{v}_i \cdot \nabla_{\boldsymbol{r}_i} f_{N_0-1} \,. \tag{2.54}$$

To do the final integral, we first split up the acceleration sum such that

$$\sum_{i=1}^{N_0} \sum_{j \neq i}^{N_0} g_{ij} = \sum_{i=1}^{N_0-1} \sum_{j \neq i}^{N_0-1} g_{ij} + \sum_{j=1}^{N_0-1} g_{N_0 j} + \sum_{i=1}^{N_0-1} g_{i N_0} \,. \tag{2.55}$$

In the double sum, the variables in the operator are independent of the variables of integration, so the order may again be reversed. The integral over the next sum,

$$\int \mathrm{d}\boldsymbol{v}_{N_0} \boldsymbol{a}_{N_0 j} \cdot \frac{\partial f_{N_0}}{\partial \boldsymbol{v}_{N_0}} = \boldsymbol{a}_{N_0 j}\, f_{N_0} \Big|_{\boldsymbol{v}_{N_0} \to -\infty}^{\boldsymbol{v}_{N_0} \to \infty} = 0 \,, \tag{2.56}$$

so I_3 becomes

$$I_3 = V \sum_{i=1}^{N_0-1} \sum_{j \neq i}^{N_0-1} \boldsymbol{a}_{ij} \cdot \nabla_{\boldsymbol{v}_i} f_{N_0-1} + \sum_{i=1}^{N_0-1} \int \mathrm{d}\boldsymbol{r}_{N_0} \mathrm{d}\boldsymbol{v}_{N_0}\, \boldsymbol{a}_{i,N_0} \cdot \nabla_{\boldsymbol{v}_i} f_{N_0} \,. \tag{2.57}$$

Combining these pieces, and dividing by V, we finally obtain

$$\frac{\partial}{\partial t} f_{N_0-1} + \sum_{i=1}^{N_0-1} \boldsymbol{v}_i \cdot \nabla_{\boldsymbol{r}_i} f_{N_0-1} + \sum_{i=1}^{N_0-1} \sum_{j \neq i}^{N_0-1} \boldsymbol{a}_{ij} \cdot \nabla_{\boldsymbol{v}_i} f_{N_0-1}$$
$$+ \frac{1}{V} \sum_{i \neq N_0}^{N_0-1} \int \mathrm{d}\boldsymbol{r}_{N_0} \mathrm{d}\boldsymbol{v}_{N_0}\, \boldsymbol{a}_{i N_0} \cdot \nabla_{\boldsymbol{v}_i} f_{N_0} = 0 \,. \tag{2.58}$$

From this expression, we first note that the equation for f_{N_0-1} depends on f_{N_0}, and secondly we note that as we have made no approximations, this is still exact. As we integrate over each set of coordinates successively, we will again obtain a similar equation, where f_k depends on f_{k+1}, leading to a chain of equations. To see how this develops, we will integrate one more time (following the pattern of 1,2,...,N), this time over $\boldsymbol{r}_{N_0-1}$ and $\boldsymbol{v}_{N_0-1}$. Noting first from equation (2.45) that

$$\begin{aligned} f_{N_0-2} &= \frac{1}{V^2}\int \mathrm{d}\boldsymbol{r}_{N_0}\mathrm{d}\boldsymbol{v}_{N_0}\mathrm{d}\boldsymbol{r}_{N_0-1}\mathrm{d}\boldsymbol{v}_{N_0-1} f_{N_0} \\ &= \frac{1}{V^2}\int \mathrm{d}\boldsymbol{r}_{N_0-1}\mathrm{d}\boldsymbol{v}_{N_0-1}\int \mathrm{d}\boldsymbol{r}_{N_0}\mathrm{d}\boldsymbol{v}_{N_0} f_{N_0} \\ &= \frac{1}{V^2}\int \mathrm{d}\boldsymbol{r}_{N_0-1}\mathrm{d}\boldsymbol{v}_{N_0-1} V f_{N_0-1}\,, \end{aligned} \tag{2.59}$$

where we have used equation (2.51), so we find

$$\int \mathrm{d}\boldsymbol{r}_{N_0-1}\mathrm{d}\boldsymbol{v}_{N_0-1} f_{N_0-1} = V f_{N_0-2}\,. \tag{2.60}$$

Using this result, we have that the integral of the first term of equation (2.58) gives $V\partial f_{N_0-2}/\partial t$.

The integral of the second term in equation (2.58) again splits into two parts as in equation (2.52), one whose sum terminates with $N_0 - 2$ terms whose variables are independent, and one where we can do the integration and evaluate at the end points and obtain a null result. The corresponding result to equation (2.54) is then

$$I_2 = V\sum_{i=1}^{N_0-2} \boldsymbol{v}_i\cdot\nabla_{\boldsymbol{r}_i} f_{N_0-2}\,. \tag{2.61}$$

The third term is broken again into a double sum and two single sums as in equation (2.55), indexing $N_0 \to N_0 - 1$, so that

$$\begin{aligned} I_3 =& V\sum_{i=1}^{N_0-2}\sum_{j\neq i}^{N_0-2} \boldsymbol{a}_{ij}\cdot\nabla_{\boldsymbol{v}_i} f_{N_0-2} + \sum_{i=1}^{N_0-2}\int \mathrm{d}\boldsymbol{r}_{N_0-1}\mathrm{d}\boldsymbol{v}_{N_0-1}\, \boldsymbol{a}_{iN_0-1}\cdot\nabla_{\boldsymbol{v}_i} f_{N_0-1} \\ &+ \sum_{j=1}^{N_0-2}\int \mathrm{d}\boldsymbol{r}_{N_0-1}\mathrm{d}\boldsymbol{v}_{N_0-1}\, \boldsymbol{a}_{N_0-1j}\cdot\nabla_{\boldsymbol{v}_{N_0-1}} f_{N_0-1}\,, \end{aligned} \tag{2.62}$$

where the last term vanishes upon integration as before.

Upon integrating the fourth term in equation (2.58), we have

$$\begin{aligned} I_4 &= \frac{1}{V}\sum_{i=1}^{N_0-1}\int \mathrm{d}\boldsymbol{r}_{N_0-1}\mathrm{d}\boldsymbol{v}_{N_0-1}\,\mathrm{d}\boldsymbol{r}_{N_0}\mathrm{d}\boldsymbol{v}_{N_0}\boldsymbol{a}_{iN_0}\cdot\nabla_{\boldsymbol{v}_i} f_{N_0} \\ &= \frac{1}{V}\sum_{i=1}^{N_0-2}\int \mathrm{d}\boldsymbol{r}_{N_0}\mathrm{d}\boldsymbol{v}_{N_0}\,\boldsymbol{a}_{iN_0}\cdot\nabla_{\boldsymbol{v}_i}\int \mathrm{d}\boldsymbol{r}_{N_0-1}\mathrm{d}\boldsymbol{v}_{N_0-1}\, f_{N_0}\,, \end{aligned} \tag{2.63}$$

where the N_0-1 term in the sum vanishes upon doing the integral over $\boldsymbol{v}_{N_0-1}$. Now the variables of integration in equation (2.63) are dummy variables, so we can exchange the subscripts N_0 and N_0-1, with $\boldsymbol{a}_{iN_0} \to \boldsymbol{a}_{iN_0-1}$. Since f_{N_0} is symmetric with respect to the interchange of coordinates, it remains the same. But from equation (2.51), the last integral is simply Vf_{N_0-1}, so we have

$$I_4 = \sum_{i=1}^{N_0-2} \int \mathrm{d}\boldsymbol{r}_{N_0-1}\mathrm{d}\boldsymbol{v}_{N_0-1}\, \boldsymbol{a}_{iN_0-1} \cdot \nabla_{\boldsymbol{v}_i} f_{N_0-1}\,. \tag{2.64}$$

This term is identical to the second term of I_3, so the final result is (after dividing by V):

$$\begin{aligned}\frac{\partial}{\partial t} f_{N_0-2} + \sum_{i=1}^{N_0-2} \Bigg(& \boldsymbol{v}_i \cdot \nabla_{\boldsymbol{r}_i} f_{N_0-2} + \sum_{j\neq i}^{N_0-2} \boldsymbol{a}_{ij} \cdot \nabla_{\boldsymbol{v}_i} f_{N_0-2} \\ & + \frac{2}{V} \int \mathrm{d}\boldsymbol{r}_{N_0-1}\mathrm{d}\boldsymbol{v}_{N_0-1}\, \boldsymbol{a}_{iN_0-1} \cdot \nabla_{\boldsymbol{v}_i} f_{N_0-1} \Bigg) = 0\,.\end{aligned} \tag{2.65}$$

Note that the equation for f_{N_0-2} involves the next higher term, f_{N_0-1}, as noted above.

Generalizing now, we can see from the pattern above what the result will be from integrating over all but k variables. The result is

$$\begin{aligned}\frac{\partial f_k}{\partial t} + \sum_{i=1}^{k} \Bigg(& \boldsymbol{v}_i \cdot \nabla_{\boldsymbol{r}_i} f_k + \sum_{j\neq i}^{k} \boldsymbol{a}_{ij} \cdot \nabla_{\boldsymbol{v}_i} f_k \\ & + \frac{(N_0-k)}{V} \int \mathrm{d}\boldsymbol{r}_{k+1}\mathrm{d}\boldsymbol{v}_{k+1}\, \boldsymbol{a}_{ik+1} \cdot \nabla_{\boldsymbol{v}_i} f_{k+1} \Bigg) = 0\,.\end{aligned} \tag{2.66}$$

This chain of equations is the **BBGKY hierarchy** (**B**ogolyubov[7], **B**orn and **G**reen[8], **K**irkwood[9], and **Y**von[10]).

2.3.2 Truncating the BBGKY hierarchy

As it stands, the BBGKY hierarchy is still exact and as impossible to treat as the Klimontovich equation or the Liouville equation because it still has N_0 equations to solve. Only if we can reduce the order of the system of equations to a reasonable number will we be able to make progress. If we start at the bottom of the chain, with $k=1$, we have

$$\frac{\partial}{\partial t} f_1(\boldsymbol{r}_1,\boldsymbol{v}_1,t) + \boldsymbol{v}_1 \cdot \nabla_{\boldsymbol{r}_1} f_1 + \frac{N_0-1}{V} \int \mathrm{d}\boldsymbol{r}_2\, \mathrm{d}\boldsymbol{v}_2\, \boldsymbol{a}_{12} \cdot \nabla_{\boldsymbol{v}_1} f_2(\boldsymbol{r}_1,\boldsymbol{v}_1,\boldsymbol{r}_2,\boldsymbol{v}_2,t) = 0\,. \tag{2.67}$$

This is coupled to the $k=2$ equation through f_2, which is coupled to the $k=3$ equation, etc. Only if we can truncate this chain can we make any real progress. In order to do this, we need to be reminded of what information is contained in each of the various f_k, and especially what the meaning of f_1, f_2 and f_3 is. From our previous discussions, we can conclude that $f_1(\boldsymbol{r}_1, \boldsymbol{v}_1, t)\mathrm{d}\boldsymbol{r}_1\mathrm{d}\boldsymbol{v}_1$ is the probability that a particle will be found in the volume between $\boldsymbol{r}_1$ and $\boldsymbol{r}_1 + \mathrm{d}\boldsymbol{r}_1$ and in the region of velocity space between $\boldsymbol{v}_1$ and $\boldsymbol{v}_1 + \mathrm{d}\boldsymbol{v}_1$. On the other hand, f_2 may be interpreted as the *joint* probability of finding particle 1 at $(\boldsymbol{r}_1, \boldsymbol{v}_1)$ *and* that particle 2 is at $(\boldsymbol{r}_2, \boldsymbol{v}_2)$ (where *at* means in the differential neighborhood of the point). Presuming still that all of the particles are of the same species, this means that $f_2 \to 0$ if $\boldsymbol{r}_1 \to \boldsymbol{r}_2$, since two particles cannot occupy the same point in space (they could have the same velocity). We also know that $f_2(\boldsymbol{r}_1, \boldsymbol{v}_1, \boldsymbol{r}_2, \boldsymbol{v}_2, t) = f_2(\boldsymbol{r}_2, \boldsymbol{v}_2, \boldsymbol{r}_1, \boldsymbol{v}_1, t)$ since the particles and coordinates were interchangeable in f_{N_0}, and the property is not lost by our integrations.

In fact, we know a little more about f_2 if we remind ourselves of the properties of conditional probability distribution functions. If one considers two separate events, each characterized by the probability $P(x)$ if taken alone, while denoted $P(x, y)$ if considered together, then we must first ask if the two events are *correlated*, that is whether the second event depends on the first. If the two events are *statistically independent* or uncorrelated, then the probability of the event x, y is characterized by the joint probability distribution

$$P(x, y) = P(x)P(y)\,, \tag{2.68}$$

or by the product of the two probabilities taken independently. For example, with a pair of fair dice, where the probability of getting any specific digit of the six possible is $1/6$, the probability of snake eyes $(1+1 = 2)$ is $(1/6)(1/6) = 1/36$, so that it is to be expected once in every 36 rolls of the pair. On the other hand, if the two events are correlated, so that one depends on the other, then we can write the more general relation as

$$P(x, y) = P(x)P(y) + \delta P(x, y)\,, \tag{2.69}$$

where $\delta P(x, y) = 0$ if the events are uncorrelated. For example, if one wants the probability of two specific cards to be dealt from a well shuffled deck in a particular order, the probability of the first is $1/52$, while the probability of the second is $1/51$ since there are only 51 cards left after the first is dealt. For this example,

$$\begin{aligned} \delta P(x, y) &= P(x, y) - P(x)P(y) \\ &= \frac{1}{52} \cdot \frac{1}{51} - \frac{1}{52} \cdot \frac{1}{52} = \frac{1}{52^2 51}\,. \end{aligned} \tag{2.70}$$

In this example, the correlation term is the order of N (=52) smaller than the uncorrelated term because the first event had only a small effect on the second event.

For our problem, we will define the correlation function in a similar way, so that

$$f_2(\boldsymbol{r}_1, \boldsymbol{v}_1, \boldsymbol{r}_2, \boldsymbol{v}_2, t) \equiv f_1(\boldsymbol{r}_1, \boldsymbol{v}_1, t) f_1(\boldsymbol{r}_2, \boldsymbol{v}_2, t) + g(\boldsymbol{r}_1, \boldsymbol{v}_1, \boldsymbol{r}_2, \boldsymbol{v}_2, t)\,, \tag{2.71}$$

so that g is the correlation function. Since there are only two particles in the correlation function, it is called a *two-particle correlation function*, or often a *two-body correlation function*, especially if the particles are of different q/m. Breaking the two-body function f_2 into these components is the first step in the *Mayer cluster expansion*. We may now insert this result into equation (2.67) to obtain

$$\frac{\partial}{\partial t} f_1(\boldsymbol{r}_1, \boldsymbol{v}_1, t) + \boldsymbol{v}_1 \cdot \nabla_{\boldsymbol{r}_1} f_1 + n_0 \int \mathrm{d}\boldsymbol{r}_2 \, \mathrm{d}\boldsymbol{v}_2 \, \boldsymbol{a}_{12} \cdot \nabla_{\boldsymbol{v}_1} [f_1(\boldsymbol{r}_1, \boldsymbol{v}_1, t) f_1(\boldsymbol{r}_2, \boldsymbol{v}_2, t) + g(\boldsymbol{r}_1, \boldsymbol{v}_1, \boldsymbol{r}_2, \boldsymbol{v}_2, t)] = 0 \,, \tag{2.72}$$

where we have approximated $(N_0 - 1)/V \approx n_0$, the average particle density, since N_0 is assumed to be large compared to 1.

If we now take the case where the correlations are small enough to be neglected, then we can reduce this even further to

$$\frac{\partial}{\partial t} f_1(\boldsymbol{r}_1, \boldsymbol{v}_1, t) + \boldsymbol{v}_1 \cdot \nabla_{\boldsymbol{r}_1} f_1 + [n_0 \int \mathrm{d}\boldsymbol{r}_2 \, \mathrm{d}\boldsymbol{v}_2 \, \boldsymbol{a}_{12} f_1(\boldsymbol{r}_2, \boldsymbol{v}_2, t)] \cdot \nabla_{\boldsymbol{v}_1} f_1(\boldsymbol{r}_1, \boldsymbol{v}_1, t) = 0, \tag{2.73}$$

where some terms have been rearranged, but the term in brackets now may be simplified. If we examine this term, it represents the integral over the coordinates of particle 2 of the acceleration weighted by the distribution function for particle 2. This is simply the ensemble averaged acceleration experienced by particle 1, so that we may introduce the notation

$$\boldsymbol{a}(\boldsymbol{r}_1, t) \equiv n_0 \int \mathrm{d}\boldsymbol{r}_2 \mathrm{d}\boldsymbol{v}_2 \, \boldsymbol{a}_{12} f_1(\boldsymbol{r}_2, \boldsymbol{v}_2, t) \,. \tag{2.74}$$

With this substitution, equation (2.73) becomes

$$\frac{\partial f_1}{\partial t} + \boldsymbol{v}_1 \cdot \nabla_{\boldsymbol{r}_1} f_1 + \boldsymbol{a} \cdot \nabla_{\boldsymbol{v}_1} f_1 = 0 \,, \tag{2.75}$$

which is the kinetic equation we encountered before. In this context, the kinetic equation must be considered the *correlationless* kinetic equation, since collisions have not been introduced in this section. We can see from the earlier discussion of dividing the particles so that $e \to 0$, $m \to 0$, etc., that this process would lead to an absence of correlations, since no particles would interact. It is likely that collisions are the most important component of the correlation function, but other processes such as the generation of plasma waves and electromagnetic radiation could affect the correlations. While the Vlasov equation is the most important equation in investigating kinetic effects in plasmas, and will receive nearly all of the attention in subsequent chapters, we would like to have some firmer handle on the collision or correlation term in order to better establish the limits of validity and the first order deviations. In order to do this, we eventually want an equation in f_1 alone, so the system of equations is closed. Finding an expression for g in terms of f_1 will be difficult, and will be approximate, but it is necessary to find some rationale for closure which depends on collisions. To do this, which means finding an expression for g, we need to examine the second equation in the hierarchy using equation (2.66) with $k = 2$, which is

$$\frac{\partial f_2}{\partial t} + (\boldsymbol{v}_1 \cdot \nabla_{\boldsymbol{r}_1} + \boldsymbol{v}_2 \cdot \nabla_{\boldsymbol{r}_2}) f_2 + (\boldsymbol{a}_{12} \cdot \nabla_{\boldsymbol{v}_1} + \boldsymbol{a}_{21} \cdot \nabla_{\boldsymbol{v}_2}) f_2 + n_0 \int \mathrm{d}\boldsymbol{r}_3 \boldsymbol{v}_3 (\boldsymbol{a}_{13} \cdot \nabla_{\boldsymbol{v}_1} + \boldsymbol{a}_{23} \cdot \nabla_{\boldsymbol{v}_2}) f_3 = 0 \,. \tag{2.76}$$

As before, we will let $f_2(12) = f_1(1)f_1(2) + g(12)$, and with the appearance of f_3 in equation (2.76), we will need to introduce the next step in the Mayer cluster expansion, or

$$f_3(123) = f_1(1)f_1(2)f_1(3) + f_1(1)g(23) + f_1(2)g(13) + f_1(3)g(12) + h(123) \,, \tag{2.77}$$

where the numbers 1,2,3 in the parenthesis refer to the coordinate set for the respective particles. In this expansion, the three-body correlation function appears, but our first truncation approximation is to neglect $h(123)$. If we can prove that g is small compared to f, then it follows that h will be small compared to g. Examining the terms of equation (2.76) separately, we find for the various terms

$$I_1 = \dot{f}_1(1)f_1(2) + \dot{f}_1(2)f_1(1) + \dot{g}(12) \,, \tag{2.78}$$

$$I_2 = f_1(2)\boldsymbol{v}_1 \cdot \nabla_{\boldsymbol{r}_1} f_1(1) + \boldsymbol{v}_1 \cdot \nabla_{\boldsymbol{r}_1} g(12) + \{1 \leftrightarrow 2\} \,, \tag{2.79}$$

$$I_3 = f_1(2)\boldsymbol{a}_{12} \cdot \nabla_{\boldsymbol{v}_1} f_1(1) + \boldsymbol{a}_{12} \cdot \nabla_{\boldsymbol{v}_1} g(12) + \{1 \leftrightarrow 2\} \,, \tag{2.80}$$

$$I_4 = n_0 \int \mathrm{d}3\boldsymbol{a}_{13} \cdot \nabla_{\boldsymbol{v}_1} [f_1(1)f_1(2)f_1(3) + f_1(1)g(23) + f_1(2)g(13) + f_1(3)g(12)] + \{1 \leftrightarrow 2\}, \tag{2.81}$$

where $\mathrm{d}3 \equiv \mathrm{d}\boldsymbol{r}_3 \, \mathrm{d}\boldsymbol{v}_3$, and $\{1 \leftrightarrow 2\}$ implies the same terms repeated with labels 1 and 2 interchanged. Now we can simplify this by using equation (2.72), if we take the first terms of I_1 and I_2 and the first and third term of I_4, we have

$$\left\{ \dot{f}_1(1) + \boldsymbol{v}_1 \cdot \nabla_{\boldsymbol{r}_1} f_1(1) + n_0 \int \mathrm{d}3\boldsymbol{a}_{13} \cdot \nabla_{\boldsymbol{v}_1} [f_1(1)f_1(3) + g(13)] \right\} f_1(2) = 0 \tag{2.82}$$

by equation (2.72). In the same fashion, the second term in I_1 combines with some of the interchange terms to vanish (using $g(12) = g(21)$), leaving us with

$$\begin{aligned} &\dot{g}(12) + (\boldsymbol{v}_1 \cdot \nabla_{\boldsymbol{r}_1} + \boldsymbol{v}_2 \cdot \nabla_{\boldsymbol{r}_2}) g(12) \\ &= -(\boldsymbol{a}_{12} \cdot \nabla_{\boldsymbol{v}_1} + \boldsymbol{a}_{21} \cdot \nabla_{\boldsymbol{v}_2}) [f_1(1)f_1(2) + g(12)] \\ &\quad - \left\{ n_0 \int \mathrm{d}3\boldsymbol{a}_{13} \cdot \nabla_{\boldsymbol{v}_1} [f_1(1)g(23) + f_1(3)g(12)] + \{1 \leftrightarrow 2\} \right\} . \end{aligned} \tag{2.83}$$

This equation, along with equation (2.72), written using the average acceleration from equation (2.74) as

$$\dot{f}_1(1) + \boldsymbol{v}_1 \cdot \nabla_{\boldsymbol{r}_1} f_1 + \boldsymbol{a} \cdot \nabla_{\boldsymbol{v}_1} f_1 = -n_0 \int \mathrm{d}2\, \boldsymbol{a}_{12} \cdot \nabla_{\boldsymbol{v}_1} g(12)\,, \qquad (2.84)$$

completes the set of two equations in two unknowns, f_1 and g. Thus, by neglecting the three-body correlations, we have closed the set. We note that equation (2.84) is essentially identical to equation (2.20).

Problem 2.1 *BBGKY Hierarchy.*
Integrate equation (2.65) over all $\boldsymbol{r}_{N_0-2}$ and $\boldsymbol{v}_{N_0-2}$ to obtain the $k{-}N_0{-}3$ equation of the BBGKY hierarchy, and compare your result to equation (2.66).

Problem 2.2 *Three-Point Correlations (Coins).*
In equation (2.77), we defined a three-point joint probability function f_3 in terms of the one-point probability f_1, the two-point correlation function g, and the three-point correlation function, h. Suppose we apply this kind of thinking to the case of three coins, each of which can come up heads (+) or tails (−). What is the meaning of f_3 in this case? Write out f_3 in the form of equation (2.77), and evaluate f_3, f_1, g, and h for each of the following cases:

1. All three coins are "honest," that is, each coin is equally likely to come up heads or tails, and each coin is unaffected by any other coin.

2. Because the coins are mysteriously locked together, in any one throw all three are heads or tails, the result changing randomly from throw to throw.

3. All three coins always come up tails.

4. The first two coins always come up heads, while the third is honest. Note that here the probability functions are not symmetric, so that, for example, $f_1(1)$ is not the same function as $f_1(3)$.

Problem 2.3 *Three-Point Correlations (Dice).*
In equation (2.77), we defined a three-point joint probability function f_3 in terms of the one-point probability f_1, the two-point correlation function g, and the three-point correlation function, h. Suppose we apply this kind of thinking to the case of three dice, each of which can take on integer values from one through six. What is the meaning of f_3 in this case? Write out f_3 in the form of equation (2.77), and evaluate f_3, f_1, g, and h for each of the following cases:

1. All three dice are "honest," that is, the value of each die is equally likely to be one through six, and is independent of the value of any other die.

2. Because the dice are mysteriously locked together, in any one throw all three always show the same value, the value changing randomly from throw to throw with all six values equally likely.

3. All of the dice always come up “five.”
4. The first two dice always come up “two,” while the third is honest.

3

THE LENARD-BALESCU EQUATION

While we have reduced the BBGKY hierarchy to two equations in two unknowns by neglecting three-body correlations, the two equations are so complicated that we can learn little from them as they are virtually unsolvable. We will thus begin making a series of approximations, which are generally quite good except for exceptional cases, and finally obtain a solution of this reduced problem. One of the fundamental concepts in the approximation has to do with our assumption that a large angle collision is the result of many small collisions, so that trajectories are only changed a small amount in a single encounter. This was dependent upon the plasma parameter being large, so the largeness of N_D (or $\Lambda = 9N_D$) will be our fundamental assumption.

3.1 Bogolyubov's hypothesis

Suppose we consider a spatially uniform ensemble of plasmas. If truly uniform, then any function of one coordinate must be independent of location, so that $f(\boldsymbol{r}_1, \boldsymbol{v}_1, t) = f(\boldsymbol{v}_1, t)$, and $\boldsymbol{a}(\boldsymbol{r}_1, t) = \boldsymbol{a}(t) = 0$. Considering two particles at a time, any function which is averaged over the ensemble must depend only on the difference between the locations of the two particles, so that

$$g(12) = g(\boldsymbol{r}_1 - \boldsymbol{r}_2, \boldsymbol{v}_1, \boldsymbol{v}_2, t)\,. \tag{3.1}$$

With these assumptions, equation (2.84) reduces to

$$\partial_t f_1(\boldsymbol{v}_1, t) = -n_0 \int \mathrm{d}\boldsymbol{r}_2\, \mathrm{d}\boldsymbol{v}_2\, \boldsymbol{a}_{12} \cdot \nabla_{\boldsymbol{v}_1} g(12)\,, \tag{3.2}$$

where we denote $\partial/\partial t \to \partial_t$. This assumption also simplifies equation (2.83), since two of the integrals are of the form

$$\left[n_0 \int \mathrm{d}3\, \boldsymbol{a}_{13} f_1(3)\right] \cdot \nabla_{\boldsymbol{v}_1} g(12) = \boldsymbol{a} \cdot \nabla_{\boldsymbol{v}_1} g(12) = 0\,. \tag{3.3}$$

The remaining terms in equation (2.83) may be written

$$\partial_t g(12) + \boldsymbol{v}_1 \cdot \nabla_{\boldsymbol{r}_1} g(12) + \boldsymbol{v}_2 \cdot \nabla_{\boldsymbol{r}_2} g(12)$$

$$+(\boldsymbol{a}_{12}\cdot\nabla_{\boldsymbol{v}_1}+\boldsymbol{a}_{21}\cdot\nabla_{\boldsymbol{v}_2})g(12)$$
$$+n_0\int \mathrm{d}3\,\boldsymbol{a}_{13}\cdot\nabla_{\boldsymbol{v}_1}f_1(1)g(23)+n_0\int \mathrm{d}3\,\boldsymbol{a}_{23}\cdot\nabla_{\boldsymbol{v}_2}f_1(2)g(13)$$
$$=-(\boldsymbol{a}_{12}\cdot\nabla_{\boldsymbol{v}_1}+\boldsymbol{a}_{21}\cdot\nabla_{\boldsymbol{v}_2})f_1(1)f_1(2)\,. \tag{3.4}$$

We now wish to estimate the order of the various terms in equation (3.4). Consider the "pulverization procedure" where $e\to 0$ and $m\to 0$ but $e/m=$ const., $n_0 e=$ const., and $v_e=$ const. This requires $n_0\to\infty$ and $T_e\to 0$ but leaves $\omega_{pe}^2\sim(n_0e)(e/m)$ and $\lambda_D\sim v_e/\omega_{pe}$ constant. Even though λ_D is constant, $N_D\to\infty$, so collisions cease. As each charge vanishes, there are no longer any correlations, since $\boldsymbol{a}\sim e^2/m\to 0$, so that $g\to 0$ as $1/N_D$. Since ω_{pe} remains constant, however, the collective effects will survive the pulverization, so that g is of order $1/N_D$ compared to f_1, as is $\boldsymbol{a}$. The integrals in equation (3.4) are multiplied by n_0, so there is the product of $n_0e^2/m=(n_0e)(e/m)$ =const., so these terms are of the same order as g. Thus all of the terms are of order N_D^{-1} except the fourth term on the left which is of order N_D^{-2}. Discarding this term, we write the equation in the form

$$\partial_t g+V_1 g+V_2 g=S\,, \tag{3.5}$$

where V_1 and V_2 are operators given by

$$V_1 g(12)=\boldsymbol{v}_1\cdot\nabla_{\boldsymbol{r}_1}g(12)+\left[n_0\int \mathrm{d}3\,\boldsymbol{a}_{13}g(23)\right]\cdot\nabla_{\boldsymbol{v}_1}f_1(1)\,, \tag{3.6}$$

$$V_2 g(12)=\boldsymbol{v}_2\cdot\nabla_{\boldsymbol{r}_2}g(12)+\left[n_0\int \mathrm{d}3\,\boldsymbol{a}_{23}g(13)\right]\cdot\nabla_{\boldsymbol{v}_2}f_1(2)\,, \tag{3.7}$$

and the source term S is given by

$$S(\boldsymbol{r}_1-\boldsymbol{r}_2,\boldsymbol{v}_1,\boldsymbol{v}_2,t)=-(\boldsymbol{a}_{12}\cdot\nabla_{\boldsymbol{v}_1}+\boldsymbol{a}_{21}\cdot\nabla_{\boldsymbol{v}_2})f_1(1)f_1(2)\,. \tag{3.8}$$

The set, Eqs. (3.2) and (3.5), are now simpler, but still very difficult to solve. With the help of the suggestion of Bogolyubov, called *Bogolyubov's hypothesis*, we can make one more significant step in making the system tractable. The suggestion is that the time scales for the relaxation of $g(t)$ and $f_1(t)$ are very different, *and may be treated independently.* Going back to our discussion of the nature of the collision process in plasmas, we expect that the particle-particle correlation function will relax on the time scale of a single, small deflection angle encounter, which is the time scale for a particle to transit through the Debye sphere of the interacting particle. This time is the limit for the interaction of two nearby particles. On the other hand, the change of the distribution function as a whole requires substantial momentum and energy changes, and a significant momentum change corresponds to a 90° deflection, so the time scale is longer by a factor of $L_{90°}/\lambda_D=6N_D/\ln\Lambda$ from equation (1.39). The fast time scale for electrons is then $\tau_f=\lambda_D/v_e=1/\omega_{pe}$ where v_e is the electron thermal speed, and this is the time scale over which

$g(12)$ relaxes. The longer time scale is the usual collision time scale, $\tau_e = 1/\nu_c$, where ν_c is given by equation (1.40). On this longer time scale, we expect g to have relaxed completely to its $t \to \infty$ limit before f_1 has begun to change significantly.

We implement this approximation by assuming $f_1(1)$ and $f_1(2)$ to be *invariant in time* in equation (3.5), so that the source term from equation (3.8) becomes

$$S(\boldsymbol{r}_1 - \boldsymbol{r}_2, \boldsymbol{v}_1, \boldsymbol{v}_2, t) = -(\boldsymbol{a}_{12} \cdot \nabla_{\boldsymbol{v}_1} + \boldsymbol{a}_{21} \cdot \nabla_{\boldsymbol{v}_2}) f_1(\boldsymbol{v}_1) f_1(\boldsymbol{v}_2) \,. \tag{3.9}$$

We then solve equation (3.5) to find the time asymptotic limit, or $g(\boldsymbol{r}_1 - \boldsymbol{r}_2, \boldsymbol{v}_1, \boldsymbol{v}_2, t \to \infty)$, and use this in equation (3.2) to find the time evolution of $f_1(\boldsymbol{v}_1, t)$. As we evolve f_1 forward in time, we must of course reevaluate $g(t \to \infty)$ with the current f_1, but the time variable in the two steps is entirely independent. To emphasize this independence, we will introduce the variable τ for the fast time scale, so that $g = g(\tau)$, and we solve equation (3.5) using the source term from equation (3.9), and then $g(\tau \to \infty)$ in equation (3.2) to find $f_1(t)$.

3.2 Solution via Fourier and Laplace transforms

In order to reduce the remaining problem to manageable proportions, we will use Fourier transforms in space and Laplace transforms in time to eliminate all but the velocity derivatives and integrals. To this end, we define the Fourier and inverse Fourier transforms by

$$\mathcal{F}[f(x)] = \tilde{f}(k) \equiv \int_{-\infty}^{\infty} \mathrm{d}x \, \mathrm{e}^{-\mathrm{i}kx} f(x) \,, \tag{3.10}$$

$$\mathcal{F}^{-1}[\tilde{f}(k)] = f(x) \equiv \int_{-\infty}^{\infty} \frac{\mathrm{d}k}{2\pi} \mathrm{e}^{\mathrm{i}kx} \tilde{f}(k) \,, \tag{3.11}$$

and the Laplace transform and its inverse by

$$\mathcal{L}[f(t)] = f_p(p) \equiv \int_0^{\infty} \mathrm{d}t \, \mathrm{e}^{-pt} f(t) \,, \tag{3.12}$$

$$\mathcal{L}^{-1}[f_p(p)] = f(t) \equiv \int_{\sigma - \mathrm{i}\infty}^{\sigma + \mathrm{i}\infty} \frac{\mathrm{d}p}{2\pi \mathrm{i}} \, \mathrm{e}^{pt} f_p(p) \,, \tag{3.13}$$

where σ ($\sigma > 0$) is to the right of all singularities of f_p. The Fourier transform is equivalent to letting

$$f(\boldsymbol{r}) = \tilde{f}(\boldsymbol{k}) \mathrm{e}^{\mathrm{i}\boldsymbol{k} \cdot \boldsymbol{r}} \,, \tag{3.14}$$

so that $\nabla_{\boldsymbol{r}_1} \to \mathrm{i}\boldsymbol{k}_1$, but the Laplace transform must be treated more carefully, since the Laplace transform of the time derivative is

$$\left(\frac{\mathrm{d}f}{\mathrm{d}t}\right)_p = pf_p + \Big[f(t)\mathrm{e}^{-pt}\Big]_{t=0}^{t\to\infty} = pf_p - f(0)\,. \tag{3.15}$$

The Fourier transform of the acceleration can be performed after writing the acceleration in terms of the difference variable, $\boldsymbol{r} = \boldsymbol{r}_1 - \boldsymbol{r}_2$, so that

$$\boldsymbol{a}_{12}(\boldsymbol{r}) = \frac{e^2}{4\pi\epsilon_0 m_e r^3}\boldsymbol{r}\,, \tag{3.16}$$

with the resulting transform

$$\tilde{\boldsymbol{a}}_{12}(\boldsymbol{k}) = \frac{-\mathrm{i}\boldsymbol{k}}{m_e}\tilde{\varphi}(k)\,, \tag{3.17}$$

where

$$\tilde{\varphi}(k) = \frac{e^2}{\epsilon_0 k^2} \tag{3.18}$$

is the Fourier transform of the electrostatic potential energy function (See problem 3.1)

$$\varphi = \frac{e^2}{4\pi\epsilon_0 r}\,. \tag{3.19}$$

We will also find useful the results of the double Fourier transforms

$$\mathcal{F}[f(\boldsymbol{r}_1 - \boldsymbol{r}_2)] = (2\pi)^3\delta(\boldsymbol{k}_1 + \boldsymbol{k}_2)\tilde{f}(\boldsymbol{k}_1)\,, \tag{3.20}$$

and

$$\int \mathrm{d}\boldsymbol{r}_1 \mathrm{e}^{-\mathrm{i}\boldsymbol{k}_1\cdot\boldsymbol{r}_1} \int \mathrm{d}\boldsymbol{r}_2 \mathrm{e}^{-\mathrm{i}\boldsymbol{k}_2\cdot\boldsymbol{r}_2} \int \mathrm{d}\boldsymbol{r}_3\, f_1(\boldsymbol{r}_1 - \boldsymbol{r}_3) f_2(\boldsymbol{r}_2 - \boldsymbol{r}_3) = (2\pi)^3\delta(\boldsymbol{k}_1 + \boldsymbol{k}_2)\tilde{f}_1(\boldsymbol{k}_1)\tilde{f}_2(-\boldsymbol{k}_1)\,, \tag{3.21}$$

where the transforms are over both $\boldsymbol{r}_1$ and $\boldsymbol{r}_2$, and

$$\int \mathrm{d}\boldsymbol{r}\, f_1(\boldsymbol{r}) f_2(\boldsymbol{r}) = (2\pi)^{-3} \int \mathrm{d}\boldsymbol{k}\, \tilde{f}_1(-\boldsymbol{k})\tilde{f}_2(\boldsymbol{k}) \tag{3.22}$$

for any functions f_1 and f_2.

Problem 3.1 *Fourier transforms of the potential and field.* Prove the Fourier transforms of Eqs. (3.17) and (3.18). (Hint: use spherical coordinates with $\boldsymbol{k}\cdot\boldsymbol{r} = kr\cos\theta$.)

Problem 3.2 *Fourier transform identities.* Prove the identities of equations (3.20) through (3.22).

3.2.1 Transforming the evolution equations

Using the results of Eqs. (3.17) and (3.22), equation (3.2) may be written as

$$\partial_t f_1(\boldsymbol{v}_1, t) = -\frac{\mathrm{i}n_0}{(2\pi)^3 m_e}\nabla_{\boldsymbol{v}_1}\cdot\int \mathrm{d}\boldsymbol{v}_2\,\mathrm{d}\boldsymbol{k}_1\,\boldsymbol{k}_1\tilde{\varphi}(k_1)\tilde{g}(\boldsymbol{k}_1, \boldsymbol{v}_1, \boldsymbol{v}_2, \tau \to \infty)\,. \quad (3.23)$$

Using the identities (3.20) and (3.21), the double Fourier transform of equation (3.5) may be written as

$$\partial_\tau \tilde{g}(\boldsymbol{k}_1, \boldsymbol{v}_1, \boldsymbol{v}_2, \tau) + V_1\tilde{g} + V_2\tilde{g} = \tilde{S}(\boldsymbol{k}_1, \boldsymbol{v}_1, \boldsymbol{v}_2)\,, \quad (3.24)$$

where the common factor, $(2\pi)^3\delta(\boldsymbol{k}_1 + \boldsymbol{k}_2)$, has been deleted, and

$$V_1\tilde{g}(12) = \mathrm{i}\boldsymbol{k}_1\cdot\boldsymbol{v}_1\tilde{g}(12) - \mathrm{i}\boldsymbol{k}_1\cdot\nabla_{\boldsymbol{v}_1} f_1(\boldsymbol{v}_1)\tilde{\varphi}(k_1)\frac{n_0}{m_e}\int \mathrm{d}\boldsymbol{v}_3\tilde{g}(\boldsymbol{k}_2, \boldsymbol{v}_3, \boldsymbol{v}_2, \tau), \quad (3.25)$$

$$V_2\tilde{g}(12) = -\mathrm{i}\boldsymbol{k}_1\cdot\boldsymbol{v}_2\tilde{g}(12) + \mathrm{i}\boldsymbol{k}_1\cdot\nabla_{\boldsymbol{v}_2} f_1(\boldsymbol{v}_2)\tilde{\varphi}(k_1)\frac{n_0}{m_e}\int \mathrm{d}\boldsymbol{v}_3\tilde{g}(\boldsymbol{k}_1, \boldsymbol{v}_3, \boldsymbol{v}_1, \tau), \quad (3.26)$$

where

$$\tilde{S}(\boldsymbol{k}_1, \boldsymbol{v}_1, \boldsymbol{v}_2) = \frac{\tilde{\phi}(k_1)}{m_e}\mathrm{i}\boldsymbol{k}_1\cdot(\nabla_{\boldsymbol{v}_1} - \nabla_{\boldsymbol{v}_2}) f_1(\boldsymbol{v}_1) f_1(\boldsymbol{v}_2)\,. \quad (3.27)$$

Our first task is to solve equation (3.24) for $g(\tau \to \infty)$, which we will accomplish using Laplace transforms over the τ time scale. Using equation (3.15), equation (3.24) becomes

$$-\tilde{g}(\boldsymbol{k}_1, \boldsymbol{v}_1, \boldsymbol{v}_2, 0) + p\tilde{g}_p(\boldsymbol{k}_1, \boldsymbol{v}_1, \boldsymbol{v}_2, p) + V_1\tilde{g}_p(12p) + V_2\tilde{g}_p(12p) = \frac{1}{p}\tilde{S}(\boldsymbol{k}_1, \boldsymbol{v}_1, \boldsymbol{v}_2). \quad (3.28)$$

In this expression, the operators V_1 and V_2 are independent of time and so independent of p, so we treat them as constants. Equation (3.28) is now an algebraic function which we may solve for $\tilde{g}_p$ to find

$$\tilde{g}_p = \frac{\tilde{g}(0) + \tilde{S}/p}{p + V_1 + V_2}\,. \quad (3.29)$$

In order to find $\tilde{g}(\tau \to \infty)$, we need the inverse Laplace transform of $\tilde{g}_p$, or

$$\tilde{g}(\tau) = \int_{\sigma-\mathrm{i}\infty}^{\sigma+\mathrm{i}\infty} \frac{\mathrm{d}p}{2\pi\mathrm{i}}\mathrm{e}^{p\tau}\frac{\tilde{g}(0) + \tilde{S}/p}{p + V_1 + V_2}\,. \quad (3.30)$$

It is apparent from the form of the integrand that there are poles at $p = 0$ and at $p = -(V_1 + V_2)$. From later discussions of Vlasov theory, we will find that stable distribution functions $f(\boldsymbol{v})$ lead to positive real parts for V_1 and V_2, so there is one pole at the origin and one (or more if V_1 or V_2 are multivalued) in the left half of the p-plane. The specified contour requires us to go to the right of all singularities, as shown in Figure 3.1, but we can just as well deform the

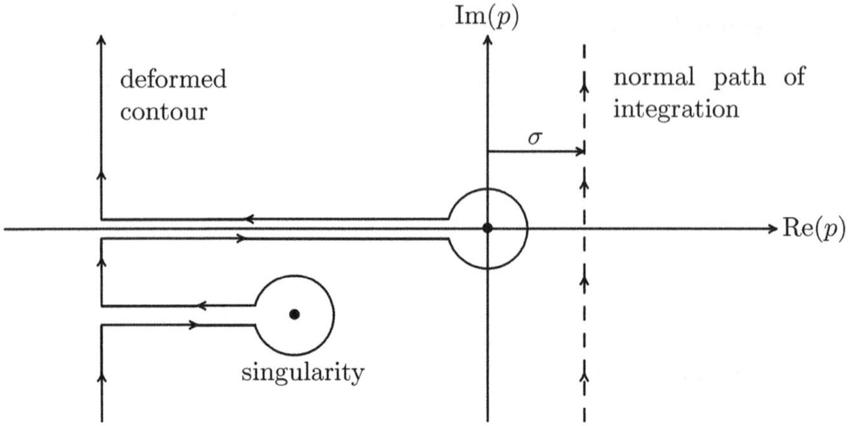

FIGURE 3.1
Deformed contour. The normal path is shifted to the left but remains to the right of all singularities and branch points.

contour to the left as long as we loop back to stay to the right of each pole (or branch point, if any).

By deforming the contour, the vertical portion of the contour may be neglected if we move it to $-\infty$, since the inverse transform in equation (3.30) has the factor $e^{p\tau}$ which vanishes as $\mathrm{Re}(p) \to -\infty$. The horizontal segments toward the poles will cancel each other out since there is one going each way. Hence, unless there is a branch cut, and we have none in this case, the integral is simply the sum of the residues. Furthermore, the pole at the origin is the most important one, as the others will decay away as $\tau \to \infty$ since the poles are in the left half of the p-plane. We then focus our attention on the pole at the origin whose residue gives

$$\tilde{g}(\tau \to \infty) = \lim_{p \to 0} \frac{\tilde{S}}{p + V_1 + V_2} . \tag{3.31}$$

We will not take the limit at this time, since the limiting behavior will be useful in what follows.

We now invoke a sneaky trick to separate the operators V_1 and V_2. Noting that we can write

$$\begin{aligned}\frac{1}{p + V_1 + V_2} &= \int_0^\infty \mathrm{d}t\, \mathrm{e}^{-(p+V_1+V_2)t} \\ &= \int_0^\infty \mathrm{d}t\, \mathrm{e}^{-pt} \oint_{C_1} \frac{\mathrm{d}p_1}{2\pi\mathrm{i}} \frac{\mathrm{e}^{p_1 t}}{p_1 + V_1} \oint_{C_2} \frac{\mathrm{d}p_2}{2\pi\mathrm{i}} \frac{\mathrm{e}^{p_2 t}}{p_2 + V_2} \\ &= \oint_{C_1} \frac{\mathrm{d}p_1}{2\pi\mathrm{i}} \oint_{C_2} \frac{\mathrm{d}p_2}{2\pi\mathrm{i}} \left(\frac{1}{p_1 + V_1}\right) \left(\frac{1}{p_2 + V_2}\right) \frac{1}{p - p_1 - p_2} , \end{aligned} \tag{3.32}$$

where the contours C_1 and C_2 must be chosen so they include the poles at $-V_1$ and $-V_2$ and so that Re(p) >Re(p_1)+Re(p_2). Using this separation technique, we can write equation (3.31) as

$$\tilde{g}(\tau \to \infty) = \lim_{p\to 0} \oint_{C_1} \frac{\mathrm{d}p_1}{2\pi \mathrm{i}} \oint_{C_2} \frac{\mathrm{d}p_2}{2\pi \mathrm{i}} \left(\frac{1}{p_2 + V_2}\right)\left(\frac{1}{p_1+V_1}\right) \frac{\tilde{S}}{p - p_1 - p_2}. \tag{3.33}$$

3.2.2 Inverting the operators

In the expressions above with V_1 and V_2 in the denominator, where V_1 and V_2 are operators, we must interpret an operation such as $(p_1 + V_1)^{-1}\tilde{S} = U_1$ as equivalent to $(p_1 + V_1)U_1 = \tilde{S}$. Hence we must find $U_1(\boldsymbol{k}_1, \boldsymbol{v}_1, \boldsymbol{v}_2)$ such that

$$\begin{aligned}\tilde{S}(\boldsymbol{k}_1, \boldsymbol{v}_1, \boldsymbol{v}_2) &= (p_1 + V_1)U_1(\boldsymbol{k}_1, \boldsymbol{v}_1, \boldsymbol{v}_2) \\ &= (p_1 + \mathrm{i}\boldsymbol{k}_1 \cdot \boldsymbol{v}_1)U_1(\boldsymbol{k}_1, \boldsymbol{v}_1, \boldsymbol{v}_2) \\ &\quad - \frac{\mathrm{i}n_0}{m_e}\boldsymbol{k}_1 \cdot \nabla_{\boldsymbol{v}_1} f_1(\boldsymbol{v}_1)\tilde{\varphi}(k_1) \int \mathrm{d}\boldsymbol{v}_3 U_1(\boldsymbol{k}_1, \boldsymbol{v}_3, \boldsymbol{v}_2)\,.\end{aligned} \tag{3.34}$$

In order to solve this integral equation, we recast equation (3.34) as

$$\begin{aligned}U_1(\boldsymbol{k}_1, \boldsymbol{v}_1, \boldsymbol{v}_2) = \frac{1}{p_1 + \mathrm{i}\boldsymbol{k}_1 \cdot \boldsymbol{v}_1} \Bigg[&\tilde{S}(\boldsymbol{k}_1, \boldsymbol{v}_1, \boldsymbol{v}_2) \\ &+ \frac{\mathrm{i}n_0}{m_e}\boldsymbol{k}_1 \cdot \nabla_{\boldsymbol{v}_1} f_1(\boldsymbol{v}_1)\tilde{\varphi}(k_1) \int \mathrm{d}\boldsymbol{v}_3 U_1(\boldsymbol{k}_1, \boldsymbol{v}_3, \boldsymbol{v}_2)\Bigg],\end{aligned} \tag{3.35}$$

and integrate over $\boldsymbol{v}_1$ to obtain

$$\begin{aligned}\int \mathrm{d}\boldsymbol{v}_1 U_1(\boldsymbol{k}_1, \boldsymbol{v}_1, \boldsymbol{v}_2) = &\int \mathrm{d}\boldsymbol{v}_1 \frac{\tilde{S}(\boldsymbol{k}_1, \boldsymbol{v}_1, \boldsymbol{v}_2)}{p_1 + \mathrm{i}\boldsymbol{k}_1 \cdot \boldsymbol{v}_1} \\ &+ \frac{\mathrm{i}n_0\tilde{\varphi}(k_1)}{m_e} \int \mathrm{d}\boldsymbol{v}_3 U_1(\boldsymbol{k}_1, \boldsymbol{v}_3, \boldsymbol{v}_2) \int \mathrm{d}\boldsymbol{v}_1 \frac{\boldsymbol{k}_1 \cdot \nabla_{\boldsymbol{v}_1} f_1(\boldsymbol{v}_1)}{p_1 + \mathrm{i}\boldsymbol{k}_1 \cdot \boldsymbol{v}_1}.\end{aligned}$$

We note, however, that $\boldsymbol{v}_3$ is just a dummy variable in the integral over $\boldsymbol{v}_3$, so that integral is equivalent to the integral on the left. Solving for this integral, we may write (with the dummy variable v)

$$\int \mathrm{d}\boldsymbol{v}\, U_1(\boldsymbol{k}_1, \boldsymbol{v}, \boldsymbol{v}_2) = \frac{1}{\epsilon(\boldsymbol{k}_1, p_1)} \int \mathrm{d}\boldsymbol{v} \frac{\tilde{S}(\boldsymbol{k}_1, \boldsymbol{v}, \boldsymbol{v}_2)}{p_1 + \mathrm{i}\boldsymbol{k}_1 \cdot \boldsymbol{v}}, \tag{3.36}$$

where we define

$$\epsilon(\boldsymbol{k}, p) \equiv 1 - \frac{\mathrm{i}n_0\tilde{\varphi}(k)}{m_e} \int \mathrm{d}\boldsymbol{v} \frac{\boldsymbol{k} \cdot \nabla_{\boldsymbol{v}} f_1(\boldsymbol{v})}{p + \mathrm{i}\boldsymbol{k} \cdot \boldsymbol{v}}, \tag{3.37}$$

which we will later show to be the dielectric constant for the plasma. Having solved for the integral of U_1, we can return to equation (3.35) to give the solution for U_1 as

$$U_1(\boldsymbol{k}_1, \boldsymbol{v}_1, \boldsymbol{v}_2) = \frac{1}{p_1 + \mathrm{i}\boldsymbol{k}_1 \cdot \boldsymbol{v}_1} \left[\tilde{S}(\boldsymbol{k}_1, \boldsymbol{v}_1, \boldsymbol{v}_2) + \frac{\mathrm{i}n_0 \boldsymbol{k}_1 \cdot \nabla_{\boldsymbol{v}_1} f_1(\boldsymbol{v}_1) \tilde{\varphi}(k_1)}{m_e \epsilon(\boldsymbol{k}_1, p_1)} \int \mathrm{d}\boldsymbol{v} \frac{\tilde{S}(\boldsymbol{k}_1, \boldsymbol{v}, \boldsymbol{v}_2)}{p_1 + \mathrm{i}\boldsymbol{k}_1 \cdot \boldsymbol{v}} \right]. \quad (3.38)$$

Unfolding equation (3.33) from the right, the next inversion involving V_2 is similar, except that this time we need $(p_2 + V_2)^{-1} U_1 = U_2$. From equation (3.23) we see that the integral $\int \mathrm{d}\boldsymbol{v}_2 \tilde{g}$ is required, and since V_2 is equivalent to V_1 except for a change in sign of $\boldsymbol{k}_1$ and an interchange of $\boldsymbol{v}_1$ and $\boldsymbol{v}_2$, we can do the inversion by analogy to equation (3.36), so that

$$\int \mathrm{d}\boldsymbol{v}_2 U_2(\boldsymbol{k}_1, \boldsymbol{v}_1, \boldsymbol{v}_2) = \frac{1}{\epsilon(-\boldsymbol{k}_1, p_2)} \int \mathrm{d}\boldsymbol{v}_2 \frac{U_1(\boldsymbol{k}_1, \boldsymbol{v}_1, \boldsymbol{v}_2)}{p_2 - \mathrm{i}\boldsymbol{k}_1 \cdot \boldsymbol{v}_2}. \quad (3.39)$$

Having done the inversion, we return to equation (3.33), and after integrating over $\boldsymbol{v}_2$, we obtain

$$\int \mathrm{d}\boldsymbol{v}_2 \tilde{g}(\boldsymbol{k}_1, \boldsymbol{v}_1, \boldsymbol{v}_2, \tau \to \infty) = \lim_{p \to 0} \oint_{C_1} \frac{dp_1}{2\pi \mathrm{i}} \oint_{C_2} \frac{dp_2}{2\pi \mathrm{i}} \frac{1}{\epsilon(-\boldsymbol{k}_1, p_2)} \times \frac{1}{p - p_1 - p_2} \int \mathrm{d}\boldsymbol{v}_2 \frac{U_1(\boldsymbol{k}_1, \boldsymbol{v}_1, \boldsymbol{v}_2)}{p_2 - \mathrm{i}\boldsymbol{k}_1 \cdot \boldsymbol{v}_2}. \quad (3.40)$$

3.2.3 Integrating over p_2

At this point, we do the integral over p_2 first, noting from the previous condition that Re(p) >Re(p_1)+ Re(p_2) and that the pole at $p_2 = p - p_1$ is to the *right* of the prescribed contour, as shown in Figure 3.2. Since the integrand varies as $1/p_2^2$ for large p_2, we may close the contour to the right and inclose only the one pole, with the result

$$\int \mathrm{d}\boldsymbol{v}_2 \tilde{g}(\boldsymbol{k}_1, \boldsymbol{v}_1, \boldsymbol{v}_2, \tau \to \infty) = \lim_{p \to 0} \oint_{C_1} \frac{\mathrm{d}p_1}{2\pi \mathrm{i}} \frac{1}{\epsilon(-\boldsymbol{k}_1, p - p_1)} \times \int \mathrm{d}\boldsymbol{v}_2 \frac{U_1(\boldsymbol{k}_1, \boldsymbol{v}_1, \boldsymbol{v}_2)}{p - p_1 - \mathrm{i}\boldsymbol{k}_1 \cdot \boldsymbol{v}_2}. \quad (3.41)$$

Expanding this expression using equation (3.38) for U_1 and equation (3.27) for $\tilde{S}$, we find

$$\int \mathrm{d}\boldsymbol{v}_2 \tilde{g}(\boldsymbol{k}_1, \boldsymbol{v}_1, \boldsymbol{v}_2, \tau \to \infty) = \lim_{p \to 0} \int \mathrm{d}\boldsymbol{v}_2 \oint_{C_1} \frac{\mathrm{d}p_1}{2\pi \mathrm{i}} \frac{T_1 + T_2 + T_3 + T_4}{\epsilon(-\boldsymbol{k}_1, p - p_1)(p - p_1 - \mathrm{i}\boldsymbol{k}_1 \cdot \boldsymbol{v}_2)(p_1 + \mathrm{i}\boldsymbol{k}_1 \cdot \boldsymbol{v}_1)}, \quad (3.42)$$

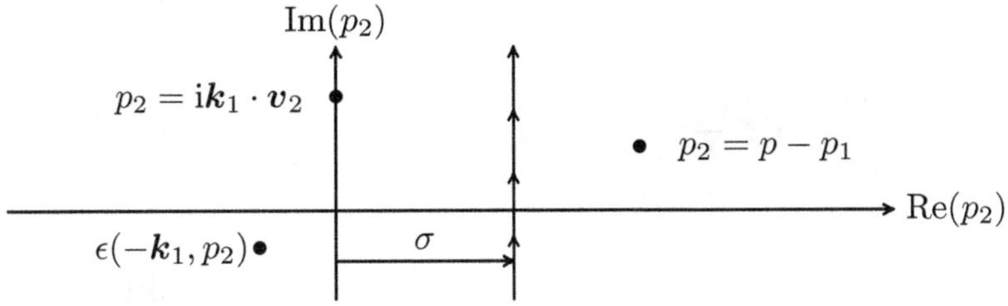

FIGURE 3.2
C_2 contour which closes to the right and incloses one pole.

where

$$T_1 = \frac{\mathrm{i}\tilde{\varphi}(k_1)}{m_e}\boldsymbol{k}_1 \cdot \nabla_{\boldsymbol{v}_1} f_1(\boldsymbol{v}_1) f_1(\boldsymbol{v}_2)\,, \tag{3.43}$$

$$T_2 = -\frac{\mathrm{i}\tilde{\varphi}(k_1)}{m_e}\boldsymbol{k}_1 \cdot \nabla_{\boldsymbol{v}_2} f_1(\boldsymbol{v}_1) f_1(\boldsymbol{v}_2)\,, \tag{3.44}$$

$$T_3 = -\frac{n_0\tilde{\varphi}^2(k_1)\boldsymbol{k}_1 \cdot \nabla_{\boldsymbol{v}_1} f_1(\boldsymbol{v}_1)}{m_e^2 \epsilon(\boldsymbol{k}_1, p_1)} \int \mathrm{d}\boldsymbol{v} \frac{\boldsymbol{k}_1 \cdot \nabla_{\boldsymbol{v}}}{p_1 + \mathrm{i}\boldsymbol{k}_1 \cdot \boldsymbol{v}} f_1(\boldsymbol{v}) f_1(\boldsymbol{v}_2)\,, \tag{3.45}$$

$$T_4 = \frac{n_0\tilde{\varphi}^2(k_1)\boldsymbol{k}_1 \cdot \nabla_{\boldsymbol{v}_1} f_1(\boldsymbol{v}_1)}{m_e^2 \epsilon(\boldsymbol{k}_1, p_1)} \int \mathrm{d}\boldsymbol{v} \frac{\boldsymbol{k}_1 \cdot \nabla_{\boldsymbol{v}_2}}{p_1 + \mathrm{i}\boldsymbol{k}_1 \cdot \boldsymbol{v}} f_1(\boldsymbol{v}) f_1(\boldsymbol{v}_2)\,. \tag{3.46}$$

Using the definition of the dielectric function in equation (3.37), the integral of T_2 over $\boldsymbol{v}_2$ is

$$\int \mathrm{d}\boldsymbol{v}_2 \frac{T_2}{p - p_1 - \mathrm{i}\boldsymbol{k}_1 \cdot \boldsymbol{v}_2} = [1 - \epsilon(-\boldsymbol{k}_1, p - p_1)]\frac{f_1(\boldsymbol{v}_1)}{n_0}\,. \tag{3.47}$$

Using equation (3.37) again for the integral over $\boldsymbol{v}$, the integral of T_3 over $\boldsymbol{v}_2$ is

$$\int \frac{T_3\,\mathrm{d}\boldsymbol{v}_2}{p - p_1 - \mathrm{i}\boldsymbol{k}_1 \cdot \boldsymbol{v}_2} = \left[\frac{1}{\epsilon(\boldsymbol{k}_1, p_1)} - 1\right] \mathrm{i}\boldsymbol{k}_1 \cdot \nabla_{\boldsymbol{v}_1} f_1(\boldsymbol{v}_1)\frac{\tilde{\varphi}}{m_e} \int \frac{f_1(\boldsymbol{v}_2)\,\mathrm{d}\boldsymbol{v}_2}{p - p_1 - \mathrm{i}\boldsymbol{k}_1 \cdot \boldsymbol{v}_2}\,. \tag{3.48}$$

The integral over the last term gives

$$\int \frac{T_4\,\mathrm{d}\boldsymbol{v}_2}{p - p_1 - \mathrm{i}\boldsymbol{k}_1 \cdot \boldsymbol{v}_2} = \frac{[1 - \epsilon(-\boldsymbol{k}_1, p - p_1)]}{\epsilon(\boldsymbol{k}_1, p_1)} \mathrm{i}\boldsymbol{k}_1 \cdot \nabla_{\boldsymbol{v}_1} f_1(\boldsymbol{v}_1)\frac{\tilde{\varphi}}{m_e} \int \frac{f_1(\boldsymbol{v})\,\mathrm{d}\boldsymbol{v}}{p_1 + \mathrm{i}\boldsymbol{k}_1 \cdot \boldsymbol{v}}\,. \tag{3.49}$$

The integral of T_1 cancels the second term in the integral of T_3, so collecting the pieces again, we have

$$\int \mathrm{d}\boldsymbol{v}_2 \tilde{g}(\boldsymbol{k}_1, \boldsymbol{v}_1, \boldsymbol{v}_2, \tau \to \infty)$$
$$= \lim_{p \to 0} \oint_{C_1} \frac{\mathrm{d}p_1}{2\pi\mathrm{i}} \frac{1}{(p_1 + \mathrm{i}\boldsymbol{k}_1 \cdot \boldsymbol{v}_1)} \left\{ \left[\frac{1}{\epsilon(-\boldsymbol{k}_1, p - p_1)} - 1\right](T_a + T_b) + T_c \right\} \tag{3.50}$$

where

$$T_a = f_1(\boldsymbol{v}_1)/n_0\,, \tag{3.51}$$

$$T_b = \frac{\mathrm{i}\boldsymbol{k}_1 \cdot \nabla_{\boldsymbol{v}_1} f_1(\boldsymbol{v}_1)}{\epsilon(\boldsymbol{k}_1, p_1)} \frac{\tilde{\varphi}}{m_e} \int \mathrm{d}\boldsymbol{v} \frac{f_1(\boldsymbol{v})}{p_1 + \mathrm{i}\boldsymbol{k}_1 \cdot \boldsymbol{v}}\,, \tag{3.52}$$

$$T_c = \frac{\mathrm{i}\boldsymbol{k}_1 \cdot \nabla_{\boldsymbol{v}_1} f_1(\boldsymbol{v}_1)}{\epsilon(-\boldsymbol{k}_1, p - p_1)\epsilon(\boldsymbol{k}_1, p_1)} \frac{\tilde{\varphi}}{m_e} \int \mathrm{d}\boldsymbol{v} \frac{f_1(\boldsymbol{v})}{p - p_1 - \mathrm{i}\boldsymbol{k}_1 \cdot \boldsymbol{v}}\,. \tag{3.53}$$

3.2.4 Integrating over p_1

Turning our attention now to the integral over the contour C_1, we note the location of the poles in the p_1-plane in Figure 3.3. The integral of the term T_a is simple, since as $|p_1| \to \infty$, the factor in square brackets in equation (3.50) varies as $1/p_1$, so we may close the contour to the left and pick up only the pole at $p_1 = -\mathrm{i}\boldsymbol{k}_1 \cdot \boldsymbol{v}_1$ [the pole due to $\epsilon(-\boldsymbol{k}_1, p - p_1)$ is to the *right* of C_1], with the result

$$\oint_{C_1} \frac{\mathrm{d}p_1}{2\pi\mathrm{i}} \frac{1}{(p_1 + \mathrm{i}\boldsymbol{k}_1 \cdot \boldsymbol{v}_1)} \left[\frac{1}{\epsilon(-\boldsymbol{k}_1, p - p_1)} - 1\right] T_a = \left[\frac{1}{\epsilon(-\boldsymbol{k}_1, p + \mathrm{i}\boldsymbol{k}_1 \cdot \boldsymbol{v}_1)} - 1\right] \frac{f_1(\boldsymbol{v}_1)}{n_0}. \tag{3.54}$$

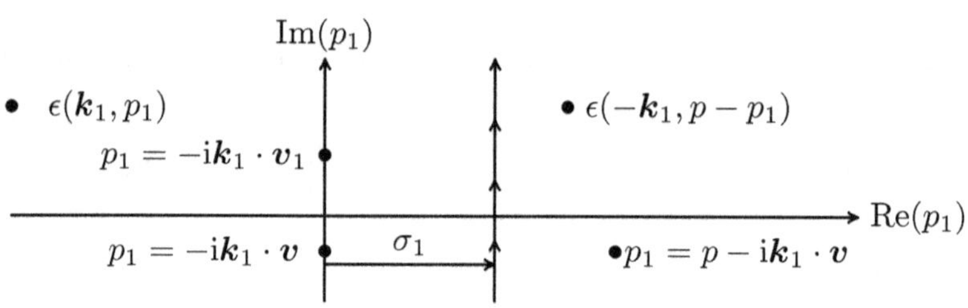

FIGURE 3.3
C_1 contour and poles.

Multiplying the T_b term by the (-1) term in the square brackets, the only poles are along the imaginary axis, so closing to the *right* will pick up no pole, so we have no contribution from that term.

Combining the remaining terms, we have

$$I = \lim_{p \to 0} \oint_{C_1} \frac{\mathrm{d}p_1}{2\pi\mathrm{i}} \frac{\mathrm{i}\boldsymbol{k}_1 \cdot \nabla_{\boldsymbol{v}_1} f_1(\boldsymbol{v}_1)\tilde{\varphi}}{m_e(p_1 + \mathrm{i}\boldsymbol{k}_1 \cdot \boldsymbol{v}_1)\epsilon(-\boldsymbol{k}_1, p - p_1)\epsilon(\boldsymbol{k}_1, p_1)} \times \int \mathrm{d}\boldsymbol{v}\, f_1(\boldsymbol{v}) \left[\frac{1}{p_1 + \mathrm{i}\boldsymbol{k}_1 \cdot \boldsymbol{v}} - \frac{1}{p_1 - p + \mathrm{i}\boldsymbol{k}_1 \cdot \boldsymbol{v}}\right]. \tag{3.55}$$

Noting that the integral is to be taken in the limit as $p \to 0$, it appears at first glance that the velocity integrand vanishes, so that we have no contribution.

We must be careful, however, since the two poles are on opposite sides of the prescribed contour, and as $p \to 0$, these two poles "pinch" the contour. Also, the pole at $p_1 = -\mathrm{i}\boldsymbol{k}_1 \cdot \boldsymbol{v}_1$ is fixed, while the other two poles move as we perform the integration over $\boldsymbol{v}$, and one passes through this fixed pole.

It is necessary at this point to introduce the Plemelj formulas,

$$\lim_{\epsilon \to 0} \frac{1}{p - a \pm \mathrm{i}\epsilon} = \wp\left(\frac{1}{p-a}\right) \pm \mathrm{i}\pi\delta(p-a)\,, \tag{3.56}$$

where the upper sign is used when a contour passes to the right of a pole, and the lower sign when it passes to the left, and $\wp$ signifies the principal part. It will be advantageous at this point to recognize from equation (3.23) that we only need the imaginary part of the integral of $\tilde{g}$, which comes from the real part of

$$\begin{aligned}
&\mathrm{Re}\lim_{p\to 0} \frac{1}{p_1 + \mathrm{i}\boldsymbol{k}_1\cdot\boldsymbol{v}_1}\left[\frac{1}{p_1 + \mathrm{i}\boldsymbol{k}_1\cdot\boldsymbol{v}} - \frac{1}{p_1 - p + \mathrm{i}\boldsymbol{k}_1\cdot\boldsymbol{v}}\right] \\
&= \mathrm{Re}\left[\wp\left(\frac{1}{p_1 + \mathrm{i}\boldsymbol{k}_1\cdot\boldsymbol{v}_1}\right) + \mathrm{i}\pi\delta(p_1 + \mathrm{i}\boldsymbol{k}_1\cdot\boldsymbol{v}_1)\right]\left[\wp\left(\frac{1}{p_1 + \mathrm{i}\boldsymbol{k}_1\cdot\boldsymbol{v}}\right)\right. \\
&\qquad \left. + \mathrm{i}\pi\delta(p_1 + \mathrm{i}\boldsymbol{k}_1\cdot\boldsymbol{v}) - \wp\left(\frac{1}{p_1 + \mathrm{i}\boldsymbol{k}_1\cdot\boldsymbol{v}}\right) + \mathrm{i}\pi\delta(p_1 + \mathrm{i}\boldsymbol{k}_1\cdot\boldsymbol{v})\right] \\
&= -2\pi^2\delta(p_1 + \mathrm{i}\boldsymbol{k}_1\cdot\boldsymbol{v}_1)\delta(p_1 + \mathrm{i}\boldsymbol{k}_1\cdot\boldsymbol{v})\,.
\end{aligned} \tag{3.57}$$

Using the first delta function to do the p_1 integral, we get

$$I = \frac{\mathrm{i}\pi\boldsymbol{k}_1\cdot\nabla_{\boldsymbol{v}_1} f_1(\boldsymbol{v}_1)}{|\epsilon(-\boldsymbol{k}_1, \mathrm{i}\boldsymbol{k}_1\cdot\boldsymbol{v}_1)|^2}\frac{\tilde{\varphi}}{m_e}\int \mathrm{d}\boldsymbol{v}\, f_1(\boldsymbol{v})\delta[\boldsymbol{k}_1\cdot(\boldsymbol{v}-\boldsymbol{v}_1)]\,, \tag{3.58}$$

where we have used the fact that $\epsilon(-\boldsymbol{k}, -p) = \epsilon^*(\boldsymbol{k}, p)$.

To complete this part of the problem, we also need the imaginary part of equation (3.54) integrated over $\boldsymbol{v}_2$ as $p \to 0$, or

$$\lim_{p\to 0}\mathrm{Im}\left[\frac{1}{\epsilon(-\boldsymbol{k}_1, \mathrm{i}\boldsymbol{k}_1\cdot\boldsymbol{v}_1)} - 1\right]\frac{f_1(\boldsymbol{v}_1)}{n_0}$$

and the -1 is real, so we need

$$\mathrm{Im}\left(\frac{1}{\epsilon_r + \mathrm{i}\epsilon_i}\right) = \frac{-\epsilon_i}{|\epsilon|^2}\,.$$

From the definition of the dielectric constant, we have

$$\epsilon(\boldsymbol{k}, p) = 1 - \frac{in_0\tilde{\varphi}(k)}{m_e}\int \mathrm{d}\boldsymbol{v}\frac{\boldsymbol{k}\cdot\nabla_{\boldsymbol{v}} f_1(\boldsymbol{v})}{p + \mathrm{i}\boldsymbol{k}\cdot\boldsymbol{v}}$$

and letting $p = \sigma - \mathrm{i}\omega$ (where σ is real and positive) and factoring out $-\mathrm{i}k$ from the denominator, we have

$$\epsilon(\boldsymbol{k}, p) = 1 + \frac{n_0\tilde{\varphi}(k)}{m_e k}\int \mathrm{d}v\frac{\boldsymbol{k}\cdot\nabla_{\boldsymbol{v}} f_1(\boldsymbol{v})}{\omega/k - v + \mathrm{i}\sigma/k}$$

$$= 1 + \frac{n_0\tilde{\varphi}(k)}{m_e k}\left[\wp\int \mathrm{d}\boldsymbol{v}\frac{\boldsymbol{k}\cdot\nabla_{\boldsymbol{v}} f_1(\boldsymbol{v})}{\omega/k - v} + \mathrm{i}\pi\boldsymbol{k}\cdot\nabla_{\boldsymbol{v}} f_1(\boldsymbol{v})|_{v=\omega/k}\right] \quad (3.59)$$

and if we represent the pole contribution (imag. part) by $\mathrm{i}\pi\boldsymbol{k}\cdot\nabla_{\boldsymbol{v}} f_1(\boldsymbol{v})\delta(p + \mathrm{i}\boldsymbol{k}\cdot\boldsymbol{v})$, then we have

$$\epsilon(-\boldsymbol{k}_1, \mathrm{i}\boldsymbol{k}_1\cdot\boldsymbol{v}_1)_i = \frac{\pi n_0\tilde{\varphi}(k_1)}{m_e}\boldsymbol{k}_1\cdot\nabla_{\boldsymbol{v}} f_1(\boldsymbol{v})\delta[\boldsymbol{k}_1\cdot(\boldsymbol{v}-\boldsymbol{v}_1)],$$

so our final result for this term is

$$\mathrm{Im}\left[\frac{1}{\epsilon(-\boldsymbol{k}_1, \mathrm{i}\boldsymbol{k}_1\cdot\boldsymbol{v}_1)} - 1\right]\frac{f_1(\boldsymbol{v})}{n_0} = -\frac{\pi\tilde{\varphi}(k_1)\boldsymbol{k}_1\cdot\nabla_{\boldsymbol{v}} f_1(\boldsymbol{v})\delta[\boldsymbol{k}_1\cdot(\boldsymbol{v}-\boldsymbol{v}_1)]}{m_e|\epsilon(-\boldsymbol{k}_1, \mathrm{i}\boldsymbol{k}_1\cdot\boldsymbol{v}_1)|^2}$$

so that combining this term with the result from equation (3.58), the combination gives

$$\mathrm{Im}\int \mathrm{d}\boldsymbol{v}_2\tilde{g}(\tau\to\infty) = \frac{\pi\tilde{\varphi}}{m_e|\epsilon(-\boldsymbol{k}_1, \mathrm{i}\boldsymbol{k}_1\cdot\boldsymbol{v}_1)|^2} \times\int \mathrm{d}\boldsymbol{v}\,\boldsymbol{k}_1\cdot(\nabla_{\boldsymbol{v}_1} - \nabla_{\boldsymbol{v}}) f_1(\boldsymbol{v}) f_1(\boldsymbol{v}_1)\delta[\boldsymbol{k}_1\cdot(\boldsymbol{v}-\boldsymbol{v}_1)]. \quad (3.60)$$

We are now ready to put this result back into equation (3.23), with the result

$$\partial_t f_1(\boldsymbol{v}_1, t) = \frac{n_0}{8\pi^2 m_e^2}\nabla_{\boldsymbol{v}_1}\cdot\int \mathrm{d}\boldsymbol{k}_1\int \mathrm{d}\boldsymbol{v}\frac{\boldsymbol{k}_1\tilde{\varphi}^2(k_1)}{|\epsilon(\boldsymbol{k}_1, -\mathrm{i}\boldsymbol{k}_1\cdot\boldsymbol{v}_1)|^2} \times\boldsymbol{k}_1\cdot(\nabla_{\boldsymbol{v}_1} - \nabla_{\boldsymbol{v}}) f_1(\boldsymbol{v}_1) f_1(\boldsymbol{v})\delta[\boldsymbol{k}_1\cdot(\boldsymbol{v}-\boldsymbol{v}_1)], \quad (3.61)$$

which we rewrite as

$$\frac{\partial f(\boldsymbol{v}, t)}{\partial t} = -\nabla_{\boldsymbol{v}}\cdot\int \mathrm{d}\boldsymbol{v}'\mathsf{Q}(\boldsymbol{v}, \boldsymbol{v}')\cdot(\nabla_{\boldsymbol{v}} - \nabla_{\boldsymbol{v}'}) f(\boldsymbol{v}) f(\boldsymbol{v}'), \quad (3.62)$$

where Q is a tensor given by

$$\begin{aligned}\mathsf{Q} &= -\frac{n_0}{4\pi m_e^2}\int\frac{\mathrm{d}\boldsymbol{k}}{2\pi}\frac{\boldsymbol{k}\boldsymbol{k}\tilde{\varphi}^2(k)}{|\epsilon(\boldsymbol{k}, -\mathrm{i}\boldsymbol{k}\cdot\boldsymbol{v})|^2}\delta[\boldsymbol{k}\cdot(\boldsymbol{v}-\boldsymbol{v}')] \\ &= -2n_0\left(\frac{e^2}{4\pi\epsilon_0 m_e}\right)^2\int \mathrm{d}\boldsymbol{k}\frac{\boldsymbol{k}\boldsymbol{k}}{k^4}\frac{\delta[\boldsymbol{k}\cdot(\boldsymbol{v}-\boldsymbol{v}')]}{|1 + (\psi/k^2\lambda_D^2)|^2}, \quad (3.63)\end{aligned}$$

where

$$\psi(\boldsymbol{k}, \boldsymbol{k}\cdot\boldsymbol{v}) = v_e^2\int \mathrm{d}\boldsymbol{v}'\frac{\boldsymbol{k}\cdot\nabla_{\boldsymbol{v}'} f(\boldsymbol{v}')}{\boldsymbol{k}\cdot(\boldsymbol{v}-\boldsymbol{v}')}, \quad (3.64)$$

since

$$\begin{aligned}\epsilon(\boldsymbol{k}, -\mathrm{i}\boldsymbol{k}\cdot\boldsymbol{v}) &= 1 - \frac{\mathrm{i}n_0 e^2}{m_e\epsilon_0 k^2}\int \mathrm{d}\boldsymbol{v}'\frac{\boldsymbol{k}\cdot\nabla_{\boldsymbol{v}'} f(\boldsymbol{v}')}{-\mathrm{i}\boldsymbol{k}\cdot\boldsymbol{v} + \mathrm{i}\boldsymbol{k}\cdot\boldsymbol{v}'} \\ &= 1 + \frac{v_e^2}{k^2\lambda_D^2}\int \mathrm{d}\boldsymbol{v}'\frac{\boldsymbol{k}\cdot\nabla_{\boldsymbol{v}'} f(\boldsymbol{v}')}{\boldsymbol{k}\cdot(\boldsymbol{v}-\boldsymbol{v}')}. \quad (3.65)\end{aligned}$$

3.2.5 Integrating over $\boldsymbol{k}$

We now wish to do the integral in $\boldsymbol{k}$-space, and to do this, we first orient $\hat{e}_{k_1}$ along the $\boldsymbol{v}-\boldsymbol{v}'$ direction, so that the Q_{ij} tensor components are given by

$$Q_{ij}(\boldsymbol{v},\boldsymbol{v}') = -2n_0\left(\frac{e^2}{4\pi\epsilon_0 m_e}\right)^2 \int \mathrm{d}k_1\,\mathrm{d}k_2\,\mathrm{d}k_3 \frac{k_i k_j}{k^4}\frac{1}{|\boldsymbol{v}-\boldsymbol{v}'|}\frac{\delta(k_1)}{|1+(\psi/k^2\lambda_D^2)|^2}\,. \tag{3.66}$$

The integral over k_1 is trivial because of the $\delta(k_1)$ factor, since if either $i=1$ or $j=1$, $Q_{ij}=0$, and k^2 is reduced to $k^2=k_2^2+k_3^2$. Using cylindrical coordinates, with $k_2=k\cos\theta$ and $k_3=k\sin\theta$, and setting the upper limit of k to be k_0, we find for Q_{33} (for example),

$$Q_{33} = -2n_0\left(\frac{e^2}{4\pi\epsilon_0 m_e}\right)^2 \frac{1}{|\boldsymbol{v}-\boldsymbol{v}'|}\int_0^{2\pi}\mathrm{d}\theta\,\sin^2\theta\int_0^{k_0}\frac{\mathrm{d}k}{k}\frac{1}{|1+(\psi/k^2\lambda_D^2)|^2}\,. \tag{3.67}$$

Now ψ is in general a function of θ, but not of k, so we can do the integral over k with the result

$$\int_0^{k_0}\frac{\mathrm{d}k}{k}\frac{1}{|1+(\psi/k^2\lambda_D^2)|^2} = \frac{1}{2}\ln\left|1+\frac{k_0^2\lambda_D^2}{\psi}\right| + \frac{\psi_r}{2\psi_i}\arg\left(1+\frac{k_0^2\lambda_D^2}{\psi}\right)\,. \tag{3.68}$$

The cutoff at $k_0 = 1/b_{\text{min}}$ (the maximum k corresponds to the minimum distance over which there is significant variation) is the same as the distance of closest approach in Section 1.7.2, and since $\Lambda = b_{\text{max}}/b_{\text{min}}$ where $b_{\text{max}}=\lambda_D$, the integral may be written

$$\begin{aligned}\int_0^{k_0}\frac{\mathrm{d}k}{k}\frac{1}{|1+(\psi/k^2\lambda_D^2)|^2} &= \frac{1}{2}\ln\left|1+\frac{\Lambda^2}{\psi}\right| - \frac{\psi_i}{2\psi_r}\arg\left(1+\frac{\Lambda^2}{\psi}\right)\\ &\simeq \ln\Lambda\,,\end{aligned} \tag{3.69}$$

since the dimensionless ψ is of order unity, and we take both Λ and $\ln\Lambda$ to be large. The integral over θ is then trivial, with the result that $Q_{23}=Q_{32}=0$, and

$$Q_{33}(\boldsymbol{v},\boldsymbol{v}') = Q_{22}(\boldsymbol{v},\boldsymbol{v}') = -2\pi n_0\left(\frac{e^2}{4\pi\epsilon_0 m_e}\right)^2\frac{1}{|\boldsymbol{v}-\boldsymbol{v}'|}\ln\Lambda\,. \tag{3.70}$$

With only these two components of the tensor nonzero, we can represent the tensor in terms of the unit tensor I, so that

$$\mathsf{Q}(\boldsymbol{v},\boldsymbol{v}') = -\frac{\omega_{pe}^4\ln\Lambda}{8\pi n_0}\frac{g^2\mathsf{I}-\boldsymbol{g}\boldsymbol{g}}{g^3}\,, \tag{3.71}$$

where $\boldsymbol{g}=\boldsymbol{v}-\boldsymbol{v}'$, recalling that $\hat{e}_{k_1}=\hat{e}_g$. This form of the tensor is known as the *Landau form* for Q.

Problem 3.3 *Analytic behavior of $\epsilon(\boldsymbol{k}, p)$.* Prove that $\epsilon(-\boldsymbol{k}, -p) = \epsilon^*(\boldsymbol{k}, p)$ if $\mathrm{Im}(p) = -\omega$ and $\mathrm{Re}(p) \to 0$, and show that

$$\mathrm{Im}[\epsilon(\boldsymbol{k}, p)] = -\mathrm{i}\pi \frac{\omega_{pe}^2}{k^2} \int \mathrm{d}\boldsymbol{v}\, \boldsymbol{k} \cdot \nabla_{\boldsymbol{v}} f_1(\boldsymbol{v}) \delta(\omega - \boldsymbol{k} \cdot \boldsymbol{v})\,. \tag{3.72}$$

Problem 3.4 *Integral over k.* Verify the integral in equation (3.68), and examine the approximations of equation (3.69) for ψ real and for ψ imaginary.

3.3 The Fokker-Planck equation

With a few manipulations and partial integrations, we can cast equation (3.62) into the form of a Fokker-Planck equation, which enables us to interpret the terms more clearly. Noting first that

$$\nabla_{\boldsymbol{v}} \nabla_{\boldsymbol{v}} g = \frac{g^2 \mathsf{I} - \boldsymbol{g}\boldsymbol{g}}{g^3}\,, \tag{3.73}$$

we can write equation (3.62) as

$$\partial_t f(\boldsymbol{v}) = Q_0 \nabla_{\boldsymbol{v}} \cdot [(\nabla_{\boldsymbol{v}} f) \cdot \nabla_{\boldsymbol{v}} \nabla_{\boldsymbol{v}} \int \mathrm{d}\boldsymbol{v}'\, f(\boldsymbol{v}') g - f(\boldsymbol{v}) \int \mathrm{d}\boldsymbol{v}'\, \nabla_{\boldsymbol{v}} \nabla_{\boldsymbol{v}} g \cdot \nabla_{\boldsymbol{v}'} f(\boldsymbol{v}')]\,, \tag{3.74}$$

where $Q_0 = \omega_{pe}^4 \ln \Lambda / 8\pi n_0 = \frac{1}{4} \nu_e v_e^3$ from equation (1.40). Integrating the second term by parts, and using $\nabla_{\boldsymbol{v}'} g = -\nabla_{\boldsymbol{v}} g$, the two terms can be written

$$Q_1 = Q_0 \nabla_{\boldsymbol{v}} \nabla_{\boldsymbol{v}} : [f(\boldsymbol{v}) \nabla_{\boldsymbol{v}} \nabla_{\boldsymbol{v}} \int \mathrm{d}\boldsymbol{v}'\, g f(\boldsymbol{v}')]\,, \tag{3.75}$$

$$Q_2 = -2 Q_0 \nabla_{\boldsymbol{v}} \cdot [f(\boldsymbol{v}) \int \mathrm{d}\boldsymbol{v}' \nabla_{\boldsymbol{v}} (\nabla_{\boldsymbol{v}} \cdot \nabla_{\boldsymbol{v}}) g f(\boldsymbol{v}')]\,, \tag{3.76}$$

where we define

$$\begin{aligned} \nabla_{\boldsymbol{v}} \nabla_{\boldsymbol{v}} : f \nabla_{\boldsymbol{v}} \nabla_{\boldsymbol{v}} h &\equiv \nabla_{\boldsymbol{v}} \cdot [(\nabla_{\boldsymbol{v}} \cdot f \nabla_{\boldsymbol{v}}) \nabla_{\boldsymbol{v}} h] \\ &= \nabla_{\boldsymbol{v}} \cdot [f (\nabla_{\boldsymbol{v}} \cdot \nabla_{\boldsymbol{v}}) \nabla_{\boldsymbol{v}} h] + \nabla_{\boldsymbol{v}} \cdot [(\nabla_{\boldsymbol{v}} f) \cdot \nabla_{\boldsymbol{v}} \nabla_{\boldsymbol{v}} h]\,, \end{aligned} \tag{3.77}$$

for any scalar functions $f(\boldsymbol{v})$ and $h(\boldsymbol{v})$. Using $\nabla_{\boldsymbol{v}}^2 g = 2/g$, the second term is

$$Q_2 = -4 Q_0 \nabla_{\boldsymbol{v}} \cdot \left[f(\boldsymbol{v}) \nabla_{\boldsymbol{v}} \int \mathrm{d}\boldsymbol{v}' \frac{f(\boldsymbol{v}')}{g} \right]. \tag{3.78}$$

Combining these, we obtain the Fokker-Planck equation,

$$\partial_t f(\boldsymbol{v}, t) = -\nabla_{\boldsymbol{v}} \cdot [\boldsymbol{A} f(\boldsymbol{v}, t)] + \tfrac{1}{2} \nabla_{\boldsymbol{v}} \nabla_{\boldsymbol{v}} : [f(\boldsymbol{v}, t) \mathsf{B}]\,, \tag{3.79}$$

where

$$\boldsymbol{A} = \nu_e v_e^3 \nabla_{\boldsymbol{v}} \int \mathrm{d}\boldsymbol{v}' \frac{f(\boldsymbol{v}', t)}{|\boldsymbol{v} - \boldsymbol{v}'|}, \tag{3.80}$$

$$\mathsf{B} = \tfrac{1}{2} \nu_e v_e^3 \nabla_{\boldsymbol{v}} \nabla_{\boldsymbol{v}} \int \mathrm{d}\boldsymbol{v}' \, |\boldsymbol{v} - \boldsymbol{v}'| f(\boldsymbol{v}', t). \tag{3.81}$$

With these coefficients, this is known as the *Landau form* of the Fokker-Planck equation, where $\boldsymbol{A}$ is the *coefficient of dynamic friction* and B is the *diffusion coefficient.*

3.4 Dynamic friction and diffusion

The Fokker-Planck equation is the basic equation which describes the relaxation of a nonequilibrium distribution function towards equilibrium. We can see from the general form of the equation that it contains a slowing down or *dynamic friction* term, which tends to keep the distribution centered about zero and pulls high velocity particles back into the distribution, and a spreading or *diffusion* term which maintains a finite width. The balance between these two effects leads to the Maxwellian distribution in steady state unless some other source or driving term is added.

Simpler forms of the Fokker-Planck equation are possible, but these invariably give a less accurate or incomplete description of the thermalization process. More complicated expressions are needed to include the effects of a magnetic field, which adds anisotropy so that the rates are different along and across the magnetic field.

An example of a simpler form for the Fokker-Planck equation is

$$\frac{\partial f}{\partial t} = \nu_e \nabla_{\boldsymbol{v}} \cdot \left[(\boldsymbol{v} - \boldsymbol{v}_0) f + \tfrac{1}{2} v_e^2 \nabla_{\boldsymbol{v}} f \right], \tag{3.82}$$

where ν_e is the electron collision frequency and $\boldsymbol{v}_0$ is a constant velocity (such as the velocity of a beam injected into the plasma). While substantially simpler, it leaves so much out that it is useful only as a first step. The principal weakness of this form is that for $v \gg v_t$, both the friction term and the diffusion term fall off as $1/v^n$ with $n \geq 2$, so it is only useful for velocities of the order of the thermal speed or less.

An even simpler form introduced in Section 1.7.3 is the *Krook model*, described by

$$\frac{\partial f}{\partial t} = -\nu(f - f_0), \tag{3.83}$$

where f_0 is the appropriate Maxwellian equilibrium distribution function. This is clearly simplistic, but tractable when added to the Vlasov equation to represent the effects of collisions on waves. In the next chapter, these two simple models will be compared for collisional corrections to Landau damping.

In the last three chapters, we will investigate the effects of collisions on the various transport coefficients, without and with a stationary magnetic field, among them the conductivity. These calculations will largely ignore the friction and diffusion effects, and will assume that a steady state distribution exists and is nearly Maxwellian. It is worthwhile to examine these assumptions when there is a steady electric field or an oscillating electric field near a cyclotron resonance, since in both of these cases, the steady state either does not exist or differs substantially from a Maxwellian.

Problem 3.5 *Dynamic friction for large velocities.* For large velocities, find approximate expressions for the integrals in equations (3.80) and (3.81), keeping the lowest and next higher order nonvanishing terms in v_e/v. Then show that the diffusion term varies as $1/v^3$.

3.4.1 Runaway electrons in a steady electric field

If one assumes a steady electric field and sets out to calculate the current, the simple dynamic friction term of equation (3.82) will not, in general, suffice. Because particles in the tail (we shall use "tail" to denote those particles with large velocities relative to the thermal velocity) of the distribution see little of the drag the main body of the distribution sees because of the rapid decline of the dynamic friction with velocity, they will not be pulled back by collisions and will continue to accelerate in the electric field. H. Dreicer[11] has shown that there exists a critical field, $E_c = m_e v_e \nu_e / e$, where a particle is accelerated from zero to the thermal velocity in one collision period. Because the time between collisions depends on both temperature and density, this critical field will vary as Joule heating due to the field will increase the temperature. In fact, there is no true steady state, since as runaway particles are drawn away from the main body of the distribution, the diffusion term in the Fokker-Planck equation will tend to fill the void and these new particles will be steadily drawn out by the electric field. In a tokamak, or other toroidal device, the runaway electrons will continue to accelerate until they lose containment, sometimes reaching relativistic energies.

To see how this comes about, we begin with the equations of motion for electrons and ions, neglecting the effects of a magnetic field, since with $\boldsymbol{E}$ perpendicular to $\boldsymbol{B}$, it simply causes a drift velocity $\boldsymbol{E} \times \boldsymbol{B}/B^2$ and for $\boldsymbol{E}$ parallel to $\boldsymbol{B}$, it has no direct effect, so we write

$$\frac{\partial f_e}{\partial t} + \boldsymbol{v} \cdot \nabla f_e - \frac{e}{m_e} \boldsymbol{E} \cdot \nabla_v f_e = \left(\frac{\partial f_e}{\partial t} \right)_c \tag{3.84}$$

$$\frac{\partial f_i}{\partial t} + \boldsymbol{v} \cdot \nabla f_i + \frac{e}{m_i} \boldsymbol{E} \cdot \nabla_v f_i = \left(\frac{\partial f_i}{\partial t} \right)_c . \tag{3.85}$$

The collision terms are represented by the Fokker-Planck equation

$$\left(\frac{\partial f_\alpha}{\partial t}\right)_c = \sum_{\beta=e,i} \left\{ -\frac{\partial}{\partial v_i}[f_\alpha \langle \Delta v_i \rangle_{\alpha\beta}] + \frac{1}{2}\frac{\partial^2}{\partial v_j \partial v_k}[f_\alpha \langle \Delta v_j \Delta v_k \rangle_{\alpha\beta}] \right\} \tag{3.86}$$

where the repeated indices, i, j, k are to be summed, and the sum over species $\beta = e, i$ indicates that each type of particle has collisions with both electrons and ions. The average velocity increments are given by the Rosenbluth H and G potentials[12] through

$$\langle \Delta v_k \rangle_{\alpha\beta} = \frac{\partial H_{\alpha\beta}}{\partial v_k} \tag{3.87}$$

$$\langle \Delta v_j \Delta v_k \rangle_{\alpha\beta} = \frac{\partial^2 G_{\alpha\beta}}{\partial v_j \partial v_k} \tag{3.88}$$

where

$$H_{\alpha\beta}(\boldsymbol{r}, \boldsymbol{v}, t) = \frac{m_\alpha + m_\beta}{m_\beta} \Gamma_\alpha \int \frac{f_\beta(\boldsymbol{r}, \boldsymbol{v}', t)}{w} \mathrm{d}^3 v' \tag{3.89}$$

$$G_{\alpha\beta}(\boldsymbol{r}, \boldsymbol{v}, t) = \Gamma_\alpha \int w f_\beta(\boldsymbol{r}, \boldsymbol{v}', t) \mathrm{d}^3 v' \tag{3.90}$$

and

$$w = |\boldsymbol{v} - \boldsymbol{v}'| \tag{3.91}$$

$$\Gamma_\alpha = 4\pi \left(\frac{e^2}{4\pi\epsilon_0 m_\alpha}\right)^2 \ln\left(\frac{\lambda_D}{p_0}\right), \tag{3.92}$$

where p_0 is the average impact parameter for a 90° Coulomb deflection. Defining the average velocity for each species by

$$\boldsymbol{v}_\alpha \equiv \frac{1}{n} \int f_\alpha \boldsymbol{v} \, \mathrm{d}^3 v \tag{3.93}$$

and n is the electron (or ion) particle density. Assuming a uniform plasma, we then take the first moments of equations (3.84) and (3.85) by multiplying by $m\boldsymbol{v}$ and integrating to obtain

$$m_e \frac{\partial \boldsymbol{v}_e}{\partial t} + e\boldsymbol{E} = \frac{m_e}{n} \int f_e(\boldsymbol{v}, t) \nabla_v H_{ei} \, \mathrm{d}^3 v, \tag{3.94}$$

$$m_i \frac{\partial \boldsymbol{v}_i}{\partial t} - e\boldsymbol{E} = \frac{m_i}{n} \int f_i(\boldsymbol{v}, t) \nabla_v H_{ie} \, \mathrm{d}^3 v. \tag{3.95}$$

These equations state that the time rate of change of momentum for each species is a balance between the electric force and the dynamical friction arising from electron-ion encounters. Encounters between like particles do not alter the momentum of the gas and therefore do not contribute to the dynamical friction. The total momentum with a steady electric field must be conserved, and this can be shown by demonstrating that the dynamical friction force obeys Newton's third law. Using equations (3.89), (3.91) and (3.92), we find

$$\int f_e(\boldsymbol{v},t)\nabla_v H_{ei}\,\mathrm{d}^3v = \frac{m_e+m_i}{m_i}\Gamma_e \iint f_e(\boldsymbol{v},t)f_i(\boldsymbol{v}',t)\nabla_v\left(\frac{1}{w}\right)\mathrm{d}^3v\,\mathrm{d}^3v'$$
$$= -\frac{m_i}{m_e}\int f_i(\boldsymbol{v}',t)\nabla_{v'}H_{ie}\,\mathrm{d}^3v'$$

and adding equations (3.94) and (3.95), we find

$$\frac{\partial \boldsymbol{v}_e}{\partial t} = -\frac{m_i}{m_e}\frac{\partial \boldsymbol{v}_i}{\partial t} \tag{3.96}$$

which shows that electrons carry nearly all of the current.

In the limit of either weak or strong electric fields, this problem simplifies somewhat. For strong fields, we may consider the effects of electron-ion collisions to be a small perturbation on the motion which electrons and ions execute in the applied electric field. To a first approximation, then, we may assume that the electrons and ions are accelerated independently and at a constant rate. The distribution functions will largely be determined by like-particle collisions and approach asymptotically a Maxwellian which is shifted by the drift velocity. If we consider *displaced* Maxwellian distributions, given by

$$f_\alpha[\boldsymbol{r},\boldsymbol{v},\boldsymbol{v}_\alpha(t)] = \frac{n(\boldsymbol{r})}{\pi^{3/2}v_{t\alpha}^3}\mathrm{e}^{-|\boldsymbol{v}-\boldsymbol{v}_\alpha(t)|^2/v_{t\alpha}^2} \tag{3.97}$$

where

$$v_{t\alpha}^2 = \frac{2k_BT_\alpha}{m_\alpha}$$

defines the thermal velocity for each species, then a more careful analysis shows that many correct results may be obtained even for weak fields. Using this form for the distribution function, the H_{ei} function may now be evaluated by substituting equation (3.97) into equation (3.89). Integrating, we find

$$H_{ei}(q) = n\Gamma_e\frac{m_e+m_i}{m_i}\frac{\mathrm{erf}(q/v_{ti})}{q} \tag{3.98}$$

where $\mathrm{erf}(x)$ is the error function

$$\mathrm{erf}(x) \equiv \frac{2}{\sqrt{\pi}}\int_0^x \mathrm{e}^{-t^2}\mathrm{d}t$$

and $q \equiv |\boldsymbol{v}-\boldsymbol{v}_i|$. In many problems, the electron thermal speed greatly exceeds the ion thermal speed and in this limit, we may approximate the ions as being at zero temperature, in which case, H_{ei} simplifies to

$$H_{ei} = n\Gamma_e/q \tag{3.99}$$

where we have neglected terms of order m_e/m_i. With this simplification, we may write equation (3.94) as

$$m_e \frac{\partial \boldsymbol{v}_e}{\partial t} + e\boldsymbol{E} = -eE_c\Psi(z) \tag{3.100}$$

where

$$\Psi(z) \equiv \frac{\text{erf}(z) - z\,\text{erf}'(z)}{z^2}$$

where the prime denotes a derivative with respect to the argument and

$$z = |\boldsymbol{v}_e - \boldsymbol{v}_i|/v_{te}$$
$$eE_c = nm_e\Gamma_e/v_{te}^2\,.$$

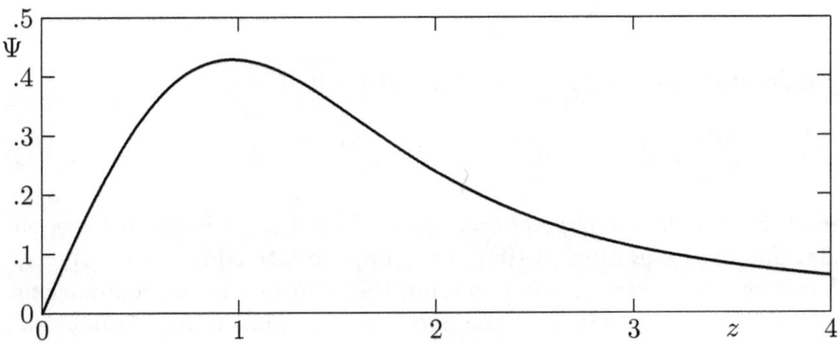

FIGURE 3.4
Variation of the dynamical friction coefficient Ψ as a function of the relative drift velocity $z = |\boldsymbol{v}_e - \boldsymbol{v}_i|/v_{te}$.

The variation of $\Psi(z)$ is shown in Figure 3.4 and it may be noted that the maximum occurs where $z = 1$. Having established the form of the dynamical friction, we now solve equation (3.94) by successive approximation. There are two types of solutions for weak and strong fields. For weak fields, such that

$$E/E_c < \Psi(1) = 0.428\,,$$

the drift velocity rises monotonically to the steady state value ($\partial/\partial t = 0$),

$$E = E_c\Psi(1)\,. \tag{3.101}$$

For stronger fields such that

$$E/E_c > \Psi(1)\,,$$

there is no steady state as the particle acceleration is always positive and both v_e and v_i increase without bound or "run away." The critical field may be expressed in terms of the collision frequency or in terms of the Debye radius, as

$$E_c = \frac{e}{4\pi\epsilon_0\lambda_D^2}\ln\Lambda\,, \tag{3.102}$$

or

$$\nu = \frac{n\Gamma_e}{v_{te}^3} = \frac{e}{m_e}\frac{E_c}{v_{te}}$$

which is related to the acceleration

$$v_{te}\nu = \frac{e}{m_e}E_c\,.$$

In the weak field limit, we may solve equation (3.94) for $v_e(t)$ using the small argument expression for $\Psi(z)$,

$$\Psi(z) \approx \frac{4}{3\sqrt{\pi}}z\,, \quad z \ll 1$$

to find the solution with initial condition $v_e(0) = 0$ to be

$$v_e(t) = -\frac{3\sqrt{\pi}v_{te}}{4}\frac{E}{E_c}\left(1 - \mathrm{e}^{-4\nu t/3\sqrt{\pi}}\right)\,. \tag{3.103}$$

Problem 3.6 *Rosenbluth H-potential.* Show that $H_{ei}(q)$ is given by equation (3.98) and that equation (3.100) is the appropriate equation for the dynamical friction for electrons, after making the appropriate approximations. (You will probably need to look at the paper by Rosenbluth, MacDonald and Judd[12].)

Problem 3.7 *Weak-field velocity.* Show that equation (3.103) is a solution of equation (3.94) with $v_e(0) = 0$ using the small argument expression for $\Psi(z)$.

3.4.2 Resonant wave heating

As our final example of using the Fokker-Planck equation to see how external fields may modify the distribution function, we examine the heating of a plasma at either the fundamental ion cyclotron frequency or its first harmonic via the fast Alfvén wave. The treatment is due to Stix[14] who begins with a two ion component plasma. The two ion components lead to resonances at each ion cyclotron fundamental, the two-ion hybrid resonance, and for minority hydrogen in deuterium, the two-ion hybrid resonance nearly coincides with the hydrogen fundamental and the deuterium harmonic since $\omega_{cH} = 2\omega_{cD}$.

One way to understand the mechanism for resonant damping is to move into a rotating reference frame at the resonant frequency in which case the particles "see" a stationary electric field depending on the wave amplitude. For the fast Alfvén wave, the effective electric field is $E_+ = \frac{1}{2}(E_x+\mathrm{i}E_y)$. The objective is to see how such a resonant wave can influence the distribution function. Because of the constant acceleration in the rotating frame, the runaway effect of the previous section tends to pull out a "tail" so that the high energy portion of the distribution is enhanced and substantially non-Maxwellian. The Fokker-Planck equation for this case includes the quasilinear diffusion coefficients in the Boltzmann equation which are given by Kennel and Engelmann[15] so that the kinetic equation takes the form

$$\frac{\partial \boldsymbol{v}(t)}{\partial t} = \lim_{V\to\infty} \sum_n \frac{\pi Z^2 e^2}{m^2} \int \frac{\mathrm{d}^3\boldsymbol{k}}{(2\pi)^3 V} L v_\perp \delta(\omega - k_\| v_\| - n\omega_{ci}) \left|\theta_{n,k}\right|^2 v_\perp L f \tag{3.104}$$

where

$$\begin{aligned}
L &\equiv \left(1 - \frac{k_\| v_\|}{\omega}\right) \frac{1}{v_\perp}\frac{\partial}{\partial v_\perp} + \frac{k_\|}{\omega}\frac{\partial}{\partial v_\|}, \\
\theta_{n,k} &= \frac{1}{2}\mathrm{e}^{\mathrm{i}\psi}(E_x - \mathrm{i}E_y)_k J_{n+1}\left(\frac{k_\perp v_\perp}{\omega_{ci}}\right) + \frac{1}{2}\mathrm{e}^{-\mathrm{i}\psi}(E_x + \mathrm{i}E_y)_k J_{n-1}\left(\frac{k_\perp v_\perp}{\omega_{ci}}\right) \\
&\quad + \frac{v_\|}{v_\perp}(E_x)_k J_n\left(\frac{k_\perp v_\perp}{\omega_{ci}}\right), \\
\boldsymbol{k} &= k_\perp \cos\psi \hat{\mathbf{e}}_x + k_\perp \sin\psi \hat{\mathbf{e}}_y + k_\| \hat{\mathbf{e}}_z\,,
\end{aligned}$$

and V is the plasma volume and the $\boldsymbol{E}_k$ are the Fourier amplitudes in the complex analysis of the wave fields. For our purposes we can drop the parallel velocity effects in the L operator and also neglect the $E_x - \mathrm{i}E_y$ and the E_z contributions to $\theta_{n,k}$, and assume $\psi = 0$ and a monochromatic spectrum. With these simplifications, carrying out the integral over $\boldsymbol{k}$, we have

$$\begin{aligned}
\frac{\partial f(\boldsymbol{v})}{\partial t} &= \frac{\pi Z^2 e^2}{8m^2 |k_\||} \left|E_x + \mathrm{i}E_y\right|^2 \\
&\quad \times \sum_n \frac{1}{v_\perp}\frac{\partial}{\partial v_\perp} v_\perp^2 \left|J_{n-1}\left(\frac{k_\perp v_\perp}{\omega_{ci}}\right)\right|^2 \delta\left(v_\| - \frac{\omega - n\omega_{ci}}{k_\|}\right) \frac{1}{v_\perp}\frac{\partial f}{\partial v_\perp}
\end{aligned} \tag{3.105}$$

where the sum is over the cyclotron harmonics. For acceleration at the cyclotron fundamental, $n = 1$, and the Bessel factor, $J_0(k_\perp v_\perp/\omega_{ci}) \to 1$, in the small Larmor radius limit.

The applicability of equation (3.105) depends on geometry, such as is found in tokamaks where this type of resonant heating is most relevant. For the validity of the quasilinear diffusion coefficient, it is assumed that the phase of the wave is random and that there is a spectrum of wave frequencies and wave numbers. We are assuming a fixed phase and a single frequency and wave number during any single pass of a resonant ion through the resonance zone, where the particle gets a small increment of velocity. The validity thus depends on particles making multiple passes through the resonance zone (the magnetic field is assumed to vary in space so the resonance zone is a finite width slice through the plasma whose width depends on the velocity spread) and "forgetting" their phase on each pass so that the phase is essentially random. This is a good approximation in tokamaks since individual particles have relatively long trajectories between passes and the dephasing effects of collisions occur more rapidly than significant deflections upon which the usual collision frequency is based.

From the form of the quasilinear diffusion through the operator L in equation (3.104) and especially in the reduced form of equation (3.105), it may appear that cyclotron heating occurs only in the perpendicular direction, but this is not so. The operator L has a gradient in velocity in the direction of a circle defined by

$$v_\perp^2 + \left(v_\parallel - \frac{\omega - n\omega_{ci}}{k_\parallel}\right)^2 = \text{constant}$$

in velocity space. This is illustrated in Figure 3.5 where the resonance condition is the dashed vertical line and the diffusion is always along a circle, so that both parallel and perpendicular diffusion occurs.

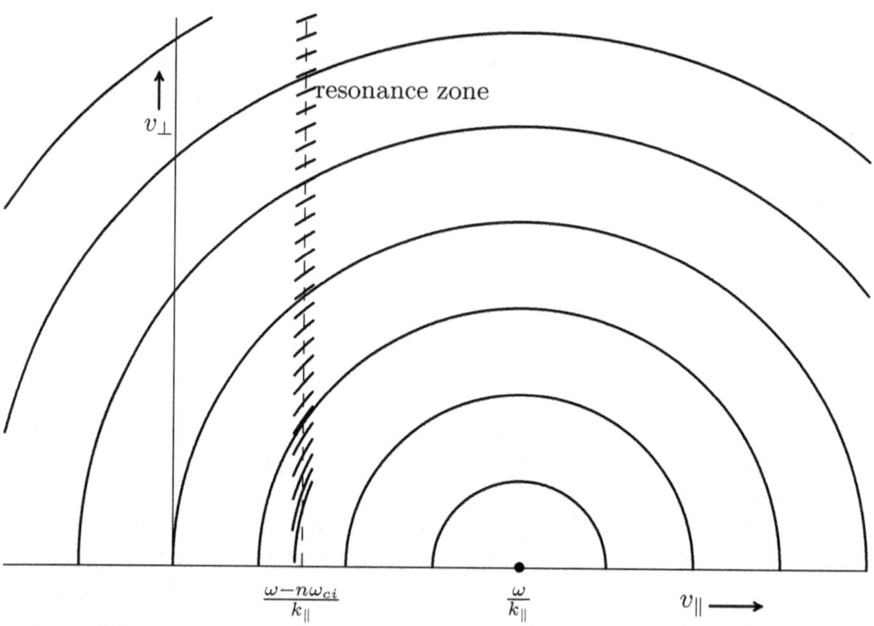

FIGURE 3.5
Contours for quasi-linear diffusion. Diffusion takes place only along the arc direction and only in the resonance zone.

We now wish to write a kinetic equation which combines the effects of quasilinear diffusion and Coulomb thermalization. A convenient form for the Fokker-Planck equation is

$$\frac{\partial f}{\partial t} = -\nabla_v \cdot (\langle \Delta \boldsymbol{v} \rangle f) + \tfrac{1}{2}\nabla_v \cdot [\nabla_v \cdot (\langle \Delta \boldsymbol{v} \Delta \boldsymbol{v} \rangle f)] , \tag{3.106}$$

the first term representing drag and the second term diffusion.

For a specific example, we choose a plasma with majority tritium and minority deuterium and examine the resonance at $\omega = \omega_{cT}$ so there is only a fundamental resonance and no two-ion hybrid resonance in the vicinity of the resonance and the nonresonant background ions may be considered Maxwellian. For the Coulomb collision effects, we use the coefficients of Spitzer[1] in velocity space given in terms of $\langle \Delta v_\parallel \rangle$, $\langle (\Delta v_\parallel)^2 \rangle$, and $\langle (\Delta v_\perp)^2 \rangle$. In this case, the parallel and perpendicular subscripts indicate directions relative to the test particle velocity $\boldsymbol{v}$ and not to the direction of $\boldsymbol{B}$. Changing from cartesian coordinates to spherical coordinates in velocity space, such that $\langle \Delta \boldsymbol{v} \rangle = \langle \Delta v_\parallel \rangle \hat{\boldsymbol{e}}_v$ and $\langle \Delta \boldsymbol{v} \Delta \boldsymbol{v} \rangle = \langle (\Delta v_\parallel)^2 \rangle \hat{\boldsymbol{e}}_v \hat{\boldsymbol{e}}_v + \langle (\Delta v_\perp)^2/2 \rangle (\hat{\boldsymbol{e}}_\theta \hat{\boldsymbol{e}}_\theta + \hat{\boldsymbol{e}}_\phi \hat{\boldsymbol{e}}_\phi)$ and then adding the quasi-linear terms to the Coulomb terms[16], the kinetic equation is

$$\frac{\partial f}{\partial t} = C(f) + Q(f) \tag{3.107}$$

where

$$\begin{aligned}
C(f) &\equiv -\frac{1}{v^2}\frac{\partial}{\partial v}\left[v^2 \left(\langle \Delta v_\parallel \rangle + \frac{1}{2v}\langle (\Delta v_\perp)^2 \rangle \right) f \right] \\
&\quad + \frac{1}{2v^2}\frac{\partial^2}{\partial v^2}\left[v^2 \langle (\Delta v_\parallel)^2 \rangle f \right] + \frac{1}{4v^2}\frac{\partial}{\partial \mu}(1-\mu^2)\frac{\partial}{\partial \mu}\left[\langle (\Delta v_\perp)^2 \rangle f \right] , \\
Q(f) &\equiv \sum_{mn} Q_{mn} \delta \left[\omega_{ci} - \frac{(\omega - k_\parallel v_\parallel)}{n} \right] R_m(f) , \\
R_m(f) &\equiv (1-\mu^2)^{m+1} \frac{1}{v^2}\frac{\partial}{\partial v} v^{2m+1} \frac{\partial}{\partial v} v f + v^{2m-2}\frac{\partial}{\partial \mu}\mu(1-\mu^2)^{m+1}\frac{\partial}{\partial \mu}\mu f \\
&\quad - v^{2m-2}\frac{\partial}{\partial \mu}\mu(1-\mu^2)^{m+1}\frac{\partial}{\partial v} v f - (1-\mu^2)^{m+1} v^{2m-2}\frac{\partial}{\partial \mu}\mu\frac{\partial}{\partial v} v f \\
&\quad - 2m(1-\mu^2)^{m+1} v^{2m-2}\frac{\partial}{\partial \mu}\mu f ,
\end{aligned}$$

where μ is the cosine of the angle between $\boldsymbol{v}$ and $\boldsymbol{B}$ and $Q_{mn}(\boldsymbol{r})$ is the coefficient of $v_\perp^{2m}$ in the power series,

$$\sum_m Q_{mn} v_\perp^{2m} = \frac{\pi Z^2 e^2}{8 m_i^2 |n|} |E_x + \mathrm{i}E_y|^2 \left| J_{n-1}\left(\frac{k_\perp v_\perp}{\omega_{ci}} \right) \right|^2 , \tag{3.108}$$

where for $n = 1$, the first few of the Q_{m1} are

$$Q_{01} = \frac{\pi Z^2 e^2}{8 m_i^2} |E_x + \mathrm{i}E_y|^2$$

$$Q_{11} = -2Q_{01}\left(\frac{k_\perp}{2\omega_{ci}}\right)^2$$
$$Q_{21} = \frac{3}{2}Q_{01}\left(\frac{k_\perp}{2\omega_{ci}}\right)^4 .$$

We can eliminate the delta function in $R_m(f)$ by averaging over a toroidal magnetic surface of minor radius r and major radius R such that

$$\frac{1}{2\pi}\int \mathrm{d}\theta Q_{mn}(x,y)\delta\delta\left[\omega_{ci} - \frac{(\omega - k_\parallel v_\parallel)}{n}\right] \approx \frac{R}{\pi\omega_{ci} r|\sin\theta_0|}Q_{mn}(x_0,|y_0|)$$

where $\tan\theta_0 = |y_0|/x_0$, and x_0, y_0 are the coordinates where the selected magnetic surface intersects the central ($v_\parallel = 0$) resonance surface. This assumes that $Q_{mn}(x,y)$ is symmetric for $y = \pm|y_0|$ and is well represented over the resonance zone by its value at x_0.

The next step in solving this kind of partial differential equation is to expand $f(r,\boldsymbol{v},t)$ in Legendre polynomials such that

$$f(r,\boldsymbol{v},t) = \sum_{\ell=0}^{\infty} g_{2\ell}(r,v,t)P_{2\ell}(\mu) . \tag{3.109}$$

This expansion is particularly useful if $Q(f) = 0$ since then the $P_k(\mu)$ are eigenfunctions of the $C(f)$ operator. For $Q(f) \neq 0$, however, the various $P_k(\mu)$ are mixed, but since the integrals over μ are simple, it is still a useful expansion. Since we have selected only the $n = 1$ case for our example, we will keep only two terms in the expansion so that

$$f(v,t) = A(v,t) + \tfrac{1}{2}B(v,t)(3\mu^2 - 1)$$

and we find from the P_0 and P_2 moments of equation (3.107)

$$\frac{\partial A}{\partial t} = \frac{1}{v^2}\frac{\partial}{\partial v}\left[-\alpha v^2 A + \frac{\partial}{\partial v}(\tfrac{1}{2}\beta v^2 A) + Kv\frac{\partial}{\partial v}v\left(A - \frac{B}{5}\right) - K\left(A + \frac{B}{5}\right)v\right] \tag{3.110}$$

$$\begin{aligned}\frac{\partial B}{\partial t} &= -\frac{1}{v^2}\frac{\partial}{\partial v}(\alpha v^2 A) + \frac{1}{2v^2}\frac{\partial^2}{\partial v^2}(\beta v^2 B) - \frac{3\gamma B}{2v^2} + \frac{K}{v^2}\frac{\partial}{\partial v}v\frac{\partial}{\partial v}v\left(-A + \frac{5}{7}B\right)\\ &\quad -\frac{K}{v^2}\left(3A + \frac{30}{7}B\right) + \frac{K}{v^2}\frac{\partial}{\partial v}v\left(4A - \frac{5}{7}B\right)\end{aligned} \tag{3.111}$$

where

$$\begin{aligned}K &\equiv \frac{2R}{3\pi\omega_{ci} r|\sin\theta_0|}Q_{01}(x_0,|y_0|) = \frac{\langle P\rangle}{3n_i m_i}\\ \alpha &\equiv \langle \Delta v_\parallel\rangle + \frac{1}{2v}\langle(\Delta v_\perp)^2\rangle\\ \beta &\equiv \langle(\Delta v_\parallel)^2\rangle\\ \gamma &\equiv \langle(\Delta v_\perp)^2\rangle\end{aligned} \tag{3.112}$$

and where

$$\begin{aligned}
\langle \Delta v_{\parallel} \rangle &= -\sum_f C_f \ell_f^2 (1 + m/m_f) G(\ell_f v) \\
\langle (\Delta v_{\parallel})^2 \rangle &= \sum_f \frac{C_f}{v} G(\ell_f v) \\
\langle (\Delta v_{\perp})^2 \rangle &= \sum_f \frac{C_f}{v} [\mathrm{erf}(\ell_f v) - G(\ell_f v)] \\
C_f &\equiv \frac{n_f Z_f^2 e^4 \ln \Lambda}{2\pi \epsilon_0^2 m^2} \\
\ell_f &\equiv 1/v_{tf} \\
G(x) &\equiv \tfrac{1}{2}\Psi(x)
\end{aligned} \tag{3.113}$$

where the subscript f denotes the background particles. A useful identity in working with these coefficients is

$$-\alpha v^2 + \frac{1}{2}\frac{\mathrm{d}}{\mathrm{d}v}(\beta v^2) = \sum_f C_f \frac{v^2}{v_{tf}^2} G(v/v_{tf}) \,. \tag{3.114}$$

Because the error function and $G(x)$ are so complicated, we introduce approximate forms so that the integrals can all be solved in closed form. We let

$$\begin{aligned}
\mathrm{erf}(x) &\approx \frac{\varepsilon x (3 + 2x^2)}{1 + 2\varepsilon x^3} \\
G(x) &\approx \frac{\varepsilon x}{1 + 2\varepsilon x^3}
\end{aligned}$$

where $\varepsilon = 2/3\sqrt{\pi}$. Some simplified forms of the Coulomb diffusion coefficients may be derived in certain regions according to Table 3.1 where Range I is for

TABLE 3.1
Ranges and simplified diffusion coefficients.

Range	α	β	γ
I	D/v	D	$2D$
II	$-\dfrac{v}{t_s}\left(1 + \dfrac{V_\alpha^3}{v^3}\right)$	$\dfrac{v_{te}}{t_s}\left(1 + \dfrac{V_\beta^3}{v^3}\right)$	$\dfrac{V_\gamma^3}{v t_s}$

$v \ll v_{ti}$ and Range II is for $v_{ti} \ll v \ll v_{te}$, and where, summing over the f species of background ions and electrons,

$$D \equiv \sum_f C_f \ell_f$$

$$= 2\sum_f \frac{n_f Z_f^2}{n_e}\left(\frac{m_f T_e}{m_e T_f}\right)^{1/2} \frac{k_B T_e}{m t_s}$$

$$t_s \equiv \frac{m_e v_{te}^3}{\varepsilon m C_e} = \frac{6.65 \cdot 10^{14} A T_{eV}^{3/2}}{Z^2 n_e \ln\Lambda} \text{ seconds}$$

and summing over the j species of field ions,

$$\frac{1}{2} m V_\alpha^2 = 1.48 \cdot 10^{-3} T_{eV} \left[\frac{A^{3/2}}{n_e}\sum_j \frac{n_j Z_j^2}{A_j}\right]^{2/3}$$

$$\frac{1}{2} m V_\beta^2 = 1.48 \cdot 10^{-3} T_{eV}^{1/3} \left[\frac{A^{3/2}}{n_e}\sum_j \frac{n_j Z_j^2 k_B T_j}{e A_j}\right]^{2/3}$$

$$\frac{1}{2} m V_\gamma^2 = 1.48 \cdot 10^{-3} T_{eV} \left[\frac{2A^{1/2}}{n_e}\sum_j n_j Z_j^2\right]^{2/3}$$

where A and A_j are the atomic masses of the test and field ions and n_e is in m^{-3}.

Problem 3.8 *Q_{m2} coefficients.* Find the first three Q_{m2} coefficients.

Problem 3.9 *Evolution of $A(v,t)$ and $B(v,t)$.* Show that equations (3.110) and (3.111) result from taking the P_0 and P_2 moments of equation (3.107) with $m = 0$ and $n = 1$.

3.4.2.1 Steady state — lowest energy range

If we assume that the wave heating is of sufficient duration that a steady state is reached, then $\partial f(v,t)/\partial t = 0$. The existence of a steady state, however, may be limited to the lowest energy range and to fundamental resonance heating, since for $n = 2$, the wave energy is primarily transfered to the tail of the distribution due to the higher order in $k_\perp v_\perp/\omega_{ci}$ in the Q_{mn} terms. This leads to a runaway condition for that part of the distribution function much as we encountered in the previous section for a constant electric field. Even in this lower energy range, significant effects can be noted, and equations (3.110) and (3.111) may be integrated to find $A(v)$ and $B(v)$. With $\partial A/\partial t = \partial B/\partial t = 0$, we can integrate equation (3.110) once immediately, and using the condition that $A \to 0$ as $v \to \infty$, and using equation (3.114) and the Range I expressions for α, β, and γ, we find

$$A(v) = \text{const. } e^{-mv^2/k_B T_{\text{eff}}}$$

$$k_B T_{\text{eff}} = \frac{D + 2K}{\varepsilon \sum_f (C_f \ell_f / k_B T_f)}$$

$$B(v) = \frac{4K}{7D + 10K}\left(\frac{mv^2}{2k_B T_{\text{eff}}}\right)^2 A\,.$$

If all the T_f are equal, such that $T_f = T_0$, then $k_B T_{\text{eff}} = (D + 2K)k_B T_0/D$. It follows that if $K \to 0$, then $T_{\text{eff}} \to T_0$ and $B = 0$. When $K \neq 0$, the resonant velocity distribution is still approximately Maxwellian with an elevated temperature. The complexity of the diffusion coefficients makes a full analysis of the angular distribution difficult, but by integrating equation (3.111) over $v^2 \mathrm{d}v$ from 0 to ∞, many of the terms vanish with the result

$$\int_0^\infty \left(\frac{\gamma(v)}{2} + \frac{10}{7}K\right) B(v)\,\mathrm{d}v = -K \int_0^\infty A(v)\,\mathrm{d}v\,, \tag{3.115}$$

which suggests that

$$B(v) \sim -KA(v) \Big/ \left(\frac{\gamma(v)}{2} + \frac{10}{7}K\right)$$

as a first estimate.

3.4.2.2 Steady state — $f(v)$ solution

If we drop the angular dependence, the problem simplifies greatly, so that the Fokker-Planck distribution is truncated at

$$f(\boldsymbol{v}) = A(v) = \tfrac{1}{2} \int f(v, \mu)\,\mathrm{d}\mu$$

which is equation (3.110) with $B = 0$, and is immediately integrable such that

$$f(v) = f(0) \exp\left[-\int_0^v \mathrm{d}v \frac{-2\alpha v^2 + (\beta v^2)'}{\beta v^2 + 2Kv^2}\right] \tag{3.116}$$

where the Coulomb diffusion coefficients are given in equations (3.112) and (3.113) and the identity of equation (3.114) provides a convenient form for the numerator in the integrand. For the background electrons, we may use $G(x_e) = \varepsilon x_e$ since $v \ll v_e$. With only one background ion species and using the approximate forms for $G(x)$, the result is

$$\ln f(v) = -\frac{E}{k_B T_e(1+\xi)}\left[1 + \frac{R_j(T_e - T_j + \xi T_e)}{T_j(1 + R_j + \xi)} H(E/E_j)\right] \tag{3.117}$$

where

$$\begin{aligned}
E &\equiv \tfrac{1}{2}mv^2 \\
\ell_j &\equiv 1/v_{tj} \\
R_j &\equiv \frac{n_j Z_j^2 \ell_j}{n_e \ell_e} \\
\xi &\equiv \frac{2K}{\varepsilon C_e \ell_e} = \frac{2\pi^{3/2}\epsilon_0^2 m \langle P\rangle v_{te}}{n_e n Z^2 e^4 \ln\Lambda}
\end{aligned}$$

$$
\begin{aligned}
E_j(\xi) &\equiv v_{tj}\left[\frac{1+R_j+\xi}{2\varepsilon(1+\xi)}\right]^{2/3} \leq E_j(0) \\
E_j(0) &= \tfrac{1}{2}mV_\beta^2 \\
H(x) &\equiv \frac{1}{x}\int_0^x \frac{\mathrm{d}u}{1+u^{3/2}} \\
&= \frac{2}{x}\left[\frac{1}{6}\ln\left(\frac{1-\sqrt{x}+x}{1+2\sqrt{x}+x}\right)+\frac{1}{\sqrt{3}}\left(\frac{\pi}{2}+\tan^{-1}\frac{2\sqrt{x}-1}{\sqrt{3}}\right)\right] \\
&\simeq 1-\tfrac{2}{5}x^{3/2}+\tfrac{1}{4}x^3-\tfrac{2}{11}x^{9/2}+\cdots, \quad |x|\ll 1 \\
&\simeq 2.4184x^{-1}-2x^{-3/2}+\tfrac{1}{2}x^{-3}-\tfrac{2}{7}x^{-9/2}+\cdots, \quad |x|\gg 1\,.
\end{aligned}
$$

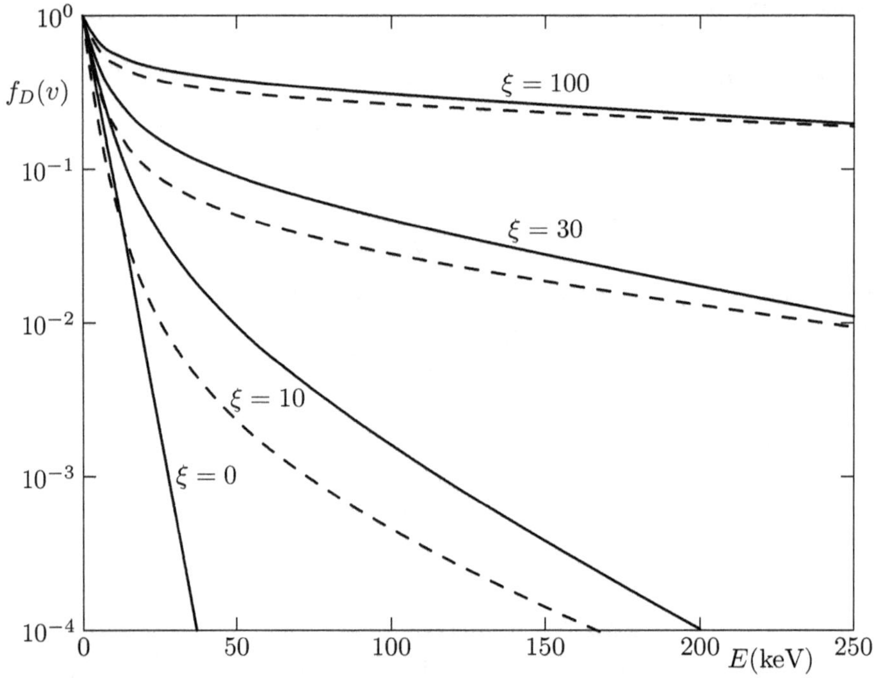

FIGURE 3.6
Plots of the distribution function versus E (in keV) for various values of ξ where $n_T = 5\cdot 10^{19}$ m^{-3}, $n_D = n_T/20$, $k_BT_e = k_BT_T = 4$ keV. The solid lines are for $f_D(v)$ and the dashed lines are for $f_D(v_\perp)$.

For a specific example, we show $f_D(v)$ as a function of E in Figure 3.6 for a deuterium minority in tritium where the deuterons are resonant. The variable ξ is proportional to the wave power absorbed per unit volume, $\langle P\rangle$.

It is evident that as the applied power is increased, the deviation from a Maxwellian ($\xi = 0$) becomes extreme with a power-law distribution on the tail.

Instead of the distribution of the velocity magnitude, we may be more interested in the distribution of the velocity perpendicular to the magnetic field since the primary driving force is in the perpendicular direction. Defining

$$f(v_\perp) = \int_{-\infty}^{\infty} f(\boldsymbol{v})\mathrm{d}v_\parallel$$

Stix[14] has shown that the steady-state Fokker-Planck equation for this case becomes

$$0 = -\frac{1}{v_\perp}\frac{\partial}{\partial v_\perp}(\alpha v_\perp f) + \frac{1}{2v_\perp}\frac{\partial^2}{\partial v_\perp^2}(\beta v_\perp f) + \frac{1}{4v_\perp}\frac{\partial}{\partial v_\perp}(\gamma f) + \frac{3}{2}\frac{K}{v_\perp}\frac{\partial}{\partial v_\perp}v_\perp\frac{\partial f}{\partial v_\perp}. \tag{3.118}$$

The steady-state solution may then be obtained by integrating twice to obtain

$$f(v_\perp) = f(0)\exp\left[-\int_0^{v_\perp}\mathrm{d}v_\perp\frac{-4\alpha v_\perp + 2(\beta v_\perp)' + \gamma}{2\beta v_\perp + 6Kv_\perp}\right].$$

Following the same steps that led to equation (3.117), we find

$$\begin{aligned} \ln f(v_\perp) &= \frac{-2E}{k_B T_e(2+3\xi)}\left\{1 + \frac{R_j[(2A+A_j)(2+3\xi)T_e - 4AT_j]}{2AT_j(2+2R_j+3\xi)}H(E/E_j)\right\} \\ E_j &= \frac{AkT_j}{A_j}\left[\frac{2+2R_j+3\xi}{2\varepsilon(2+3\xi)}\right]^{2/3}. \end{aligned} \tag{3.119}$$

It has been tacitly assumed to this point that the wave dispersion relation is not extremely important in estimating the distribution function. Since, however, the two-ion hybrid resonance is close to the minority fundamental, and in an inhomogeneous plasma the two-ion hybrid resonance is a mode conversion layer where the incident wave may be transmitted, reflected, or mode converted to a Bernstein wave, the details of the mode conversion analysis make a huge difference. For the parameters in Figure 3.6, there is virtually no tunneling, and with $k_\parallel = 0$ there is no absorption at all. Both the reflection coefficient and the conversion coefficients are strongly dependent on $k_\parallel$, with the reflection coefficient varying as $|R|^2 \sim \exp(-\alpha k_\parallel^2)$[17] where α depends on the local plasma parameters and requires an integral equation to be solved numerically. The conversion coefficient is even more complicated. As the power is applied with a given ξ, the distribution function approaches a steady state value given by equation (3.117), and the effective temperature increases in time. A peculiarity of the two-ion hybrid resonance is that as the minority temperature increases, the resonance moves so that reflection vanishes and the wave energy not absorbed is mode-converted, eventually heating electrons[18]. For this example, the transition occurs near $\xi \sim 45$. The amount of incident energy absorbed ($\langle P \rangle$), and hence ξ, then will change as the process changes in time, complicating estimates of the heating of the minority and the majority as the two species relax via collisions.

4

THE VLASOV-MAXWELL EQUATIONS

The first and simplest approximation including the effects of kinetic theory in equation (2.84) is to neglect any effects due to correlations. This zero order equation (in $1/N_D$) along with the Maxwell equations are called the Vlasov equations, although frequently this collisionless Boltzmann equation alone is referred to as the Vlasov equation. In the singular, we prefer to refer to this fundamental equation either as the collisionless Boltzmann equation or as the kinetic equation, and refer to the system of equations as the Vlasov-Maxwell equations or simply as the Vlasov equations (plural).

The Vlasov equations, then, are comprised of the set

$$\frac{\partial f_j}{\partial t} + \boldsymbol{v} \cdot \nabla f_j + \frac{q_j}{m_j}(\boldsymbol{E} + \boldsymbol{v} \times \boldsymbol{B}) \cdot \nabla_{\boldsymbol{v}} f_j = 0 \tag{4.1}$$

for each species, and the Maxwell equations, with

$$\rho(\boldsymbol{r}, t) = \sum_j q_j \int \mathrm{d}^3 v\, f_j \tag{4.2}$$

$$\boldsymbol{J}(\boldsymbol{r}, t) = \sum_j q_j \int \mathrm{d}^3 v\, \boldsymbol{v} f_j \tag{4.3}$$

as the sources and $\boldsymbol{E} = \boldsymbol{E}(\boldsymbol{r}, t)$ and $\boldsymbol{B} = \boldsymbol{B}(\boldsymbol{r}, t)$. This set of equations is nonlinear and its solutions in the linear, quasilinear, and nonlinear approximations comprise the majority of kinetic theory. The other major topic in kinetic theory is to assess the effects of collisions or correlations, which are ignored in the Vlasov equations, on the distribution functions, waves and transport coefficients such as electrical conductivity, viscosity, thermal diffusion, etc.

4.1 Electrostatic waves in an unmagnetized plasma

In the unmagnetized plasma, the preferred direction is the $\boldsymbol{k}$ direction, and motions of particles parallel and perpendicular to that direction will have different effects. We will focus particular attention on the classic problem that serves to illustrate the most important effects of thermal plasmas on waves. This classic case is the $K_{zz} = P = 0$ solution of cold plasma, or the Bohm-Gross solution in the fluid plasma, namely an electrostatic wave near the plasma frequency.

We are first going to try a simple way to find the solution following the 1945 analysis of Vlasov[19], which will lead us to a dilemma because the method is not well posed. We will then back up and start again with a more carefully posed problem that will provide a recipe to be used with the simpler method so that it may be used as a foundation for the study of more complicated cases.

4.1.1 Vlasov method

We first linearize the Vlasov equations by separating out zero and first order terms that are assumed to vary as

$$\begin{aligned} f_j &= f_{0j}(\boldsymbol{v}) + f_{1j}(\boldsymbol{v})\mathrm{e}^{\mathrm{i}(\boldsymbol{k}\cdot\boldsymbol{r}-\omega t)} \\ \boldsymbol{E} &= \qquad\qquad \boldsymbol{E}_1\mathrm{e}^{\mathrm{i}(\boldsymbol{k}\cdot\boldsymbol{r}-\omega t)} \\ \boldsymbol{B} &= \qquad\qquad \boldsymbol{B}_1\mathrm{e}^{\mathrm{i}(\boldsymbol{k}\cdot\boldsymbol{r}-\omega t)} \end{aligned} \tag{4.4}$$

and $\boldsymbol{v}$ is now an independent variable so is not linearized. This is equivalent to having Fourier transformed in both space and time. We choose the normalization of the velocity distribution function so that $\int \mathrm{d}^3v\, f_j = n_j$. Using these results in Eqs. (4.1) through (4.3) along with the Maxwell equations, and assuming only a single species of ions plus electrons so that charge neutrality requires $n_i = n_e = n_0$, the equations to be solved are

$$-\mathrm{i}(\omega - \boldsymbol{k}\cdot\boldsymbol{v})f_{1j} + \frac{q_j}{m_j}\boldsymbol{E}_1\cdot\nabla_{\boldsymbol{v}} f_{0j} = 0\,, \quad j = e, i \tag{4.5}$$

$$\rho = e\int \mathrm{d}^3v\,(f_{1i} - f_{1e}) \tag{4.6}$$

$$\boldsymbol{J} = e\int \mathrm{d}^3v\,\boldsymbol{v}(f_{1i} - f_{1e})\,, \tag{4.7}$$

$$\mathrm{i}\boldsymbol{k}\times\boldsymbol{E}_1 = \mathrm{i}\omega\boldsymbol{B}_1 \tag{4.8}$$

$$\mathrm{i}\boldsymbol{k}\times\boldsymbol{B}_1 = \mu_0\boldsymbol{J} - \frac{\mathrm{i}\omega}{c^2}\boldsymbol{E}_1\,. \tag{4.9}$$

There is no $\boldsymbol{B}_0$ term by assumption and no $\boldsymbol{B}_1$ term since for an electrostatic wave, $\boldsymbol{k} \parallel \boldsymbol{E}$ so that from Maxwell's equations, $\boldsymbol{B}_1 = \boldsymbol{k}\times\boldsymbol{E}/\omega = 0$.

From $\boldsymbol{E}_1 = -\nabla\varphi$, Poisson's equation gives

$$-\nabla^2\varphi = k^2\varphi = \frac{\rho}{\epsilon_0} = \frac{e}{\epsilon_0}\int \mathrm{d}^3v\,(f_{1i} - f_{1e})\,. \tag{4.10}$$

Solving equation (4.10) for the potential and equation (4.5) for f_{1j}, we find

$$\varphi(\boldsymbol{k},\omega) = -\frac{\varphi}{\epsilon_0 k^2}\sum_j \int \mathrm{d}^3v\,\frac{q_j^2}{m_j}\frac{\boldsymbol{k}\cdot\nabla_{\boldsymbol{v}} f_{0j}}{(\omega - \boldsymbol{k}\cdot\boldsymbol{v})}\,,$$

but since φ appears on both sides, this requires

$$1 + \sum_j \frac{q_j^2}{\epsilon_0 k^2 m_j} \int \mathrm{d}^3 v \, \frac{\boldsymbol{k} \cdot \nabla_{\boldsymbol{v}} f_{0j}}{(\omega - \boldsymbol{k} \cdot \boldsymbol{v})} = 0 \,. \tag{4.11}$$

We are free to align $\boldsymbol{k}$ with v_z, and then the integrals over v_x and v_y are trivial, leaving the remaining integrals over $u = v_z/v_j$, where $v_j^2 \equiv 2k_B T_j/m_j$, which may be written as

$$1 + \sum_j \frac{\omega_{pj}^2}{k^2 v_j^2} \frac{1}{\sqrt{\pi}} \int_{-\infty}^{\infty} \frac{\frac{\mathrm{d}}{\mathrm{d}u} \mathrm{e}^{-u^2} \, \mathrm{d}u}{(\omega/k v_j) - u} \,, \tag{4.12}$$

where we have introduced the Maxwellian distribution (in one dimension),

$$f_{0j}(v_z) = \frac{n_{0j}}{v_j \sqrt{\pi}} \mathrm{e}^{-v_z^2/v_j^2} \,.$$

Now the difficulties begin. It is clear that the integral over u has a pole along the path of integration, at $u = \omega/k v_j$, and is hence undefined unless we specify that the path should always go above or below the pole. At this point in the problem, Vlasov chose to take the principal part of the integral, or the average of the two paths above and below the pole[19]. This provides symmetry in time, but ignores some of the physics. We shall examine one of the two paths, and try to determine the implications of making one choice or the other after we see the effects of our choice. Let us assume first that $\mathrm{Im}(\omega) > 0$, so that ω has a small positive imaginary part which puts the pole just above the path of integration. This choice corresponds to "turning on" the perturbation slowly from $t \to -\infty$, and happens to coincide with the definition of the Plasma Dispersion function which is a tabulated function[20] defined by

$$Z(\zeta) \equiv \frac{1}{\sqrt{\pi}} \int_{-\infty}^{\infty} \frac{e^{-\xi^2}}{\xi - \zeta} \, d\xi \,, \quad \mathrm{Im}(\zeta) > 0 \,, \tag{4.13}$$

and whose properties are listed in Appendix A. In terms of this function we find

$$1 - \sum_j \frac{\omega_{pj}^2}{k^2 v_j^2} Z'\left(\frac{\omega}{k v_j}\right) = 0 \,. \tag{4.14}$$

The asymptotic forms of the Plasma Dispersion function, listed in Appendix A, vary for real argument as

$$\begin{aligned} Z(\zeta) &= \mathrm{i}\sqrt{\pi}\mathrm{e}^{-\zeta^2} - \frac{1}{\zeta}\left(1 + \frac{1}{2\zeta^2} + \frac{3}{4\zeta^4} + \cdots\right) \\ Z'(\zeta) &= -2\mathrm{i}\sqrt{\pi}\zeta e^{-\zeta^2} + \frac{1}{\zeta^2}\left(1 + \frac{3}{2\zeta^2} + \frac{15}{4\zeta^4} + \cdots\right) . \end{aligned} \tag{4.15}$$

Using the expansion for Z' for the electron term (the ion terms are of order m_e/m_i unless $T_e \gg T_i$), the dispersion relation becomes

$$1 - \frac{\omega_{pe}^2}{\omega^2}\left[1 + \frac{3}{2}\left(\frac{v_e}{v_p}\right)^2 + \mathcal{O}\left(\frac{v_e}{v_p}\right)^4\right] + 2i\sqrt{\pi}\frac{\omega_{pe}^2}{k^2 v_e^2}\frac{v_p}{v_e}\exp\left[-\left(\frac{v_p}{v_e}\right)^2\right] = 0\,, \tag{4.16}$$

or since $v_e/v_p \ll 1$,

$$\omega \simeq \omega_{pe}\left\{1 + \frac{3}{4}\left(\frac{v_e}{v_p}\right)^2 - i\sqrt{\pi}\left(\frac{v_p}{v_e}\right)^3 \exp\left[-\left(\frac{v_p}{v_e}\right)^2\right]\right\}, \tag{4.17}$$

where we see that $\omega \sim \omega_{pe}$ and $\mathrm{Im}(\omega) < 0$! Unfortunately, our assumption about the imaginary part of ω being positive has led us to the conclusion that the imaginary part is negative. It is not difficult to show that if we had chosen the imaginary part to be negative, then the analysis would have led to a positive imaginary part! There is no consistent solution, because the problem is ill-posed. While this seems to justify the approach of Vlasov in keeping only the principal part, so that there would be no imaginary part to ω, the question of whether there is or is not any damping of the wave was left unresolved. It did not stay unresolved for long, however, since the solution was provided within a year by Landau, although the result was disputed for over 20 years until unequivocally verified by experiment.

Problem 4.1 *Properties of the Plasma Dispersion function.*

1. Prove from the definition of equation (4.13) that the Plasma Dispersion function satisfies the differential equation

$$Z'(\zeta) = -2[1 + \zeta Z(\zeta)]\,. \tag{4.18}$$

2. Expand the denominator of equation (4.13) to obtain the asymptotic expansion of the Plasma Dispersion function ($\zeta \to \infty$). Compare with the result in Appendix B and discuss why this simple expansion fails to get the imaginary part right.

3. Derive the power series expansion for the Plasma Dispersion function from its definition and its differential equation, and show that the series may be grouped into two series, one of which may be summed to get an analytic expression.

4. If you were to make a numerical subroutine to evaluate the Plasma Dispersion function for real argument, using only the power series and the asymptotic series:

   (a) Show how to pick the crossover between the power series and asymptotic series for optimum accuracy. The error in a numerical representation of a power series usually comes from subtraction errors between successive (largest) terms rather than from the smallest (last) term. The error for an asymptotic series comes from the last term kept because the series must be truncated when the terms no longer decrease.

(b) Find the optimum crossover point for an 8-digit computer and for a 14-digit computer, and estimate the relative accuracy obtainable on the two machines.

4.1.2 Landau solution

In 1946, Landau[21] recognized that the difficulty could be resolved by treating the problem as an initial value problem rather than using the Fourier transforms of Vlasov. Because of the importance of this and other related problems, we follow the Landau development closely, restricting our attention to longitudinal plasma oscillations only and consider only the initial value problem in time rather than include the time harmonic antenna problem (which is treated, however, in Landau's original paper). We will also neglect ion motions, since they play little role (unless $T_e \gg T_i$ as noted above), and since the ion terms are so similar in form to the electron terms.

For the electrostatic case, $\boldsymbol{k} \parallel \boldsymbol{E}$, so $\boldsymbol{k} \times \boldsymbol{E} = 0$ and $\boldsymbol{E} = -\nabla\varphi$. The equations to be solved are then

$$\frac{\partial f_1}{\partial t} + \boldsymbol{v}\cdot\nabla f_1 - \frac{e}{m_e}\boldsymbol{E}\cdot\nabla_{\boldsymbol{v}} f_0 = 0 \tag{4.19}$$

$$\nabla\cdot\boldsymbol{E} = -\nabla^2\varphi = -\frac{e}{\epsilon_0}\int \mathrm{d}^3v\, f_1 = -\frac{\rho}{\epsilon_0}\,. \tag{4.20}$$

Taking the Fourier transform in space,

$$f_1(\boldsymbol{r},\boldsymbol{v},t) = \tilde{f}(\boldsymbol{k},\boldsymbol{v},t)\mathrm{e}^{\mathrm{i}\mathbf{k}\cdot\mathbf{r}} \tag{4.21}$$

$$\varphi(\boldsymbol{r},t) = \tilde{\varphi}(\boldsymbol{k},t)\mathrm{e}^{\mathrm{i}\mathbf{k}\cdot\mathbf{r}} \tag{4.22}$$

the equations become, with $\boldsymbol{k} = k\hat{e}_z$,

$$\frac{\partial \tilde{f}}{\partial t} + \mathrm{i}kv_z\tilde{f} + \frac{e}{m_e}\mathrm{i}k\tilde{\varphi}\frac{\partial f_0}{\partial v_z} = 0 \tag{4.23}$$

$$k^2\tilde{\varphi} = -\frac{e}{\epsilon_0}\int \mathrm{d}^3v\,\tilde{f} \tag{4.24}$$

where $f_0(\boldsymbol{v})$ is given and $\tilde{f}(\boldsymbol{v},0) \equiv g(\boldsymbol{v})$ is the given initial perturbation.

For the initial value problem in time, it is convenient to use the Laplace transform of the time variable

$$X_p(\boldsymbol{v},p) = \int_0^\infty \mathrm{e}^{-pt}X(\boldsymbol{v},t)\,\mathrm{d}t \tag{4.25}$$

and its inverse

$$X(\boldsymbol{v},t) = \int_{\sigma-\mathrm{i}\infty}^{\sigma+\mathrm{i}\infty} \mathrm{e}^{pt}X_p(\boldsymbol{v},p)\,\frac{\mathrm{d}p}{2\pi\mathrm{i}} \tag{4.26}$$

where σ is real ($\sigma > 0$) and the path is to the right of all singularities of X_p. The Laplace transform of the time derivative is

$$\left(\frac{\mathrm{d}X}{\mathrm{d}t}\right)_p = pX_p + \left[X\mathrm{e}^{-pt}\right]_{t=0}^{t\to\infty} .$$

The condition on σ comes from the assumption that $|\tilde{f}(\boldsymbol{v},t)| < |Me^{\gamma t}|$, i. e., that the growth of $\tilde{f}$ is bounded, and that $\mathrm{Re}(p) > |\gamma|$. The Laplace transforms of equations (4.23) and (4.24) are

$$(p + \mathrm{i}kv_z)\tilde{f}_p + \frac{e}{m_e}\mathrm{i}k\tilde{\varphi}_p\frac{\partial f_0}{\partial v_z} = \tilde{f}(\boldsymbol{v},0) \equiv g(\boldsymbol{v}) \tag{4.27}$$

$$k^2\tilde{\varphi}_p = -\frac{e}{\epsilon_0}\int \mathrm{d}^3v\,\tilde{f}_p \tag{4.28}$$

from which we deduce

$$\tilde{f}_p(k,\boldsymbol{v},p) = \frac{g(\boldsymbol{v}) - \frac{e}{m_e}\mathrm{i}k\tilde{\varphi}_p\frac{\partial f_0}{\partial v_z}}{p + \mathrm{i}kv_z} \tag{4.29}$$

$$\begin{aligned}\tilde{\varphi}_p &= -\frac{e}{\epsilon_0 k^2}\int \mathrm{d}^3v\,\frac{g(\boldsymbol{v}) - \frac{e}{m_e}\mathrm{i}k\tilde{\varphi}_p\frac{\partial f_0}{\partial v_z}}{p + \mathrm{i}kv_z}\\ &= \frac{-\frac{e}{\epsilon_0}\int \mathrm{d}^3v\,\frac{g(\boldsymbol{v})}{p + \mathrm{i}kv_z}}{k^2\left[1 + \frac{e^2}{m_e\epsilon_0\mathrm{i}k}\int \mathrm{d}^3v\,\frac{\frac{\partial f_0}{\partial v_z}}{p + \mathrm{i}kv_z}\right]} .\end{aligned} \tag{4.30}$$

We integrate first over v_x and v_y, using the notation

$$\begin{aligned}g(u) &= \int_{-\infty}^{\infty}\int_{-\infty}^{\infty} g(\boldsymbol{v})\,\mathrm{d}v_x\mathrm{d}v_y\\ \frac{\mathrm{d}f_0(u)}{\mathrm{d}u} &= \int_{-\infty}^{\infty}\int_{-\infty}^{\infty}\frac{\partial f_0(\boldsymbol{v})}{\partial v_z}\,\mathrm{d}v_x\mathrm{d}v_y\end{aligned}$$

and we let $v_z \to u$. The remaining pair of equations is now one-dimensional,

$$\tilde{f}_p(k,u,p) = \frac{g(u) - \frac{e}{m_e}\mathrm{i}k\tilde{\varphi}_p\frac{\mathrm{d}f_0(u)}{\mathrm{d}u}}{p + \mathrm{i}ku} \tag{4.31}$$

$$\tilde{\varphi}_p(k,p) = \frac{-\frac{e}{\epsilon_0}\int_{-\infty}^{\infty}\frac{g(u)\,\mathrm{d}u}{p + \mathrm{i}ku}}{k^2\left[1 + \frac{e^2}{m_e\epsilon_0\mathrm{i}k}\int_{-\infty}^{\infty}\frac{\frac{\mathrm{d}f_0(u)}{\mathrm{d}u}\mathrm{d}u}{p + \mathrm{i}ku}\right]} . \tag{4.32}$$

The inverse transformation is

$$\varphi(z,t) = \int_{\sigma-\mathrm{i}\infty}^{\sigma+\mathrm{i}\infty}\frac{\mathrm{d}p}{2\pi\mathrm{i}}\int_{-\infty}^{\infty}\frac{\mathrm{d}k}{2\pi}\,\mathrm{e}^{\mathrm{i}kz+pt}\tilde{\varphi}_p(k,p) . \tag{4.33}$$

The normal path of integration as defined in the complex p-plane is to the right of all singularities. If, however, we deform the contour far enough to the left, the large negative real part of p eliminates the contribution from the vertical portion of the contour. Landau proposed moving the path to the left, but keeping to the right of all singularities and around all branch cuts as in Figure 4.1. Since the vertical portion of the contour no longer contributes, only the singularities (residues) and branch cuts need be evaluated. Consider the contribution from singularities, assuming there are no branch cuts. Then,

$$\tilde{\varphi}(k,t) = \int_{\substack{\text{deformed}\\ \text{contour}}} \frac{\mathrm{d}p}{2\pi \mathrm{i}} \mathrm{e}^{pt} \tilde{\varphi}_p(k,p) \tag{4.34}$$

$$= \sum_n \mathrm{e}^{p_n t} [(p - p_n)\tilde{\varphi}_p(k,p)]_{p=p_n} \, . \tag{4.35}$$

Due to the $\mathrm{e}^{p_n t}$ factor, after a short time, only the rightmost pole in the p-plane will contribute to $\tilde{\varphi}(k,t)$, so the sum collapses to a single term.

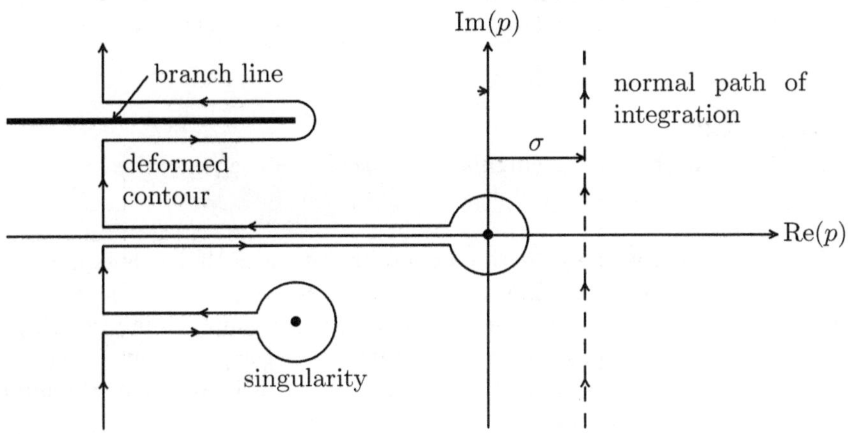

FIGURE 4.1
Landau contour. The normal path is shifted to the left but remains to the right of all singularities and branch points.

In order to evaluate this remaining expression, we need to know $\tilde{\varphi}_p$ in a region where it was not defined, since it was defined for $\mathrm{Re}(p) > |\gamma|$, and we need to know $\tilde{\varphi}_p$ for $\mathrm{Re}(p) < |\gamma|$. It is precisely this point which led to the difficulties in the Vlasov method. Here, however, we can study the analytic continuation of $\tilde{\varphi}_p$ as we deform the contour. Since $\tilde{\varphi}_p$ is the ratio of two terms, we must investigate the numerator and denominator separately. In the numerator of the expression for $\tilde{\varphi}_p$ in equation (4.31), we may take $g(u)$ to be an entire (analytic) function of u (no poles in the complex u-plane), so that the integral

$$G\left(\frac{\mathrm{i}p}{k}\right) = -\frac{\mathrm{i}ek}{\epsilon_0}\int_{-\infty}^{\infty}\frac{g(u)\,\mathrm{d}u}{u - \mathrm{i}p/k} \tag{4.36}$$

can be evaluated by the residue theorem

$$G\left(\frac{\mathrm{i}p}{k}\right)_{(1)} - G\left(\frac{\mathrm{i}p}{k}\right)_{(2)} = 2\pi\mathrm{i}[g(u)]_{u=\mathrm{i}p/k}$$

where the paths (1) and (2) are shown in Figure 4.2.

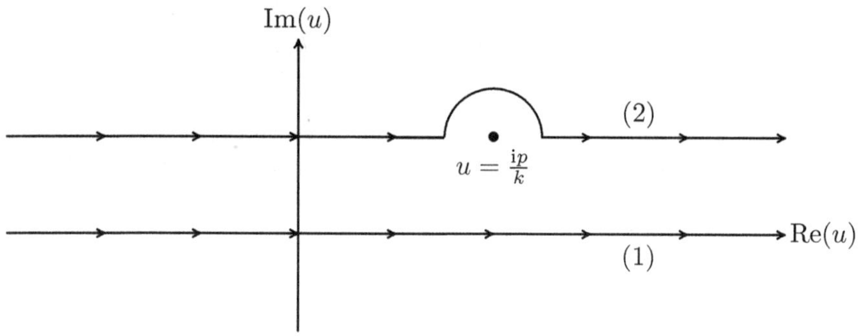

FIGURE 4.2
Paths above and below the singularity. Path (1) is the prescribed path.

The contour (1) is the prescribed contour, but note that it is equal to the contour (2) plus a contribution from the pole. As $\mathrm{Re}(p)$ goes to negative values, contour (2) remains analytic, and since g was assumed analytic, $G(\frac{\mathrm{i}p}{k})_{(1)}$ is also always analytic provided we integrate *under* the pole (sometimes called the Nautilus convention) as in Figure 4.3. With this convention, the numerator is always an entire function of p, or analytic.

Similar arguments apply to the denominator of $\tilde{\varphi}_p$ since $f(u)$ is also assumed to be an entire function of u, or analytic. Thus the only poles of $\tilde{\varphi}_p$ occur at the zeros of the denominator, and there are no branch cuts.

The value of p that makes the denominator vanish is the value such that

$$1 + \frac{e^2}{m_e\epsilon_0\mathrm{i}k}\int_{-\infty\,(1)}^{\infty}\frac{\frac{\mathrm{d}f_0(u)}{\mathrm{d}u}}{p + \mathrm{i}ku}\,\mathrm{d}u = 0,\ \mathrm{Re}(p) > 0$$

$$1 + \frac{e^2}{m_e\epsilon_0\mathrm{i}k}\int_{-\infty(2)}^{\infty}\frac{\frac{\mathrm{d}f_0(u)}{\mathrm{d}u}}{p + \mathrm{i}ku}\,\mathrm{d}u + \frac{e^2 2\pi\mathrm{i}}{m_e\epsilon_0(\mathrm{i}k)^2}\left.\frac{\mathrm{d}f_0(u)}{\mathrm{d}u}\right|_{u=\mathrm{i}p/k} = 0,\ \mathrm{Re}(p) < 0$$

where both integrals are along the real axis.

The principal value of an integral through an isolated singular point is the average of the two integrals along paths just to either side of the point. We

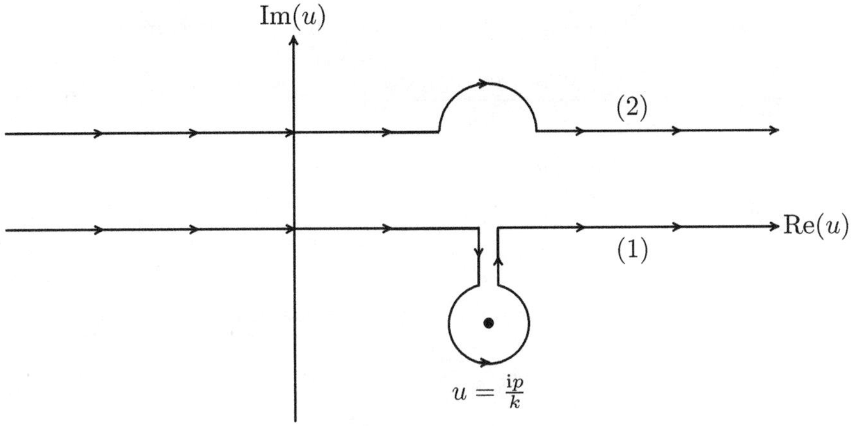

FIGURE 4.3
Analytic continuation by deforming the path to remain below the pole when $\mathrm{Re}(p) < 0$.

can use this concept to combine the two equations above into one that is valid for all p by writing

$$1 + \frac{e^2}{m_e \epsilon_0 \mathrm{i}k} \wp \int_{-\infty}^{\infty} \frac{\frac{\mathrm{d}f_0(u)}{\mathrm{d}u}}{p + \mathrm{i}ku} \mathrm{d}u - \frac{\pi \mathrm{i} e^2}{m_e \epsilon_0 k^2} \left. \frac{\mathrm{d}f_0(u)}{\mathrm{d}u} \right|_{u=\mathrm{i}p/k} = 0 . \tag{4.37}$$

If we now let $\mathrm{i}p = \omega$ and use equation (4.16) to approximate the principal part, we may write this as

$$1 - \frac{\omega_{pe}^2}{\omega^2} \left[1 + \frac{3}{2} \left(\frac{v_e}{v_p} \right)^2 + \cdots \right] - \mathrm{i}\epsilon_i \simeq 0 , \tag{4.38}$$

where

$$\epsilon_i = \frac{\pi \omega_{pe}^2}{k^2} \left. \frac{\mathrm{d}f_0(u)}{\mathrm{d}u} \right|_{u=\omega/k} = 0 \tag{4.39}$$

if we normalize $\int f_0 \, \mathrm{d}u = 1$ (instead of $\int f_0 \, \mathrm{d}u = n_0$). Then, breaking up ω into real and imaginary parts, equation (4.38) becomes

$$(\omega_r + \mathrm{i}\gamma)^2 \simeq \omega_{pe}^2 [1 + \mathcal{O}(v_e/v_p)^2] + \mathrm{i}\omega^2 \epsilon_i$$

so if $\gamma \ll \omega_r$, then the imaginary part is given by

$$\gamma = \frac{1}{2} \omega_r \epsilon_i = \omega_r \frac{\pi \omega_{pe}^2}{2k^2} \left. \frac{\mathrm{d}f_0(u)}{\mathrm{d}u} \right|_{u=\omega_r/k} , \tag{4.40}$$

and the sign of the imaginary part depends on the slope of the distribution function at the phase velocity.

If we take f_0 to be Maxwellian, then equation (4.37) becomes

$$1+\frac{\omega_{pe}^2}{k^2 v_e}\wp\frac{1}{\sqrt{\pi}}\int_{-\infty}^{\infty}\frac{\frac{\mathrm{d}}{\mathrm{d}u}[\exp(-u^2/v_e^2)]\,\mathrm{d}u}{\frac{\omega}{k}-u}+2\sqrt{\pi}\mathrm{i}\frac{\omega_{pe}^2\omega}{k^3 v_e^3}\mathrm{e}^{-(\omega/kv_e)^2}=0\,. \quad (4.41)$$

Integrating by parts, this becomes

$$1-\frac{\omega_{pe}^2}{k^2 v_e^2}\wp\frac{1}{\sqrt{\pi}}\int_{-\infty}^{\infty}\frac{\exp(-u^2/v_e^2)\,\mathrm{d}u}{\left(u-\frac{\omega}{k}\right)^2}+2\sqrt{\pi}\mathrm{i}\frac{\omega_{pe}^2\omega}{k^3 v_e^3}\mathrm{e}^{-(\omega/kv_e)^2}=0\,. \quad (4.42)$$

For this case, the Landau damping rate is the imaginary part of ω, given by

$$\gamma=-\sqrt{\pi}\frac{\omega_{pe}^2\omega_r^2}{k^3 v_e^3}\mathrm{e}^{-(\omega_r/kv_e)^2}\,, \quad (4.43)$$

and the dispersion relation is equivalent to

$$1-\frac{\omega_{pe}^2}{k^2 v_e^2}Z'\left(\frac{\omega}{kv_e}\right)=0 \quad (4.44)$$

which is the same result we got by following the Vlasov method when we assumed $\mathrm{Im}(\omega)>0$.

Up to this point, we have assumed that $k>0$. If this is not the case, then we must change our prescription. If $k<0$, then the pole lies below the path of integration, so we must integrate *over* the pole (Byrd convention)*. Only the sign of the imaginary part changes, so that sometimes the Plasma Dispersion function is written

$$Z\left(\frac{\omega}{kv_e}\right)=\frac{1}{\sqrt{\pi}}\wp\int_{-\infty}^{\infty}\frac{\mathrm{e}^{-\xi^2}\,\mathrm{d}\xi}{\xi-\omega/kv_e}+\left(\frac{k}{|k|}\right)\mathrm{i}\sqrt{\pi}\mathrm{e}^{-(\omega/kv_e)^2}\,. \quad (4.45)$$

It may seem surprising that the same result is obtained by both the Vlasov method and the Landau method, but actually the Landau solution serves to justify the assumption we made that ω should have a small positive imaginary part in the Vlasov method. The initial value problem and the assumption that the perturbation was "turned on slowly" from infinitely long ago both result in the recipe that the velocity integral should go under the pole. The fact that ω has a negative imaginary part (for $k>0$) is now seen as resulting from the initial value problem, guaranteeing that disturbances decay away if there are no sources of free energy, and is called Landau damping. Having done this problem both ways, we can now choose whichever is most convenient in the future as the meaning of the recipe is now clear. When magnetic field effects are included, it will be much simpler to use the Vlasov method with Fourier transforms in both time and space with $\mathrm{Im}(\omega)>0$ than to do the initial value problem.

*Just as the Nautilus submarine first sailed under the North Pole, Admiral Byrd first flew over the pole.

Problem 4.2 *Landau damping with a Lorentzian distribution.*

1. Find the normalization constant A for the Lorentzian velocity distribution function $f_0(u) = A/(v_e^2 + u^2)$.

2. Find closed form expressions for both $\tilde{f}_p(k, u, p)$ and $\tilde{\varphi}_p(k, p)$ with $g(u) = f_0(u)\Delta u$ with Δ a constant.

3. Do the inverse Laplace transform for both $\tilde{f}(k, u, t)$ and $\tilde{\varphi}(k, t)$ and show that the potential $\tilde{\varphi}$ decays in time but that $\tilde{f}$ has a term that does not decay in time.

Problem 4.3 *The Ordinary wave in a hot unmagnetized plasma.* For the cold ordinary wave, it was shown in problem 1.4 that $v_p v_g = c^2$ so that the phase velocity always exceeds the velocity of light. In terms of the physical picture of Landau damping, what does this imply about the Landau damping of the ordinary wave?

4.2 Effects of collisions on Landau damping

The analytic treatment of collisions in the context of Landau damping has long been problematical because of the difficulty of representing the collision frequency in a kinetic treatment. Coulomb collisions have a significant range of velocities where the cross section varies as v^{-3}, and such behavior is generally intractable in solving the kinetic equation. If one considers electron-neutral collisions or weak Coulomb collisions, however, an analytic collision operator can be formed, and even solved. For this analysis, we assume at the outset that we are looking for waves that have the dependence $\exp[\mathrm{i}(kx - \omega t)]$, and our equations to be solved are the linearized kinetic equation and Poisson's equation,

$$\frac{\partial f}{\partial t} + v\frac{\partial f}{\partial x} - \frac{e}{m_e}\frac{\partial f_0}{\partial v}E = \nu\frac{\partial}{\partial v}\left(vf + \frac{k_B T_e}{m_e}\frac{\partial f}{\partial v}\right) \tag{4.46}$$

$$\frac{\partial E}{\partial x} = -\frac{e}{\epsilon_0}\int_{-\infty}^{\infty} f\,\mathrm{d}v\,. \tag{4.47}$$

The collision term comes from equation (3.82) and has the properties that it conserves particles and derives from a Fokker-Planck treatment of collisions. This set of equations was first analyzed by Ng, Bhattachargee, and Skiff[22], but we shall follow the development of Short and Simon[23] except that the notation is changed to be consistent with previous calculations.

We first change to dimensionless variables such that $u \equiv v/v_t$, $\zeta \equiv \omega/kv_t$, $g \equiv v_t f/n_0$, $g_0 \equiv \pi^{-1/2}\mathrm{e}^{-u^2}$, $\eta \equiv \alpha(\partial g_0/\partial u)$, $\alpha \equiv \omega_{pe}^2/k^2v_t^2$, and $\mu = \nu/kv_t$.

With these changes of variable, equations (4.46) and (4.47) become

$$(u-\zeta)g(u)-\eta(u)\int_{-\infty}^{\infty} g(u')\,\mathrm{d}u' = -\mathrm{i}\mu\frac{\partial}{\partial u}\left(ug+\frac{1}{2}\frac{\partial g}{\partial u}\right). \tag{4.48}$$

We next Fourier transform in velocity so that

$$G(w)=\frac{1}{\sqrt{2\pi}}\int_{-\infty}^{\infty} \mathrm{e}^{\mathrm{i}wu}g(u)\,\mathrm{d}u$$

so that Eq. (4.48) becomes

$$\mathrm{i}(1+\mu w)\frac{\mathrm{d}G}{\mathrm{d}w}+\left(\zeta+\frac{\mathrm{i}\mu}{2}w^2\right)G=\frac{\mathrm{i}\alpha}{2}w\mathrm{e}^{-w^2/4}G(0)\,. \tag{4.49}$$

Changing variables so that $x \equiv 1+\mu w$, Eq. (4.49) may be written (with $G(w)\rightarrow H(x)$) as

$$\begin{aligned}\frac{\mathrm{d}H}{\mathrm{d}x}+\left[\frac{1}{2x}\left(\frac{x-1}{\mu}\right)^2-\frac{\mathrm{i}\zeta}{\mu x}\right]H &= \frac{\mathrm{d}H}{\mathrm{d}x}+\left[b(\mu)(x-2)+\frac{a(\zeta,\mu)}{x}\right]H\\ &= \alpha b(\mu)\left(1-\frac{1}{x}\right)\mathrm{e}^{-(x-1)^2/4\mu^2}H(1) \end{aligned} \tag{4.50}$$

where

$$\begin{aligned} b(\mu) &= \frac{1}{2\mu^2}\\ a(\zeta,\mu) &= \frac{1}{2\mu^2}-\frac{\mathrm{i}\zeta}{\mu}\,. \end{aligned}$$

Equation (4.50) is a first-order ordinary differential equation with solution

$$H(x)=\mathrm{e}^{b(2x-x^2/2)}x^{-a}\left[C+\alpha b\mathrm{e}^{-b/2}H(1)\int_0^x x'^a\left(1-\frac{1}{x'}\right)\mathrm{e}^{-bx'}\mathrm{d}x'\right] \tag{4.51}$$

where C is a constant of integration. We find $g(u)$ from the inverse Fourier transformation,

$$g(u)=\frac{1}{\sqrt{2\pi}}\int_{-\infty}^{\infty}\mathrm{e}^{-\mathrm{i}wu}G(w)\,\mathrm{d}w=\frac{1}{\sqrt{2\pi}\mu}\mathrm{e}^{\mathrm{i}u/\mu}\int_{-\infty}^{\infty}\mathrm{e}^{-\mathrm{i}ux/\mu}H(x)\,\mathrm{d}x\,.$$

Note that the first term in $H(x)$ blows up as $x\rightarrow 0$ as $1/x^b$ with $b>0$. This means that for small μ (large b), the inverse Fourier transform will not converge, requiring that we set $C=0$. Evaluating the remaining expression for $H(x)$ at $x=1$ and dividing by $H(1)$, we have the consistency relation

$$1=\alpha b\mathrm{e}^b\int_0^1(x^a-x^{a-1})\mathrm{e}^{-bx}\mathrm{d}x \tag{4.52}$$

which is our dispersion relation. The integrals may be expressed in terms of the Incomplete Gamma function $\gamma(a,x) = \int_0^x \mathrm{e}^{-t} t^{a-1}\,\mathrm{d}t$, such that

$$\int_0^1 (x^a - x^{a-1})\mathrm{e}^{-bx}\mathrm{d}x = [b^{-a}(a-b)\gamma(a,b) - \mathrm{e}^{-b}]/b$$

where the recursion formula from Appendix A has been used. The dispersion relation may then be written as

$$1 + \alpha\left\{1 + \frac{\mathrm{i}\zeta}{\mu}(2\mu^2)^{a(\zeta,\mu)}\mathrm{e}^{b(\mu)}\gamma[a(\zeta,\mu), b(\mu)]\right\} = 0 \qquad (4.53)$$

where the dependence of a and b on ζ and μ are again displayed. A form more amenable to calculation may be obtained from the relation $\gamma(a,x) = \Gamma(a) - \Gamma(a,x)$, where $\Gamma(a,x) = \int_x^\infty \mathrm{e}^{-t} t^{a-1}\,dt$ and may be evaluated by the continued fraction given in Appendix A.

For very weak collisions, μ is very small, and the numerical evaluation of the roots becomes more and more difficult. It is possible to expand the dispersion relation in a power series in μ of the form

$$1 + \alpha\left[1 + \sum_{n=0}^{\infty} \mu^n f_n(\zeta)\right] = 0 \qquad (4.54)$$

where Short and Simon[23] find

$$f_0(\zeta) = \zeta Z(\zeta) \qquad (4.55)$$

$$f_1(\zeta) = \mathrm{i}\zeta[2(1-\zeta^2) + \zeta Z(\zeta)(3 - 2\zeta^2)]/3\,. \qquad (4.56)$$

A table of roots for $\alpha = 9$ for several values of μ is given in Table 4.1, where the last entry with $\mu = 0$ is the Landau result. It is apparent that the damping rate for the electric field increases as collisions increase. These values agree with those of Ng, Bhattachargee, and Skiff[22] who used a different numerical scheme, except for the real part of ζ for $\mu = 0.01$, which has a typographical error (corrected here).

In problem 4.2, it was shown that the potential decays at the Landau rate but that the distribution function has a component that does not decay away. In the collisional environment, the perturbation to the distribution function will also decay, but at a slower rate than the potential. The calculation of this rate is beyond the scope of this book, but an analysis of the decay rate in space away from a localized antenna has been investigated by Short and Simon[23]. The differential decay rates are crucial for the observation of plasma wave echoes where the potential of one antenna decays away and the potential from a second antenna at a different frequency and location also decays away, after which an echo is observed at yet another frequency, providing the collisions have not damped the waves away.

TABLE 4.1
Roots of the dispersion relation as a function of μ, the collisions term. Exact roots from equation (4.53) are shown in the first column for $\alpha = 9$. The middle column of roots are from equation (4.54), and the third column of roots are from equation (4.14) with $\omega \to \omega + i\nu$. The bottom root for each case is the Landau root, ζ_L. In the third column, one may notice that $\mathrm{Re}(\zeta)$ is unchanged from $\mathrm{Re}(\zeta_L)$, and $\mathrm{Im}(\zeta)$ is simply $\mathrm{Im}(\zeta_L) - \mu$.

μ	ζ [Eq. (4.53)]	ζ [Eq. (4.54)]	ζ [Eq. (4.14)]
10^{-1}	2.5177324−0.1270101i	2.5160641−0.1381947i	2.5458154−0.1548864i
10^{-2}	2.5428465−0.0622458i	2.5428231−0.0623408i	2.5458154−0.0648864i
10^{-3}	2.5455167−0.0556237i	2.5455164−0.0556246i	2.5458154−0.0558864i
10^{-4}	2.5457849−0.0549602i	2.5457855−0.0549601i	2.5458154−0.0549864i
10^{-5}	2.5458124−0.0548937i	2.5458124−0.0548937i	2.5458154−0.0548964i
0	2.5458154−0.0548864i	2.5458154−0.0548864i	2.5458154−0.0548864i

Problem 4.4 *Collisional dispersion relation.*

1. Show that $H(x)$ in equation (4.51) is a solution to equation (4.49).
2. Fill in the steps leading from equation (4.51) to equation (4.53).

4.3 The Debye potential

4.3.1 Potential of a stationary test charge

The original Debye potential was derived from rather elementary considerations with little regard for the complexities of kinetic theory. It is interesting to note that an examination of the static potential for a test charge may be derived from kinetic theory. A test charge, by definition, has a small enough charge that the background plasma may be described by a plasma dielectric function which does not include the test charge, but the plasma will still have some response. We begin with the Maxwell equation,

$$\nabla \cdot \boldsymbol{D} = \rho\,, \tag{4.57}$$

where $\boldsymbol{D} = \epsilon \boldsymbol{E}$, $\boldsymbol{E} = -\nabla\varphi(\boldsymbol{r}, t)$, and $\rho = q\delta(\boldsymbol{r})$. Taking Fourier transforms in space and time, equation (4.57) becomes

$$k^2 \epsilon(k,\omega)\tilde{\varphi}(k) = q\,, \tag{4.58}$$

where for the unmagnetized plasma,

$$\epsilon(k,\omega) = \epsilon_0 \left[1 - \sum_{j=e,i} \frac{\omega_{pj}^2}{k^2 v_j^2} Z'\left(\frac{\omega}{k v_j}\right) \right].$$

We are looking for the potential for a stationary test charge, so $\omega = 0$. Solving for $\tilde{\varphi}(k)$, we find

$$\tilde{\varphi}(k) = \frac{q}{\epsilon_0 k^2[1 + (k_{De}^2 + k_{Di}^2)/k^2]} = \frac{q}{\epsilon_0(k^2 + k_D^2)}, \tag{4.59}$$

where $k_{Dj}^2 = 2\omega_{pj}^2/v_j^2$ and $k_D^2 \equiv k_{De}^2 + k_{Di}^2$, having used $Z'(0) = -2$. It follows that the potential is given by

$$\varphi(r) = \frac{q}{\epsilon_0}\int\int\int \frac{\mathrm{d}^3k}{(2\pi)^3}\frac{\mathrm{e}^{\mathrm{i}\boldsymbol{k}\cdot\boldsymbol{r}}}{(k^2 + k_D^2)}. \tag{4.60}$$

In order to execute the inverse Fourier transform, we will align the k_z axis with $\boldsymbol{r}$, and use spherical coordinates in k space. The potential then becomes

$$\varphi(r) = \frac{q}{(2\pi)^3\epsilon_0}\int_0^{2\pi}\mathrm{d}\phi\int_0^{\pi}\mathrm{d}\theta\,\sin\theta\int_0^{\infty}\mathrm{d}k\frac{k^2\mathrm{e}^{\mathrm{i}kr\cos\theta}}{k^2 + k_D^2}.$$

The integral over ϕ is trivial, and if we let $\mu = \cos\theta$, the angular integral is

$$\int_0^{\pi}\sin\theta\mathrm{e}^{\mathrm{i}kr\cos\theta}\mathrm{d}\theta = \int_{-1}^{1}\mathrm{e}^{\mathrm{i}kr\mu}\mathrm{d}\mu = \frac{1}{\mathrm{i}kr}(\mathrm{e}^{\mathrm{i}kr} - \mathrm{e}^{-\mathrm{i}kr}) = 2\frac{\sin kr}{kr}.$$

The potential is therefore given by

$$\begin{aligned}\varphi(r) &= \frac{4\pi q}{(2\pi)^3\epsilon_0 r}\int_0^{\infty}\frac{k\sin kr\,\mathrm{d}k}{k^2 + k_D^2}\\ &= \frac{q}{(2\pi)^2\epsilon_0 r}\int_{-\infty}^{\infty}\frac{k\sin kr\,\mathrm{d}k}{k^2 + k_D^2}\\ &= \frac{\pi q}{(2\pi)^3\epsilon_0\mathrm{i}r}(I_1 - I_2),\end{aligned}$$

where we used the evenness of the integrand to extend the range of the integral and where

$$I_1 = \int_{-\infty}^{\infty}\frac{k\mathrm{e}^{\mathrm{i}kr}\mathrm{d}k}{(k - \mathrm{i}k_D)(k + \mathrm{i}k_D)},$$

$$I_2 = \int_{-\infty}^{\infty}\frac{k\mathrm{e}^{-\mathrm{i}kr}\mathrm{d}k}{(k - \mathrm{i}k_D)(k + \mathrm{i}k_D)}.$$

The integral I_1 may be closed above in Figure 4.4, picking up the pole at $k = \mathrm{i}k_D$, so that $I_1 = \pi\mathrm{i}\mathrm{e}^{-k_D r}$. Similarly, the integral I_2 may be closed below, picking up the pole at $k = -\mathrm{i}k_D$, so that $I_2 = -\pi\mathrm{i}\mathrm{e}^{-k_D r}$. The result is that

$$\varphi(r) = \frac{q}{4\pi\epsilon_0 r}\mathrm{e}^{-k_D r}, \tag{4.61}$$

which is the Debye potential.

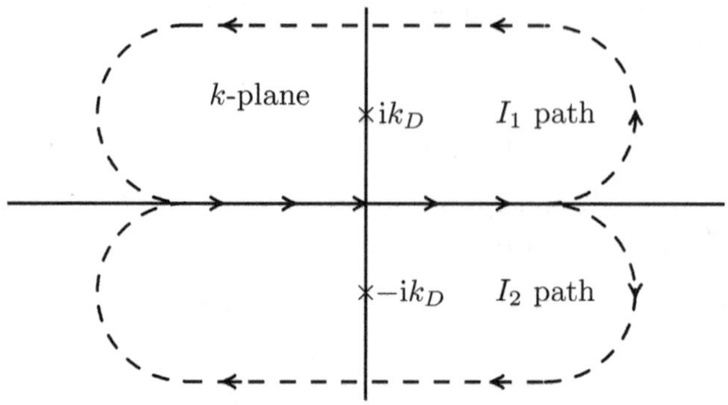

FIGURE 4.4
Poles in the k plane for integrals I_1 and I_2.

4.3.2 Average potential of thermal test charges

For the potential of a moving test charge, whose location is given by $\boldsymbol{r} = \boldsymbol{x} - \boldsymbol{v}t$, the corresponding potential is given by[24]

$$\varphi(\boldsymbol{r}, \boldsymbol{v}) = q \int\int\int \frac{\mathrm{d}^3 k}{(2\pi)^3} \frac{\mathrm{e}^{\mathrm{i}\boldsymbol{k}\cdot\boldsymbol{r}}}{k^2 \epsilon(\boldsymbol{k}, \boldsymbol{k}\cdot\boldsymbol{v})} . \tag{4.62}$$

The only difference from the static case is that $\omega = 0 \to \omega = \boldsymbol{k} \cdot \boldsymbol{v}$.

This potential is not symmetric, and studies have shown that there is a kind of bow shock ahead of the moving test charge and an elongated tail following along behind[26, 27]. Furthermore, the moving charge launches plasma waves. When a test charge moves very fast, however, the plasma particles do not have time to respond to the charge and create very little shielding, so that the only significant effect is to launch plasma waves. One would thus expect that since electrons traveling at their thermal velocity are moving much faster than the typical ion, the slower moving ions would be ineffective at shielding the fast moving electrons, suggesting that when considering electrons, $k_D \sim k_{De}$. It also seems that since the typical electron is not at rest, the average self-shielding of electrons might be reduced from the value estimated for a particle at rest. It is of interest to estimate, therefore, what the *average* shielding is for a *distribution of test charges*. For this example, we will assume that the number density of test charges is insignificant relative to the overall plasma density, so that they do not shield one another. We will assume them to have a thermal velocity, v_t, independent of the velocities of the background plasma. We also can argue from symmetry that the asymmetric potential of a single moving test charge will result in a spherically symmetric potential when averaged over the test particle distribution. This average potential, designated $\Psi(r)$, is described by[25]

$$\Psi(r) = \int \varphi(\boldsymbol{r}, \boldsymbol{v}) f_t(v) \mathrm{d}^3 v \,, \tag{4.63}$$

where $\varphi(\boldsymbol{r}, \boldsymbol{v})$ is given by equation (4.62), and $f_t(v)$ is the Maxwellian

$$f_t(v) = \frac{1}{\pi^{3/2} v_t^3} \mathrm{e}^{-v^2/v_t^2} \,,$$

with $v_t^2 = 2k_B T_t / m_t$ for the test particles. We now orient the z axis in velocity space along $\boldsymbol{k}$, integrate over v_y and v_z, and define $u \equiv v_z / v_t$. We also introduce the quantity

$$k_0^2(u) = -\sum_{j=e,i} \frac{\omega_{pj}^2}{v_j^2} Z'\left(\frac{u v_t}{v_j}\right) ,$$

so that

$$\Psi(r) = \frac{q}{\sqrt{\pi}\epsilon_0} \int_{-\infty}^{\infty} \mathrm{d}u \iiint \frac{\mathrm{d}^3 k}{(2\pi)^3} \frac{\mathrm{e}^{-u^2} \mathrm{e}^{\mathrm{i}\boldsymbol{k}\cdot\boldsymbol{r}}}{[k^2 + k_0^2(u)]} = \frac{1}{\sqrt{\pi}} \int_{-\infty}^{\infty} \mathrm{d}u \, \psi(u, r) \mathrm{e}^{-u^2} ,$$

where

$$\psi(u, r) = \frac{q}{\epsilon_0} \iiint \frac{\mathrm{d}^3 k}{(2\pi)^3} \frac{\mathrm{e}^{\mathrm{i}\boldsymbol{k}\cdot\boldsymbol{r}}}{[k^2 + k_0^2(u)]} = \frac{q}{4\pi\epsilon_0 r} \mathrm{e}^{-k_0(u) r} .$$

Now from the form of $\psi(u, r)$, we know

$$\nabla^2 \psi = -k_0^2(u) \psi \,, \quad r \neq 0 \,,$$

so that

$$\nabla^2 \Psi = -\frac{1}{\sqrt{\pi}} \int_{-\infty}^{\infty} \mathrm{d}u \, \mathrm{e}^{-u^2} k_0^2(u) \psi \,.$$

Thus we may write

$$\nabla^2 \Psi + k_D^2(r) \Psi = \frac{q}{\epsilon_0} \delta(\boldsymbol{r}) \,,$$

where

$$\begin{aligned} k_D^2(r) &\equiv \frac{\int_{-\infty}^{\infty} \mathrm{d}u \, \mathrm{e}^{-u^2} k_0^2(u) \psi(u, r)}{\int_{-\infty}^{\infty} \mathrm{d}u \, \mathrm{e}^{-u^2} \psi(u, r)} \\ &= \frac{\int_{-\infty}^{\infty} \mathrm{d}u \, k_0^2(u) \mathrm{e}^{-u^2 - k_0(u) r}}{\int_{-\infty}^{\infty} \mathrm{d}u \, \mathrm{e}^{-u^2 - k_0(u) r}} \,. \end{aligned} \tag{4.64}$$

We now assume that $k_D(r)$ is slowly varying, so that we may make a WKB approximation for $\Psi(r)$ to find

$$\Psi(r) \simeq \frac{q}{4\pi\epsilon_0 r} \left[\frac{k_0(0)}{k_0(r)}\right]^{1/2} \exp\left[-\int_0^r k_D(r') \, \mathrm{d}r'\right] .$$

We need the quantity $\sqrt{k_D^2(r)}$, so we expand about $r = 0$, finding

$$e^{-k_0(u)r} = 1 - k_0(u)r + \tfrac{1}{2}k_0^2(u)r^2 + \cdots$$

and integrate so that

$$k_D^2(r) = \langle k_0^2 \rangle - (\langle k_0^3 \rangle - \langle k_0^2 \rangle \langle k_0 \rangle) r + \cdots$$

where

$$\langle k_0^n \rangle = \frac{1}{\sqrt{\pi}} \int_{-\infty}^{\infty} du\, e^{-u^2} k_0^n(u)\,,$$

and

$$\langle k_0^2 \rangle = -\frac{1}{\sqrt{\pi}} \int_{-\infty}^{\infty} du\, e^{-u^2} \sum_{j=e,i} \frac{\omega_{pj}^2}{v_j^2} Z'\left(\frac{uv_t}{v_j}\right) = \sum_{j=e,i} \frac{2\omega_{pj}^2}{v_t^2 + v_j^2}\,. \tag{4.65}$$

Since the principal behavior of the potential in the near region is determined by $k_D(0) = \langle k_0^2 \rangle^{1/2}$, we examine $\langle k_0^2 \rangle$ from equation (4.65) more carefully. We note three cases in particular:

1. With $v_t = 0$, the result from equation (4.65) is identical to the result for a stationary test charge.

2. If the test charges are taken to be electrons with the same temperature as the background electrons, then the effective Debye wavenumber is given by
$$k_D^2(0) \sim \tfrac{1}{2}k_{De}^2 + \mathcal{O}(m_e/m_i)\,, \tag{4.66}$$
so the ions don't screen the electrons at all, as one would expect since the electrons are going much too fast for the ions to respond. We also note that the electrons do not fully screen themselves since the slower moving electrons are unable to screen the higher velocity electrons. Numerical calculations of $k_D(r)$ from equation (4.64) indicate that at $k_{De}r = 1$, $k_D(1/k_{De}) \sim \frac{1}{2}[k_D(0) + k_{De}]$, so it lies midway between the classical wavenumber and that from equation (4.66), while for $k_{De}r = 2$, $k_D(2/k_{De}) \sim k_{De}$, so that $k_D(r)$ varies smoothly between $k_D(0)$ and k_{De} over the first two Debye lengths.

3. If the test charges are taken to be ions of the same temperature as the background ions, then the effective Debye wavenumber is given by
$$k_D^2(0) \sim k_{De}^2 + \tfrac{1}{2}k_{Di}^2 + \mathcal{O}(m_e/m_i)\,, \tag{4.67}$$
so the ions are fully screened by the electrons (except for a term of the order m_e/m_i), and the ion contribution is halved because the slower moving ions are unable to screen the faster moving ions. Numerical calculations for this case show that $k_D(r) \sim k_D(0)$ from equation (4.67) within a few percent for the first several Debye lengths. Even at $k_{Di}r = 4$, $k_D(4/k_{Di})$ is much closer to $k_D(0)$ than to k_{Di}.

The general conclusion is that in an average sense, $k_D(0)$ from either equation (4.66) for electrons or from equation (4.67) for ions is a better representations of the effective Debye length than the classical result which is appropriate only for a test particle at rest. For large r, numerical calculations show that the averaged potential falls off faster than the Debye potential, eventually oscillating due to the excitation of plasma waves. This "average Debye wavenumber" has little practical use, but it helps us to better understand the shielding process from a kinetic theory analysis.

5

WAVES IN A MAGNETIZED HOT PLASMA

5.1 The hot plasma dielectric tensor

The calculation of the response of a hot plasma in a magnetic field to a wave is considerably more formidable than the unmagnetized case. For this case, all nine of the dielectric tensor components are nonzero when thermal effects are included, and we shall find the symmetries of the tensor will be even more involved when the full kinetic effects are included. This difficulty relates to the additional effect that the zero order motions of electrons and ions in a uniform magnetic field are spirals, drifting uniformly parallel to the field while they execute circular motion at the cyclotron frequency with their individual Larmor radii around a field line.

The technique we shall use is due to J. E. Drummond[28], R. Z. Sagdeev and V. D. Shafranov[29], and M. N. Rosenbluth and N. Rostoker[30], but we will follow the development of Stix[31] most closely. The idea of the method is to find the perturbation of the distribution function due to the wave by integrating along the unperturbed orbits. This is called the method of characteristics, and we have effectively used it already, except in a trivial fashion, since up to this point the unperturbed orbits were straight lines.

5.1.1 The evolution of the distribution function

We begin by describing a zero-order trajectory by

$$\boldsymbol{R}(t) = \boldsymbol{R}[\boldsymbol{r}(t), \boldsymbol{v}(t), t]$$

and calculate the rate of change of the distribution function along this trajectory by

$$\left.\frac{\mathrm{d}f}{\mathrm{d}t}\right|_{\mathbf{R}} = \frac{\partial f}{\partial t} + \frac{\partial f}{\partial \boldsymbol{r}} \cdot \frac{\mathrm{d}\boldsymbol{r}}{\mathrm{d}t} + \frac{\partial f}{\partial \boldsymbol{v}} \cdot \frac{\mathrm{d}\boldsymbol{v}}{\mathrm{d}t} \tag{5.1}$$

where $\mathrm{d}\boldsymbol{r}/\mathrm{d}t = \boldsymbol{v}$ and $\mathrm{d}\boldsymbol{v}/\mathrm{d}t = \boldsymbol{a}$ where $\boldsymbol{a}$ is the acceleration along the zero-order trajectory

$$\boldsymbol{a} = \frac{\mathrm{d}\boldsymbol{v}}{\mathrm{d}t} = \frac{q}{m}\boldsymbol{v} \times \boldsymbol{B}_0\,. \tag{5.2}$$

We then write equation (5.1) as

$$\left.\frac{\mathrm{d}f}{\mathrm{d}t}\right|_{\mathbf{R}} = \frac{\partial f}{\partial t} + \boldsymbol{v}\cdot\nabla f + \frac{q}{m}\boldsymbol{v}\times\boldsymbol{B}_0\cdot\nabla_v f\,. \tag{5.3}$$

The zero order distribution function, f_0, is, of course, independent of $\boldsymbol{r}$ and t, so

$$\frac{\mathrm{d}f_0}{\mathrm{d}t} = \left.\frac{\mathrm{d}f_0}{\mathrm{d}t}\right|_{\mathbf{R}} = \frac{q}{m}\boldsymbol{v}\times\boldsymbol{B}_0\cdot\nabla_v f_0 = 0\,. \tag{5.4}$$

The most general form of f_0 that satisfies equation (5.4) is

$$f_0(\boldsymbol{v}) = f_0(v_\perp, v_z) \tag{5.5}$$

where $v_\perp^2 = v_x^2 + v_y^2$. Adding and subtracting the wave field terms in equation (5.3) leads to

$$\begin{aligned}\left.\frac{\mathrm{d}f}{\mathrm{d}t}\right|_{\mathbf{R}} &= \left\{\frac{\partial f}{\partial t} + \boldsymbol{v}\cdot\nabla f + \frac{q}{m}[\boldsymbol{E}_1 + \boldsymbol{v}\times(\boldsymbol{B}_0+\boldsymbol{B}_1)]\cdot\nabla_v f\right\}\\ &\quad -\frac{q}{m}(\boldsymbol{E}_1 + \boldsymbol{v}\times\boldsymbol{B}_1)\cdot\nabla_v f\end{aligned} \tag{5.6}$$

and the term in curly brackets vanishes due to the collisionless Boltzmann equation. If we now separate the distribution function into $f = f_0 + f_1$, then

$$\left.\frac{\mathrm{d}f}{\mathrm{d}t}\right|_{\mathbf{R}} = \left.\frac{\mathrm{d}f_0}{\mathrm{d}t}\right|_{\mathbf{R}} + \left.\frac{\mathrm{d}f_1}{\mathrm{d}t}\right|_{\mathbf{R}} = -\frac{q}{m}(\boldsymbol{E}_1 + \boldsymbol{v}\times\boldsymbol{B}_1)\cdot\nabla_v f \tag{5.7}$$

where the zero-order term vanishes by equation (5.4). This leaves us with a total derivative of f_1, so if we integrate equation (5.7) along $\boldsymbol{r}$, we obtain

$$f_1(\boldsymbol{r},\boldsymbol{v},t) = -\frac{q}{m}\int_{t_0}^{t}[\boldsymbol{E}_1(\boldsymbol{r}',t') + \boldsymbol{v}'\times\boldsymbol{B}_1(\boldsymbol{r}',t')]\cdot\nabla_{v'}f_0(\boldsymbol{v}')\,\mathrm{d}t' + f_1(\boldsymbol{r},\boldsymbol{v},t_0)\,. \tag{5.8}$$

The recipe we developed from the Vlasov-Landau analysis that ω should have a positive imaginary part corresponds to growing waves in time, but it guarantees that the waves vanish as $t_0 \to -\infty$, so if we change the lower limit in equation (5.8) to $-\infty$, we may neglect the effects of the initial conditions. This is effectively equivalent to the Landau prescription of the initial value problem, but will be easier to manipulate. The perturbed distribution function is then described by

$$f_1(\boldsymbol{r},\boldsymbol{v},t) = -\frac{q}{m}\int_{-\infty}^{t}[\boldsymbol{E}_1(\boldsymbol{r}',t') + \boldsymbol{v}'\times\boldsymbol{B}_1(\boldsymbol{r}',t')]\cdot\nabla_{v'}f_0(\boldsymbol{v}')\,\mathrm{d}t' \tag{5.9}$$

where we are to integrate along the trajectories, $\boldsymbol{r}(\boldsymbol{r}',\boldsymbol{v}',t')$, that end at $\boldsymbol{r}(\boldsymbol{r},\boldsymbol{v},t)$ when $t' \to t$.

Problem 5.1 *Zero-order distribution function.* Prove that any zero-order distribution function having the form of equation (5.5) will satisfy equation (5.4) as long as it is differentiable.

5.1.2 Integrating along the unperturbed orbits

In order to evaluate the integral of equation (5.9), we shall assume that the wave electric and magnetic fields are of the form

$$\boldsymbol{E}_1 = \boldsymbol{E}e^{i(\boldsymbol{k}\cdot\boldsymbol{r}'-\omega t')} \tag{5.10}$$

$$\boldsymbol{B}_1 = \boldsymbol{B}e^{i(\boldsymbol{k}\cdot\boldsymbol{r}'-\omega t')} \tag{5.11}$$

so we may use the Maxwell equations to obtain

$$\boldsymbol{B}_1 = \frac{\boldsymbol{k}\times\boldsymbol{E}_1}{\omega} = \frac{\boldsymbol{k}\times\boldsymbol{E}}{\omega}e^{i(\boldsymbol{k}\cdot\boldsymbol{r}'-\omega t')} \tag{5.12}$$

so that equation (5.9) may be written

$$f_1(\boldsymbol{r},\boldsymbol{v},t) = -\frac{q}{m}\int_{-\infty}^{t} dt'\boldsymbol{E}\left(1+\frac{\cdot\,\boldsymbol{v}'\,\boldsymbol{k}-\boldsymbol{v}'\cdot\boldsymbol{k}}{\omega}\right)\cdot\nabla_{v'}f_0(\boldsymbol{v}')e^{i(\boldsymbol{k}\cdot\boldsymbol{r}'-\omega t')}\,. \tag{5.13}$$

The trajectory that reaches $\boldsymbol{r}' = \boldsymbol{r}$ when $t' = t$ is governed by the equation of motion from equation (5.2)

$$\frac{d\boldsymbol{v}'}{dt'} = \epsilon\boldsymbol{v}'\times\omega_c\hat{e}_z \tag{5.14}$$

where $\epsilon = q/|q|$ and we suppress all other species specific notation until we begin to combine each species' contribution to the total current. The solution of equation (5.14) that reaches $\boldsymbol{v}' = \boldsymbol{v}$ at $t' = t$ is

$$\begin{aligned} v'_x &= v_x\cos\omega_c\tau - \epsilon v_y\sin\omega_c\tau \\ v'_y &= \epsilon v_x\sin\omega_c\tau + v_y\cos\omega_c\tau \\ v'_z &= v_z \end{aligned} \tag{5.15}$$

where $\tau = t - t'$. Integrating these to find the zero-order trajectory that ends at $\boldsymbol{r}' = \boldsymbol{r}$ at $t' = t$ we find

$$\begin{aligned} x' &= x - \frac{v_x}{\omega_c}\sin\omega_c\tau + \frac{\epsilon v_y}{\omega_c}(1-\cos\omega_c\tau) \\ y' &= y - \frac{\epsilon v_x}{\omega_c}(1-\cos\omega_c\tau) - \frac{v_y}{\omega_c}\sin\omega_c\tau \\ z' &= z - v_z\tau\,. \end{aligned} \tag{5.16}$$

The phase factor in equation (5.13) becomes

$$\begin{aligned} i\boldsymbol{k}\cdot\boldsymbol{r}' - i\omega t' = i\boldsymbol{k}\cdot\boldsymbol{r} - i\omega t &+ \frac{iv_x}{\omega_c}[-k_x\sin\omega_c\tau - \epsilon k_y(1-\cos\omega_c\tau)] \\ &+ \frac{iv_y}{\omega_c}[-k_y\sin\omega_c\tau + \epsilon k_x(1-\cos\omega_c\tau)] + i(\omega - k_z v_z)\tau\,. \end{aligned} \tag{5.17}$$

Because $v_\perp$ and v_z are constants of the motion, we know that $f_0(v'_\perp, v'_z) = f_0(v_\perp, v_z)$. If we now define

$$\frac{\partial f_0}{\partial v_\perp} \equiv f_{0\perp} \tag{5.18}$$

$$\frac{\partial f_0}{\partial v_z} \equiv f_{0z} \tag{5.19}$$

then

$$\frac{\partial f_0}{\partial v_x} = \frac{v_x}{v_\perp} f_{0\perp}\,, \qquad \frac{\partial f_0}{\partial v_y} = \frac{v_y}{v_\perp} f_{0\perp}\,.$$

Using these definitions, the remaining factor in equation (5.13) may be written

$$\begin{aligned}
&\boldsymbol{E}\left(1 + \frac{\cdot\, \boldsymbol{v}'\,\boldsymbol{k} - \boldsymbol{v}'\cdot\boldsymbol{k}}{\omega}\right)\cdot\nabla_{v'} f_0(\boldsymbol{v}') \\
&= (E_x v'_x + E_y v'_y)\left[\frac{f_{0\perp}}{v_\perp} + \frac{k_z}{\omega}\left(f_{0z} - \frac{v'_z}{v_\perp} f_{0\perp}\right)\right] \\
&\quad + E_z\left[f_{0z} - \frac{k_x v'_x + k_y v'_y}{\omega}\left(f_{0z} - \frac{v'_z}{v_\perp} f_{0\perp}\right)\right] \\
&= (v_x \cos\omega_c\tau - \epsilon v_y \sin\omega_c\tau)\left[\frac{E_x f_{0\perp}}{v_\perp} + \frac{E_x k_z - E_z k_x}{\omega}\left(f_{0z} - \frac{v_z}{v_\perp} f_{0\perp}\right)\right] \\
&\quad + (\epsilon v_x \sin\omega_c\tau + v_y \cos\omega_c\tau)\left[\frac{E_y f_{0\perp}}{v_\perp} + \frac{E_y k_z - E_z k_y}{\omega}\left(f_{0z} - \frac{v_z}{v_\perp} f_{0\perp}\right)\right] \\
&\quad + E_z f_{0z}\,.
\end{aligned} \tag{5.20}$$

We complete the variable change by writing the integral in the form

$$f_1(\boldsymbol{r}, \boldsymbol{v}, t) = \int_{-\infty}^{t} \mathrm{d}t' \cdots = \mathrm{e}^{\mathrm{i}(\boldsymbol{k}\cdot\boldsymbol{r} - \omega t)} \int_0^\infty \mathrm{d}\tau \cdots$$

where we have factored out the leading terms of the phase factor and will subsequently suppress this factor, interpreting the remaining integral as the Fourier amplitude of the distribution function.

We conclude this section by noting that we need only integrate over τ and average the current density over velocity for some particular $f_0(v_\perp, v_z)$ to obtain the mean current density from

$$\boldsymbol{j} = q \int \mathrm{d}^3 v \;\; \boldsymbol{v} f_1\,. \tag{5.21}$$

From this current density, we can construct the effective dielectric tensor and obtain the dispersion relation.

5.1.3 General $f_0(v_\perp, v_z)$

It is possible to execute the integral over τ without specifying the zero-order distribution function, $f_0(v_\perp, v_z)$. This is done most conveniently by using polar coordinates for the velocity and wave vector such that

$$\begin{aligned} v_x &= v_\perp \cos\phi \quad & k_x &= k_\perp \cos\psi \\ v_y &= v_\perp \sin\phi \quad & k_y &= k_\perp \sin\psi \end{aligned}$$

so that the phase factor of equation (5.17) may be written,

$$\begin{aligned} \mathrm{e}^{\mathrm{i}(\boldsymbol{k}\cdot\boldsymbol{r}'-\omega t')} &= \mathrm{e}^{\mathrm{i}(\boldsymbol{k}\cdot\boldsymbol{r}-\omega t)}\mathrm{e}^{-\mathrm{i}b[\sin(\phi-\psi+\epsilon\omega_c\tau)-\sin(\phi-\psi)]+\mathrm{i}a\tau} \\ &= \mathrm{e}^{\mathrm{i}(\boldsymbol{k}\cdot\boldsymbol{r}-\omega t)}\sum_{m,n=-\infty}^{\infty} J_m(b)J_n(b)\mathrm{e}^{\mathrm{i}(m-n)(\phi-\psi)}\mathrm{e}^{\mathrm{i}(\omega-n\epsilon\omega_c-k_z v_z)\tau}, \end{aligned} \tag{5.22}$$

where $a = (\omega - k_z v_z)$, $b = \epsilon k_\perp v_\perp/\omega_c$, and we have used the Bessel identity

$$\mathrm{e}^{\mathrm{i}b\sin\theta} = \sum_{n=-\infty}^{\infty} J_n(b)\mathrm{e}^{\mathrm{i}n\theta}\,. \tag{5.23}$$

Assuming that ω has a positive imaginary part, the integral over τ can now be done immediately with the result (again suppressing the $\mathrm{e}^{\mathrm{i}(\boldsymbol{k}\cdot\boldsymbol{r}-\omega t)}$ so that the result is the Fourier amplitude)

$$\begin{aligned} f_1(\boldsymbol{k}, \boldsymbol{v}, \omega) = -\frac{\mathrm{i}q}{m}\sum_{m=-\infty}^{\infty}\sum_{n=-\infty}^{\infty} &\frac{J_m(b)\mathrm{e}^{\mathrm{i}(m-n)(\phi-\psi)}}{(\omega - n\epsilon\omega_c - k_z v_z)} \\ \times\Bigg\{ &\frac{nJ_n(b)}{b}\left[f_{0\perp} + \frac{k_z}{\omega}(v_\perp f_{0z} - v_z f_{0\perp})\right](E_x\cos\psi + E_y\sin\psi) \\ &+\mathrm{i}J_n'(b)\left[f_{0\perp} + \frac{k_z}{\omega}(v_\perp f_{0z} - v_z f_{0\perp})\right](-E_x\sin\psi + E_y\cos\psi) \\ &+J_n(b)\left[f_{0z} - \frac{n\epsilon\omega_c}{\omega}\left(f_{0z} - \frac{v_z}{v_\perp}f_{0\perp}\right)\right]E_z\Bigg\}\,. \end{aligned} \tag{5.24}$$

In obtaining equation (5.24), we have used the Bessel identities

$$\begin{aligned} J_{\ell-1}(b) + J_{\ell+1}(b) &= \frac{2\ell}{b}J_\ell(b) \\ J_{\ell-1}(b) - J_{\ell+1}(b) &= 2J_\ell'(b) \end{aligned}$$

and have let $n \pm 1 \to \ell$ so that, for example, with $\psi = 0$,

$$\begin{aligned} &\sum_{n=-\infty}^{\infty} J_n(b)\mathrm{e}^{\mathrm{i}(m-n)\phi-\mathrm{i}n\epsilon\omega_c\tau}\cos(\phi+\epsilon\omega_c\tau) \\ &= \sum_{n=-\infty}^{\infty}\frac{J_n(b)}{2}\left[\mathrm{e}^{\mathrm{i}(m+1-n)\phi-\mathrm{i}(n-1)\epsilon\omega_c\tau} + \mathrm{e}^{\mathrm{i}(m-1-n)\phi-\mathrm{i}(n+1)\epsilon\omega_c\tau}\right] \end{aligned}$$

$$= \sum_{\ell=-\infty}^{\infty} \left[\frac{J_{\ell+1}(b) + J_{\ell-1}(b)}{2}\right] \mathrm{e}^{\mathrm{i}(m-\ell)\phi - \mathrm{i}\ell\epsilon\omega_c\tau}$$

$$= \sum_{\ell=-\infty}^{\infty} \frac{\ell}{b} J_\ell(b) \mathrm{e}^{\mathrm{i}(m-\ell)\phi - \mathrm{i}\ell\epsilon\omega_c\tau} .$$

Then for $\psi \neq 0$,

$$\begin{aligned} \cos(\phi - \psi + \epsilon\omega_c\tau) &\rightarrow \frac{n}{b} J_n(b) \\ \cos(\phi + \epsilon\omega_c\tau) &\rightarrow \frac{n}{b} J_n(b)\cos\psi - \mathrm{i}J_n'(b)\sin\psi \\ \sin(\phi + \epsilon\omega_c\tau) &\rightarrow \frac{n}{b} J_n(b)\sin\psi + \mathrm{i}J_n'(b)\cos\psi . \end{aligned} \tag{5.25}$$

In order to complete the integrals over velocity to obtain the mean current density, we note that the volume element of the integral is

$$\int \mathrm{d}^3 v = \int_0^{2\pi} \mathrm{d}\phi \int_0^\infty v_\perp\, \mathrm{d}v_\perp \int_{-\infty}^{\infty} \mathrm{d}v_z$$

and it is convenient to use the orthogonality integral over ϕ which takes the form (for $\langle v_x \rangle, \langle v_y \rangle$ and $\langle v_z \rangle$, respectively)

$$\sum_{m=-\infty}^{\infty} J_m(b) \int_0^{2\pi} \mathrm{d}\phi \mathrm{e}^{\mathrm{i}(m-n)(\phi-\psi)} \begin{Bmatrix} \cos\phi \\ \sin\phi \\ 1 \end{Bmatrix} = 2\pi \begin{Bmatrix} \dfrac{nJ_n(b)}{b}\cos\psi + \mathrm{i}J_n'(b)\sin\psi \\ \dfrac{nJ_n(b)}{b}\sin\psi - \mathrm{i}J_n'(b)\cos\psi \\ J_n(b) \end{Bmatrix} . \tag{5.26}$$

Using these elements, the effective dielectric tensor may be expressed as

$$\mathsf{K} = \begin{pmatrix} K_1 + \sin^2\psi K_0 & K_2 - \cos\psi\sin\psi K_0 & \cos\psi K_4 + \sin\psi K_5 \\ -K_2 - \cos\psi\sin\psi K_0 & K_1 + \cos^2\psi K_0 & \sin\psi K_4 - \cos\psi K_5 \\ \cos\psi K_6 - \sin\psi K_7 & \sin\psi K_6 + \cos\psi K_7 & K_3 \end{pmatrix} \tag{5.27}$$

where

$$K_0 = \sum_j \frac{\omega_{pj}^2}{\omega} \sum_{n=-\infty}^{\infty} \int \mathrm{d}^2 v \frac{[b_j J_n'(b_j)]^2 - n^2 J_n^2(b_j)}{b_j^2(\omega - n\epsilon_j\omega_{cj} - k_z v_z)} F_\perp \tag{5.28}$$

$$K_1 = 1 + \sum_j \frac{\omega_{pj}^2}{\omega} \sum_{n=-\infty}^{\infty} \int \mathrm{d}^2 v \frac{n^2 J_n^2(b_j)}{b_j^2(\omega - n\epsilon_j\omega_{cj} - k_z v_z)} F_\perp \tag{5.29}$$

$$K_2 = \mathrm{i} \sum_j \frac{\omega_{pj}^2}{\omega} \sum_{n=-\infty}^{\infty} \int \mathrm{d}^2 v \frac{n J_n(b_j) J_n'(b_j)}{b_j(\omega - n\epsilon_j\omega_{cj} - k_z v_z)} F_\perp \tag{5.30}$$

$$K_3 = 1 + \sum_j \frac{\omega_{pj}^2}{\omega} \sum_{n=-\infty}^{\infty} \int \mathrm{d}^2v \frac{J_n^2(b_j)}{(\omega - n\epsilon_j\omega_{cj} - k_z v_z)} F_z \tag{5.31}$$

$$K_4 = \sum_j \frac{\omega_{pj}^2}{\omega} \sum_{n=-\infty}^{\infty} \int \mathrm{d}^2v \frac{nJ_n^2(b_j)}{b_j(\omega - n\epsilon_j\omega_{cj} - k_z v_z)} \frac{v_\perp}{v_z} F_z \tag{5.32}$$

$$K_5 = \mathrm{i} \sum_j \frac{\omega_{pj}^2}{\omega} \sum_{n=-\infty}^{\infty} \int \mathrm{d}^2v \frac{J_n(b_j)J_n'(b_j)}{(\omega - n\epsilon_j\omega_{cj} - k_z v_z)} \frac{v_\perp}{v_z} F_z \tag{5.33}$$

$$K_6 = \sum_j \frac{\omega_{pj}^2}{\omega} \sum_{n=-\infty}^{\infty} \int \mathrm{d}^2v \frac{nJ_n^2(b_j)}{b_j(\omega - n\epsilon_j\omega_{cj} - k_z v_z)} \frac{v_z}{v_\perp} F_\perp \tag{5.34}$$

$$K_7 = \mathrm{i} \sum_j \frac{\omega_{pj}^2}{\omega} \sum_{n=-\infty}^{\infty} \int \mathrm{d}^2v \frac{J_n(b_j)J_n'(b_j)}{(\omega - n\epsilon_j\omega_{cj} - k_z v_z)} \frac{v_z}{v_\perp} F_\perp \tag{5.35}$$

where $\int \mathrm{d}^2v = 2\pi \int_{-\infty}^{\infty} \mathrm{d}v_z \int_0^{\infty} v_\perp \, \mathrm{d}v_\perp$ and

$$F_\perp = v_\perp \left[\frac{\partial f_{0j}}{\partial v_\perp} \left(1 - \frac{k_z v_z}{\omega} \right) + \frac{k_z v_\perp}{\omega} \frac{\partial f_{0j}}{\partial v_z} \right]$$

$$F_z = v_z \left[\frac{\partial f_{0j}}{\partial v_z} + \frac{n\epsilon_j\omega_{cj}}{\omega} \left(\frac{v_z}{v_\perp} \frac{\partial f_{0j}}{\partial v_\perp} - \frac{\partial f_{0j}}{\partial v_z} \right) \right].$$

When the distribution function is isotropic ($v_\perp \partial F/\partial v_z = v_z \partial F/\partial v_\perp$), then $K_6 = K_4$ and $K_7 = K_5$. Also, only the K_1, K_2, and K_3 components survive in the cold plasma limit, so all of the others are first order or higher in the temperature.

Problem 5.2 *Polar coordinates in velocity space.* Fill in the steps leading to equations (5.22) and (5.25).

Problem 5.3 *General tensor elements.* Fill in the steps leading to any (except $K_{zz} = K_3$) of the composite tensor elements K_{ij} in equation (5.27).

Problem 5.4 *Sum rules.*

1. Using the Newberger sum rule[32],

$$\sum_{n=-\infty}^{\infty} \frac{J_n(z)J_{n-m}(z)}{a-n} = \frac{(-1)^m \pi}{\sin \pi a} J_{m-a}(z) J_a(z) , \qquad m \geq 0 , \tag{5.36}$$

prove the identities

$$\sum_{n=-\infty}^{\infty} \frac{n^2 J_n^2(z)}{a-n} = \frac{\pi a^2}{\sin \pi a} J_a(z) J_{-a}(z) - a \tag{5.37}$$

$$\sum_{n=-\infty}^{\infty} \frac{[J_n'(z)]^2}{a-n} = \frac{\pi}{\sin \pi a} J_a'(z) J_{-a}'(z) + \frac{a}{z^2} \tag{5.38}$$

$$\sum_{n=-\infty}^{\infty} \frac{nJ_n(z)J_n'(z)}{a-n} = \frac{\pi a}{\sin \pi a} J_a(z)J_{-a}'(z) + \frac{a}{z} \tag{5.39}$$

$$\sum_{n=-\infty}^{\infty} \frac{J_n(z)J_n'(z)}{a-n} = \frac{\pi}{\sin \pi a} J_a(z)J_{-a}'(z) + \frac{1}{z} \tag{5.40}$$

$$\sum_{n=-\infty}^{\infty} \frac{nJ_n^2(z)}{a-n} = \frac{\pi a}{\sin \pi a} J_a(z)J_{-a}(z) - 1 \tag{5.41}$$

$$\sum_{n=-\infty}^{\infty} \frac{J_n^2(z)}{a-n} = \frac{\pi}{\sin \pi a} J_a(z)J_{-a}(z) . \tag{5.42}$$

2. Show that the dielectric tensor elements of equations (5.28) through (5.35) can be written in terms of eight alternative integrals without any Bessel function sums by use of equations (5.37) through (5.42), the first four of which are

$$A_0 = \sum_j \frac{\epsilon_j \omega_{pj}^2}{\omega \omega_{cj}} \int d^2v \frac{\pi J_{a_j}'(b_j) J_{-a_j}'(b_j)}{\sin \pi a_j} F_\perp \tag{5.43}$$

$$A_1 = \sum_j \frac{\epsilon_j \omega_{pj}^2}{\omega \omega_{cj}} \int d^2v \frac{\pi a_j^2 J_{a_j}(b_j) J_{-a_j}(b_j)}{b_j^2 \sin \pi a_j} F_\perp \tag{5.44}$$

$$A_2 = \sum_j \frac{\epsilon_j \omega_{pj}^2}{\omega \omega_{cj}} \int d^2v \frac{\pi a_j J_{a_j}(b_j) J_{-a_j}'(b_j)}{b_j \sin \pi a_j} F_\perp \tag{5.45}$$

$$A_\perp = \sum_j \frac{\epsilon_j \omega_{pj}^2}{\omega \omega_{cj}} \int d^2v \frac{a_j}{b_j^2} F_\perp \tag{5.46}$$

with $a_j = (\omega - k_z v_z)/\epsilon_j \omega_{cj}$.

3. Show that four of the dielectric tensor elements may be alternatively represented by

$$K_{xx} = 1 + A_0 \sin^2\psi + A_1 \cos^2\psi - A_\perp \cos 2\psi \tag{5.47}$$
$$K_{yy} = 1 + A_0 \cos^2\psi + A_1 \sin^2\psi + A_\perp \cos 2\psi \tag{5.48}$$
$$K_{xy} = iA_2 + \tfrac{1}{2}(A_1 - A_0)\sin 2\psi + A_\perp(i - \sin 2\psi) \tag{5.49}$$
$$K_{yx} = -iA_2 + \tfrac{1}{2}(A_1 - A_0)\sin 2\psi - A_\perp(i + \sin 2\psi) . \tag{5.50}$$

Problem 5.5 *Isotropic distribution function.* Prove that $K_4 = K_6$ and $K_5 = K_7$ if the distribution function is isotropic.

5.1.4 Maxwellian distributions

When the distribution function is Maxwellian, the integrals over the perpendicular and parallel velocities can be done in closed form (although an infinite sum remains). We shall treat first only the perpendicular form of the distribution function, leaving the parallel distribution function until later, so that

$$f_0(v_\perp, v_z) = \frac{F(v_z)}{\pi v_t^2} \mathrm{e}^{-v_\perp^2/v_t^2} \tag{5.51}$$

where $v_t^2 = 2k_B T_\perp/m$ denotes the *transverse* thermal speed (we shall introduce v_ℓ, the *longitudinal* thermal speed later), and $T_\perp$ is the perpendicular temperature. It is unlikely that a plasma will ever be truly Maxwellian with different perpendicular and parallel temperatures, but it is not so unlikely that the perpendicular and parallel distributions will differ, especially if a wave is preferentially heating one or the other, as is often the case. The deviation from equilibrium will occasionally lead to instabilities, as we shall show in a subsequent section.

5.1.4.1 Integrating over perpendicular velocities

It is possible to evaluate the integrals of equations (5.28) through (5.35), but the preferred method is to return to an earlier step and integrate over the perpendicular velocities *before* we integrate over τ. With the distribution function of equation (5.51), the Fourier amplitude of f_1 may be expressed as

$$\begin{aligned} f_1(\boldsymbol{k}, \boldsymbol{v}, \omega) = -\frac{q}{m\pi v_t^2} \int_0^\infty & \mathrm{d}\tau (A_x v_x + A_y v_y + \alpha_z) \\ & \times \exp\left[-\mathrm{i}a_x v_x - \frac{v_x^2}{v_t^2} - \mathrm{i}a_y v_y - \frac{v_y^2}{v_t^2} + \mathrm{i}(\omega - k_z v_z)\tau\right] \end{aligned} \tag{5.52}$$

where

$$a_x = \frac{1}{\omega_c}[k_x \sin\omega_c\tau + \epsilon k_y(1 - \cos\omega_c\tau)] \tag{5.53}$$

$$a_y = \frac{1}{\omega_c}[k_y \sin\omega_c\tau - \epsilon k_x(1 - \cos\omega_c\tau)] \tag{5.54}$$

$$A_x = \alpha_x \cos\omega_c\tau + \epsilon\alpha_y \sin\omega_c\tau \tag{5.55}$$

$$A_y = \alpha_y \cos\omega_c\tau - \epsilon\alpha_x \sin\omega_c\tau \tag{5.56}$$

$$\alpha_x = -\frac{2F}{v_t^2}E_x + \left(F' + \frac{2v_z}{v_t^2}F\right)\left(\frac{k_z E_x - k_x E_z}{\omega}\right) \tag{5.57}$$

$$\alpha_y = -\frac{2F}{v_t^2}E_y + \left(F' + \frac{2v_z}{v_t^2}F\right)\left(\frac{k_z E_y - k_y E_z}{\omega}\right) \tag{5.58}$$

$$\alpha_z = F' E_z\,. \tag{5.59}$$

Since we need to calculate the mean current density from equation (5.21), we will require integrals over the perpendicular velocities of the type

$$G_n(a) = \frac{1}{\sqrt{\pi} v_t} \int_{-\infty}^{\infty} v^n \mathrm{e}^{-\mathrm{i}av - v^2/v_t^2}\, \mathrm{d}v \tag{5.60}$$

which by completing the square are

$$G_0(a) = \mathrm{e}^{-a^2 v_t^2/4} \tag{5.61}$$

$$G_1(a) = \mathrm{e}^{-a^2 v_t^2/4}\left(-\frac{iav_t^2}{2}\right) \tag{5.62}$$

$$G_2(a) = \mathrm{e}^{-a^2 v_t^2/4}\left(\frac{v_t^2}{2} - \frac{a^2 v_t^4}{4}\right). \tag{5.63}$$

Now each integral is of the form $G_n(a_x)G_m(a_y)$, so they all have the common exponential factor $\exp[-(a_x^2 + a_y^2)v_t^2/4]$, which may be written

$$(a_x^2 + a_y^2)\frac{v_t^2}{4} = \lambda(1 - \cos\omega_c\tau) \tag{5.64}$$

where $\lambda = \frac{1}{2}k_\perp^2 \rho_L^2$ and $\rho_L = v_t/\omega_c$ is the Larmor radius.

The pertinent integrals then lead to

$$\langle f_1 \rangle_\perp = \frac{q}{m}\int_0^\infty \mathrm{d}\tau\, e^\phi \left[\frac{\mathrm{i}v_t^2}{2}(A_x a_x + A_y a_y) - \alpha_z\right] \tag{5.65}$$

$$\langle v_x f_1 \rangle_\perp = \frac{q}{m}\int_0^\infty \mathrm{d}\tau\, e^\phi \left[\frac{v_t^4}{4}(A_x a_x + A_y a_y)a_x + \frac{v_t^2}{2}(\mathrm{i}a_x\alpha_z - A_x)\right] \tag{5.66}$$

$$\langle v_y f_1 \rangle_\perp = \frac{q}{m}\int_0^\infty \mathrm{d}\tau\, e^\phi \left[\frac{v_t^4}{4}(A_x a_x + A_y a_y)a_y + \frac{v_t^2}{2}(\mathrm{i}a_y\alpha_z - A_y)\right] \tag{5.67}$$

where now $\phi = \mathrm{i}(\omega - k_z v_z)\tau - \lambda(1 - \cos\omega_c\tau)$, and

$$A_x a_x + A_y a_y = \frac{1}{\omega_c}[(\alpha_x k_x + \alpha_y k_y)\sin\omega_c\tau + \epsilon(\alpha_y k_x - \alpha_x k_y)(1 - \cos\omega_c\tau)].$$

Problem 5.6 *Maxwellian distribution.*

1. Fill in the steps leading to equation (5.52).
2. Fill in the steps leading to equations (5.53) through (5.59).

Problem 5.7 *Integrating over the perpendicular velocities.*

1. Verify equations (5.61) through (5.63).
2. Verify equation (5.64).

5.1.4.2 Integrating over time

In order to integrate over τ, it will be convenient to introduce another Bessel identity,

$$\mathrm{e}^{\lambda\cos\omega_c\tau} = \sum_{n=-\infty}^{\infty} I_n(\lambda)\mathrm{e}^{\mathrm{i}n\omega_c\tau} \tag{5.68}$$

where $I_n(\lambda)$ is the modified Bessel function of the first kind. We may use this identity in a similar fashion to the other Bessel identity and its use in the orthogonality relation of equation (5.26) to obtain

$$\sum_{n=-\infty}^{\infty} \int_0^{\infty} \mathrm{d}\tau I_n(\lambda) \mathrm{e}^{\mathrm{i}(\omega+n\omega_c-k_z v_z)\tau} \left\{ \begin{array}{c} 1 \\ \cos\omega_c\tau \\ \sin\omega_c\tau \\ \sin\omega_c\tau\cos\omega_c\tau \\ \sin^2\omega_c\tau \end{array} \right\}$$

$$= \sum_{n=-\infty}^{\infty} \left\{ \begin{array}{c} \mathrm{i}I_n(\lambda) \\ \mathrm{i}I_n'(\lambda) \\ \frac{n}{\lambda} I_n(\lambda) \\ \frac{n}{\lambda^2}[\lambda I_n'(\lambda) - I_n(\lambda)] \\ \frac{\mathrm{i}}{\lambda^2}[\lambda I_n'(\lambda) - n^2 I_n(\lambda)] \end{array} \right\} \frac{1}{(\omega + n\omega_c - k_z v_z)} . \qquad (5.69)$$

Using these relations in equation (5.65) through equation (5.67) accomplishes the integral over time, and the results may be summarized as

$$\langle f_1 \rangle_\perp = \frac{\mathrm{i}v_t^2 q\,\mathrm{e}^{-\lambda}}{2\omega_c m} \sum_{n=-\infty}^{\infty} \frac{\kappa_+ n I_n/\lambda + \mathrm{i}\epsilon\kappa_-(I_n - I_n') - 2\alpha_z\omega_c I_n/v_t^2}{\omega + n\omega_c - k_z v_z} \qquad (5.70)$$

$$\langle v_x f_1 \rangle_\perp = \frac{\mathrm{i}v_t^2 q\,\mathrm{e}^{-\lambda}}{2\omega_c m} \sum_{n=-\infty}^{\infty} \frac{\kappa_x n I_n + (\mathrm{i}\epsilon\kappa_y - \kappa_- k_y v_t^2/\omega_c)\lambda(I_n - I_n')}{\lambda(\omega + n\omega_c - k_z v_z)} \qquad (5.71)$$

$$\langle v_y f_1 \rangle_\perp = \frac{\mathrm{i}v_t^2 q\,\mathrm{e}^{-\lambda}}{2\omega_c m} \sum_{n=-\infty}^{\infty} \frac{\kappa_y n I_n - (\mathrm{i}\epsilon\kappa_x - \kappa_- k_x v_t^2/\omega_c)\lambda(I_n - I_n')}{\lambda(\omega + n\omega_c - k_z v_z)} \qquad (5.72)$$

where $\kappa_+ = \alpha_x k_x + \alpha_y k_y$, $\kappa_- = \alpha_y k_x - \alpha_x k_y$, $\kappa_x = \alpha_x n\omega_c - \alpha_z k_x$, and $\kappa_y = \alpha_y n\omega_c - \alpha_z k_y$.

5.1.4.3 Integrating over the parallel velocity

In order to integrate over the parallel velocity distribution, we must specify the form of the distribution, and we shall choose a shifted Maxwellian, or equivalently, a Maxwellian with a drift velocity such that

$$F(v_z) = \frac{1}{\sqrt{\pi}v_\ell} \exp\left[-\frac{(v_z - v_0)^2}{v_\ell^2}\right] \qquad (5.73)$$

where v_ℓ is the longitudinal thermal speed given by $v_\ell^2 \equiv 2k_B T_\parallel/m$. Since α_x, α_y, and α_z may be expressed in terms of $F, v_z F$, and F', and since

$$F'(v_z) = -\frac{2(v_z - v_0)}{v_\ell^2} F(v_z)$$

the required integrals are all of the form

$$F_m = \frac{1}{\sqrt{\pi}v_\ell} \int_{-\infty}^{\infty} \mathrm{d}v_z \frac{v_z^m F(v_z)}{\omega + n\omega_c - k_z v_z} \qquad m = 0, 1, 2 . \qquad (5.74)$$

Changing variables to $u = (v_z - v_0)/v_\ell$, these moments of F may be expressed as

$$F_m(\zeta_n) = -\frac{1}{k_z v_\ell \sqrt{\pi}} \int_{-\infty}^{\infty} \frac{(v_0 + u v_\ell)^m \mathrm{e}^{-u^2}\, du}{u - \zeta_n} \tag{5.75}$$

where

$$\zeta_n = \frac{\omega + n\omega_c - k_z v_0}{k_z v_\ell}\,. \tag{5.76}$$

Using the definition of the Plasma Dispersion function of equation (4.13), the required moments are

$$F_0(\zeta_n) = -\frac{1}{k_z v_\ell} Z(\zeta_n) \tag{5.77}$$

$$F_1(\zeta_n) = \frac{1}{k_z}\left[\frac{1}{2} Z'(\zeta_n) - \frac{v_0}{v_\ell} Z(\zeta_n)\right] \tag{5.78}$$

$$F_2(\zeta_n) = \frac{v_\ell}{k_z}\left[\left(\frac{\zeta_n}{2} + \frac{v_0}{v_\ell}\right) Z'(\zeta_n) - \frac{v_0^2}{v_\ell^2} Z(\zeta_n)\right] \tag{5.79}$$

and using these, the integrals involving F' are

$$\int_{-\infty}^{\infty} \mathrm{d}v_z \frac{F'(v_z)}{\omega + n\omega_c - k_z v_z} = -\frac{1}{k_z v_\ell^2} Z'(\zeta_n) \tag{5.80}$$

$$\int_{-\infty}^{\infty} \mathrm{d}v_z \frac{v_z F'(v_z)}{\omega + n\omega_c - k_z v_z} = -\frac{1}{k_z v_\ell}\left(\zeta_n + \frac{v_0}{v_\ell}\right) Z'(\zeta_n)\,. \tag{5.81}$$

The total current density is finally constructed from

$$\boldsymbol{j} = \sum_j n_j q_j \int_{-\infty}^{\infty} \mathrm{d}v_z [\langle v_x f_{1j}\rangle_\perp \hat{e}_x + \langle v_y f_{1j}\rangle_\perp \hat{e}_y + v_z \langle f_{1j}\rangle_\perp \hat{e}_z]\,. \tag{5.82}$$

Problem 5.8 *Parallel velocity integrals.* Verify equations (5.78) – (5.81).

Problem 5.9 *Lorentzian distribution.* Evaluate the integrals corresponding to equations (5.77) through (5.81) for the Lorentzian distribution function:

$$F(v_z) = A/[(v_z - v_0)^2 + v_\ell^2]\,.$$

5.1.5 The dielectric tensor

From the current density of equation (5.82), all of the dielectric tensor elements may be constructed from the mobility tensor, M, where $\langle \boldsymbol{v}_j \rangle = \mathsf{M}_j \cdot \boldsymbol{E}$ by

$$\mathsf{K} = \mathsf{I} + \sum_j \frac{n_j q_j}{-\mathrm{i}\omega\epsilon_0} \mathsf{M}_j\,. \tag{5.83}$$

The final forms are not unique, since

$$\sum_{n=-\infty}^{\infty} nI_n = \sum_{n=-\infty}^{\infty} (I_n - I_n') = 0\,, \tag{5.84}$$

so certain terms can be added or subtracted. The general components may be expressed as

$$\begin{aligned} K_0 = 2\sum_j \frac{\omega_{pj}^2 \mathrm{e}^{-\lambda_j}}{\omega k_z v_{\ell j}} \sum_{n=-\infty}^{\infty} \lambda_j (I_n - I_n') \Bigg[\left(1 - \frac{k_z v_{0j}}{\omega}\right) Z(\zeta_{nj}) \\ + \frac{k_z v_{\ell j}}{\omega}\left(1 - \frac{T_{\perp j}}{T_{\parallel j}}\right) \frac{Z'(\zeta_{nj})}{2}\Bigg] \end{aligned} \tag{5.85}$$

$$\begin{aligned} K_1 = 1 + \sum_j \frac{\omega_{pj}^2 \mathrm{e}^{-\lambda_j}}{\omega k_z v_{\ell j}} \sum_{n=-\infty}^{\infty} \frac{n^2 I_n}{\lambda_j} \Bigg[\left(1 - \frac{k_z v_{0j}}{\omega}\right) Z(\zeta_{nj}) \\ + \frac{k_z v_{\ell j}}{\omega}\left(1 - \frac{T_{\perp j}}{T_{\parallel j}}\right) \frac{Z'(\zeta_{nj})}{2}\Bigg] \end{aligned} \tag{5.86}$$

$$\begin{aligned} K_2 = \mathrm{i}\sum_j \frac{\epsilon_j \omega_{pj}^2 \mathrm{e}^{-\lambda_j}}{\omega k_z v_{\ell j}} \sum_{n=-\infty}^{\infty} n(I_n - I_n') \Bigg[\left(1 - \frac{k_z v_{0j}}{\omega}\right) Z(\zeta_{nj}) \\ + \frac{k_z v_{\ell j}}{\omega}\left(1 - \frac{T_{\perp j}}{T_{\parallel j}}\right) \frac{Z'(\zeta_{nj})}{2}\Bigg] \end{aligned} \tag{5.87}$$

$$\begin{aligned} K_3 = 1 - \sum_j \frac{\omega_{pj}^2 \mathrm{e}^{-\lambda_j}}{\omega k_z v_{\ell j}} \sum_{n=-\infty}^{\infty} I_n \left(\frac{\omega + n\omega_{cj}}{k_z v_{\ell j}}\right) \\ \times \Bigg\{\left[1 + \frac{n\omega_{cj}}{\omega}\left(1 - \frac{T_{\parallel j}}{T_{\perp j}}\right)\right] Z'(\zeta_{nj}) \\ + \frac{2n\omega_{cj} T_{\parallel j} v_{0j}}{\omega T_{\perp j} v_{\ell j}} \left[Z(\zeta_{nj}) + \frac{k_z v_{\ell j}}{\omega + n\omega_{cj}}\right]\Bigg\} \end{aligned} \tag{5.88}$$

$$\begin{aligned} K_4 = \sum_j \frac{k_\perp \omega_{pj}^2 \mathrm{e}^{-\lambda_j}}{k_z \omega \omega_{cj}} \sum_{n=-\infty}^{\infty} \frac{nI_n}{\lambda_j} \Bigg\{\frac{n\omega_{cj} v_{0j}}{\omega v_{\ell j}} Z(\zeta_{nj}) \\ + \left[\frac{T_{\perp j}}{T_{\parallel j}} - \frac{n\omega_{cj}}{\omega}\left(1 - \frac{T_{\perp j}}{T_{\parallel j}}\right)\right] \frac{Z'(\zeta_{nj})}{2}\Bigg\} \end{aligned} \tag{5.89}$$

$$\begin{aligned} K_5 = \mathrm{i}\sum_j \frac{k_\perp \epsilon_j \omega_{pj}^2 \mathrm{e}^{-\lambda_j}}{k_z \omega \omega_{cj}} \sum_{n=-\infty}^{\infty} (I_n - I_n') \Bigg\{\frac{n\omega_{cj} v_{0j}}{\omega v_{\ell j}} Z(\zeta_{nj}) \\ + \left[\frac{T_{\perp j}}{T_{\parallel j}} - \frac{n\omega_{cj}}{\omega}\left(1 - \frac{T_{\perp j}}{T_{\parallel j}}\right)\right] \frac{Z'(\zeta_{nj})}{2}\Bigg\} \end{aligned} \tag{5.90}$$

and $K_6 = K_4$ and $K_7 = K_5$, so the hot plasma dielectric tensor of equation (5.27) reduces to the form

$$\mathsf{K} = \begin{pmatrix} K_1 + \sin^2\psi K_0 & K_2 - \cos\psi\sin\psi K_0 & \cos\psi K_4 + \sin\psi K_5 \\ -K_2 - \cos\psi\sin\psi K_0 & K_1 + \cos^2\psi K_0 & \sin\psi K_4 - \cos\psi K_5 \\ \cos\psi K_4 - \sin\psi K_5 & \sin\psi K_4 + \cos\psi K_5 & K_3 \end{pmatrix} \tag{5.91}$$

where $k_x = k_\perp \cos\psi$ and $k_y = k_\perp \sin\psi$. The proof that $K_6 = K_4$ and $K_7 = K_5$, where K_6 and K_7 come from $\langle v_z \rangle$ and K_4 and K_5 come from either $\langle v_x \rangle$ or $\langle v_y \rangle$ is only apparent when the Bessel identities of equation (5.84) are used.

The dielectric tensor, which is *not*, in general, hermitian, does have the symmetry property that $K_{ij}(B_0) = K_{ji}(-B_0)$ $(\omega_{cj} \to -\omega_{cj})$, since $K_2(-B_0) = -K_2(B_0)$ and $K_5(-B_0) = -K_5(B_0)$ while the other components are invariant, and this is a general result from the Onsager relations.

It is customary to set $\psi = 0$ so $k_x = k_\perp$ and $k_y = 0$, which can be accomplished by merely rotating the coordinate system, but the full symmetry is more apparent in this presentation.

5.1.5.1 Special Case: isotropic Maxwellian without drifts

When $v_{0j} = 0$ and $T_{\perp j} = T_{\parallel j}$, then the tensor components simplify significantly, and may be represented by

$$K_0 = 2\sum_j \frac{\omega_{pj}^2 e^{-\lambda_j}}{\omega k_z v_j} \sum_{n=-\infty}^{\infty} \lambda_j (I_n - I_n') Z(\zeta_{nj}) \tag{5.92}$$

$$K_1 = 1 + \sum_j \frac{\omega_{pj}^2 e^{-\lambda_j}}{\omega k_z v_j} \sum_{n=-\infty}^{\infty} \frac{n^2 I_n}{\lambda_j} Z(\zeta_{nj}) \tag{5.93}$$

$$K_2 = i\sum_j \frac{\epsilon_j \omega_{pj}^2 e^{-\lambda_j}}{\omega k_z v_j} \sum_{n=-\infty}^{\infty} n(I_n - I_n') Z(\zeta_{nj}) \tag{5.94}$$

$$K_3 = 1 - \sum_j \frac{\omega_{pj}^2 e^{-\lambda_j}}{\omega k_z v_j} \sum_{n=-\infty}^{\infty} I_n \zeta_{nj} Z'(\zeta_{nj}) \tag{5.95}$$

$$K_4 = \sum_j \frac{k_\perp \omega_{pj}^2 e^{-\lambda_j}}{2 k_z \omega \omega_{cj}} \sum_{n=-\infty}^{\infty} \frac{n I_n}{\lambda_j} Z'(\zeta_{nj}) \tag{5.96}$$

$$K_5 = i\sum_j \frac{k_\perp \epsilon_j \omega_{pj}^2 e^{-\lambda_j}}{2 k_z \omega \omega_{cj}} \sum_{n=-\infty}^{\infty} (I_n - I_n') Z'(\zeta_{nj})\,. \tag{5.97}$$

Problem 5.10 *Cold plasma limits.* For $v_{0j} = 0$,

1. Calculate the six dielectric tensor elements as $T_\perp \to 0$, $T_\parallel \neq 0$.

2. Calculate the six dielectric tensor elements as $T_\parallel \to 0$, $T_\perp \neq 0$.

3. Calculate the six dielectric tensor elements as $T_\parallel, T_\perp \to 0$. Show that this reduces to the cold plasma dielectric tensor. Does the order in which these limits are taken matter?

5.1.6 The hot plasma dispersion relation

The vector wave equation when $k_y \neq 0$ takes the form

$$\begin{pmatrix} \kappa_{xx} - k_z^2 - k_y^2 & \kappa_{xy} + k_x k_y & \kappa_{xz} + k_x k_z \\ \kappa_{yx} + k_y k_x & \kappa_{yy} - k_z^2 - k_x^2 & \kappa_{yz} + k_y k_z \\ \kappa_{zx} + k_z k_x & \kappa_{zy} + k_z k_y & \kappa_{zz} - k_\perp^2 \end{pmatrix} \begin{pmatrix} E_x \\ E_y \\ E_z \end{pmatrix} = 0 \qquad (5.98)$$

where the various tensor elements are given in equation (5.91) and we define $\kappa_j \equiv (\omega^2/c^2)K_j$. The dispersion relation is given by setting the determinant of coefficients to zero, and may be written either in terms of K_{ij}, n_x, n_y, and n_z (dimensionless quantities), or in terms of κ_{ij}, k_x, k_y, and k_z as written here.

While it is apparent that the components of the wave equation depend on ψ, the hot plasma dispersion relation (HPDR) does not, and may be written as

$$\begin{aligned} &[\gamma(\gamma - \kappa_0 + k_\perp^2) + \kappa_2^2]\kappa_3 + k_\perp^2[(\gamma - \kappa_0 + k_\perp^2)\kappa_1 - \kappa_2^2] \\ &+\kappa_4(\gamma - \kappa_0 + k_\perp^2)(2k_\perp k_z + \kappa_4) - \kappa_5[\gamma\kappa_5 + 2\kappa_2(k_\perp k_z + \kappa_4)] = 0 \end{aligned} \qquad (5.99)$$

where we have introduced $\gamma \equiv k_z^2 - \kappa_1$.

Problem 5.11 *Hot plasma dispersion relation.* Show that the determinant of coefficients of equation (5.98) results in the hot plasma dispersion relation given by equation (5.99).

5.1.7 Examples of hot plasma wave effects

5.1.7.1 Parallel propagation

For parallel propagation, $k_\perp = \lambda = 0$ and the tensor elements simplify since the infinite sums reduce to either one or two terms. In addition, we find $K_0 = K_4 = K_5 = 0$, so the dispersion relation reduces to

$$(\gamma^2 + \kappa_2^2)\kappa_3 = 0 \qquad (5.100)$$

so the roots are $\kappa_3 = 0$, and $k_z^2 = \kappa_1 \pm i\kappa_2$. The first root is the plasma wave which is unaffected by the magnetic field, so it is the case treated in Section 4.1 with the dispersion relation of equation (4.44). The other two roots are the R-wave and the L-wave, whose dispersion relations reduce to

$$n_{R,L}^2 = K_1 \pm iK_2$$

$$= 1 + \sum_j \frac{\omega_{pj}^2}{\omega k_z v_{\ell j}} \left[Z_{1j} \left(\frac{1 \pm \epsilon_j}{2} \right) + Z_{-1j} \left(\frac{1 \mp \epsilon_j}{2} \right) \right] \tag{5.101}$$

where

$$Z_{\pm 1} = \left(1 - \frac{k_z v_0}{\omega} \right) Z(\zeta_{\pm 1}) + \frac{k_z v_\ell}{2\omega} \left(1 - \frac{T_\perp}{T_\parallel} \right) Z'(\zeta_{\pm 1}) \,. \tag{5.102}$$

The structure of this dispersion relation confirms the general character we had observed with the cold plasma waves except that now resonance has a different meaning. The R-wave is here seen to be a function of $Z(\zeta_{1i})$ and $Z(\zeta_{-1e})$, and assuming that $\omega \gg k_z v_{\ell j}$, then there is virtually no ion damping associated with the R-wave since $|\zeta_{1i}| \gg 1$ and the ion damping is exponentially small. Near the electron cyclotron resonance, however, $|\zeta_{-1e}| \simeq 0$, and in this limit $Z(\zeta_{-1e}) \simeq \mathrm{i}\sqrt{\pi}$ so there is no longer any resonance at the electron cyclotron frequency, but now there is strong damping. If we neglect drifts and anisotropic temperature effects, and assume that $|\zeta_{-1e}|$ is large, but not too large, then the R-wave dispersion relation reduces to

$$\frac{k_z^2 c^2}{\omega^2} \approx 1 - \frac{\omega_{pe}^2}{\omega(\omega - \omega_{ce})} + \frac{\mathrm{i}\sqrt{\pi}\omega_{pe}^2}{\omega k_z v_e} \exp\left[- \left(\frac{\omega - \omega_{ce}}{k_z v_e} \right)^2 \right] . \tag{5.103}$$

Assuming weak damping such that $\omega_i \ll \omega_r$, the damping is approximately

$$\frac{\omega_i}{\omega_r} \sim - \frac{\sqrt{\pi}\omega_{pe}^2}{\omega_r k_z v_e [2 + \omega_{pe}^2 \omega_{ce} / \omega_r (\omega_r - \omega_{ce})^2]} \exp\left[- \left(\frac{\omega_r - \omega_{ce}}{k_z v_e} \right)^2 \right] , \tag{5.104}$$

which indicates the damping is exponentially small far from resonance, but as resonance approaches, the exponential term grows but the denominator also grows, suggesting a maximum value of the damping rate before resonance is reached. This is misleading, however, since the growing denominator depended on the weak damping assumption and this is no longer valid where this growing term dominates. In fact, it may be shown that for $\omega_r = \omega_{ce}$ that $\omega_i \gg \omega_r$. Except near this resonance, the damping is weak, so the cold plasma dispersion relation is relevant except near resonance.

The transition to an electron-acoustic wave is much more questionable, however, since the transition does not occur until the influence of the resonance has slowed the phase velocity to the neighborhood of the thermal velocity. In order to see the difficulty more clearly, we write the dispersion relation including the next higher order term in the expansion of the Plasma Dispersion function. This leads to

$$\frac{k_z^2 c^2}{\omega^2} = 1 - \frac{\omega_{pe}^2}{\omega(\omega - \omega_{ce})} \left[1 + \frac{k_z^2 v_e^2}{2(\omega - \omega_{ce})^2} \right] + \frac{\mathrm{i}\sqrt{\pi}\omega_{pe}^2}{\omega k_z v_e} \exp\left[- \left(\frac{\omega - \omega_{ce}}{k_z v_e} \right)^2 \right] \tag{5.105}$$

where now we can see that for the thermal term to become important, which is a necessary condition for the acoustic branch, we require $k_z v_e/(\omega - \omega_{ce}) > 1$ and before this happens we have entered or passed through the strong cyclotron damping region. For waves dominated by the derivative of the Plasma Dispersion function, $Z'(\zeta)$, rather than $Z(\zeta)$, as was the case for the ion-acoustic wave in Section 4.1, there is a weakly damped wave on the other side when $|\zeta| \ll 1$, but such is not the case for the R-wave or L-wave with $T_\perp = T_\parallel$.

Problem 5.12 *Damping rate at resonance.* Show that $\omega_i \gg \omega_r$ for the R-wave at $\omega_r = \omega_{ce}$.

5.1.7.2 Finite Larmor orbit effects

The addition of magnetic field effects to the hot plasma adds another effect that is entirely independent of the Landau or cyclotron damping we have discovered in the unmagnetized plasma or in the case of parallel propagation in a magnetized plasma. These effects are generally called finite Larmor orbit (FLR) effects and introduce two new general kinds of effects, one of which is the addition of higher cyclotron harmonic effects, indicated by the infinite sums in each of the dielectric tensor components, and of a class of electrostatic waves that have no counterpart in the cold plasma, and differ dramatically from the warm plasma electrostatic waves we have already encountered. These latter waves, commonly called Bernstein modes after I. Bernstein because of his analysis of the hot plasma electrostatic dispersion relation as $k_z \to 0$ [33], will be treated in the next section. In this section, we investigate the lowest order FLR effects by treating $\lambda_j = \frac{1}{2}k_\perp^2 \rho_{Lj}^2$ as a small parameter.

Before expanding the tensor elements in λ_j, we first discuss the physics of this new phenomenon. When the wavelength perpendicular to the magnetic field is infinite ($k_\perp = 0$), then as each particle executes its circular orbit, it maintains the same phase relation relative to the driving wave since the driving wave has no spatial dependence across the orbit. If we consider a case where $\omega = 2\omega_c$ and $k_\perp = 0$, and follow an ion making a counterclockwise orbit beginning at point a in Figure 5.1(a) where the vectors indicate the direction of the wave electric field at various points along the orbit, it is clear that the particle gains energy from the wave at a where the motion and the field are parallel. Following it around, it is moving perpendicular to the wave field at b, so it is neither accelerated nor decelerated there. It is moving antiparallel to the wave field at c so it is being slowed down there, and the point d is essentially equivalent to b, where there is no effect on the orbit. We can see that for this case, on the average, there is no net effect at this frequency, so the particles and wave show no special effects. If, however, we consider a case where the wavelength across the orbit is twice the orbit diameter, the situation is as shown in Figure 5.1(b) (with the wave traveling to the right).

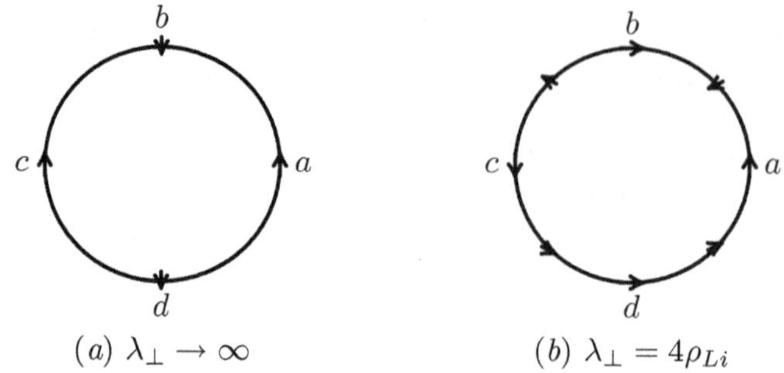

FIGURE 5.1
L-Wave electric fields on an ion as it follows its orbit counterclockwise with $\omega = 2\omega_{ci}$. (***a***) With $k_\perp = 0$. (***b***) with $k_\perp \rho_{Li} = \pi/2$.

For this case, the phase is chosen so that the electric field is again parallel to the motion at a, but now it is antiparallel at b, and parallel again at both c and d. From the indicated directions for the wave field at the intermediate points, it may be seen that the interaction nearly cancels on the upper half of the orbit, while the ion is being accelerated continuously on the lower half of the orbit. This difference is due to the fact that in the upper half of the orbit, the particle is moving to the left and the wave is moving to the right, so the phase changes rapidly, while in the lower half, the particle is moving in the same direction as the wave and the particle nearly stays in phase with the wave field. For any finite $k_\perp$ there is a nonvanishing contribution at *every* harmonic when averaged over the distribution, but for small $k_\perp \rho_L$, the interaction is progressively weaker as the harmonic number increases.

Keeping only first order terms in the expansion parameter λ_j (not to be confused with the perpendicular wavelength $\lambda_\perp$ in Figure 5.1), the isotropic temperature tensor components of equation (5.92) through equation (5.97) are

$$K_0 = \sum_j \frac{2\omega_{pj}^2 \lambda_j}{\omega k_z v_j} \{Z(\zeta_0) - \tfrac{1}{2}[Z(\zeta_1) + Z(\zeta_{-1})]\}_j \tag{5.106}$$

$$K_1 = 1 + \sum_j \frac{\omega_{pj}^2}{2\omega k_z v_j} \{[Z(\zeta_{-1}) + Z(\zeta_1)](1-\lambda) + \lambda[Z(\zeta_{-2}) + Z(\zeta_2)]\}_j \tag{5.107}$$

$$K_2 = \mathrm{i} \sum_j \frac{\epsilon_j \omega_{pj}^2}{2\omega k_z v_j} \{[Z(\zeta_{-1}) - Z(\zeta_1)](1-2\lambda) + \lambda[Z(\zeta_{-2}) - Z(\zeta_2)]\}_j \tag{5.108}$$

$$K_3 = 1 - \sum_j \frac{\omega_{pj}^2}{\omega k_z v_j} \{\zeta_0 Z'(\zeta_0)(1-\lambda) + \frac{\lambda}{2}[\zeta_{-1} Z'(\zeta_{-1}) + \zeta_1 Z'(\zeta_1)]\}_j \tag{5.109}$$

$$K_4 = \sum_j \frac{\omega_{pj}^2 \sqrt{\lambda_j}}{2\sqrt{2}\omega k_z v_j} [Z'(\zeta_1) - Z'(\zeta_{-1})]_j \tag{5.110}$$

$$K_5 = \mathrm{i}\sum_j \frac{\epsilon_j \omega_{pj}^2 \sqrt{\lambda_j}}{\sqrt{2}\omega k_z v_j}\{Z'(\zeta_0) - \tfrac{1}{2}[Z'(\zeta_1) + Z'(\zeta_{-1})]\}_j \,. \tag{5.111}$$

From these expressions, several things are immediately evident. To zero order in λ_j, it is apparent that only K_1, K_2, and K_3 are nonzero, and they have no cyclotron interactions above the fundamental resonance (which we shall call the *first* harmonic so that $n = 1$ is the first harmonic, $n = 2$ is the second harmonic, etc. Other authors sometimes use $n = 1$ as the fundamental, $n = 2$ as the *first* harmonic, etc.). From these it is easy to recover the cold plasma dielectric tensor by using the large argument expansion of the Plasma Dispersion function. It is also apparent that through first order, only K_1 and K_2 have an interaction at the second harmonic ($n = \pm 2$). Thus if one wanted to examine effects near the second harmonic, it would be appropriate to neglect all other first order terms in λ_j except the harmonic terms, because the harmonic resonant terms can be taken to be large near their resonance (a large term times a small term could be considered zero order, while all other first order terms would be small by comparison). It follows that near these harmonics, we only need K_1, K_2, and K_3.

Some comments about the order of K_4 and K_5 are in order since they appear to be of order $\sqrt{\lambda_j}$ in equations (5.110) and (5.111). That these are properly considered as first order in λ_j is evident first by noting from the hot plasma dispersion relation, equation (5.99), that both K_4 and K_5 appear multiplied by $k_z k_\perp$ or one another, so their terms appear *in the dispersion relation* as first order in λ_j. If we use the large argument expansion for $Z(\zeta)$ in equation (5.110) for K_4, for example, we find

$$K_4 \simeq -\sum_j \frac{\omega_{pj}^2 k_z k_\perp v_j^2}{(\omega^2 - \omega_{cj}^2)^2}$$

and the expression for K_5 is similar. This means that multiplying either by $k_z k_\perp$ or by one another produces a term of the order of λ_j in the HPDR. Hence we have not included any higher order terms for these components.

Problem 5.13 *Third harmonic.* Find the tensor elements corresponding to equations (5.106) through (5.111) near the third harmonic. (Neglect $n \pm 2$ terms, but go to order λ^2 for $n \pm 3$).

5.2 Electrostatic waves

5.2.1 Electrostatic dispersion relation

For hot plasmas, where the tensor elements are so formidable individually and the dispersion relation virtually defies any analytic analysis for any but the simplest cases, the simplifications of the electrostatic approximation make it even more attractive than it was in either the cold or warm plasma approximations. The general hot plasma electrostatic dispersion relation is

$$k_\perp^2 \kappa_1 + 2k_\perp k_z \kappa_4 + k_z^2 \kappa_3 = 0\,. \tag{5.112}$$

While it is possible to combine the terms in the expressions for K_1, K_3, and K_4 to simplify this dispersion relation, it will be useful for a large $k_\perp$ approximation to begin again with the integrals over the unperturbed orbits. We start with the electrostatic restriction that $\omega \boldsymbol{B}_1 = \boldsymbol{k} \times \boldsymbol{E}_1 = 0$ so the Fourier transform of the electric field may be represented by $\boldsymbol{E} = -i\boldsymbol{k}\varphi$. Then equation (5.65) reduces to

$$\langle f_1 \rangle_\perp = -\frac{q\varphi}{m} \int_0^\infty \mathrm{d}\tau \mathrm{e}^\phi \left(\frac{F k_\perp^2}{\omega_c} \sin \omega_c \tau - i k_z F' \right). \tag{5.113}$$

Integrating over time, this becomes

$$\langle f_1 \rangle_\perp = -\frac{q\varphi \mathrm{e}^{-\lambda}}{m} \sum_{n=-\infty}^{\infty} \left(\frac{nFk_\perp^2}{\lambda \omega_c} + k_z F' \right) \frac{I_n(\lambda)}{\omega + n\omega_c - k_z v_z} \tag{5.114}$$

and finally, integrating over the parallel velocity distribution, this becomes

$$\langle f_1 \rangle = \frac{2q\varphi \mathrm{e}^{-\lambda}}{m v_\ell^2} \sum_{n=-\infty}^{\infty} \left[1 + \frac{\omega + n\omega_c (1 - T_\parallel / T_\perp) - k_z v_0}{k_z v_\ell} Z(\zeta_n) \right] I_n(\lambda)\,. \tag{5.115}$$

Then we use Poisson's equation,

$$\nabla^2 \varphi = -k^2 \varphi = -\frac{\rho}{\epsilon_0} = -\frac{1}{\epsilon_0} \sum_j n_{0j} q_j \langle f_{1j} \rangle$$

to obtain the hot plasma electrostatic dispersion relation

$$k^2 + \sum_j \frac{2\omega_{pj}^2 \mathrm{e}^{-\lambda_j}}{v_{\ell j}^2} \sum_{n=-\infty}^{\infty} \left[1 + \frac{\omega + n\omega_{cj} \left(1 - \frac{T_{\parallel j}}{T_{\perp j}} \right) - k_z v_{0j}}{k_z v_{\ell j}} Z(\zeta_{nj}) \right] I_n(\lambda_j) = 0\,. \tag{5.116}$$

Problem 5.14 *The hot plasma electrostatic dispersion relation.*

Show that the hot plasma electrostatic dispersion relation, equation (5.116) (with $T_\perp = T_\parallel$ and $v_{0j} = 0$), can be obtained from the general hot plasma dielectric tensor elements. First show that $\boldsymbol{k} \cdot \mathsf{K} \cdot \boldsymbol{k} = 0$ leads to equation (5.112). Then use the tensor elements from equations (5.92) through (5.97) for $T_\perp = T_\parallel$ and $v_{0j} = 0$. (*Hint*: A Bessel identity is required.)

5.2.2 Perpendicular propagation – Bernstein modes

We have seen a variety of hot plasma effects due to Landau and cyclotron damping, but another important result has no damping associated with it at all in the absence of collisions, and this is the case for perpendicularly propagating electrostatic waves. As $k_z \to 0$, the dispersion relation reduces to

$$k_\perp^2 = \sum_j 2k_{Dj}^2 \mathrm{e}^{-\lambda_j} \sum_{n=1}^{\infty} I_n \frac{n^2}{\nu_j^2 - n^2}, \tag{5.117}$$

where again $k_{Dj}^2 = 2\omega_{pj}^2/v_j^2$ is the Debye wavenumber and $\nu_j \equiv \omega/\omega_{cj}$. This dispersion relation has a resonance at every harmonic of both cyclotron frequencies but the "strength" of the resonance indicated by $I_n(\lambda_j)$ becomes small for large n. We also note that there is no absorption here to damp out the wave at resonance. Within the framework of the collisionless theory outlined in this chapter, these resonances remain unresolved.

Examining equation (5.117) for small λ_i (which means $\lambda_e \ll 1$ since $\lambda_e/\lambda_i = T_e m_e/T_i m_i \ll 1$), we shall approximate $I_n(\lambda) \simeq (\lambda/2)^n/n!$ and consider cold electrons. In this case, the dispersion relation can be approximated by

$$\frac{V_A^2}{c^2} = \frac{1}{\nu^2 - 1} + \frac{\lambda_i}{\nu^2 - 4} + \frac{3\lambda_i^2}{8(\nu^2 - 9)} + \cdots \tag{5.118}$$

where we have neglected m_e/m_i and taken $\mathrm{e}^{-\lambda_i} = 1$. If we investigate the behavior near $\nu \simeq 2$, letting $\nu = 2$ in the nonresonant terms, then to lowest order,

$$\nu = 2 - \frac{3\lambda_i}{4(1 - 3V_A^2/c^2)}, \tag{5.119}$$

so the wave propagates below the second harmonic, and for large λ_i, the dispersion relation approaches the fundamental. This is shown in Figure 5.2(a) where the dispersion relation falls away initially from $\nu = 2$ with linear slope for small λ_i.

Using the same technique near $\nu \simeq 3$, the result to lowest order is

$$\nu = 3 - \frac{\lambda_i^2}{2(1 - 8V_A^2/c^2)}, \tag{5.120}$$

so this wave begins at $\lambda_i = 0$ with zero slope in Figure 5.2(a) and then falls toward $\nu = 2$ as λ_i gets large. We can generalize this analysis to

$$\nu = n - \frac{(n^2 - 1)\lambda_i^{n-1}}{2^n (n-1)! [1 - (n^2 - 1)V_A^2/c^2]}, \tag{5.121}$$

so for the higher harmonics, ν deviates less and less from the resonance for fixed λ_i, but eventually approaches the next lower harmonic.

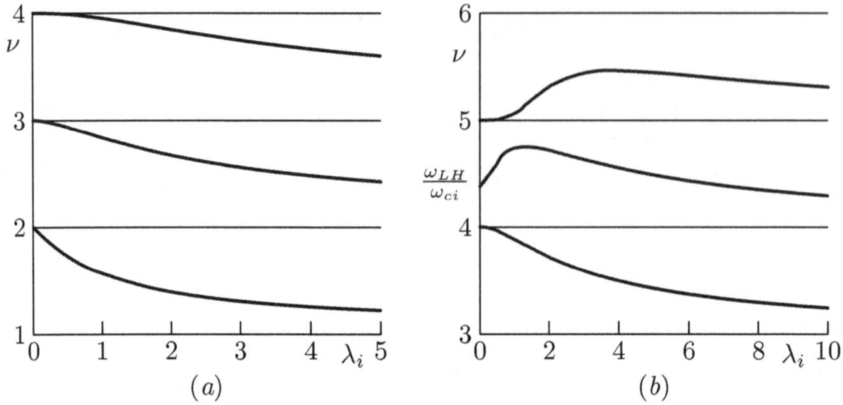

FIGURE 5.2
Ion Bernstein wave dispersion relations. (***a***) The first few modes with $\omega \ll \omega_{LH}$ $(\omega_{pe}^2/\omega_{ce}^2 = 10)$. (***b***) Higher order modes near ω_{LH} $(\omega_{pe}/\omega_{ce} = 0.1)$.

For sufficiently high harmonics, or for sufficiently low density, however, the character of these dispersion curves changes, since for $\lambda_i = 0$, equation (5.117) reduces to

$$1 = \frac{\omega_{pe}^2}{\omega^2 - \omega_{ce}^2} + \frac{\omega_{pi}^2}{\omega^2 - \omega_{ci}^2}, \tag{5.122}$$

which is the condition for the hybrid resonances. Hence one curve begins at $\nu = \omega_{LH}/\omega_{ci}$ for $\lambda_i = 0$ and then falls to the next lower harmonic as $\lambda_i \to \infty$. The wave whose dispersion curve starts at the next higher resonance then lies *above* the resonance, rising to a maximum at some finite λ_i (but below the next higher resonance) and then falls back to the same resonance as $\lambda_i \to \infty$. This behavior is illustrated in Figure 5.2(b).

5.3 Velocity space instabilities

5.3.1 Anisotropic temperature

While it is often difficult to do anything more than estimate the damping or growth rate when the imaginary part of ω is much smaller than the real part, there is one case where we can calculate exactly the threshold condition, or the marginal stability condition. For either the R-wave or the L-wave, only one species contributes any damping or growth, and the condition that ω be exactly real is that Z_{-1} from equation (5.102) have no imaginary part. The imaginary part of Z_{-1} may be written

$$\left[1 - \frac{k_z v_0}{\omega} - \left(1 - \frac{T_\perp}{T_\parallel}\right) \frac{\omega - k_z v_0 - \omega_c}{\omega}\right] \mathrm{i}\sqrt{\pi}\mathrm{e}^{-\zeta_{-1}^2} = 0\,. \tag{5.123}$$

With $v_0 = 0$, since we are considering only the anisotropic temperature effects here, this leads to the marginal stability condition

$$\omega_m = \left(1 - \frac{T_\parallel}{T_\perp}\right) \omega_c\,. \tag{5.124}$$

Instability occurs when $\mathrm{Im}(Z_{-1}) < 0$ or whenever $\omega < \omega_m$, provided that k_z is real. To check this, we note that for $\omega = \omega_m$,

$$Z_{-1} = \frac{k_z v_\ell}{\omega_c} \frac{T_\perp}{T_\parallel} \tag{5.125}$$

which is purely real, and the dispersion relation for the R-wave, from equation (5.101) (neglecting the ion term), is

$$k_z^2 c^2 = \omega_{ce}^2 \left(1 - \frac{T_\parallel}{T_\perp}\right)^2 + \omega_{pe}^2 \left(\frac{T_\perp}{T_\parallel} - 1\right). \tag{5.126}$$

Thus, with $T_\perp > T_\parallel$, the wave is propagating and it does go unstable beyond the marginal condition. This implies that for any $T_\perp > T_\parallel$, the plasma is unstable, but unless $\omega \sim \omega_c$, the growth rate is very small as the exponent is very large. In practical terms, the anisotropy must be very large, in which case the marginal frequency may approach the cyclotron frequency closely enough to have significant effects.

Another way in which this effect of marginal stability may be perceived is the occurrence of a transition from strong absorption to transparency in plasma heating experiments using either electron or ion cyclotron waves. In this scenario, we imagine, for example, an L-wave propagating toward resonance in a very slowly decreasing magnetic field that is termed a "magnetic beach"[34]. If we imagine that any wave energy absorbed leads to increasing $T_\perp$ only, then the wave is absorbed very weakly far from resonance, but more strongly as resonance is approached. This stronger absorption increases $T_\perp$ locally, raising ω_m to ω whereupon the plasma absorbs no further wave energy. Of course, as ω approaches ω_c, the transparency condition requires $T_\perp \to \infty$, so some absorption must always occur. It is clear from this example that this type of wave heating could never lead to an instability unless the plasma with higher $T_\perp$ drifted back towards the source at higher magnetic field, a result that is unlikely due to the magnetic mirror effect which would confine the higher $T_\perp$ plasma particles to the lower field region. In an interesting experiment on the Model C Stellarator, an L-wave was launched in a predominantly hydrogen plasma with a deuterium minority that was locally resonant in a narrow depression in the magnetic field ($\omega \simeq \omega_{cH}/2$ so no significant hydrogen absorption occurred). Since the mirror was only a few percent deep, it was possible to raise the $T_{D\perp}$ to nearly a hundred times the average temperature, trapping this small population in the mirror in the process.[35]

Problem 5.15 *Anisotropic temperature instability.*

1. Fill in the steps leading to equations (5.124) and (5.126).
2. Estimate the maximum k_{zi}/k_{zr} with $\omega_{ce}/k_{zr}v_e = 3$, $\omega_{pe}/\omega_{ce} = 2$, for $T_\perp/T_\parallel = 2$ and $T_\perp/T_\parallel = 10$.

5.3.2 Bump-on-the-tail instability

As another example of an instability due to the velocity distribution function, we discuss a double-humped distribution that has a background Maxwellian distribution with another shifted Maxwellian added as if there were a beam of streaming particles with a thermal distribution passing through the background plasma. We found in equation (4.40) that the sign of the imaginary part of ω depended on the slope of the distribution function through the relation,

$$\gamma = \frac{1}{2}\omega_r \epsilon_i = \omega_r \frac{\pi \omega_{pe}^2}{2k^2} \left. \frac{\mathrm{d}f_0(u)}{\mathrm{d}u} \right|_{u=\omega_r/k} .$$

In Figure 5.3, such a distribution function is illustrated, and the slope of the distribution function is positive between v_1 and v_2, indicating that $\gamma > 0$ in this velocity range, or unstable. This type of instability is sometimes called the beam-plasma instability or if the beam intensity is small and the beam temperature is high enough that the bump is "gentle," it is referred to as a *weak* bump on the tail distribution. The amplitude of the wave begins to grow so that the linear approximation soon fails. With a finite but not too large an amplitude, this case can be analyzed using quasilinear theory. It is shown in this theory that as the wave amplitude grows, a wave-driven diffusion begins to develop, consisting of two parts, a resonant quasilinear diffusion that occurs in the range where the slope of the distribution function is positive, and a nonresonant quasilinear diffusion that spreads the entire distribution and drives it back towards equilibrium.

The first phase, due to resonant diffusion, works rapidly to eliminate the positive slope, and leads to what is known as a plateau region, illustrated in Figure 5.4. When this plateau is reached, growth ceases, but the wave amplitude is still large enough that nonresonant diffusion will continue until equilibrium is reached. The area where the dashed line (original distribution) is above the plateau is equal the area where the dashed line is below the plateau. This conserves energy, but not momentum, so the figure is only indicative of the distribution function.

In the final phase, nonresonant diffusion begins to smooth out the distribution and relax toward equilibrium. This is illustrated in Figure 5.5 where

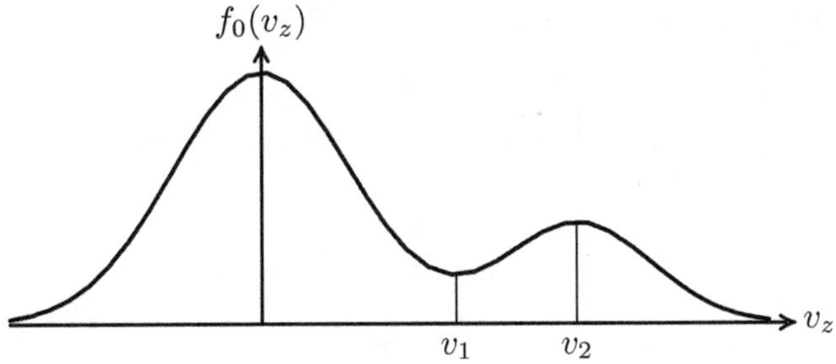

FIGURE 5.3
Bump-on-the-tail distribution showing initial unstable region between v_1 and v_2.

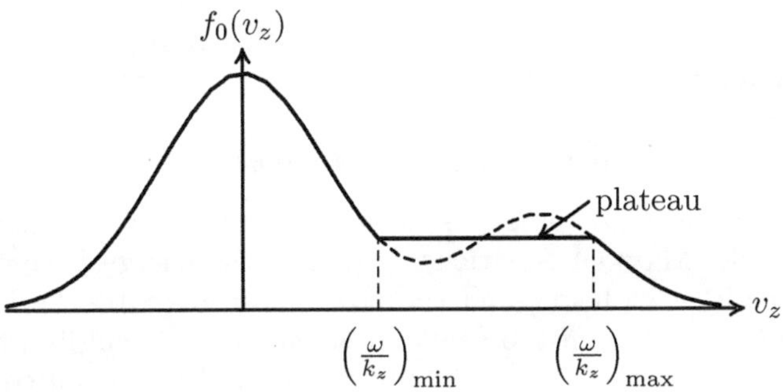

FIGURE 5.4
Quasilinear plateau between end points of k_z spectrum.

the smoothing is evident and the slight shift in the maximum to higher energy is also shown. This process occurs on a much longer time scale than the resonant diffusion, but ultimately, a true equilibrium requires collisions as the wave amplitude decays away too rapidly for quasilinear diffusion to accomplish it alone. A discussion of quasiliinear theory and more details about this example may be found in chapter 7 of Swanson[5].

Problem 5.16 *Weak bump-on-the-tail instability.* For the distribution function,

$$f(v) = A_0 e^{-v^2/v_0^2} + A_b e^{-(v-v_d)^2/v_b^2},$$

with $A_0 = 1$, $v_0 = 1$, $A_b = .4$, $v_d = 2$, and $v_b = .6$,

1. find v_1 and v_2, the bounds of the velocity range where the distribution is unstable.

2. find the equilibrium distribution function if this system relaxes completely to equilibrium and sketch both the original and equilibrium distribution functions in the same figure.

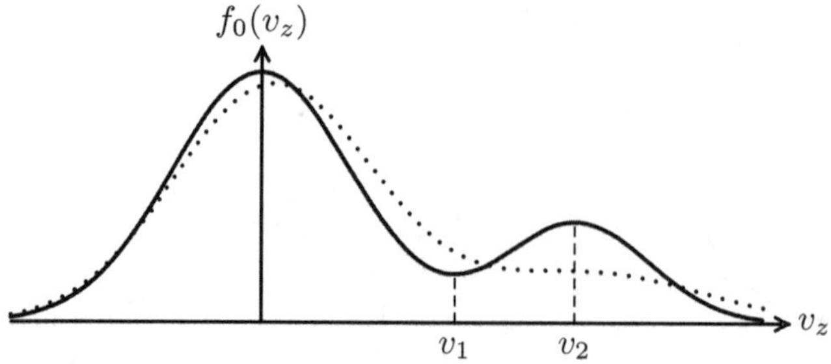

FIGURE 5.5
Original bump-on-the-tail distribution(solid) and form of $f_0(v,t)$(dotted) after only nonresonant decay remains.

5.4 Conservation of energy and power flow

In vacuum, we saw in Chapter 1 that energy conservation and power flow were related by the Maxwell equations, and that the energy was stored in the electromagnetic wave fields, and the power flow given by the Poynting vector was dependent only on the fields. In Chapter 6 we will find that the plasma particles contribute both to the power flow and to the stored energy. In this section, we use a more general formalism to include the effects of the plasma and the effects of dissipation through an antihermitian component of the dielectric tensor and/or the effects of a complex frequency. We treat first the temporal problem, where we obtain the stored energy density and the effects of dissipation. We then examine in more detail the concept of group velocity, and obtain the kinetic flux component of the power flow.

5.4.1 Poynting's theorem for kinetic waves

When the angular frequency ω is assumed to have an imaginary part, especially in the case where this imaginary part may vary slowly in time to represent the slow turning on of the wave, the steady state results of Chapter 1 are not the most fruitful in understanding energy density and power flow. Using the more general representation of a wave field amplitude by

$$\boldsymbol{A}(t) = \mathrm{Re}\left[\hat{A}(\omega)\mathrm{e}^{-\mathrm{i}\phi(t)}\right], \quad \phi(t) = \int_{-\infty}^{t} \omega(t')\,\mathrm{d}t', \tag{5.127}$$

the general product of two vectors may be represented by

$$[\boldsymbol{A}][\boldsymbol{B}] = \tfrac{1}{4}\left\{[\hat{A}][\hat{B}]\mathrm{e}^{-2\mathrm{i}\phi(t)} + ([\hat{A}][\hat{B}^*] + [\hat{A}^*][\hat{B}])\mathrm{e}^{2\phi_i(t)} + [\hat{A}^*][\hat{B}^*]\mathrm{e}^{2\mathrm{i}\phi^*(t)}\right\}. \tag{5.128}$$

If we now take $\omega_i \ll \omega_r$ and integrate over a period $T = 2\pi/\omega_r$, then

$$\frac{1}{T}\int_0^T \exp\left[-2\mathrm{i}\int_{-\infty}^t \omega(t')\,\mathrm{d}t'\right]\,\mathrm{d}t \simeq \frac{-\mathrm{i}\omega_i(0)}{\omega_r(0)}\mathrm{e}^{-2\mathrm{i}\phi(0)}\,.$$

This means both the first and the last terms in equation (5.128) may be neglected so that a general product of this type reduces to

$$[\boldsymbol{A}][\boldsymbol{B}] = \tfrac{1}{4}([\hat{A}][\hat{B}^*] + [\hat{A}^*][\hat{B}])\mathrm{e}^{2\phi_i(t)}. \tag{5.129}$$

Using these expressions, the complex Poynting vector is represented by

$$\boldsymbol{P} = \tfrac{1}{4}(\hat{E}\times\hat{H}^* + \hat{E}^*\times\hat{H})\mathrm{e}^{2\phi_i(t)} \tag{5.130}$$

and the conservation law from the Maxwell equations,

$$\nabla\cdot(\boldsymbol{E}\times\boldsymbol{H}) = -\left(\boldsymbol{H}\cdot\frac{\partial\boldsymbol{B}}{\partial t} + \boldsymbol{E}\cdot\frac{\partial\boldsymbol{D}}{\partial t}\right)$$

becomes

$$\nabla\cdot\boldsymbol{P} = -\frac{\partial W}{\partial t} \tag{5.131}$$

where

$$\begin{aligned}\frac{\partial W}{\partial t} &= \tfrac{1}{4}[\hat{H}\cdot(-\mathrm{i}\omega\hat{B})^* + \hat{H}^*\cdot(-\mathrm{i}\omega\hat{B}) + \hat{E}\cdot(-\mathrm{i}\omega\epsilon_0\mathsf{K}\cdot\hat{E})^* \\ &\quad +\hat{E}^*\cdot(-\mathrm{i}\omega\epsilon_0\mathsf{K}\cdot\hat{E})]\mathrm{e}^{2\phi_i(t)} \\ &= \tfrac{1}{4}[2\omega_i\mu_0\hat{H}\cdot\hat{H}^* + \omega_i\epsilon_0\hat{E}^*\cdot(\mathsf{K}+\mathsf{K}^\dagger)\cdot\hat{E} + \omega_r\epsilon_0\hat{E}^*\cdot(\mathrm{i}\mathsf{K}^\dagger - \mathrm{i}\mathsf{K})\cdot\hat{E}] \\ &\quad \times\mathrm{e}^{2\phi_i(t)}\end{aligned} \tag{5.132}$$

where we have used $\hat{E}\cdot(\mathsf{K}\cdot\hat{E})^* = \hat{E}^*\cdot\mathsf{K}^\dagger\cdot\hat{E}$ and $\mathsf{K}^\dagger$ is the hermitian adjoint of K given by $\mathsf{K}^\dagger = \tilde{\mathsf{K}}^*$. We define the hermitian and antihermitian portions of K by

$$\mathsf{K}_h \equiv \frac{1}{2}(\mathsf{K}+\mathsf{K}^\dagger) \quad \text{and} \quad \mathsf{K}_a \equiv \frac{1}{2\mathrm{i}}(\mathsf{K}-\mathsf{K}^\dagger)\,.$$

It is apparent from equation (5.132) that if ω is real and K is hermitian, there is no loss, so any dissipation must arise from the antihermitian portion of the dielectric tensor or the imaginary part of ω.

If we now expand the dielectric tensor about the real part of ω, we find

$$\mathsf{K}(\omega) = \mathsf{K}(\omega_r) + \left.\frac{\partial\mathsf{K}}{\partial\omega}\right|_{\omega_r}\mathrm{i}\omega_i + \cdots$$

so that

$$\mathsf{K}_h(\omega) = \mathsf{K}_h(\omega_r) + \left.\frac{\partial\mathsf{K}_h}{\partial\omega}\right|_{\omega_r}\mathrm{i}\omega_i + \cdots \tag{5.133}$$

$$-\mathrm{i}\mathsf{K}(\omega) + \mathrm{i}\mathsf{K}^\dagger(\omega^*) = 2\mathsf{K}_a(\omega_r) + 2\omega_i\left.\frac{\partial\mathsf{K}_h}{\partial\omega}\right|_{\omega_r} + \cdots \tag{5.134}$$

and the energy term of equation (5.132) becomes

$$\frac{\partial W}{\partial t} = \frac{1}{4}\left[2\omega_i\left(\mu_0\hat{H}\cdot\hat{H}^* + \epsilon_0\hat{E}^*\cdot\frac{\partial}{\partial\omega}(\omega\mathsf{K}_h)\bigg|_{\omega_r}\cdot\hat{E}\right)\right.$$
$$\left. + 2\omega_r\epsilon_0\hat{E}^*\cdot\mathsf{K}_a(\omega_r)\cdot\hat{E}\right]\mathrm{e}^{2\phi_i(t)} \tag{5.135}$$

and we can identify the total energy as being comprised of the stored energy,

$$W_0 = \frac{1}{4}\left[\mu_0\hat{H}\cdot\hat{H}^* + \epsilon_0\hat{E}^*\cdot\frac{\partial}{\partial\omega}(\omega\mathsf{K}_h)\bigg|_{\omega_r}\cdot\hat{E}\right] \tag{5.136}$$

and a dissipative term associated with the antihermitian part of K. We can also see that some of the stored energy is electromagnetic or electrostatic, and some is in the particle kinetic energy, even in the cold plasma.

It is often useful to relate these two terms through the quality factor, which is given by the ratio of the stored energy to the energy lost per cycle, or

$$Q = \frac{\omega_r W_0}{\partial W/\partial t(\text{loss})} = \frac{\mu_0\hat{H}\cdot\hat{H}^* + \epsilon_0\hat{E}^*\cdot\partial(\omega\mathsf{K}_h)/\partial\omega|_{\omega_r}\cdot\hat{E}}{2\epsilon_0\hat{E}^*\cdot\mathsf{K}_a(\omega_r)\cdot\hat{E}} \tag{5.137}$$

so that a high Q indicates that the stored energy lasts many cycles and a low Q may mean that there is very little energy circulating, as if it were all absorbed on a single pass, or nearly so. Since in subsequent discussions, we deal nearly universally with the amplitudes, we will delete the hat from now on, but recall the recipe when questions about power or energy are desired.

Problem 5.17 *Power and energy.*

1. Fill in the steps leading to equation (5.129).

2. Fill in the steps leading to equation (5.135) and justify equation (5.136).

Problem 5.18 *Stored energy.* Show that for simple cold plasma oscillations ($\omega = \omega_p$) that the electrostatic stored energy is equal to the particle kinetic energy.

5.4.2 Group velocity and kinetic flux

The discussion of group velocity in the context of a hot plasma is much more complicated than in a cold plasma, since now we need to include the power flow due the particles, which we call the kinetic flux. For this discussion, we restrict ourselves to a lossfree plasma and introduce the Maxwell operator, M, which represents the Maxwell wave equation, such that

$$\mathsf{M}\cdot\boldsymbol{E} = \frac{1}{4\mu_0\omega}\boldsymbol{k}\times(\boldsymbol{k}\times\boldsymbol{E}) + \frac{\omega\epsilon_0}{4}\mathsf{K}_h\cdot\boldsymbol{E} \tag{5.138}$$

and the wave equation is then simply

$$\mathsf{M} \cdot \boldsymbol{E} = 0 \,. \tag{5.139}$$

The hermitian adjoint of equation (5.139) is

$$\boldsymbol{E}^* \cdot \mathsf{M}^\dagger = \boldsymbol{E}^* \cdot \mathsf{M} = 0 \tag{5.140}$$

where the last equality holds for real ω and real $\boldsymbol{k}$ in which case M is hermitian.

The idea at this point is to make small displacements in ω, $\boldsymbol{k}$, and in the plasma parameters, or what is essentially a variational calculation, and the extremum will yield the group velocity. The perturbed wave equation is

$$\mathsf{M}' \cdot \boldsymbol{E}' = 0 \tag{5.141}$$

but if the perturbations are small, then we can expand

$$\mathsf{M}' = \mathsf{M} + \delta\omega \frac{\partial \mathsf{M}}{\partial \omega} + \delta\boldsymbol{k} \cdot \frac{\partial \mathsf{M}}{\partial \boldsymbol{k}} + \delta \mathsf{M} \,.$$

We now take the scalar product of $\boldsymbol{E}^*$ with equation (5.141) and use equation (5.140) to obtain

$$\boldsymbol{E}^* \cdot \left(\delta\omega \frac{\partial \mathsf{M}}{\partial \omega} + \delta\boldsymbol{k} \cdot \frac{\partial \mathsf{M}}{\partial \boldsymbol{k}} + \delta \mathsf{M} \right) \cdot \boldsymbol{E}' = 0$$

where $\partial/\partial\boldsymbol{k} = \nabla_k$. Since the perturbation is small, the differences between $\boldsymbol{E}$ and $\boldsymbol{E}'$ lead to second order corrections, so we may write to first order

$$\boldsymbol{E}^* \cdot \left(\delta\omega \frac{\partial \mathsf{M}}{\partial \omega} + \delta\boldsymbol{k} \cdot \frac{\partial \mathsf{M}}{\partial \boldsymbol{k}} + \delta \mathsf{M} \right) \cdot \boldsymbol{E} = 0 \,. \tag{5.142}$$

Examining each of the terms in equation (5.142) separately, we find

$$\begin{aligned} \boldsymbol{E}^* \cdot \frac{\partial \mathsf{M}}{\partial \omega} \cdot \boldsymbol{E} &= \boldsymbol{E}^* \cdot \frac{-1}{4\omega^2 \mu_0} \boldsymbol{k} \times (\boldsymbol{k} \times \boldsymbol{E}) + \frac{\epsilon_0}{4} \boldsymbol{E}^* \cdot \frac{\partial}{\partial \omega} (\omega \mathsf{K}_h) \cdot \boldsymbol{E} \\ &= \frac{1}{4} \left[\mu_0 \boldsymbol{H}^* \cdot \boldsymbol{H} + \epsilon_0 \boldsymbol{E}^* \cdot \frac{\partial}{\partial \omega} (\omega \mathsf{K}_h) \cdot \boldsymbol{E} \right] \\ &= W_0 \end{aligned} \tag{5.143}$$

$$\begin{aligned} \boldsymbol{E}^* \cdot \frac{\partial \mathsf{M}}{\partial \boldsymbol{k}} \cdot \boldsymbol{E} &= -\frac{1}{4} (\boldsymbol{E}^* \times \boldsymbol{H} + \boldsymbol{E} \times \boldsymbol{H}^*) + \boldsymbol{E}^* \cdot \frac{\omega \epsilon_0}{4} \frac{\partial}{\partial \boldsymbol{k}} (\mathsf{K}_h) \cdot \boldsymbol{E} \\ &= -\boldsymbol{P} - \boldsymbol{T} \end{aligned} \tag{5.144}$$

where

$$\boldsymbol{T} \equiv -\frac{\omega \epsilon_0}{4} \boldsymbol{E}^* \cdot \frac{\partial}{\partial \boldsymbol{k}} (\mathsf{K}_h) \cdot \boldsymbol{E} \,. \tag{5.145}$$

The vector $\boldsymbol{T}$ is the kinetic flux for the hot plasma, and represents the generalization of the kinetic flux of a cold plasma. From the form of equation (5.145), it is apparent that $\boldsymbol{T} = 0$ in a cold plasma or any other case where K does not depend on $\boldsymbol{k}$. In the thermal or streaming plasma, this term represents the power flow carried by the particles themselves, and is essential for the conservation of energy.

The final term in equation (5.142) is given by

$$\begin{aligned}\boldsymbol{E}^* \cdot \delta\mathsf{M} \cdot \boldsymbol{E} &= \boldsymbol{E}^* \cdot \frac{\omega}{4}\delta\mathsf{K}_h \cdot \boldsymbol{E} \\ &= \boldsymbol{E}^* \cdot \frac{\omega\epsilon_0}{8}[\delta\mathsf{K}_h - (\delta\mathsf{K}_h)^\dagger] \cdot \boldsymbol{E} + \boldsymbol{E}^* \cdot \frac{\omega\epsilon_0}{8}[\delta\mathsf{K}_h + (\delta\mathsf{K}_h)^\dagger] \cdot \boldsymbol{E}\end{aligned}$$

where we have broken the right hand side into antihermitian and hermitian components. Referring back to equation (5.132), we have

$$\boldsymbol{E}^* \cdot \delta\mathsf{M} \cdot \boldsymbol{E} = \frac{\mathrm{i}}{2}\delta\left(\frac{\partial W}{\partial t}\right)_{\text{loss}} + \text{hermitian terms,} \tag{5.146}$$

and for this term the changes are due only to variations in the plasma parameters. Since the hermitian terms represent reactive or oscillating terms, and not energy flow, we will neglect them and write equation (5.142) as

$$W_0\delta\omega - (\boldsymbol{P} + \boldsymbol{T}) \cdot \delta\boldsymbol{k} + \frac{\mathrm{i}}{2}\delta\left(\frac{\partial W}{\partial t}\right)_{\text{loss}} = 0\,. \tag{5.147}$$

For the discussion of group velocity, we are not concerned with losses, provided they are small, in which case the last term in equation (5.147) may be ignored and the result written as

$$\boldsymbol{v}_g = \frac{\delta\omega}{\delta\boldsymbol{k}} = \frac{\boldsymbol{P} + \boldsymbol{T}}{W_0} = \frac{\text{energy flux}}{\text{energy density}}\,. \tag{5.148}$$

If, on the other hand, one wishes to consider losses, then another relationship can be derived from equation (5.147) for the temporal decay of the wave ($\delta\boldsymbol{k} = 0$) where

$$\delta\omega = -\frac{\mathrm{i}}{2W_0}\delta\left(\frac{\partial W}{\partial t}\right)_{\text{loss}} \tag{5.149}$$

and for the spatial decay, we have the corresponding relationship ($\delta\omega = 0$)

$$(\boldsymbol{P} + \boldsymbol{T}) \cdot \delta\boldsymbol{k} = \frac{\mathrm{i}}{2}\delta\left(\frac{\partial W}{\partial t}\right)_{\text{loss}}\,. \tag{5.150}$$

We find then from equation (5.147) the three basic components involved in energy conservation, namely the transport of energy in the direction of the group velocity, given in equation (5.148), and the temporal and spatial decay of wave energy through dissipation, indicated by equation (5.149) and equation (5.150). Together, these give a good picture of the transport of energy, and the expressions for W_0, $\boldsymbol{P}$, and $\boldsymbol{T}$ give the balance between electromagnetic stored energy and power flow and the kinetic components of each.

5.5 Collisional effects

The effects of collisions in a magnetized plasma are more difficult to analyze than in an unmagnetized plasma. The most common collisional models are either simplistic or intractable. We shall note two cases.

5.5.1 Collisions via the Krook model

In Section 1.7.3, it was noted that for charge-neutral collisions in a partially ionized plasma, the Krook model was useful because the collisions were often large angle events so the drag and diffusion terms of the Fokker-Planck model were less important. This model is simple enough that the dielectric tensor modifications are relatively simple, such that the arguments of the Plasma Dispersion function are changed so that

$$\zeta_{ne} = \frac{\omega + n\omega_{ce}}{k_\parallel v_{te}} \rightarrow \frac{\omega + \mathrm{i}\nu_e + n\omega_{ce}}{k_\parallel v_{te}}$$
$$\zeta_{ni} = \frac{\omega + n\omega_{ci}}{k_\parallel v_{ti}} \rightarrow \frac{\omega + \mathrm{i}\nu_i + n\omega_{ci}}{k_\parallel v_{ti}} .$$

It may appear that the change is simply $\omega \rightarrow \omega + \mathrm{i}\nu$, but this is not the case since any ω coming from the Maxwell equations is *not* changed. While this may be adequate for collisions between charged particles and neutrals, it ignores both the drag and diffusion effects. Because it is so simple to include, it is often used as a first approximation.

5.5.2 Collisions via a Fokker-Planck model

A more appropriate model for collisions between charged particles includes dynamic friction and diffusion through a slight modification of equation (4.46) that was used for an unmagnetized plasma. We may write our kinetic equation as a modification of equation (5.6)

$$\left.\frac{\mathrm{d}f}{\mathrm{d}t}\right|_{\mathbf{R}} = \nu\nabla_v \cdot \left[(\boldsymbol{v} - \boldsymbol{u})f + \tfrac{1}{2}v_t^2\nabla_v f\right] - \frac{q}{m}(\boldsymbol{E}_1 + \boldsymbol{v} \times \boldsymbol{B}_1) \cdot \nabla_v f ,$$

where we have replaced the term in curly brackets (which vanished when we used the collisionless Boltzmann equation) with the the collision operator that includes both drag and diffusion. Even this term is probably too simplistic because the diffusion is likely to be different parallel and perpendicular to the magnetic field which would require a tensor coefficient for the diffusion term. This form of the collision operator has been used by Dougherty[36] to assess these effects, but the various integrals that result require numerical evaluation and no numerical results have been included.

Dougherty examines some limits, especially when $k_\parallel = 0$ where waves are undamped with many cyclotron harmonic resonances without collisions. The usual limit for weak collisional damping is $0 < \nu_j \ll \omega_{cj}$, but for this perpendicular propagation case, the more pertinent condition is $k_\perp^2 v_{tj}^2 \nu_j / \omega_{cj}^3 \ll 1$ or

$$\nu_j \ll \frac{\omega_{cj}^3}{k_\perp^2 v_{tj}^2}$$

which means that if

$$\left(\frac{k_\perp v_{tj}}{\omega_{cj}}\right)^2 \gg 1\,,$$

even with $\nu_j \ll \omega_{cj}$ the resonances may be damped out.

5.6 Relativistic plasma effects

While some plasmas are relatively hot, in that thermal effects beyond those included through the pressure term are important, few plasmas are truly relativistic such that the mean thermal speed begins to approach the speed of light. On the other hand, some relativistic effects occur at relatively low temperatures when one encounters cyclotron fundamental and harmonic resonances and $k_\parallel \to 0$. In fact, the typical argument of the Plasma Dispersion function, $\zeta_n = (\omega - n\omega_c)/k_\parallel v_\parallel$, is indeterminate as both the numerator and denominator may approach zero at cyclotron harmonics with $k_\parallel = 0$. Since the behavior of the Plasma Dispersion function is dramatically different as $\zeta_n \to 0$ or $\zeta_n \to \infty$, we need to consider the physics more carefully, since the mathematics suggests some discontinuous behavior in this region of parameter space, and we suspect there is no such discontinuity in the physical world.

The fundamental weakness in our model is that we have taken the cyclotron frequency to be a simple constant, independent of velocity, whereas from special relativity, we know that this is not so, since we should write $\omega_c = qB_0/\gamma m$. In the integral over velocity, then, the singularity in the denominator is significantly changed, and the numerator of ζ_n effectively never vanishes. Thus the indeterminate nature of the appropriate limit to take as $k_\parallel \to 0$ is resolved, since it is possible to impose $k_\parallel = 0$ externally, but the cyclotron harmonic resonances are broadened to some finite (though frequently small) extent and prevent the argument of the Plasma Dispersion function from blowing up.

It would seem from the above discussion that since hot plasma theory has already indicated that $k_\parallel = 0$ implies no absorption, even at cyclotron harmonics, and the relativistic corrections appear to lead to large arguments for the Plasma Dispersion function in this limit, that there is still no absorption. This, however, is not the case, and in fact the absorption at $k_\parallel = 0$

is considerably stronger than one might guess, since nontrivial absorption is encountered at the electron cyclotron fundamental and harmonics at temperatures of only a few keV. In order to see how this comes about, we will review the development of the relativistic dielectric tensor, including only electrons, and eventually pay particular attention to the $k_{\|} = 0$ case.

5.6.1 The relativistic dielectric tensor

In this section, we follow the development of Trubnikov[37] for the general development of the dielectric tensor, and present an outline of the derivation, noting the similarities to and differences from the hot plasma derivation.

We begin with the relativistic collisionless Boltzmann equation, and shall assume again that we may Fourier transform in both time and space, taking ω to have a small positive imaginary part when it becomes necessary to resolve the singularity in the momentum integrals and guarantee convergence. The kinetic equation is

$$\frac{\partial f}{\partial t} + \boldsymbol{v} \cdot \nabla f + q[\boldsymbol{E}_1 + \boldsymbol{v} \times (\boldsymbol{B}_0 + \boldsymbol{B}_1)] \cdot \nabla_p f = 0 \tag{5.151}$$

where the zero-order distribution is an equilibrium distribution given by

$$f_0 = A\mathrm{e}^{-\mathcal{E}/k_B T} \qquad \int f_0(p)\,\mathrm{d}^3p = n_0 \qquad \mathcal{E} = \sqrt{p^2c^2 + m^2c^4}\,. \tag{5.152}$$

We note that the Boltzmann equation is unchanged except that now the distribution function is a function of momentum rather than velocity. The distribution function is of the standard form, except that now the energy is the relativistic total energy (or relativistic kinetic energy by redefining the constant A).

At this point, we take the usual coordinate system, with $\boldsymbol{B}_0 = B_0\hat{e}_z$, and choose $k_\perp = k_x$ ($k_y = 0$) and use cylindrical coordinates in momentum space such that $p_x \equiv p_\perp \cos\phi$, $p_y \equiv p_\perp \sin\phi$. We then choose to write the first order distribution function in terms of another function, defined by

$$f_1(\boldsymbol{r}, \boldsymbol{p}, t) \equiv \mathrm{e}^{\mathrm{i}(\boldsymbol{k}\cdot\boldsymbol{r} - \omega t)} f_0(p)\Phi(\boldsymbol{p}) \tag{5.153}$$

so that with $\boldsymbol{p} = \gamma m \boldsymbol{v}$, the first order Boltzmann equation becomes

$$\begin{aligned} -\mathrm{i}\omega f_1 + \frac{\mathrm{i}(k_\| p_\| + k_\perp p_\perp \cos\phi)}{\gamma m} f_1 + q(\boldsymbol{E}_1 + \boldsymbol{v} \times \boldsymbol{B}_1) \cdot \nabla_p f_0 \\ + \frac{q}{\gamma m}(\boldsymbol{p} \times \boldsymbol{B}_0) \cdot \nabla_p f_1 = 0 \end{aligned} \tag{5.154}$$

since $(\boldsymbol{p} \times \boldsymbol{B}_0) \cdot \nabla_p f_0 = 0$ along the unperturbed orbit. Then we note that

$$\nabla_p f_0 = -\frac{f_0 c^2 \boldsymbol{p}}{k_B T \mathcal{E}} \qquad \text{so} \qquad (\boldsymbol{v} \times \boldsymbol{B}_1) \cdot \nabla_p f_0 = 0$$

and

$$\nabla_p f_1 = -\frac{c^2 \boldsymbol{p}}{k_B T \mathcal{E}} f_1 + \frac{f_1}{\Phi} \nabla_p \Phi$$

so

$$\frac{q}{\gamma m}(\boldsymbol{p} \times \boldsymbol{B}_0) \cdot \nabla_p f_1 = -\frac{\epsilon \omega_c}{\gamma} f_0 \mathrm{e}^{\mathrm{i}(\boldsymbol{k}\cdot\boldsymbol{r}-\omega t)} \frac{\partial \Phi}{\partial \phi}$$

(where $\epsilon = -1$ for electrons as before) and

$$q\boldsymbol{E}_1 \cdot \nabla_p f_0 = -\frac{f_0 q \boldsymbol{E} \cdot \boldsymbol{p}}{\gamma m k_B T}$$

since $\mathcal{E} = \gamma m c^2$ (and we drop the subscript on $\boldsymbol{E}$). Using these relations, equation (5.154) may be written as

$$\mathrm{i}\left(\frac{\gamma\omega}{\epsilon\omega_c} - \frac{k_\parallel p_\parallel + k_\perp p_\perp \cos\phi}{m\epsilon\omega_c}\right)\Phi + \frac{\partial \Phi}{\partial \phi} = -\frac{\boldsymbol{p}\cdot\boldsymbol{E}}{B_0 k_B T}\,. \tag{5.155}$$

This is a first order differential equation whose solution may be written as

$$\Phi(\boldsymbol{p}) = -\frac{1}{B_0 k_B T} \mathrm{e}^{\mathrm{i}(a\phi - b\sin\phi)} \int_\phi^{\phi_0} \mathrm{e}^{-\mathrm{i}(a\psi - b\sin\psi)} (\boldsymbol{p}\cdot\boldsymbol{E})_\psi \, \mathrm{d}\psi \tag{5.156}$$

with $a = \gamma\omega/\epsilon\omega_c - k_\parallel p_\parallel / m\epsilon\omega_c$, $b = k_\perp p_\perp / \epsilon m \omega_c$, and $(\boldsymbol{p}\cdot\boldsymbol{E})_\psi \equiv p_\parallel E_z + p_\perp(E_x \cos\psi + E_y \sin\psi)$. We may now let $\phi_0 \to -\infty$ and be guaranteed of convergence since we have assumed that ω has a positive imaginary part. Using the variable change $\psi = \phi - \xi$, this result may also be written as

$$\Phi(\boldsymbol{p}) = \frac{1}{B_0 k_B T} \int_0^\infty \mathrm{e}^{\mathrm{i}a\xi - \mathrm{i}b[\sin(\xi-\phi)+\sin\phi]} (\boldsymbol{p}\cdot\boldsymbol{E})_{\phi-\xi} \, \mathrm{d}\xi\,. \tag{5.157}$$

The current is then given by

$$\boldsymbol{J}(\boldsymbol{k},\omega) = \sum_j q_j \int \boldsymbol{v} f_{1j}(\boldsymbol{k},\omega)\, \mathrm{d}^3 p = \sum_j \frac{q_j}{m_j} \int \frac{f_0(p)\Phi_j(\boldsymbol{p})}{\gamma_j} \boldsymbol{p}\, \mathrm{d}^3 p\,. \tag{5.158}$$

We may now proceed in either of the two directions we discussed in Section 5.1.3, where we integrated over τ first, or in Section 5.1.4 where we did some of the velocity integrals first. In this case we integrate over ξ first and find that $\Phi(\boldsymbol{p})$ may be written as

$$\Phi(\boldsymbol{p}) = \frac{1}{B_0 k_B T}\left[p_\perp I_x(\phi) E_x + p_\perp I_y(\phi) E_y + p_\parallel I_z(\phi) E_z\right] \tag{5.159}$$

with

$$I_x(\phi) = \sum_{m,n=-\infty}^{\infty} \frac{\mathrm{i} J_m(b) n J_n(b)}{b(a-n)} \mathrm{e}^{\mathrm{i}(n-m)\phi} \tag{5.160}$$

$$I_y(\phi) = \sum_{m,n=-\infty}^{\infty} \frac{J_m(b) J_n'(b)}{(a-n)} \mathrm{e}^{\mathrm{i}(n-m)\phi} \tag{5.161}$$

$$I_z(\phi) = \sum_{m,n=-\infty}^{\infty} \frac{\mathrm{i} J_m(b) J_n(b)}{a-n} \mathrm{e}^{\mathrm{i}(n-m)\phi} \tag{5.162}$$

where $b = \bar{p}_\perp \nu_\perp$ with $\bar{p} \equiv p/m_j c$ so that $\gamma_j = (1+\bar{p}^2)^{1/2}$ and $\nu_\perp \equiv n_\perp \omega/\epsilon\omega_c$. We next integrate over ϕ, but because of the complexity of the several dielectric tensor components, we will calculate only K_{xx} and then list the results for the complete tensor. This component is given by

$$\begin{aligned}
K_{xx} &= 1 + \frac{\mathrm{i}\sigma_{xx}}{\omega\epsilon_0} \\
&= 1 + \sum_j \frac{\mathrm{i}q_j n_{0j}}{\epsilon_0 m_j \omega 4\pi m_j^2 c (k_B T_j)^2 K_2(\mu_j) B_0} \int_{-\infty}^{\infty} \mathrm{d}p_\parallel \int_0^\infty p_\perp \,\mathrm{d}p_\perp \int_0^{2\pi} \mathrm{d}\phi \\
&\quad \sum_{m,n=-\infty}^{\infty} \frac{\mathrm{e}^{-\mu_j\gamma_j}}{\gamma_j} p_\perp^2 \cos\phi \, \frac{\mathrm{i}J_m(b) n J_n(b)}{b(a-n)} \mathrm{e}^{\mathrm{i}(m-n)\phi} \\
&= 1 + \sum_j \frac{\omega_{pj}^2}{\omega\epsilon_j\omega_{cj}} \frac{\mu_j^2}{4\pi K_2(\mu_j)} \int_{-\infty}^{\infty} \mathrm{d}\bar{p}_\parallel \int_0^\infty \bar{p}_\perp \,\mathrm{d}\bar{p}_\perp \frac{\mathrm{e}^{-\mu_j\gamma_j}}{\gamma_j} \\
&\quad \sum_{n=-\infty}^{\infty} \frac{\bar{p}_\perp^2 n J_n(b)}{b(n-a)} \sum_{m=-\infty}^{\infty} J_m(b) \int_0^{2\pi} \mathrm{e}^{\mathrm{i}(m-n)\phi} \cos\phi \,\mathrm{d}\phi \\
&= 1 + \sum_j \frac{\omega_{pj}^2}{\omega\epsilon_j\omega_{cj}} \frac{\mu_j^2}{2K_2(\mu_j)} \int_{-\infty}^{\infty} \mathrm{d}\bar{p}_\parallel \int_0^\infty \bar{p}_\perp \,\mathrm{d}\bar{p}_\perp \frac{\mathrm{e}^{-\mu_j\gamma_j}}{\gamma_j} \\
&\quad \sum_{n=-\infty}^{\infty} \left(\frac{\bar{p}_\perp n J_n(b)}{b}\right)^2 \frac{1}{n-a}
\end{aligned} \tag{5.163}$$

where $\mu_j = m_j c^2/k_B T_j$ and we have used

$$\begin{aligned}
\sum_{m=-\infty}^{\infty} J_m(b) \int_0^{2\pi} \mathrm{e}^{\mathrm{i}(m-n)\phi} \cos\phi \,\mathrm{d}\phi &= \pi[J_{n-1}(b) + J_{n+1}(b)] \\
&= \frac{2\pi n J_n(b)}{b} .
\end{aligned} \tag{5.164}$$

The remaining components may be obtained with the integrals

$$\sum_{m=-\infty}^{\infty} J_m(b) \int_0^{2\pi} \mathrm{e}^{\mathrm{i}(m-n)\phi} \sin\phi \,\mathrm{d}\phi = -\mathrm{i}\pi[J_{n-1}(b) - J_{n+1}(b)] = -2\pi\mathrm{i}J_n'(b) \tag{5.165}$$

$$\sum_{m=-\infty}^{\infty} J_m(b) \int_0^{2\pi} \mathrm{e}^{\mathrm{i}(m-n)\phi} \,\mathrm{d}\phi = 2\pi J_n(b) \tag{5.166}$$

so that including only the electron component of the dielectric tensor (only electrons are assumed to be relativistic), we may write

$$\begin{aligned}
K_{ij} = \delta_{ij} - \frac{\omega_{pe}^2}{\omega^2} \frac{\mu_e^2}{2K_2(\mu_e)} \int_{-\infty}^{\infty} \mathrm{d}\bar{p}_\parallel \int_0^\infty \mathrm{d}\bar{p}_\perp \bar{p}_\perp \frac{\mathrm{e}^{-\mu_e\gamma_e}}{\gamma_e} \\
\sum_{n=-\infty}^{\infty} \frac{P_{ij}^n}{\gamma_e - n_\parallel \bar{p}_\parallel + n\omega_{ce}/\omega}
\end{aligned} \tag{5.167}$$

where

$$P^n_{xx} = \frac{n^2}{\nu_\perp^2} J_n^2(\nu_\perp \bar{p}_\perp) \tag{5.168}$$

$$P^n_{xy} = -P^n_{yx} = \frac{\mathrm{i}\bar{p}_\perp n}{\nu_\perp} J_n(\nu_\perp \bar{p}_\perp) J_n'(\nu_\perp \bar{p}_\perp) \tag{5.169}$$

$$P^n_{xz} = P^n_{zx} = -\frac{\bar{p}_\parallel n}{\nu_\perp} J_n^2(\nu_\perp \bar{p}_\perp) \tag{5.170}$$

$$P^n_{yy} = \bar{p}_\perp^2 {J_n'}^2(\nu_\perp \bar{p}_\perp) \tag{5.171}$$

$$P^n_{yz} = -P^n_{zy} = \mathrm{i}\bar{p}_\parallel \bar{p}_\perp J_n(\nu_\perp \bar{p}_\perp) J_n'(\nu_\perp \bar{p}_\perp) \tag{5.172}$$

$$P^n_{zz} = \bar{p}_\parallel^2 J_n^2(\nu_\perp \bar{p}_\perp) \,. \tag{5.173}$$

This representation is equivalent to that of Brambilla[38], except for notation such that his $\Omega_c = \epsilon\omega_c = -\omega_{ce}$ for electrons. If we change the definition of $\nu_\perp$ to be positive so that $\nu_\perp = n_\perp \omega/\omega_{ce}$, then we must change the signs of P^n_{xz} and P^n_{yz} in equations (5.170) and (5.172), respectively.

Problem 5.19 *Normalization constant.* Find the normalization constant A for $f_0(p)$.
Ans. $A = n_0\mu/4\pi(mc)^3 K_2(\mu)$.

Problem 5.20 *Calculating* $\Phi(\phi)$. Integrate equation (5.157) over ξ using the Bessel identity (5.23) and show that the result is given by equation (5.159) along with equations (5.160) through (5.162).

5.6.2 The relativistic dielectric tensor without sums

We may use the Newberger sum rules from equations (5.36) and (5.37) – (5.42) to eliminate the sums and cast the sums of the P^n_{ij} into the form

$$\sum_{n=-\infty}^{\infty} \frac{P^n_{xx}}{a+n} = \frac{1}{\nu_\perp^2}\left[\frac{\pi a^2}{\sin \pi a} J_a(\nu_\perp \bar{p}_\perp) J_{-a}(\nu_\perp \bar{p}_\perp) - a\right] \tag{5.174}$$

$$\sum_{n=-\infty}^{\infty} \frac{P^n_{xy}}{a+n} = -\frac{\mathrm{i}\bar{p}_\perp}{\nu_\perp}\left[\frac{\pi a}{\sin \pi a} J_a(\nu_\perp \bar{p}_\perp) J_{-a}'(\nu_\perp \bar{p}_\perp) + \frac{a}{\nu_\perp \bar{p}_\perp}\right] \tag{5.175}$$

$$\sum_{n=-\infty}^{\infty} \frac{P^n_{xz}}{a+n} = \frac{\bar{p}_\parallel}{\nu_\perp}\left[\frac{\pi a}{\sin \pi a} J_a(\nu_\perp \bar{p}_\perp) J_{-a}(\nu_\perp \bar{p}_\perp) - 1\right] \tag{5.176}$$

$$\sum_{n=-\infty}^{\infty} \frac{P^n_{yy}}{a+n} = \bar{p}_\perp^2\left[\frac{\pi}{\sin \pi a} J_a'(\nu_\perp \bar{p}_\perp) J_{-a}'(\nu_\perp \bar{p}_\perp) + \frac{a}{\nu_\perp^2 \bar{p}_\perp^2}\right] \tag{5.177}$$

$$\sum_{n=-\infty}^{\infty} \frac{P^n_{yz}}{a+n} = \mathrm{i}\bar{p}_\parallel \bar{p}_\perp\left[\frac{\pi}{\sin \pi a} J_a(\nu_\perp \bar{p}_\perp) J_{-a}'(\nu_\perp \bar{p}_\perp) + \frac{1}{\nu_\perp \bar{p}_\perp}\right] \tag{5.178}$$

$$\sum_{n=-\infty}^{\infty} \frac{P^n_{zz}}{a+n} = \bar{p}_\parallel^2 \frac{\pi}{\sin \pi a} J_a(\nu_\perp \bar{p}_\perp) J_{-a}(\nu_\perp \bar{p}_\perp) \tag{5.179}$$

where now $a = (\omega/\omega_{ce})(\gamma_e - n_\parallel \bar{p}_\parallel)$ and $\nu_\perp = n_\perp \omega/\omega_{ce}$. The dielectric tensor then assumes the form

$$K_{ij} = \delta_{ij} - \frac{\omega_{pe}^2}{\omega^2}\frac{\mu_e^2}{2K_2(\mu_e)}\int_{-\infty}^{\infty} \mathrm{d}\bar{p}_\parallel \int_0^\infty \mathrm{d}\bar{p}_\perp\, \bar{p}_\perp \mathrm{e}^{-\mu_e\gamma_e}\Pi_{ij} + \text{ ion terms} \quad (5.180)$$

where

$$\Pi_{xx} = \frac{a}{n_\perp^2(a + n_\parallel \bar{p}_\parallel \omega/\omega_{ce})}\left[\frac{\pi a}{\sin \pi a}J_a(b)J_{-a}(b) - 1\right] \quad (5.181)$$

$$\Pi_{xy} = -\Pi_{yx} = -\frac{\mathrm{i}a}{n_\perp^2(a + n_\parallel \bar{p}_\parallel \omega/\omega_{ce})}\left[\frac{\pi b}{\sin \pi a}J_a(b)J'_{-a}(b) + 1\right] \quad (5.182)$$

$$\Pi_{xz} = \Pi_{zx} = -\frac{\omega \bar{p}_\parallel}{\omega_{ce} n_\perp(a + n_\parallel \bar{p}_\parallel \omega/\omega_{ce})}\left[\frac{\pi a}{\sin \pi a}J_a(b)J_{-a}(b) - 1\right] \quad (5.183)$$

$$\Pi_{yy} = \frac{1}{n_\perp^2(a + n_\parallel \bar{p}_\parallel \omega/\omega_{ce})}\left[\frac{\pi b^2}{\sin \pi a}J'_a(b)J'_{-a}(b) + a\right] \quad (5.184)$$

$$\Pi_{yz} = -\Pi_{zy} = -\frac{\mathrm{i}\omega \bar{p}_\parallel}{\omega_{ce} n_\perp(a + n_\parallel \bar{p}_\parallel \omega/\omega_{ce})}\left[\frac{\pi b}{\sin \pi a}J_a(b)J'_{-a}(b) + 1\right] \quad (5.185)$$

$$\Pi_{zz} = \frac{\omega^2 \bar{p}_\parallel^2}{\omega_{ce}^2(a + n_\parallel \bar{p}_\parallel \omega/\omega_{ce})}\frac{\pi}{\sin \pi a}J_a(b)J_{-a}(b) \quad (5.186)$$

and where now $b = \nu_\perp \bar{p}_\perp$.

Problem 5.21 *Off-diagonal tensor elements.* Pick one of the off-diagonal dielectric tensor elements, and verify it is given by equation (5.180) along with either (5.182), (5.183), or (5.185), starting from equation (5.158).

5.6.3 The weakly relativistic dielectric tensor

In this section, we return to equation (5.158) and integrate over the momenta first instead of the phase. We can obtain another expression via the identity,

$$I(s, \boldsymbol{r}) = \frac{1}{4\pi}\int \frac{\mathrm{d}p}{\sqrt{1+p^2}}\mathrm{e}^{-s\sqrt{1+p^2} - \mathrm{i}\boldsymbol{r}\cdot\boldsymbol{p}} = \frac{K_1(\sqrt{s^2+r^2})}{\sqrt{s^2+r^2}} \quad (5.187)$$

and its derivative where K_n is the modified Bessel function of the second kind of order n. The dielectric tensor is then given by

$$\mathsf{K} = \mathsf{I} + \sum_j \frac{\mathrm{i}\omega_{pj}^2}{\epsilon_j \omega\omega_{cj}}\frac{\mu^2}{K_2(\mu)}\int_0^\infty \mathrm{d}\xi \left[\frac{K_2(\sqrt{R})}{R}\mathsf{T}_1 - \frac{K_3(\sqrt{R})}{R^{3/2}}\mathsf{T}_2\right], \quad (5.188)$$

where

$$R = \left(\mu - \mathrm{i}\xi\frac{\omega}{\epsilon_j\omega_{cj}}\right)^2 + 2\nu_\perp^2(1-\cos\xi) + n_\parallel^2\xi^2,$$

where now $n_\| = n_\|\omega/\epsilon_j\omega_{cj}$, and

$$\mathsf{T}_1 = \begin{pmatrix} \cos\xi & -\sin\xi & 0 \\ \sin\xi & \cos\xi & 0 \\ 0 & 0 & 1 \end{pmatrix}, \tag{5.189}$$

and

$$\mathsf{T}_2 = \begin{pmatrix} \nu_\perp^2 \sin^2\xi & -\nu_\perp^2 \sin\xi(1-\cos\xi) & \nu_\perp\nu_\|\xi\sin\xi \\ \nu_\perp^2 \sin\xi(1-\cos\xi) & -\nu_\perp^2(1-\cos\xi)^2 & \nu_\perp\nu_\|\xi(1-\cos\xi) \\ \nu_\perp\nu_\|\xi\sin\xi & -\nu_\perp\nu_\|\xi(1-\cos\xi) & \nu_\|^2\xi^2 \end{pmatrix}. \tag{5.190}$$

Up to this point, the analysis is exact, but is valuable only for numerical integration of the tensor components. In the weakly relativistic limit, we follow the development of Shkarofsky[39] and take $\mu_e \gg 1$ (μ_i is generally so large that relativistic effects are negligible, so that the ion contributions will be ignored from this point on and we shall delete the subscript on μ). Since this quantity appears in the argument of the modified Bessel function, we can use the asymptotic limit so that $K_n(x) \simeq \sqrt{\pi/2x}\mathrm{e}^{-x}$. If we also take the limit of small $\lambda = \frac{1}{2}k_\perp^2\rho_L^2 = \nu_\perp^2/\mu$, then we can simplify the expression for R such that

$$R^{\frac{1}{2}} = \mu\left[\left(1+\frac{\mathrm{i}\xi\omega}{\mu\omega_{ce}}\right)^2 + \left(\frac{\nu_\|}{\mu}\right)^2\xi^2\right]^{\frac{1}{2}} + \Lambda(1-\cos\xi)$$

where

$$\Lambda \equiv \lambda\left[\left(1+\frac{\mathrm{i}\xi\omega}{\mu\omega_{ce}}\right)^2 + \left(\frac{\nu_\|}{\mu}\right)^2\xi^2\right]^{-\frac{1}{2}}.$$

Now we will need to keep the $\Lambda(1-\cos\xi)$ term in the exponential of the K_n terms, because the oscillating phase in the exponent is important, but it is safe to neglect it otherwise since $\mu \gg \Lambda$. This approximation leads to $k_\perp^2c^2/R^{1/2}\omega_{ce}^2 = \Lambda$ except in the exponent. For the oscillating exponential term, we will use equation (5.68) to write

$$\mathrm{e}^{\Lambda\cos\xi} = \sum_{n=-\infty}^{\infty} I_n(\Lambda)\mathrm{e}^{-\mathrm{i}n\xi} \tag{5.191}$$

$$\cos\xi\mathrm{e}^{\Lambda\cos\xi} = \sum_{n=-\infty}^{\infty} I_n'(\Lambda)\mathrm{e}^{-\mathrm{i}n\xi} \tag{5.192}$$

and other similar results as in equation (5.69). Then by changing variables to $t = -\xi\omega/\mu\omega_{ce}$ so that $\sqrt{R}$ becomes

$$\sqrt{R} = \mu[(1-\mathrm{i}t)^2 + n_\|^2t^2]^{\frac{1}{2}} + \Lambda(1-\cos\xi)$$

and $\exp(-in\xi) \to \exp(in\nu_n t)$ with $\nu_n = n\mu\omega_{ce}/\omega$, then the weakly relativistic dielectric tensor may be written as

$$\mathsf{K} = \mathsf{I} + \mathrm{i}\frac{\omega_p^2}{\omega^2}\mu \sum_{n=-\infty}^{\infty} \int_0^\infty \mathsf{T}_3 \frac{\mathrm{e}^{-\Lambda}\exp\{\mu - \mu[(1-\mathrm{i}t)^2 + n_\parallel^2 t^2]^{\frac{1}{2}} + \mathrm{i}\nu_n t\}}{[(1-\mathrm{i}t)^2 + n_\parallel^2 t^2]^{\frac{7}{4}}}\,\mathrm{d}t \tag{5.193}$$

where

$$\mathsf{T}_3 = [(1-\mathrm{i}t)^2 + n_\parallel^2 t^2]^{\frac{1}{2}}\mathsf{T}_3^{(a)} + \frac{k_\perp k_\parallel c^2}{\omega\omega_c}\mathsf{T}_3^{(b)}\frac{\partial}{\partial\nu_n} \tag{5.194}$$

with

$$\mathsf{T}_3^{(a)} = \begin{pmatrix} \frac{n^2 I_n}{\Lambda} & -\mathrm{i}n(I_n' - I_n) & 0 \\ \mathrm{i}n(I_n' - I_n) & \frac{n^2 I_n}{\Lambda} + 2\Lambda(I_n - I_n') & 0 \\ 0 & 0 & I_n\left(1 + k_\parallel\frac{\partial}{\partial k_\parallel}\right) \end{pmatrix} \tag{5.195}$$

$$\mathsf{T}_3^{(b)} = \begin{pmatrix} 0 & 0 & \frac{nI_n}{\Lambda} \\ 0 & 0 & \mathrm{i}(I_n' - I_n) \\ \frac{nI_n}{\Lambda} & -\mathrm{i}(I_n' - I_n) & 0 \end{pmatrix}. \tag{5.196}$$

For small Λ, we can write this in terms of the $\mathcal{F}_q$ function that is defined by

$$\mathcal{F}_q(\nu_n, n_\parallel) \equiv -\mathrm{i}\int_0^\infty \mathrm{d}t \frac{\exp\{\mu - \mu[(1-\mathrm{i}t)^2 + n_\parallel^2 t^2]^{\frac{1}{2}} + \mathrm{i}\nu_n t\}}{[(1-\mathrm{i}t)^2 + n_\parallel^2 t^2]^{\frac{q}{2}}} \tag{5.197}$$

although the most common definition is a further approximation in the smallness of $n_\parallel^2$ given by

$$\mathcal{F}_q(z_n, a) \equiv -\mathrm{i}\int_0^\infty \mathrm{d}t \frac{\exp[\mathrm{i}z_n t - at^2/(1-\mathrm{i}t)]}{(1-\mathrm{i}t)^q} \tag{5.198}$$

where $z_n = \mu + \nu_n = \mu(\omega + n\omega_{ce})/\omega$ and $a = \frac{1}{2}\mu n_\parallel^2$. This generalized weakly relativistic dispersion function is real for $z_n \geq a$ and complex for $z_n < a$. Some of its characteristics are shown in Figure 5.6 for $a = 1$ and in Figure 5.7 for $a = 5$. Its mathematical properties are given in Appendix A in Section A.2.2.

Then for the case with small Λ, assuming the same dispersion function for each power of λ, the dielectric tensor elements are given by

$$K_{xx} \simeq 1 - \frac{\omega_p^2}{\omega^2}\mu\mathrm{e}^{-\lambda}\sum_{n=1}^{\infty}\frac{n^2 I_n}{\lambda}\left[\mathcal{F}_{n+3/2}(z_n, a) + \mathcal{F}_{n+3/2}(z_{-n}, a)\right] \tag{5.199}$$

$$K_{xy} \simeq \mathrm{i}\frac{\omega_p^2}{\omega^2}\mu\mathrm{e}^{-\lambda}\sum_{n=1}^{\infty} n(I_n - I_n')\left[\mathcal{F}_{n+3/2}(z_n, a) - \mathcal{F}_{n+3/2}(z_{-n}, a)\right] \tag{5.200}$$

$$K_{zz} \simeq 1 - \frac{\omega_p^2}{\omega^2}\mu\mathrm{e}^{-\lambda}\sum_{n=-\infty}^{\infty} I_n\left\{(1-4a)\mathcal{F}_{n+5/2}(z_n, a)\right.$$

$$+2a[\mathcal{F}_{n+3/2}(z_n,a)+\mathcal{F}_{n+7/2}(z_n,a)]\} \tag{5.201}$$

$$K_{yy} \simeq K_{xx} + \frac{\omega_p^2}{\omega^2}\mu e^{-\lambda}2\lambda\left\{(I_0-I_1)\mathcal{F}_{5/2}(\mu,a)\right.$$

$$\left.+\sum_{n=1}^{\infty}(I_n-I_n')\left[\mathcal{F}_{n+5/2}(z_n,a)+\mathcal{F}_{n+5/2}(z_{-n},a)\right]\right\} \tag{5.202}$$

$$K_{xz} \simeq -\frac{\omega_p^2}{\omega\omega_c}\mu e^{-\lambda}n_\parallel\nu_\perp\sum_{n=1}^{\infty}\frac{nI_n}{\lambda}\left[\mathcal{F}_{n+5/2}(z_n,a)-\mathcal{F}_{n+3/2}(z_n,a)\right.$$

$$\left.-\mathcal{F}_{n+5/2}(z_{-n},a)+\mathcal{F}_{n+3/2}(z_{-n},a)\right] \tag{5.203}$$

$$K_{yz} \simeq -i\frac{\omega_p^2}{\omega\omega_c}\mu n_\parallel\nu_\perp e^{-\lambda}\left\{(I_1-I_0)\mathcal{F}_{7/2}(\mu,a)\right.$$

$$\left.+\sum_{n=1}^{\infty}(I_n'-I_n)\left[\mathcal{F}_{n+5/2}(z_n,a)+\mathcal{F}_{n+5/2}(z_{-n},a)\right]\right\} \tag{5.204}$$

where $K_{yx}=-K_{xy}$, $K_{zx}=K_{xz}$, $K_{zy}=-K_{yz}$, and $z_\pm=\mu(\omega\pm n\omega_{ce})/\omega$ and now the argument of I_n is λ (which is independent of t) instead of Λ (which is a function of t).

5.6.4 Moderately relativistic expressions

When one keeps higher order terms in the Bessel function expansions, higher order dispersion functions should be used for those terms since Λ appeared *inside* the integral of equation (5.193). In fact, if we define the function of λ associated with K_{xx} as $f_{xx}(n,\lambda)$ such that

$$f_{xx}(n,\lambda)=\frac{n^2e^{-\lambda}I_n(\lambda)}{\lambda}=\sum_{k=0}^{\infty}a_{xx,n}^{(k)}\lambda^k\,, \tag{5.205}$$

then we could write a more nearly precise expression for K_{xx} as

$$K_{xx}=1-\frac{\omega_p^2}{\omega^2}\mu\sum_{n=1}^{\infty}\sum_{k=0}^{\infty}a_{xx,n}^{(k)}\lambda^k\left[\mathcal{F}_{k+5/2}(z_n,a)+\mathcal{F}_{k+5/2}(z_{-n},a)\right]. \tag{5.206}$$

Although this is a doubly infinite sum, usually only an $n=1$ and perhaps one other n needs to be included, and $k=0$ through $k=3$ will usually suffice unless λ is approaching unity, in which case an exact treatment is needed. Extending this same type of expansion to the other dielectric tensor terms leads to the *Moderately Relativistic* approximation.

As an example of the moderately relativistic expressions, the expression for K_{xx} through order λ^3 is given by

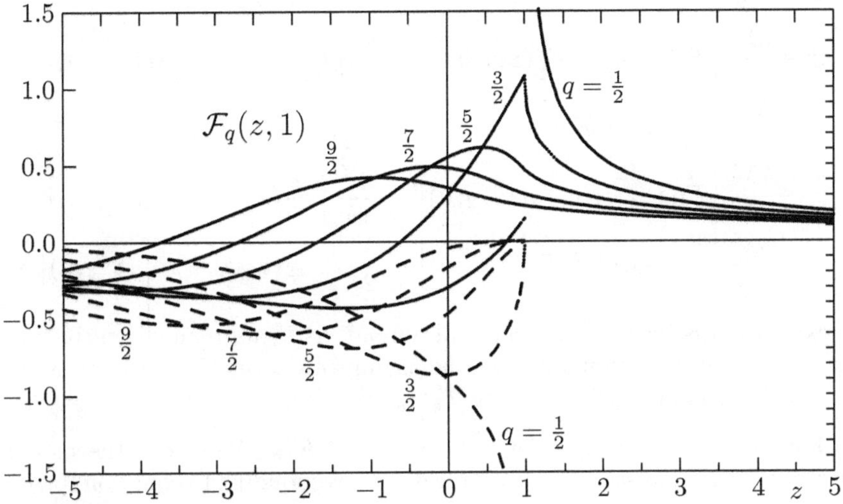

FIGURE 5.6
Generalized weakly relativistic dispersion function, $\mathcal{F}_q(z,a)$ for half-integral q and $a = 1$, showing both real (solid) and imaginary (dashed) parts.

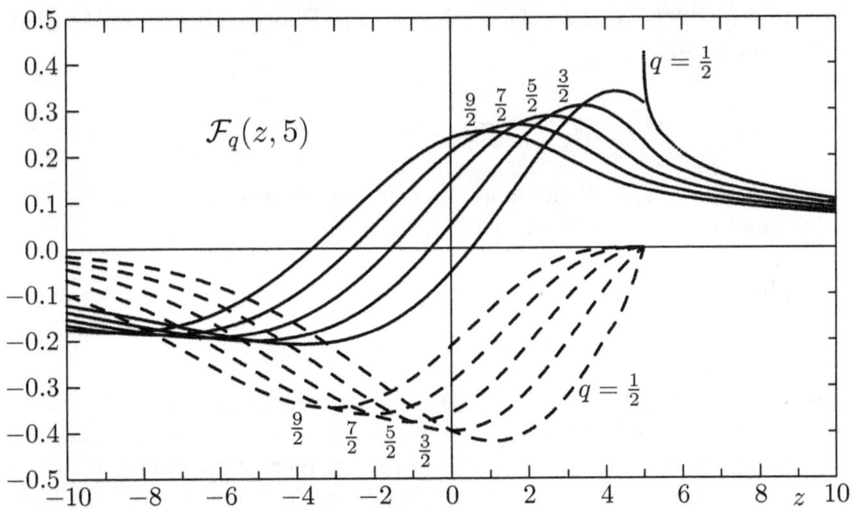

FIGURE 5.7
Generalized weakly relativistic dispersion function, $\mathcal{F}_q(z,a)$ for half-integral q and $a = 5$, showing both real (solid) and imaginary (dashed) parts.

$$\begin{aligned}
K_{xx} = 1 - \frac{\omega_p^2}{\omega^2}\mu \Bigg\{ & \frac{1}{2}\left[\mathcal{F}_{\frac{5}{2}}(z_1,a) + \mathcal{F}_{\frac{5}{2}}(z_{-1},a)\right] - \frac{\lambda}{2}\left[\mathcal{F}_{\frac{7}{2}}(z_1,a) + \mathcal{F}_{\frac{7}{2}}(z_{-1},a)\right] \\
& + \frac{5\lambda^2}{16}\left[\mathcal{F}_{\frac{9}{2}}(z_1,a) + \mathcal{F}_{\frac{9}{2}}(z_{-1},a)\right] - \frac{7\lambda^3}{48}\left[\mathcal{F}_{\frac{11}{2}}(z_1,a) + \mathcal{F}_{\frac{11}{2}}(z_{-1},a)\right] \\
& + \frac{\lambda}{2}\left[\mathcal{F}_{\frac{7}{2}}(z_2,a) + \mathcal{F}_{\frac{7}{2}}(z_{-2},a)\right] - \frac{\lambda^2}{2}\left[\mathcal{F}_{\frac{9}{2}}(z_2,a) + \mathcal{F}_{\frac{9}{2}}(z_{-2},a)\right] \qquad (5.207) \\
& + \frac{7\lambda^3}{24}\left[\mathcal{F}_{\frac{11}{2}}(z_2,a) + \mathcal{F}_{\frac{11}{2}}(z_{-2},a)\right] + \frac{3\lambda^2}{16}\left[\mathcal{F}_{\frac{9}{2}}(z_3,a) + \mathcal{F}_{\frac{9}{2}}(z_{-3},a)\right] \\
& - \frac{3\lambda^3}{16}\left[\mathcal{F}_{\frac{11}{2}}(z_3,a) + \mathcal{F}_{\frac{11}{2}}(z_{-3},a)\right] + \frac{\lambda^3}{24}\left[\mathcal{F}_{\frac{11}{2}}(z_4,a) + \mathcal{F}_{\frac{11}{2}}(z_{-4},a)\right]\Bigg\}.
\end{aligned}$$

This expression is better than the weakly relativistic approximation, but not exact. This approximation still uses the asymptotic form of $K_2(\mu)$ and is still based on the smallness of $n_\|^2$ for the $\mathcal{F}_q(z_n,a)$.

Problem 5.22 *Moderately relativistic K_{xy} and K_{yy}.* Work out the moderately relativistic expression for K_{xy} and K_{yy} corresponding to equation (5.207) (through order λ^3).

A further simplification we examine for both the weakly relativistic and the moderately relativistic approximations is the $n_\| \to 0$ limit (or the $a \to 0$ limit), where

$$\mathcal{F}_q(z,0) = F_q(z) \equiv -\mathrm{i}\int_0^\infty \frac{\mathrm{e}^{\mathrm{i}zt}}{(1-\mathrm{i}t)^q}\,\mathrm{d}t \qquad (5.208)$$

and for $q = \frac{1}{2}$, we have the relationship to the Plasma Dispersion function (see equation (A.31))

$$\mathrm{i}F_{\frac{1}{2}} = \int_0^\infty \frac{\mathrm{e}^{\mathrm{i}zt}}{(1-\mathrm{i}t)^{\frac{1}{2}}}\,\mathrm{d}t = \frac{1}{\sqrt{z}}Z(\mathrm{i}\sqrt{z})\,. \qquad (5.209)$$

This weakly relativistic dispersion function, often referred to as the Dnestrovskii function, is illustrated in Figure 5.8. Other properties are listed in Section A.2.1.2.

The higher order functions may be obtained from the recursion formula,

$$(q-1)F_q(z) = 1 - zF_{q-1}(z) \qquad (5.210)$$

and the analytic continuation for $\mathrm{Im}(\omega) < 0$ is given by the properties of the Plasma Dispersion function. Also, in this limit, $K_{xz} = K_{zx} = K_{yz} = K_{zy} = 0$, so the leading terms above are the only terms, not merely the dominant terms.

5.6.5 Exact expressions with $n_\| = 0$

It is possible to obtain the exact dielectric tensor expressions in terms of a few relatively simple integrals when $n_\| = 0$. For this analysis, we return to the

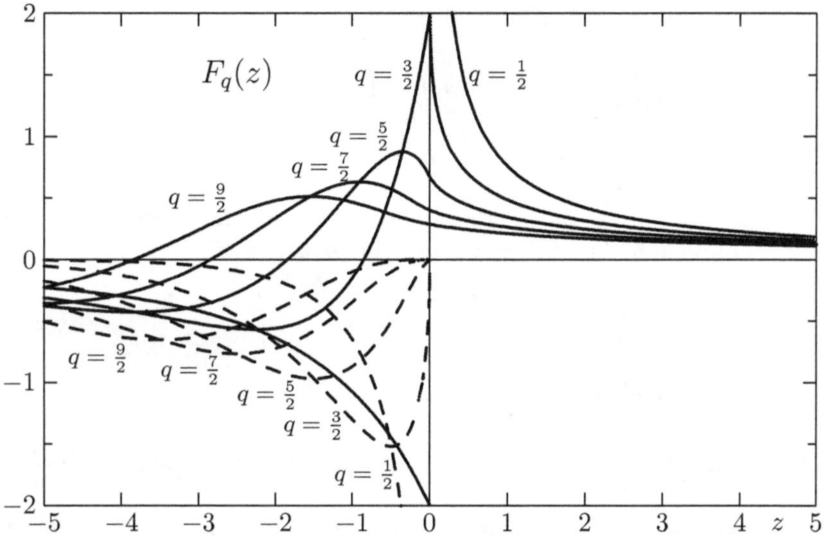

FIGURE 5.8
Weakly relativistic dispersion function, $F_q(z)$ for half-integral q, showing both real (solid) and imaginary (dashed) parts.

dielectric tensor elements without sums given in Section 5.6.2. Although the dielectric tensor terms are complicated, since the order of the Bessel functions is not integral in the general case, by changing into polar coordinates in momentum space, the integrals over the angle can be done analytically, leaving us with either a single integral for each n or even with a single integral for the sum over n. The derivation for K_{xx} will be worked out in detail, while the others will be left as an exercise.

With $n_\parallel = 0$, we may write an exact expression for K_{xx} as

$$K_{xx} = 1 - \frac{\omega_p^2}{\omega^2}\sum_{n=1}^{\infty} F_{xx}^{(n)}$$

where each term in the sum is given by

$$F_{xx}^{(n)} = \frac{2\mu^2}{K_2(\mu)}\frac{n^2}{\nu_\perp^2}\int_0^\pi \mathrm{d}\theta\,\sin\theta\int_0^\infty \mathrm{d}\bar{p}\,\bar{p}^2\mathrm{e}^{-\mu\gamma}\frac{J_n^2(\nu_\perp\bar{p}\sin\theta)}{\gamma^2 - n^2\omega_c^2/\omega^2}\,. \tag{5.211}$$

After integrating over the angle using equation (A.58), equation (5.211) may be expressed as

$$F_{xx}^{(n)} = \frac{\mu^2}{K_2(\mu)}\frac{2n^2}{\nu_\perp^2(2n+1)!}\int_0^\infty \mathrm{d}\bar{p}\,\frac{\bar{p}^2\mathrm{e}^{-\mu\gamma}}{\bar{p}^2 - p_n^2}b^{2n}{}_1F_2\left(n+\tfrac{1}{2}; n+\tfrac{3}{2}, 2n+1; -b^2\right) \tag{5.212}$$

where $\gamma(\bar{p}) = \sqrt{1+\bar{p}^2}$, $p_n^2 = n^2\omega_c^2/\omega^2 - 1$, $b^2 = \mu\lambda\bar{p}^2$, since $\nu_\perp = \nu_\perp\omega/\omega_c = \sqrt{\mu\lambda}$. Alternatively, using the Newberger sum rule, we have a single expression for the sum as

$$F_{xx} = \frac{\mu^2}{2K_2(\mu)} \int_0^{\pi} \mathrm{d}\theta \, \sin\theta \int_0^{\infty} \mathrm{d}\bar{p}\,\bar{p}^2 \frac{\mathrm{e}^{-\mu\gamma}}{\gamma} \frac{\omega a}{\omega_c \nu_\perp^2} \times \left[\frac{\pi a}{\sin \pi a} J_a(\nu_\perp \bar{p} \sin\theta) J_{-a}(\nu_\perp \bar{p} \sin\theta) - 1 \right] \tag{5.213}$$

where $a = \omega\gamma/\omega_c$. Integrating this expression over the angle (see equation (A.59) in appendix A), the result may be expressed as

$$K_{xx} = 1 - \frac{\omega_p^2}{\omega^2} \frac{\mu}{\nu_\perp^2} [I_1(z, \lambda, \mu) - 1] \tag{5.214}$$

where I_1 is the integral (*not* a Bessel function)

$$I_1(z, \lambda, \mu) = \frac{\mu}{K_2(\mu)} \int_0^{\infty} \mathrm{d}p\, p^2 \mathrm{e}^{-\mu\gamma} {}_2F_3(\tfrac{1}{2}, 1; \tfrac{3}{2}, 1-a, 1+a; -\lambda\mu p^2) \,. \tag{5.215}$$

The dependence on $z = z_{-n} \equiv \mu(1 - n\omega_c/\omega)$ is through $a(z) = n\gamma/(1 - z/\mu)$. The integrands utilize the hypergeometric functions that are defined by equations (A.56) and (A.57).

Problem 5.23 *Remaining exact tensor elements.* Show that the remaining dielectric tensor elements with $n_\parallel = 0$ may be written as

$$K_{xy} = \mathrm{i} \frac{\omega_p^2}{\omega^2} \frac{\mu}{2n_\perp^2} [I_2(z, \lambda, \mu) - I_3(z, \lambda, \mu) + 1] \tag{5.216}$$

$$K_{zz} = 1 - \frac{\omega_p^2}{\omega^2} \mu I_4(z, \lambda, \mu) \tag{5.217}$$

$$K_{yy} = K_{xx} - \frac{\omega_p^2}{\omega^2} \frac{2\mu}{n_\perp^2} [1 + I_5(z, \lambda, \mu) - I_2(z, \lambda, \mu)] \tag{5.218}$$

where

$$I_2(z, \lambda, \mu) = \frac{\mu}{K_2(\mu)} \int_0^{\infty} \mathrm{d}p\, p^2 \mathrm{e}^{-\mu\gamma} {}_2F_3(\tfrac{1}{2}, 1; \tfrac{3}{2}, 1-a, a; -\lambda\mu p^2) \tag{5.219}$$

$$I_3(z, \lambda, \mu) = \frac{\mu}{K_2(\mu)} \int_0^{\infty} \mathrm{d}p\, p^2 \mathrm{e}^{-\mu\gamma} {}_2F_3(\tfrac{1}{2}, 1; \tfrac{3}{2}, -a, 1+a; -\lambda\mu p^2) \tag{5.220}$$

$$I_4(z, \lambda, \mu) = \frac{\mu}{K_2(\mu)} \int_0^{\infty} \mathrm{d}p\, \frac{p^4 \mathrm{e}^{-\mu\gamma}}{3(1+p^2)} {}_2F_3(\tfrac{1}{2}, 1; \tfrac{5}{2}, 1-a, 1+a; -\lambda\mu p^2) \tag{5.221}$$

$$I_5(z, \lambda, \mu) = \frac{\lambda\mu^2}{K_2(\mu)} \int_0^{\infty} \mathrm{d}p\, \frac{p^4 \mathrm{e}^{-\mu\gamma}}{3a(a-1)} {}_2F_3(\tfrac{1}{2}, 2; \tfrac{5}{2}, 2-a, a; -\lambda\mu p^2) \,. \tag{5.222}$$

Problem 5.24 *Hypergeometric function identity.* Prove that $I_6 = I_2 - 1$ where

$$I_6(z, \lambda, \mu) = \frac{\lambda\mu^2}{K_2(\mu)} \int_0^{\infty} \mathrm{d}p\, \frac{p^4 \mathrm{e}^{-\mu\gamma}}{3a(a-1)} {}_2F_3(\tfrac{3}{2}, 1; \tfrac{5}{2}, 2-a, 1+a; -\lambda\mu p^2) \,.$$

The imaginary parts of the tensor elements may be obtained rather directly by examining the poles of the integrands. In this section, we simplify the notation by dropping the electron subscript, and note that the pole in equation (5.212) is at $\bar{p} = \bar{p}_n$. Evaluating the integral at this pole yields

$$\mathrm{Im}[F_{xx}^{(n)}] = -\frac{\mu^2}{K_2(\mu)}\frac{\pi n^2 \mathrm{e}^{-\mu n\omega_c/\omega} b_n^{2n+1}}{(2n+1)!(\mu\lambda)^{3/2}}\,{}_1F_2\left(n+\tfrac{1}{2}; n+\tfrac{3}{2}, 2n+1; -b_n^2\right) \tag{5.223}$$

where $b_n^2 = \mu\lambda(n^2\omega_c^2/\omega^2 - 1)$. This expression is exact, but in order to compare it with the weakly relativistic approximation, we take $e^{\mu}K_2(\mu) \sim \sqrt{\pi/2\mu}$, so that the exponential term becomes e^{z_n}. If we write the corresponding weakly relativistic expression as $W_{xx}^{(n)}$, and examine the imaginary part, we find, using equation (A.36),

$$\mathrm{Im}[W_{xx}^{(n)}] = -\frac{\mu n^2 \mathrm{e}^{-\lambda} I_n(\lambda)}{\lambda}\frac{\pi(-z)^{q-1}\mathrm{e}^z}{\Gamma(q)}$$

where $q = n + \frac{1}{2}$, then expanding both the Bessel function terms and the hypergeometric function, the ratio of the exact to the weakly relativistic expression may be written as

$$\frac{\mathrm{Im}[F_{xx}^{(n)}]}{\mathrm{Im}[W_{xx}^{(n)}]} = \frac{1 + \lambda z/q + \cdots}{1 - \lambda + q\lambda^2/(2q-1) + \cdots}. \tag{5.224}$$

From this ratio, we observe first that the exact expression is a function of the product, λz, not a function of λ times a function of z as the weakly relativistic expression indicates. Secondly, we see that the higher order terms in the Bessel function expansion do not improve the accuracy, so this justifies the truncation of the Bessel functions to the lowest order terms in equations (5.199) through (5.204). There is another interesting feature observable in comparing these imaginary parts in Figure 5.9 for two different values of λ. After factoring out some leading factors, it is apparent that the peak of the exact expressions always exceeds the peak of the weak expression, but for large negative z, the weak expression exceeds the exact expression. In fact, if one approximates $b_n^2 = -\lambda z(2 - z/\mu) \simeq -2\lambda z$, then the integral

$$\int_0^\infty \mathrm{d}x\,\mathrm{e}^{-x}x^{1/2}\int_0^\pi \mathrm{d}\theta\, J_\nu^2(\sqrt{2\lambda x}\sin\theta)\sin\theta = \sqrt{\pi}\mathrm{e}^{-\lambda}I_\nu(\lambda) \tag{5.225}$$

guarantees that each curve has equal area. Similar identities exist for the other dielectric terms. It is also apparent that the moderately relativistic approximation (with terms through λ^5) is much better than the weakly relativistic case with $\lambda = 1$, but the difference is small for small λ.

Problem 5.25 *Weakly relativistic identities.*

1. Use equation (5.225) to prove that the integrals over z for $\mathrm{Im}[K_{xx}(z)]$ for the exact result and the weakly relativistic result are identical.

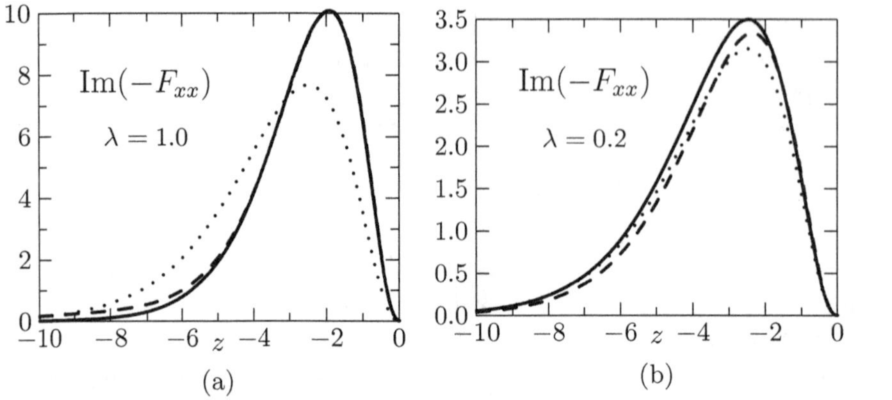

FIGURE 5.9
$\mathrm{Im}(-F_{xx})$ vs. z with $\mu = 50$ for the exact (solid), moderately relativistic (dashed), and weakly relativistic case (dotted) with $n_\| = 0$ for (a) $\lambda = 1.0$ and (b) $\lambda = 0.2$.

2. Use the identity

$$2\int_0^\infty \mathrm{d}x\, \mathrm{e}^{-x} x^{3/2} \int_0^\pi \mathrm{d}\theta\, J_\nu^2(\sqrt{2\lambda x}\sin\theta)\cos^2\theta\sin\theta = \sqrt{\pi}\mathrm{e}^{-\lambda} I_\nu(\lambda) \tag{5.226}$$

to prove that the integrals over z for $\mathrm{Im}[K_{zz}(z)]$ for the exact result and the weakly relativistic result are identical.

When μ is not so large and λ is of order unity or greater, the deviations from the weakly relativistic approximation are significant and the differences between K_{xx} and K_{xy} are no longer ignorable. A strongly relativistic case where $\mu = 20$ and $\lambda = 1$ is illustrated in Fig. 5.10(a) for F_{xx} in the exact and two approximations and in Fig. 5.10(b) for F_{xx} and F_{yy} for the exact case only. It is apparent that the exact and moderately relativistic cases are still close, but that the weakly relativistic approximation is no longer reliable. Figure 5.10(a) also shows the effects from the third and fourth harmonics which occur at $z = \mu/3 = 6.67$ for $n = 3$ and at $z = \mu/2 = 10$ for $n = 4$.

5.6.6 The relativistic X-wave

The exact X-wave dispersion relation with $n_\| = 0$ is

$$n_\perp^2 = \frac{K_{xx}K_{yy} + K_{xy}^2}{K_{xx}}\,. \tag{5.227}$$

This case leads to no damping at all in the nonrelativistic theory, so the relativistic effects are especially apparent. Figure 5.11 shows an example dispersion relation with $\mu = 50$ ($T_e \sim 10$ keV) illustrating the differences between using the exact relativistic tensor components and the weakly and moderately relativistic expressions.

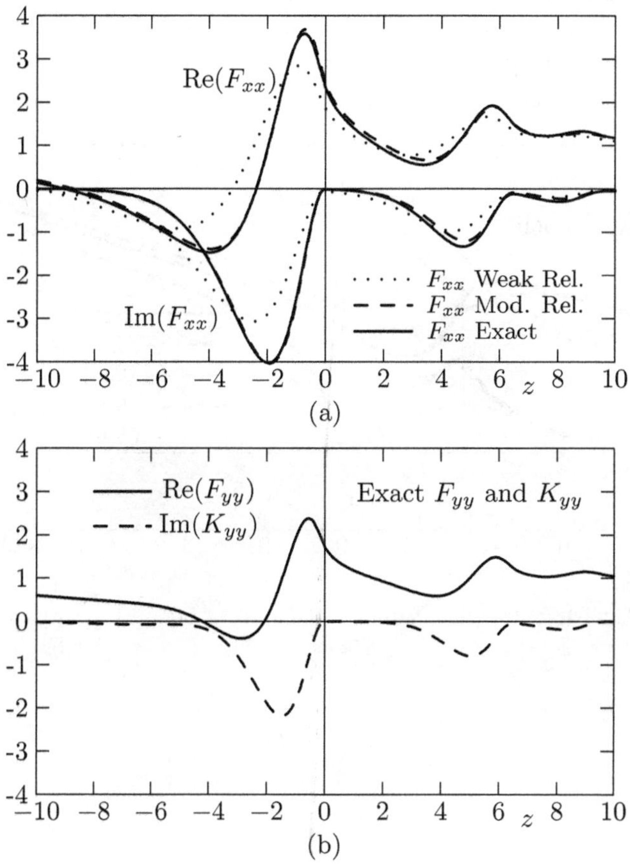

FIGURE 5.10
Plots of (a) the real and imaginary parts of F_{xx} for the exact, moderately relativistic, and weakly relativistic approximations for $\mu = 20$ and $\lambda = 1$ and (b) the exact real part of F_{yy} and imaginary part of K_{yy}.

For this comparison, we let $\omega^2/\omega_p^2 = 4$ and solve for λ in the neighborhood of the second harmonic ($z = 0$ corresponds to $\omega = 2\omega_c$), assuming that the variation in z is due to changes in the magnetic field only. The dispersion is plotted for five separate cases. The cold plasma result is a simple curve showing no significant features near the second harmonic, while the nonrelativistic hot plasma case shows no absorption, but it does indicate a mode conversion region. The exact result and the moderately relativistic cases nearly overlay one another while the weakly relativistic case shows that even with μ large and λ small, the differences are significant in the dispersion relation. The peak of the imaginary part of λ for the weakly relativistic case is approximately half the value for the exact case while the peak imaginary parts of F_{xx}, F_{yy}, and F_{xy} are about 3.7% low for $\lambda = 0.05$. For the moderately relativistic case, the peak imaginary part of λ is only 5.5% below the exact value while the individual component differences are less than 2.5% low for $\lambda = 0.05$.

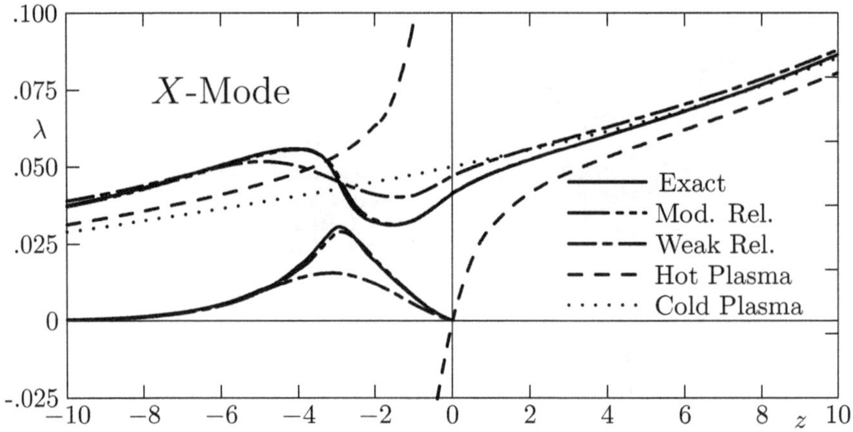

FIGURE 5.11
Variation of λ with z for the X-mode with $\mu = 50$ and $\omega^2/\omega_p^2 = 4$ (from reference [40]).

6

MOMENT EQUATIONS AND FLUID PLASMAS

The analysis of hot plasma waves in the previous chapter is so complicated that even the simplest results are very formidable to obtain. In this chapter, we shall reduce the complexity, keeping only the lowest order pressure terms to include some finite temperature effects, but none of the kinetic effects such as Landau damping. First, we shall examine two different moment expansions in a simple electron plasma with no magnetic field in order to illustrate the method of using moment expansions. Then we shall examine the more general fluid equations for each species including the effects of a uniform magnetic field, and will examine several different approximation methods for studying wave propagation, including the derivations of the low frequency dispersion relation (LFDR), the electrostatic dispersion relation (ESDR), and the warm plasma dispersion relation (WPDR). Each of these dispersion relations has something different to offer and each has its own limitations, but their advantage is that we can come to understand some of the effects of finite temperature without all of the complexity of the hot plasma dispersion relation.

We begin with the *Fluid Equations*, which come from a moment expansion of the Boltzmann equation, equation (2.20),

$$\frac{\partial f}{\partial t} + \boldsymbol{v} \cdot \nabla f + \boldsymbol{A} \cdot \nabla_v f = \left. \frac{\mathrm{d}f}{\mathrm{d}t} \right|_{\mathrm{coll.}} , \tag{6.1}$$

where $\boldsymbol{A}$ is the acceleration due to electric and magnetic fields through the Lorentz force.

Whereas in the kinetic equation we talked about motions of individual particles via a distribution function and averaged the motions to find the currents, in the fluid plasma we describe the motion of a fluid element which is an average over many particles of the same species and we assume that the separate fluid elements for each species move freely among one another, except as collisions exchange momentum between them. The averaging process we shall use here is to expand the kinetic equation for each species in a velocity moment expansion, truncating the expansion at some suitable level, depending on the particular problem. The effects of collisions will be included in subsequent chapters.

6.1 Moments of the distribution function

6.1.1 The simple moment equations

If we neglect collisions for the time being, then equation (6.1) can also be written as

$$\frac{\partial f}{\partial t} + \nabla \cdot (\boldsymbol{v} f) + \nabla_v \cdot (\boldsymbol{A} f) = 0\,, \tag{6.2}$$

since $f\nabla \cdot \boldsymbol{v} = 0$ and $f\nabla_v \cdot \boldsymbol{A} = 0$ since $\boldsymbol{r}$ and $\boldsymbol{v}$ are independent variables. We then introduce a scalar function of velocity, $Q(\boldsymbol{v})$, and define the moment process by an average over the velocity as

$$\begin{aligned} \langle Q(\boldsymbol{v})\rangle &= Q(\boldsymbol{v}) \quad \text{averaged over velocity} \\ &= \frac{\int Qf\,\mathrm{d}^3v}{\int f\,\mathrm{d}^3v} = \frac{1}{n}\int Qf\,\mathrm{d}^3v\,, \end{aligned} \tag{6.3}$$

where $n(\boldsymbol{r}) = \int f\,\mathrm{d}^3v$ is the density in configuration space.

If we now multiply equation (6.2) by Q and integrate over velocity, we have

$$\int Q\frac{\partial f}{\partial t}\,\mathrm{d}^3v + \int Q\nabla \cdot (\boldsymbol{v} f)\,\mathrm{d}^3v + \int Q\nabla_v \cdot (\boldsymbol{A} f)\,\mathrm{d}^3v = 0\,.$$

Since Q is a function of $\boldsymbol{v}$ only, this becomes

$$\frac{\partial}{\partial t}\int Qf\,\mathrm{d}^3v + \nabla \cdot \int Q\boldsymbol{v} f\,\mathrm{d}^3v + \int Q\nabla_v \cdot \boldsymbol{A} f\,\mathrm{d}^3v = 0\,. \tag{6.4}$$

The first term of equation (6.4) is simply $\frac{\partial}{\partial t}(n\langle Q\rangle)$ while the second is $\nabla \cdot (n\langle Q\boldsymbol{v}\rangle)$. The third term expands to

$$\begin{aligned} \int Q\nabla_v \cdot \boldsymbol{A} f\,\mathrm{d}^3v &= \int [\nabla_v \cdot (Q\boldsymbol{A} f) - f\boldsymbol{A} \cdot \nabla_v Q]\,\mathrm{d}^3v \\ &= \oint_{S_v} (Q\boldsymbol{A} f) \cdot d\boldsymbol{S}_v - \int f\boldsymbol{A} \cdot \nabla_v Q\,\mathrm{d}^3v\,, \end{aligned}$$

and the surface integral in velocity space vanishes because we assume the distribution function vanishes as $|v| \to \infty$. We can then write equation (6.4) as

$$\frac{\partial}{\partial t}(n\langle Q\rangle) + \nabla \cdot (n\langle Q\boldsymbol{v}\rangle) - n\langle \boldsymbol{A} \cdot \nabla_v Q\rangle = 0\,. \tag{6.5}$$

6.1.1.1 0th moment.

Let $Q = 1$. Then $\langle Q\rangle = 1$ and $\langle Q\boldsymbol{v}\rangle = \langle \boldsymbol{v}\rangle \equiv \boldsymbol{u}$ where $\boldsymbol{u}$ is the mean or average velocity of the fluid element. Then equation (6.5) leads to the *Continuity equation*

$$\frac{\partial n}{\partial t} + \nabla \cdot (n\boldsymbol{u}) = 0\,. \tag{6.6}$$

6.1.1.2 1st moment.

Let $Q = mv_x$. Then $\langle Q \rangle = mu_x$ and $\nabla_v Q = m\hat{e}_{v_x}$ so $\langle \boldsymbol{A} \cdot \nabla_v Q \rangle = m\langle a_x \rangle$. This leads to

$$\frac{\partial}{\partial t}(nmu_x) + \nabla \cdot (nm\langle \boldsymbol{v} v_x \rangle) - nm\langle a_x \rangle = 0\,.$$

Now we let $\boldsymbol{v} = \boldsymbol{u} + \boldsymbol{w}$, where $\boldsymbol{w}$ measures the perturbation from the average velocity ($\langle \boldsymbol{w} \rangle = 0$). Then

$$\langle \boldsymbol{v} v_x \rangle = \langle (\boldsymbol{u} + \boldsymbol{w})(u_x + w_x) \rangle = \boldsymbol{u} u_x + \langle \boldsymbol{w} w_x \rangle\,.$$

We then define $nm\langle \boldsymbol{w}\boldsymbol{w} \rangle \equiv \boldsymbol{\Psi}$, the stress tensor. The last term is

$$\begin{aligned} nm\langle a_x \rangle &= q \int (\boldsymbol{E} + \boldsymbol{v} \times \boldsymbol{B})_x f \, \mathrm{d}^3 v \\ &= nq(\boldsymbol{E} + \boldsymbol{u} \times \boldsymbol{B})_x\,, \end{aligned}$$

so that by taking all three such component equations from letting $Q = mv_x$, mv_y, and mv_z, we obtain the first moment equation

$$\frac{\partial}{\partial t}(nm\boldsymbol{u}) + \nabla \cdot (nm\boldsymbol{u}\boldsymbol{u}) + \nabla \cdot \boldsymbol{\Psi} - nq(\boldsymbol{E} + \boldsymbol{u} \times \boldsymbol{B}) = 0\,. \qquad (6.7)$$

This equation is generally written, using equation (6.6) and the identity

$$\nabla \cdot (nm\boldsymbol{u}\boldsymbol{u}) = nm(\boldsymbol{u} \cdot \nabla)\boldsymbol{u} + m\boldsymbol{u}\nabla \cdot (n\boldsymbol{u})\,,$$

as the *Momentum equation*

$$nm\left[\frac{\partial \boldsymbol{u}}{\partial t} + (\boldsymbol{u} \cdot \nabla)\boldsymbol{u}\right] + \nabla \cdot \boldsymbol{\Psi} - nq(\boldsymbol{E} + \boldsymbol{u} \times \boldsymbol{B}) = 0\,. \qquad (6.8)$$

This equation is the first moment of the Boltzmann equation, and represents the conservation of momentum in the plasma:

$$\frac{\partial}{\partial t}(\text{momentum density}) + \text{flux of momentum density} = \text{force density}\,.$$

6.1.1.3 Higher moments.

Equation (6.6) gave the evolution of n as a function of $\boldsymbol{u}$, equation (6.8) gave the evolution of $\boldsymbol{u}$ as a function of $\boldsymbol{\Psi}$, and to find the evolution of $\boldsymbol{\Psi}$, one needs the next higher moment, etc. This process must be truncated in order to use the moment equations. If we assume the higher rank tensor (which would appear in the second moment equation) vanishes, then we may represent the stress tensor by a scalar pressure so that $\nabla \cdot \boldsymbol{\Psi} = \nabla p = \nabla(nk_BT)$.

6.1.2 Plasma oscillations

In order to illustrate the moment expansion method, we shall investigate plasma oscillations in an unmagnetized plasma. For this case, we assume $\boldsymbol{k}$ is parallel to $\boldsymbol{E}$, $\boldsymbol{E} = E\hat{e}_x$ so that $\boldsymbol{k} = k\hat{e}_x$, and $\boldsymbol{B}_0 = 0$. The moment expansion for the αth moment is obtained by multiplying the collisionless Boltzmann equation by v^α and integrating(for electrons only):

$$\frac{\partial}{\partial t}\int_{-\infty}^{\infty} v^\alpha f\,\mathrm{d}v + \frac{\partial}{\partial x}\int_{-\infty}^{\infty} v^{\alpha+1} f\,\mathrm{d}v - \frac{e}{m}E\int_{-\infty}^{\infty} v^\alpha \frac{\partial f}{\partial v}\,\mathrm{d}v = 0\,.$$

Integrating the last term by parts leads to

$$\frac{\partial}{\partial t}\int_{-\infty}^{\infty} v^\alpha f\,\mathrm{d}v + \frac{\partial}{\partial x}\int_{-\infty}^{\infty} v^{\alpha+1} f\,\mathrm{d}v + \frac{\alpha e E}{m}\int_{-\infty}^{\infty} v^{\alpha-1} f\,\mathrm{d}v = 0\,. \tag{6.9}$$

We then define the first few moments in terms of $v = u + w$ where u is the average velocity and $w = v - u$ is a measure of the deviations from average due to thermal processes.

$$\begin{aligned}
n &= \int_{-\infty}^{\infty} f\,\mathrm{d}v\,, & nu &= \int_{-\infty}^{\infty} fv\,\mathrm{d}v\,,\\
\frac{P}{m} &= \int_{-\infty}^{\infty} fw^2\,\mathrm{d}v\,, & \frac{Q}{m} &= \int_{-\infty}^{\infty} fw^3\,\mathrm{d}v\,,\\
\frac{R}{m} &= \int_{-\infty}^{\infty} fw^4\,\mathrm{d}v\,, & \vdots &= \vdots\,, \quad \textit{etc.}
\end{aligned}$$

These lead to the set of simple moment equations for $\alpha = 0, 1, 2, 3, \ldots$

$$\begin{aligned}
&\frac{\partial n}{\partial t} + \frac{\partial (nu)}{\partial x} = 0\\
&\frac{\partial (nu)}{\partial t} + \frac{\partial}{\partial x}\left(nu^2 + \frac{P}{m}\right) + \frac{eE}{m}n = 0\\
&\frac{\partial}{\partial t}\left(nu^2 + \frac{P}{m}\right) + \frac{\partial}{\partial x}\left(nu^3 + 3u\frac{P}{m} + \frac{Q}{m}\right) + \frac{2eE}{m}nu = 0\\
&\frac{\partial}{\partial t}\left(nu^3 + 3u\frac{P}{m} + \frac{Q}{m}\right) + \frac{\partial}{\partial x}\left(nu^4 + 6u^2\frac{P}{m} + 4u\frac{Q}{m} + \frac{R}{m}\right)\\
&\qquad + \frac{3eE}{m}\left(nu^2 + \frac{P}{m}\right) = 0
\end{aligned}$$

where we linearize the variables such that

$$
\begin{aligned}
E &= \qquad \tilde{E}e^{i(kx-\omega t)} \\
n &= \bar{n} + \tilde{n}e^{i(kx-\omega t)} \\
u &= \qquad \tilde{u}e^{i(kx-\omega t)} \\
P &= \bar{P} + \tilde{P}e^{i(kx-\omega t)} \\
Q &= \qquad \tilde{Q}e^{i(kx-\omega t)} \\
R &= \bar{R} + \tilde{R}e^{i(kx-\omega t)} \\
\vdots &= \vdots
\end{aligned}
$$

and note that only the even order moments have a zero-order value. Using these moments and keeping only zero and first-order terms, the moment equations reduce to

$$-i\omega\tilde{n} + ik\bar{n}\tilde{u} = 0 \tag{6.10}$$

$$-i\omega\bar{n}\tilde{u} + ik\frac{\tilde{P}}{m} + \frac{e}{m}\bar{n}\tilde{E} = 0 \tag{6.11}$$

$$-i\omega\frac{\tilde{P}}{m} + 3ik\frac{\bar{P}}{m}\tilde{u} + ik\frac{\tilde{Q}}{m} = 0 \tag{6.12}$$

$$-3i\omega\frac{\bar{P}}{m}\tilde{u} - i\omega\frac{\tilde{Q}}{m} + ik\frac{\tilde{R}}{m} + \frac{3e}{m}\frac{\bar{P}}{m}\tilde{E} = 0 \tag{6.13}$$

$$\vdots$$

We then solve Poisson's equation,

$$ik\tilde{E} = -\frac{e}{\epsilon_0}\tilde{n}\,, \tag{6.14}$$

using successive approximations, truncating the equations with successively higher values of P, Q, R, etc. The first step is to neglect all thermal motions so that $\bar{P} = \tilde{P} = 0$ along with all the higher order terms. The set of equations (6.10) through (6.13) reduces to the set

$$
\begin{aligned}
-i\omega\tilde{n} + ik\bar{n}\tilde{u} &= 0 \\
-i\omega\bar{n}\tilde{u} + \frac{e}{m}\bar{n}\tilde{E} &= 0 \\
ik\tilde{E} &= -e\tilde{n}/\epsilon_0
\end{aligned}
$$

with solution $\omega^2 = \omega_{pe}^2$ which is the cold plasma result. To include thermal results, we add a Maxwellian distribution, $f(v) = A\exp(-w^2/v_t^2)$ with $A = (\bar{n}/\sqrt{\pi}v_t)$ so that $f(v)$ is normalized to the number density, and we find

$$\frac{\bar{P}}{m} = \int_{-\infty}^{\infty} w^2 A\exp(-w^2/v_t^2)\,dw = \frac{\bar{n}v_t^2}{2} = \frac{\bar{n}k_BT}{m}\,,$$

where we define the thermal speed as $v_t \equiv (2k_BT/m)^{1/2}$. The successive stages are first from equation (6.13),

$$\frac{\tilde{Q}}{m} = \frac{3}{2}\left(\frac{e\tilde{E}}{\mathrm{i}\omega m} - \tilde{u}\right)\bar{n}v_t^2\,,$$

so that from equation (6.12) we find

$$\frac{\tilde{P}}{m} = \frac{3k}{2\omega}\left(\frac{e\tilde{E}}{\mathrm{i}\omega m}\right)\bar{n}v_t^2\,,$$

and from equation (6.11) we find

$$\tilde{u} = \left(1 + \frac{3k^2v_t^2}{2\omega^2}\right)\frac{e\tilde{E}}{\mathrm{i}\omega m}\,,$$

and finally from equation (6.11) and equation (6.14) we find

$$\tilde{n} = \left(1 + \frac{3k^2v_t^2}{2\omega^2}\right)\frac{k\bar{n}e\tilde{E}}{\mathrm{i}\omega^2 m} = -\frac{\mathrm{i}k\epsilon_0\tilde{E}}{e}\,,$$

and equating the last two terms, we have the Bohm-Gross dispersion relation,

$$\omega^2 = \omega_p^2\left(1 + \frac{3k^2v_t^2}{2\omega^2}\right). \tag{6.15}$$

Keeping terms through the fifth moment, the result is

$$\omega^2 = \omega_p^2\left(1 + \frac{3k^2v_t^2}{2\omega^2} + \frac{15k^4v_t^4}{4\omega^4}\right),$$

so it is evident that keeping higher order terms leads to an asymptotic series for ω^2 in the expansion parameter $(v_t/v_p)^2$ where $v_p = \omega/k$ is the phase velocity.

Problem 6.1 *Simple moment expansions.*

1. *Second moment.* If we take $P \neq 0$ and $Q = 0$, show that

$$\omega^2 = \omega_p^2\left[1 + \frac{3k^2v_t^2}{2\omega^2} + \mathcal{O}\left(\frac{k^4v_t^4}{\omega^4}\right)\right].$$

2. *Fourth moment.* If we take $R \neq 0$ and $S = 0$, show that

$$\omega^2 = \omega_p^2\left[1 + \frac{3k^2v_t^2}{2\omega^2} + \frac{15k^4v_t^4}{4\omega^4} + \mathcal{O}\left(\frac{k^6v_t^6}{\omega^6}\right)\right].$$

3. *Fifth moment.* Show that if we let $S \neq 0$ that the term, $\mathcal{O}(v_t^6/v_p^6)$, disappears.

6.1.3 The kinetic moment equations

It is possible to extend the moment equations to contain some of the information in the collisionless Boltzmann equation by defining the various moment integrals as

$$F^{(\alpha)} = \int v^\alpha f(v)\,dv\,, \tag{6.16}$$

where the velocity $v = u + w$ is comprised of the average velocity, $\langle v\rangle \equiv u$, and the random part, w. If we look at the first few moments, we find

$$\begin{aligned}
F^{(0)} &= n\\
F^{(1)} &= nu\\
F^{(2)} &= nu^2 + M^{(2)}\\
F^{(3)} &= nu^3 + 3uM^{(2)} + M^{(3)}\\
F^{(4)} &= nu^4 + 6u^2M^{(2)} + 4uM^{(3)} + M^{(4)}
\end{aligned}$$

where we have introduced modified moments depending only on the random part of v,

$$M^{(\alpha)} \equiv \int w^\alpha f(v)\,dv\,. \tag{6.17}$$

The first few of these moments are $M^{(0)} = n$, $M^{(1)} = 0$, and $mM^{(2)} = p$. Since we wish to linearize the moment equations, and since u is a first-order quantity, we may write the linearized moments as

$$\begin{aligned}
F^{(2k)} &= \bar{M}^{(2k)} + \tilde{M}^{(2k)}\\
&= \frac{(2k-1)!!\bar{n}v_t^{2k}}{2^k} + \tilde{M}^{(2k)}\,,
\end{aligned} \tag{6.18}$$

$$\begin{aligned}
F^{(2k+1)} &= (2k+1)\bar{M}^{(2k)}\tilde{u} + \tilde{M}^{(2k+1)}\\
&= \frac{(2k+1)!!\bar{n}v_t^{2k}}{2^k}\tilde{u} + \tilde{M}^{(2k+1)}\,,
\end{aligned} \tag{6.19}$$

where $\tilde{u} = \tilde{u}e^{i(kz-\omega t)}$, etc. and we have used a Maxwellian distribution to obtain

$$\bar{M}^{(\alpha)} = \frac{\bar{n}}{\sqrt{\pi}v_t}\int_{-\infty}^{\infty} w^\alpha e^{-w^2/v_t^2}\,dw = \frac{\bar{n}v_t^\alpha(\alpha-1)!!}{2^{\alpha/2}} \tag{6.20}$$

for even α with $v_t = \sqrt{2kT/m}$ and $(-1)!! = 1$. It is evident from equations (6.18) and (6.19) that for α even, there is a zero order and a first order term, while for α odd, there are only first order terms.

We then write equation (6.9) in terms of the $F^{(\alpha)}$ with the result

$$\frac{\partial}{\partial t}F^{(\alpha)} + \frac{\partial}{\partial z}F^{(\alpha+1)} - \frac{\alpha q}{m}E_zF^{(\alpha-1)} = 0\,. \tag{6.21}$$

Assuming E is first order and that n has both zero and first order terms, the linearized moment equations for $\alpha = 0$ through $\alpha = 4$ are (with $\partial/\partial z \to \mathrm{i}k$ and $\partial/\partial t \to -\mathrm{i}\omega$)

$$-\mathrm{i}\omega\tilde{n} + \mathrm{i}k\bar{n}\tilde{u} = 0, \qquad \alpha = 0 \tag{6.22}$$

$$-\mathrm{i}\omega\bar{n}\tilde{u} + \mathrm{i}k\tilde{M}^{(2)} - \frac{q}{m}\bar{n}\tilde{E} = 0, \qquad \alpha = 1 \tag{6.23}$$

$$-\mathrm{i}\omega\tilde{M}^{(2)} + \left(\frac{3\bar{n}v_t^2}{2}\tilde{u} + \tilde{M}^{(3)}\right) = 0, \qquad \alpha = 2 \tag{6.24}$$

$$-\mathrm{i}\omega\left(\frac{3\bar{n}v_t^2}{2}\tilde{u} + \tilde{M}^{(3)}\right) + \mathrm{i}k\tilde{M}^{(4)} - \frac{3q\bar{n}v_t^2}{2m}\tilde{E} = 0, \qquad \alpha = 3 \tag{6.25}$$

$$-\mathrm{i}\omega\tilde{M}^{(4)} + \mathrm{i}k\left(\frac{5!!\bar{n}v_t^4}{4}\tilde{u} + \tilde{M}^{(5)}\right) = 0, \qquad \alpha = 4\,. \tag{6.26}$$

The subsequent terms can be obtained from the two general expressions for α even and α odd:

$$-\mathrm{i}\omega\tilde{M}^{(\alpha)} + \mathrm{i}k\left[\frac{(\alpha+1)!!\bar{n}v_t^\alpha}{2^{\alpha/2}}\tilde{u} + \tilde{M}^{(\alpha+1)}\right] = 0, \qquad \alpha \text{ even}, \tag{6.27}$$

$$-\mathrm{i}\omega\left[\frac{\alpha!!\bar{n}v_t^{\alpha-1}}{2^{(\alpha-1)/2}}\tilde{u} + \tilde{M}^{(\alpha)}\right] + \mathrm{i}k\tilde{M}^{(\alpha+1)} - \frac{q\alpha!!\bar{n}v_t^{\alpha-1}}{m2^{(\alpha-1)/2}}\tilde{E} = 0, \qquad \alpha \text{ odd}. \tag{6.28}$$

We may then use equation (6.22), written as $\tilde{u} = (\omega/k)\tilde{n}/\bar{n}$, to eliminate $\tilde{u}$ in equation (6.27) so that the pair may be written as

$$-\mathrm{i}\omega\left[\tilde{M}^{(\alpha)} - \frac{(\alpha+1)!!v_t^\alpha}{2^{\alpha/2}}\tilde{n}\right] + \mathrm{i}k\tilde{M}^{(\alpha+1)} = 0, \qquad \alpha \text{ even}, \tag{6.29}$$

$$-\mathrm{i}\omega\tilde{M}^{(\alpha)} + \mathrm{i}k\tilde{M}^{(\alpha+1)} + \frac{\alpha!!\bar{n}v_t^{\alpha-1}}{2^{(\alpha-1)/2}}\left(-\mathrm{i}\omega\tilde{u} - \frac{q}{m}\tilde{E}\right) = 0, \qquad \alpha \text{ odd}. \tag{6.30}$$

By taking these equations two at a time, we may generate higher and higher order approximations in a systematic manner. Defining

$$\tilde{f} \equiv -\mathrm{i}\omega\tilde{u} - \frac{q}{m}\tilde{E} \tag{6.31}$$

and using equation (6.30) with $\alpha = 1$, we first obtain

$$\tilde{f} = -\frac{\mathrm{i}k}{\bar{n}}\tilde{M}^{(2)}\,. \tag{6.32}$$

Then taking the $\alpha = 2$ and $\alpha = 3$ cases together,

$$-\mathrm{i}\omega\tilde{M}^{(2)} + \mathrm{i}k\tilde{M}^{(3)} + \mathrm{i}\omega\frac{3!!v_t^2}{2}\tilde{n} = 0$$

$$-\mathrm{i}\omega\tilde{M}^{(3)} + \mathrm{i}k\tilde{M}^{(4)} + \frac{3!!\bar{n}v_t^2}{2}\tilde{f} = 0$$

and eliminating $\tilde{M}^{(3)}$ between them, we find

$$\tilde{f}\left(1+\frac{3!!}{2\zeta^2}\right) = -\frac{ik}{\bar{n}}\left(\frac{3!!v_t^2}{2}\tilde{n}+\frac{k^2}{\omega^2}\tilde{M}^{(4)}\right), \tag{6.33}$$

where $\zeta \equiv \omega/kv_t$. Continuing on two at a time, the general result may be written

$$S_m\tilde{f} = -\frac{ik}{\bar{n}}\left[v_t^2\zeta^2(S_m-1)\tilde{n}+\left(\frac{k}{\omega}\right)^{2m}\tilde{M}^{(2m+2)}\right], \tag{6.34}$$

where equation (6.32) is the result for $m=0$, equation (6.33) is the result for $m=1$, and

$$S_m(\zeta) = \sum_{j=0}^{m}\frac{(2j+1)!!}{2^j\zeta^{2j}} \qquad \zeta \gg 1\,. \tag{6.35}$$

Equation (6.34) must now be solved along with Poisson's equation,

$$ik\tilde{E} = \frac{q\tilde{n}}{\epsilon_0}\,. \tag{6.36}$$

Using equation (6.36), we may write $\tilde{f}$ as

$$\tilde{f} = -\frac{i}{k\bar{n}}(\omega^2-\omega_p^2)\tilde{n}\,.$$

Using this result along with equation (6.34) results in

$$\omega^2 = \omega_p^2 S_m(\zeta) + k^2\left(\frac{k}{\omega}\right)^{2m}\frac{\tilde{M}^{(2m+2)}}{\tilde{n}}\,. \tag{6.37}$$

This is an asymptotic series in ζ with the remainder term set to zero for some m. No matter how many terms are kept, there will never be any imaginary part. As more terms are kept, ζ must be larger and larger since each term must be smaller than the previous term.

The lowest nontrivial approximation is to let $\tilde{M}^{(4)} = 0$ to obtain

$$\omega^2 = \omega_{pe}^2\left[1+\frac{3}{2}\left(\frac{v_e}{v_p}\right)^2\right], \tag{6.38}$$

which again is the Bohm-Gross Dispersion Relation[41] with $v_e^2 \equiv 2k_BT_e/m_e$. Keeping terms through the fifth moment, the result is

$$\omega^2 = \omega_{pe}^2\left[1+\frac{3}{2}\left(\frac{v_e}{v_p}\right)^2+\frac{15}{4}\left(\frac{v_e}{v_p}\right)^4\right], \tag{6.39}$$

as before, so it is apparent that *the moment expansion is an expansion in the ratio of the thermal velocity to the phase velocity.* The Bohm-Gross dispersion relation now resolves the pure oscillation in cold plasma theory, since this dispersion relation describes a wave with a cutoff at ω_{pe} which propagates near the electron thermal speed for high frequencies.

We have thus learned from this example that the fluid equations which are based on moment expansions are valid as long as the phase velocity is large compared to the thermal speed. When this approximation fails, then we must solve the equations without expansion, as was done with the Vlasov equation example. It should be noted that using the real part of the asymptotic expansion of the Plasma Dispersion function to solve for the dispersion relation for this case, the result corresponds exactly to the result found here by the moment expansion. Furthermore, if one examines the smallest term in the asymptotic series (for a given value of ζ), which is always a good estimate of the error in the series, that smallest term is of the same order as the pole term in the Plasma Dispersion function.

Problem 6.2 *Kinetic moment expansions.*

1. Zero order moments. Verify equation (6.20).
2. Moment pairs. Fill in the steps leading to equations (6.29) and (6.30).
3. Fifth moment. Continue with the $\alpha = 4$ and $\alpha = 5$ terms beyond equation (6.33) and verify that equation (6.34) is valid for $m = 2$.
4. Using Poisson's equation, complete the steps from equation (6.34) to equation (6.37).

6.2 The fluid equations

Having used the notation in the previous section that $\boldsymbol{v}$ is an independent variable and that $\boldsymbol{u}$ is the velocity of a fluid element, we now denote the fluid velocity by $\boldsymbol{u}_j$ since now it relates to a specific species j. We shall also use the definition of the mass density as $\rho_j = n_j m_j$ which is the mass per unit volume for a fluid element of species j. With these definitions, the fluid equations are

$$\frac{\partial \rho_j}{\partial t} + \nabla \cdot (\rho_j \boldsymbol{u}_j) = Z_j \,, \qquad \text{Continuity equation,} \tag{6.40}$$

where Z_j is the ionization rate for species j and

$$\rho_j \left[\frac{\partial \boldsymbol{u}_j}{\partial t} + (\boldsymbol{u}_j \cdot \nabla)\boldsymbol{u}_j \right] = q_j n_j (\boldsymbol{E} + \boldsymbol{u}_j \times \boldsymbol{B}) - \nabla p_j \,, \quad \text{Momentum equation,} \tag{6.41}$$

where collisions have been neglected for now. We normally take $p = nk_BT$ but for a magnetized plasma, we might want to take $p_\perp \neq p_\parallel$. Then we use the equations of state, $p_{\|j} = \gamma_{\|j}\rho_{1j}$ and $p_{\perp j} = \gamma_{\perp j}\rho_{1j}$. As $B_0 \to \infty$, there is only one degree of freedom due to the parallel motion, so $\gamma_{\|} = (f+2)/f = 3$. For transverse motions, there is one degree of freedom for rotational motion and one for parallel motion, so $\gamma_{\perp} = (f+2)/f = 2$. For weaker fields and collisions, these constraints on the degrees of freedom may be relaxed for one or both species, in which case $\gamma_{\perp} = \gamma_{\|} = 5/3$.

For particle conservation, the ionization rates among the various species (including neutrals) must satisfy

$$\sum_j Z_j = 0\,.$$

The linearized fluid equations are

$$\frac{\partial \rho_{1j}}{\partial t} + \rho_{0j}\nabla\cdot\boldsymbol{u}_{1j} = 0\,, \qquad \text{Continuity equation,} \tag{6.42}$$

where ionization has been neglected and

$$\rho_{0j}\frac{\partial \boldsymbol{u}_{1j}}{\partial t} = q_j n_{0j}(\boldsymbol{E}_1 + \boldsymbol{u}_{1j}\times\boldsymbol{B}_0) - \nabla p_{1j}\,, \quad \text{Momentum equation,} \tag{6.43}$$

since all zero order quantities are assumed to be constant in space and time and here we assume $\boldsymbol{E}_0 = \boldsymbol{u}_{0j} = 0$, although these restrictions can be relaxed to study beams and drift waves in inhomogeneous plasmas.

These equations with a separate set of equations for each species form the basis for the two-fluid equations (*multi*-fluid equations for multiple ion species) which are joined by including the Maxwell equations.

Problem 6.3 *Fluid equations with zero-order drifts and fields.* Write the zero-order and first-order continuity and momentum equations when $\boldsymbol{u}_{0j} \neq 0$ and $\boldsymbol{E}_0 \neq 0$.

6.3 Low frequency waves

The waves from the fluid equations conveniently break up into two regions when $m_e \ll m_i$. For the high frequency waves, ion motions are completely neglected, while in the low frequency region, we neglect terms of order m_e/m_i.

6.3.1 The low frequency dispersion relation

Beginning with the linearized fluid equations for electrons and one species only of singly charged ions, following the development of Stringer[42], we write for each species the continuity equation,

$$\frac{\partial \rho_{1j}}{\partial t} + \rho_{0j}\nabla\cdot\boldsymbol{v}_{1j} = 0\,, \tag{6.44}$$

and the momentum equation,

$$\rho_{0j}\frac{\partial \boldsymbol{v}_{1j}}{\partial t} = q_j n_{0j}(\boldsymbol{E} + \boldsymbol{v}_{1j} \times \boldsymbol{B}_0) - \nabla p_{1j} \,. \tag{6.45}$$

By adding the momentum equations for electrons and ions, we obtain the result,

$$\rho_0 \frac{\partial \boldsymbol{v}}{\partial t} = \boldsymbol{j} \times \boldsymbol{B}_0 - \nabla p \,, \tag{6.46}$$

where $\rho_0 = \rho_{0i} + \rho_{0e}$, $p = p_i + p_e$, and

$$\boldsymbol{v} = \frac{\rho_{0i}\boldsymbol{v}_{1i} + \rho_{0e}\boldsymbol{v}_{1e}}{\rho_{0i} + \rho_{0e}} \,, \qquad \boldsymbol{j} = n_0 e(\boldsymbol{v}_{1i} - \boldsymbol{v}_{1e}) \,.$$

The other fluid equation is obtained by multiplying each of the momentum equations by q_j/m_j with $q_i = e$, $q_e = -e$, and adding, whereby the result may be expressed as

$$\frac{m_e}{n_0 e^2}\frac{\partial \boldsymbol{j}}{\partial t} = \boldsymbol{E} + \boldsymbol{v} \times \boldsymbol{B}_0 - \frac{m_i}{e}\frac{\partial \boldsymbol{v}}{\partial t} - \frac{1}{n_0 e}\nabla p_i \,, \tag{6.47}$$

where terms of order m_e/m_i have been neglected.

We next assume that all first order quantities vary as $\exp(\mathrm{i}\boldsymbol{k}\cdot\boldsymbol{r} - \mathrm{i}\omega t)$ so that the Maxwell equations lead to the wave equation,

$$k^2\left(1 - \frac{\omega^2}{k^2c^2}\right)\boldsymbol{E} - \boldsymbol{k}(\boldsymbol{k}\cdot\boldsymbol{E}) = \mathrm{i}\omega\mu_0\boldsymbol{j} \,. \tag{6.48}$$

Introducing the thermal speeds, defined by $c_j^2 \equiv \gamma_j k_B T_j/m_j$, where γ_j is the ratio of specific heats for species j and comes from the equation of state (an alternative to truncating the moment equations), we may relate the pressure and density as $p_j = c_j^2\rho_{1j}$ so the continuity equations give

$$p_i = \frac{n_0 m_i c_i^2(\boldsymbol{k}\cdot\boldsymbol{v}_i)}{\omega} = \frac{g n_0 m_i c_s^2(\boldsymbol{k}\cdot\boldsymbol{v}_i)}{\omega} \simeq \frac{g n_0 m_i c_s^2}{\omega}\boldsymbol{k}\cdot\boldsymbol{v} \,, \tag{6.49}$$

$$\begin{aligned} p &= \frac{n_0 c_s^2(m_i + m_e)}{\omega}\boldsymbol{k}\cdot\boldsymbol{v} - \frac{(1-g)c_s^2 m_i}{\omega e}\left(1 - \frac{c_i^2}{c_e^2}\right)\boldsymbol{k}\cdot\boldsymbol{j} \\ &\simeq \frac{n_0 c_s^2 m_i}{\omega}\boldsymbol{k}\cdot\boldsymbol{v} - \frac{(1-g)c_s^2 m_i}{\omega e}\boldsymbol{k}\cdot\boldsymbol{j} \,, \end{aligned} \tag{6.50}$$

where

$$g \equiv \frac{\gamma_i T_i}{\gamma_i T_i + \gamma_e T_e} \,, \quad \text{and} \quad c_s^2 \equiv \frac{\gamma_e k_B T_e + \gamma_i k_B T_i}{m_e + m_i} \,, \tag{6.51}$$

and c_s is called the ion-acoustic speed.

If we now solve equation (6.46) for $\boldsymbol{v}$, using equation (6.50) to eliminate p we obtain

$$\begin{aligned} \boldsymbol{v} &= \frac{1}{\omega^2\mu_0\rho_0}\left[k^2\left(1 - \frac{\omega^2}{k^2c^2}\right)(\boldsymbol{E}\times\boldsymbol{B}_0) - (\boldsymbol{k}\cdot\boldsymbol{E})(\boldsymbol{k}\times\boldsymbol{B}_0)\right] \\ &= \quad + \frac{c_s^2}{\omega^2}(\boldsymbol{k}\cdot\boldsymbol{v})\boldsymbol{k} - \frac{i\epsilon_0(1-g)c_s^2}{n_0 e\omega}(\boldsymbol{k}\cdot\boldsymbol{E})\boldsymbol{k} \,, \end{aligned}$$

but if $\boldsymbol{v} = \boldsymbol{u} + \alpha(\boldsymbol{k}\cdot\boldsymbol{v})\boldsymbol{k}$, where $\boldsymbol{u}$ is *any* vector, then $\boldsymbol{v} \equiv \boldsymbol{u} + \alpha(\boldsymbol{k}\cdot\boldsymbol{u})\boldsymbol{k}/(1-\alpha k^2)$, so we may write

$$\boldsymbol{v} = \frac{1}{\omega^2\mu_0\rho_0}\left[k^2\left(1+\frac{c_s^2\boldsymbol{k}\,\boldsymbol{k}\cdot}{\omega^2-k^2c_s^2}\right)(\boldsymbol{E}\times\boldsymbol{B}_0)-(\boldsymbol{k}\cdot\boldsymbol{E})(\boldsymbol{k}\times\boldsymbol{B}_0)\right] - \frac{\mathrm{i}\omega\epsilon_0(1-g)c_s^2}{n_0e(\omega^2-k^2c_s^2)}(\boldsymbol{k}\cdot\boldsymbol{E})\boldsymbol{k}\,, \tag{6.52}$$

where we have neglected ω^2/k^2c^2 compared to unity. If we make this same approximation in equation (6.48), then $\boldsymbol{k}\cdot\boldsymbol{j}=0$ and the last term in equation (6.52) disappears, with the result

$$\boldsymbol{v} = \frac{1}{\omega^2\mu_0\rho_0}\left[k^2\left(1+\frac{c_s^2\boldsymbol{k}\,\boldsymbol{k}\cdot}{\omega^2-k^2c_s^2}\right)(\boldsymbol{E}\times\boldsymbol{B}_0)-(\boldsymbol{k}\cdot\boldsymbol{E})(\boldsymbol{k}\times\boldsymbol{B}_0)\right]. \tag{6.53}$$

We then use this equation for $\boldsymbol{v}$ and equation (6.48) for $\boldsymbol{j}$ and equation (6.49) for p_i in equation (6.47) to obtain

$$\begin{aligned}&\left(1+\frac{k^2c^2}{\omega_{pe}^2}-\frac{k^2V_A^2}{\omega^2}\right)\boldsymbol{E}+\frac{\mathrm{i}V_A^2}{\omega\omega_{ci}}[k^2(\boldsymbol{E}\times\hat{e}_z)-(\boldsymbol{k}\cdot\boldsymbol{E})(\boldsymbol{k}\times\hat{e}_z)]\\ &\quad+\left(\frac{V_A^2}{\omega^2}-\frac{c^2}{\omega_{pe}^2}\right)(\boldsymbol{k}\cdot\boldsymbol{E})\boldsymbol{k}+\frac{V_A^2}{\omega^2}[k^2E_z-(\boldsymbol{k}\cdot\boldsymbol{E})k_z]\hat{e}_z\\ &-\frac{k^2c_s^2}{\omega^2-k^2c_s^2}\left[\frac{V_A^2}{\omega^2}\boldsymbol{E}\cdot(\boldsymbol{k}\times\hat{e}_z)(\boldsymbol{k}\times\hat{e}_z)+\frac{\mathrm{i}V_A^2(1-g)}{\omega\omega_{ci}}\boldsymbol{E}\cdot(\boldsymbol{k}\times\hat{e}_z)\boldsymbol{k}\right]=0\,,\end{aligned} \tag{6.54}$$

where V_A is the Alfvén speed and where we have chosen $\boldsymbol{B}_0 = B_0\hat{e}_z$. We can now multiply equation (6.54) successively by $\boldsymbol{k}$, $\hat{e}_z$, and $\boldsymbol{k}\times\hat{e}_z$ to form the set of equations whose determinant of coefficients set to zero,

$$\begin{vmatrix} \dfrac{k^2V_A^2}{\omega^2}k_z & -\dfrac{\mathrm{i}k^2V_A^2(\omega^2-gk^2c_s^2)}{\omega\omega_{ci}(\omega^2-k^2c_s^2)} & 1-\dfrac{k_z^2V_A^2}{\omega^2} \\ 1+\dfrac{k^2c^2}{\omega_{pe}^2} & -\dfrac{\mathrm{i}k_zV_A^2(1-g)k^2c_s^2}{\omega\omega_{ci}(\omega^2-k^2c_s^2)} & -\dfrac{k_zc^2}{\omega_{pe}^2} \\ -\dfrac{\mathrm{i}k^2V_A^2}{\omega\omega_{ci}}k_z & 1+\dfrac{k^2c^2}{\omega_{pe}^2}-\dfrac{k^2V_A^2}{\omega^2-k^2c_s^2}\left(1-\dfrac{k_z^2c_s^2}{\omega^2}\right) & \dfrac{\mathrm{i}k_z^2V_A^2}{\omega\omega_{ci}} \end{vmatrix}=0\,, \tag{6.55}$$

leads to the low frequency dispersion relation (LFDR),

$$\left(1-\frac{\omega^2}{k^2V_A^2}-\frac{\omega^2}{\omega_{ce}\omega_{ci}}+\frac{k^2c_s^2\sin^2\theta}{\omega^2-k^2c_s^2}\right)\left(\cos^2\theta-\frac{\omega^2}{k^2V_A^2}-\frac{\omega^2}{\omega_{ce}\omega_{ci}}\right)=\frac{\omega^2\cos^2\theta}{\omega_{ci}^2}\,, \tag{6.56}$$

where θ is the angle between the direction of propagation and the static magnetic field. This dispersion relation is equivalent to that given by Stringer[42] and Braginskii[43]. If one lets $y=\omega/\omega_{ci}$ and $x=kc_s/\omega_{ci}$, then the dispersion relation may be written as a cubic in y^2 as

$$a_3y^6+a_2y^4+a_1y^2+a_0=0\,, \tag{6.57}$$

where

$$
\begin{aligned}
a_0 &= -x^6 \cos^4 \theta \\
a_1 &= [1 + 2(\beta + \epsilon x^2) + x^2]x^4 \cos^2 \theta \\
a_2 &= -[(x^2 + \beta + \epsilon x^2) \cos^2 \theta + (\beta + \epsilon x^2)(1 + \beta + \epsilon x^2)]x^2 \\
a_3 &= (\beta + \epsilon x^2)^2
\end{aligned}
\tag{6.58}
$$

where the only constants are $\epsilon \equiv m_e/m_i$, $\beta \equiv c_s^2/V_A^2$, and the angle. The β as defined here is the same as the plasma pressure divided by the magnetic pressure if $\gamma_i = \gamma_e = 2$. The Stringer diagrams are usually plotted with several constants listed, but equations (6.57) and (6.58) show that only three are needed to specify the warm plasma dispersion relation. This form of the equation is useful because Stringer diagrams are usually plots of $\log_{10} y$ vs. $\log_{10} x$.

Problem 6.4 *Fluid equations.* Show that equation (6.47) may be written exactly as

$$
\frac{m_e}{n_0 e^2 (1+\epsilon)} \frac{\partial \boldsymbol{j}}{\partial t} = \boldsymbol{E} + \boldsymbol{v} \times \boldsymbol{B}_0 - \frac{m_i}{e}(1-\epsilon)\frac{\partial \boldsymbol{v}}{\partial t} - \frac{1}{n_0 e(1+\epsilon)} \nabla(p_i + \epsilon p_e)
$$

where $\epsilon = m_e/m_i$, so the approximation includes neglecting ϵp_e compared to p_i as well as neglecting ϵ compared to unity.

Problem 6.5 *Determinant of coefficients.* Show that equation (6.54) can be written as

$$
\begin{pmatrix} a_{11} & a_{12} & a_{13} \\ a_{21} & a_{22} & a_{23} \\ a_{31} & a_{32} & a_{33} \end{pmatrix}
\begin{pmatrix} E_x \\ E_y \\ E_z \end{pmatrix} = 0
$$

and find the coefficients. Then show that this determinant of coefficients is equivalent to that of equation (6.55).

Problem 6.6 *Derivation of the low frequency dispersion relation.* Show that equation (6.55) leads to equation (6.56).

Problem 6.7 *Simplified form of the low frequency dispersion relation.* Show that equation (6.56) can be written as equation (6.57).

6.3.2 Stringer diagrams

Plots of the low frequency dispersion relation which exhibit the character of the various waves in the low frequency region have been given by Stringer[42]. In these plots, ω/ω_{ci} is plotted against kc_s/ω_{ci} on logarithmic scales so that regions of constant phase velocity appear as straight lines.

6.3.2.1 Variation with angle and temperature

In order to illustrate some of the features of the LFDR, we examine Figure 6.1 where the figures on the left are high-β (or overdense) cases for different angles and on the right the figures represent the same cases for a low-β (or underdense) plasma. One feature of these figures is that there appear to be places where the roots cross, indicated with a box labeled either as A, B, or C. On this scale, the upper two branches appear to cross at boxes A and B while the lower two branches appear to cross at boxes C. In fact, there are no crossings for $0° < \theta < 90°$ as is shown in Figure 6.2 which are expansions of the boxes in Figure 6.1. Only when $\theta = 0°$ does crossing occur. This is because in between these limits, the polarizations of the two wave fields always have a component parallel near the apparent crossings which causes the different waves to couple and prevent the crossing. When $\theta = 0$, the polarization of the branches are orthogonal with one longitudinal and one transverse, so no coupling occurs and there is a true crossing. It may also be noted that in the neighborhood of the near crossings, β makes little difference as there is only one figure for box A and for box B in Figure 6.2 since the cases with different β are within a line width of one another. For box C, the difference is still small as shown in box C of Figure 6.2.

6.4 High frequency waves

In the derivation of the LFDR, we made approximations based on the phase velocity relative to the speed of light and assumed low frequency, even though the dispersion relation was carried to high frequencies. In this section, we make no assumption about the phase velocity or the frequency, but do keep first order terms in pressure to include finite temperature effects, leading to the Warm Plasma Dispersion Relation (WPDR).

6.4.1 Warm plasma dispersion relation

For high frequencies, we may use the WPDR, where we neglect ion motions entirely. We do this by effectively letting $m_i \to \infty$. Using only the fluid equations for electrons and deleting the subscript denoting the species (except for c_e), the linearized time and space harmonic fluid equations are

$$-\mathrm{i}\omega\rho_1 + \rho_0\mathrm{i}\boldsymbol{k}\cdot\boldsymbol{v}_1 = 0 \tag{6.59}$$

and

$$-\mathrm{i}\omega\rho_0\boldsymbol{v}_1 = -en_0(\boldsymbol{E} + \boldsymbol{v}_1 \times \boldsymbol{B}_0) - \mathrm{i}\boldsymbol{k}p_1\,, \tag{6.60}$$

along with the equations of state, $p_{1\parallel} = \gamma_\parallel c_e^2\rho_1$ and $p_{1\perp} = \gamma_\perp c_e^2\rho_1$, where $c_e^2 \equiv k_BT/m_e$. These may be combined to give an expression for the velocity

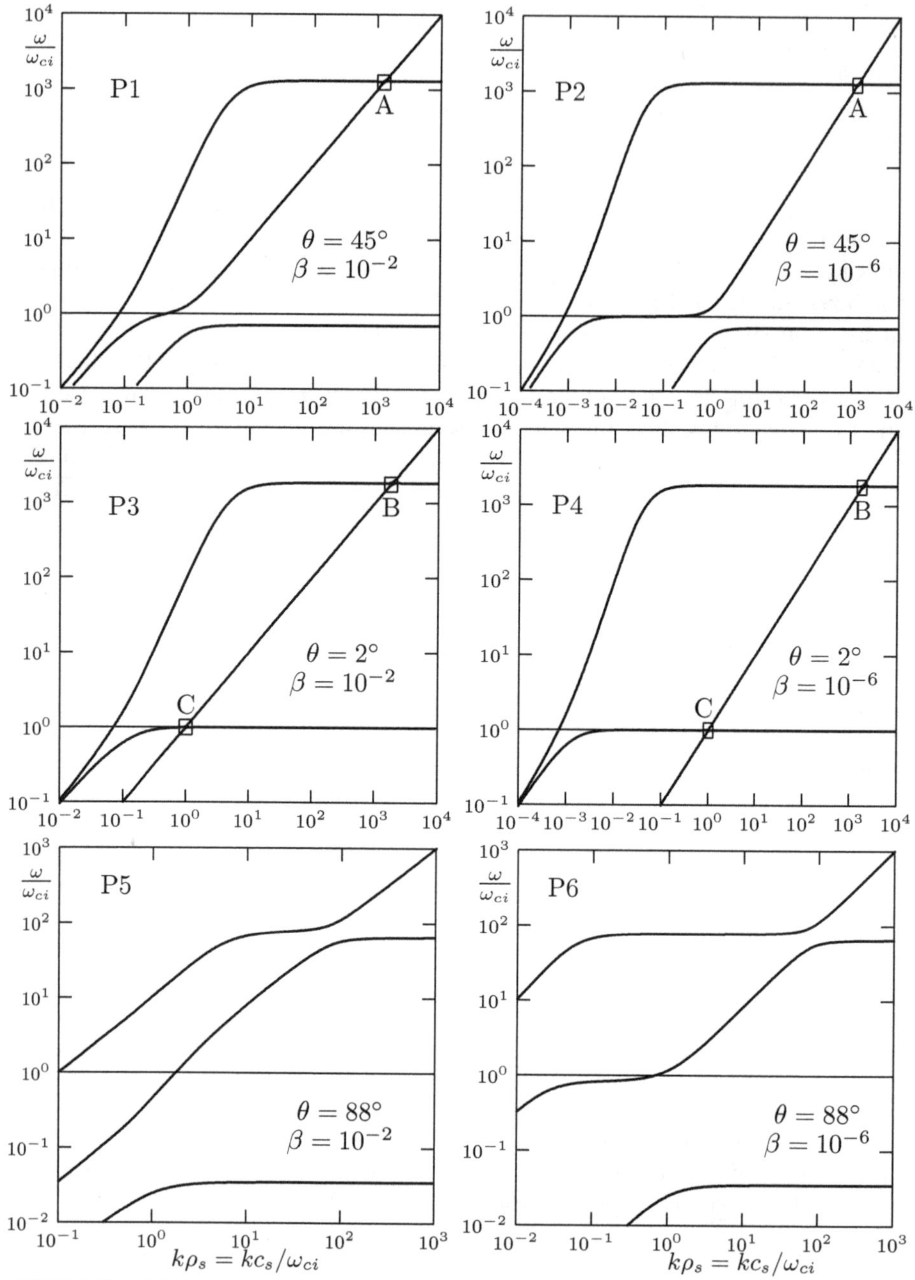

FIGURE 6.1

Dispersion curves from the low frequency dispersion relation for an overdense ($\beta = 10^{-2}$) and underdense ($\beta = 10^{-6}$) for several angles.

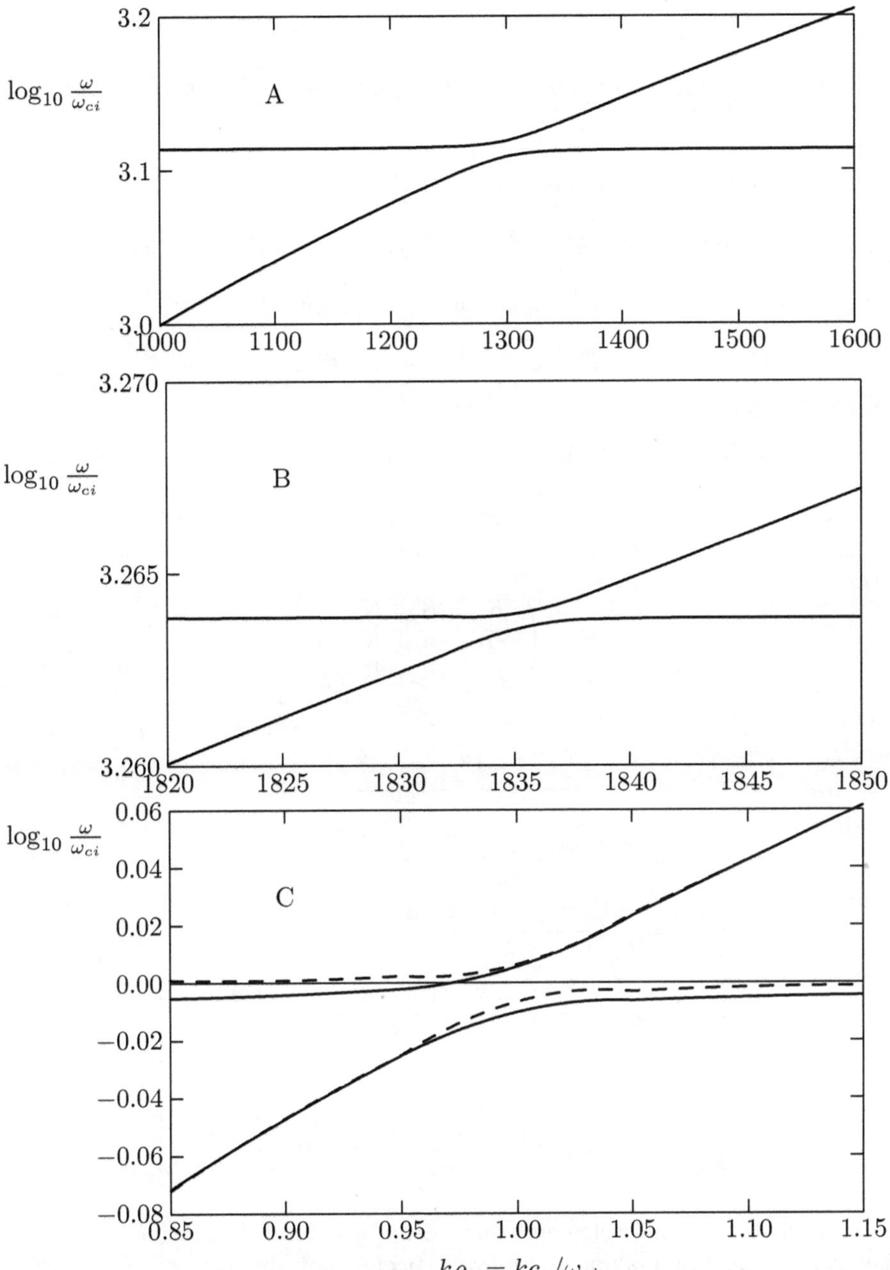

FIGURE 6.2
Expanded dispersion curves from boxes in Figure 6.1. For box C, the solid lines have $\beta = 10^{-2}$, and the dashed lines have $\beta = 10^{-6}$. For the other boxes, β does not make a discernible difference.

as

$$\boldsymbol{v}_1 = \frac{e}{\mathrm{i}\omega m}(\boldsymbol{E} + \boldsymbol{v}_1 \times \boldsymbol{B}_0) + \frac{c_e^2}{\omega^2}(\boldsymbol{k}\cdot\boldsymbol{v}_1)(\gamma_\| k_z \hat{\boldsymbol{e}}_z + \gamma_\perp k_x \hat{\boldsymbol{e}}_x)\,, \tag{6.61}$$

where $k_x = k\sin\theta$, $k_z = k\cos\theta$, and $k_y = 0$. Solving for the components of $\boldsymbol{v}_1$, we find, after some tedious algebra,

$$\begin{aligned}
v_{1x} &= \frac{e}{imD}\left[(\omega E_x - i\omega_{ce}E_y)(\omega^2 - k^2c_e^2\gamma_\|\cos^2\theta) + \omega E_z k^2 c_e^2\gamma_\perp\cos\theta\sin\theta\right] \\
v_{1y} &= \frac{e}{imD}\left[(\omega E_y(\omega^2 - k^2c_e^2\gamma_\|\cos^2\theta - k^2c_e^2\gamma_\perp\sin^2\theta)\right. \\
&\quad \left. +\mathrm{i}\omega_{ce}E_x(\omega^2 - k^2c_e^2\gamma_\|\cos^2\theta) + \mathrm{i}\omega_{ce}E_z k^2c_e^2\gamma_\perp\cos\theta\sin\theta\right] \\
v_{1z} &= \frac{e}{imD}\left[(\omega E_z(\omega^2 - \omega_{ce}^2 - k^2c_e^2\gamma_\perp\sin^2\theta)\right. \\
&\quad \left. +k^2c_e^2\gamma_\|\cos\theta\sin\theta(\omega E_x - \mathrm{i}\omega_{ce}E_y)\right]
\end{aligned}$$

where $D = \omega^2(\omega^2 - \omega_{ce}^2) + \omega_{ce}^2k^2c_e^2\gamma_\|\cos^2\theta - \omega^2k^2c_e^2(\gamma_\perp\sin^2\theta + \gamma_\|\cos^2\theta)$. Using $\boldsymbol{j} = -n_0e\boldsymbol{v}_1 = \boldsymbol{\sigma}\cdot\boldsymbol{E}$ and $\mathsf{K} = \mathsf{I} - \boldsymbol{\sigma}/\mathrm{i}\omega\epsilon_0$, the dielectric tensor has the form

$$\mathsf{K} = \begin{pmatrix} K_{xx} & K_{xy} & K_{xz} \\ -K_{xy} & K_{yy} & K_{yz} \\ K_{zx} & K_{xy} & K_{zz} \end{pmatrix} \tag{6.62}$$

where

$$\begin{aligned}
K_{xx} &= 1 - \frac{\omega_{pe}^2(\omega^2 - k^2c_e^2\gamma_\|\cos^2\theta)}{D} \\
K_{xy} &= \frac{\mathrm{i}\omega_{ce}\omega_{pe}^2(\omega^2 - k^2c_e^2\gamma_\|\cos^2\theta)}{\omega D} \\
K_{xz} &= -\frac{\omega_{pe}^2k^2c_e^2\gamma_\perp\cos\theta\sin\theta}{D} = \frac{\gamma_\perp}{\gamma_\|}K_{zx} \\
K_{yy} &= 1 - \frac{\omega_{pe}^2[\omega^2 - k^2c_e^2(\gamma_\|\cos^2\theta + \gamma_\perp\sin^2\theta)]}{D} \\
K_{yz} &= -\frac{\mathrm{i}\omega_{ce}\omega_{pe}^2k^2c_e^2\gamma_\perp\cos\theta\sin\theta}{\omega D} = -\frac{\gamma_\perp}{\gamma_\|}K_{zy} \\
K_{zz} &= 1 - \frac{\omega_{pe}^2(\omega^2 - \omega_{ce}^2 - k^2c_e^2\gamma_\perp\sin^2\theta)}{D}\,.
\end{aligned} \tag{6.63}$$

These dielectric tensor elements may easily be extended to include ions (letting $\omega_{ce} \to -\omega_{ci}$) by making a sum over species, but are very complicated for general use, except that they are valid in both electrostatic and electromagnetic regions. For cases where more than one ion species is involved, however, both the one-fluid model and the WPDR are very complicated with simple dispersion relations difficult to obtain, but the straightforwardness of the dielectric tensor method may be preferred for numerical work. In Figure 6.3, both ion and electron terms are included for comparison with the LFDR using the WPDR with $\gamma_\perp = \gamma_\parallel = 2$, equation (6.65), indicated by the dot-dash lines.

With this dielectric tensor, which has all nine components (although only six are independent), the wave equation generalizes to

$$\begin{pmatrix} K_{xx} - n^2\cos^2\theta & K_{xy} & n^2\cos\theta\sin\theta + K_{xz} \\ -K_{xy} & K_{yy} - n^2 & K_{yz} \\ n^2\cos\theta\sin\theta + K_{xz} & -K_{yz} & K_{zz} - n^2\sin^2\theta \end{pmatrix} \begin{pmatrix} E_x \\ E_y \\ E_z \end{pmatrix} = 0 \qquad (6.64)$$

where $n = kc/\omega$. The determinant of coefficients of equation (6.64) gives the WPDR, which is considerably more complicated than the cold plasma dispersion relation (where $K_{xz} = K_{yz} = 0$), but may be written as

$$\begin{aligned} &\big[(K_{xx} - n^2\cos^2\theta)(K_{yy} - n^2) + K_{xy}^2\big]K_{zz} - n^2\sin^2\theta\big[K_{xx}(K_{yy} - n^2) + K_{xy}^2\big] \\ &\quad + K_{xy}(K_{yz}K_{zx} - K_{zy}K_{xz}) + K_{xy}n^2\cos\theta\sin\theta(K_{yz} - K_{zy}) \\ &-(K_{xx} - n^2\cos^2\theta)K_{yz}K_{zy} - (K_{yy} - n^2)\big[n^2\cos\theta\sin\theta(K_{zx} + K_{xz}) + K_{xz}K_{zx}\big] \\ &= 0\,. \end{aligned} \qquad (6.65)$$

This warm plasma dispersion relation (WPDR) is very formidable, but simplifies in certain limits. Its general form is identical to that of the hot plasma dispersion relation, which included finite temperature effects through kinetic theory and was the subject of Chapter 5. It may be noted that K_{yz} as given here does *not* have the same form as the K_{yz} from the hot plasma dielectric tensor when only first order terms in temperature are kept, although the other tensor elements do agree to terms of that order. This is attributed to the use here of the equation of state to close the system of moment equations where we use the Vlasov-Maxwell equations for the hot plasma case. It may also be noted that, effectively, $\gamma_\perp = \gamma_\parallel = 2$ in the Vlasov-Maxwell equations.

Problem 6.8 *Derivation of the warm plasma dispersion relation.* Fill in the steps from equation (6.59) to equations (6.63) and (6.65).

Problem 6.9 The warm plasma dielectric tensor elements are similar, but not equal to, the hot plasma dielectric tensor elements. Compare K_{xz} and K_{yz} for each case for electrons only, keeping only the lowest order terms in v_e^2, and setting $\gamma_\perp = \gamma_\parallel = 2$ for the warm plasma terms.

6.4.1.1 Parallel propagation

When $\theta = 0$, the tensor elements simplify greatly, such that $K_{xz} = K_{yz} = 0$ and

$$K_{xx} = K_{yy} = 1 - \frac{\omega_{pe}^2}{\omega^2 - \omega_{ce}^2}$$

$$K_{xy} = \frac{\mathrm{i}\omega_{ce}\omega_{pe}^2}{\omega(\omega^2 - \omega_{ce}^2)}$$

$$K_{zz} = 1 - \frac{\omega_{pe}^2}{\omega^2 - k^2 c_e^2 \gamma_\parallel}$$

so except for K_{zz}, waves are unchanged from the cold plasma results (with $m_i \to \infty$ for the high frequency cases described here) by the fluid pressure terms. There is only one change, namely that the $K_{zz} = 0$ root leads now to

$$\omega^2 = \omega_{pe}^2 + k^2 c_e^2 \gamma_\parallel \tag{6.66}$$

which is equivalent to the Bohm-Gross dispersion relation for plasma waves if we take $\gamma_\parallel = 3$ and the thermal speed is small compared to the phase velocity.

6.4.1.2 Perpendicular propagation

When $\theta \to \pi/2$, we have $K_{xz} = K_{yz} = 0$ and now K_{zz} takes its cold plasma form so the ordinary wave is unchanged, but

$$\begin{aligned} K_{xx} &= 1 - \frac{\omega_{pe}^2}{\omega^2 - \omega_{ce}^2 - k^2 c_e^2 \gamma_\perp} \\ K_{yy} &= 1 - \frac{\omega_{pe}^2(\omega^2 - k^2 c_e^2 \gamma_\perp)}{\omega^2(\omega^2 - \omega_{ce}^2 - k^2 c_e^2 \gamma_\perp)} \\ K_{xy} &= \frac{\mathrm{i}\omega_{ce}\omega_{pe}^2}{\omega(\omega^2 - \omega_{ce}^2 - k^2 c_e^2 \gamma_\perp)} \end{aligned} \tag{6.67}$$

so that the extraordinary wave (X-wave) is now given by

$$n_X^2 = \frac{K_{xx}K_{yy} + K_{xy}^2}{K_{xx}} = \frac{(\omega^2 - \omega_{pe}^2)(\omega^2 - \omega_e^2 - k^2 c_e^2 \gamma_\perp) - \omega_{pe}^2 \omega_{ce}^2}{\omega^2(\omega^2 - \omega_e^2 - k^2 c_e^2 \gamma_\perp)} \tag{6.68}$$

where $\omega_e^2 \equiv \omega_{pe}^2 + \omega_{ce}^2$, and it is apparent the upper hybrid resonance has been shifted by the thermal term. The upper branch of the extraordinary wave then propagates from its cutoff up to the upper hybrid resonance as in the cold plasma X-wave, but when it reaches $v_p \sim kc_e$, it couples to the electron sound wave.

6.4.1.3 Arbitrary angle

The general WPDR is very complicated, but it may be approximated in certain regions. The cutoffs, of course, do not depend on the angle. Away from the cutoffs, we may approximate the high frequency branch of the R-X-wave by

$$\omega^2 \simeq \omega_{ce}^2 \left(1 + \frac{\omega_{pe}^2(1 + \cos^2\theta)}{\omega_{ce}^2 - \omega_{pe}^2 - k^2c^2} \right) + k^2 c_e^2 \sin^2\theta \gamma_\perp \,, \tag{6.69}$$

provided the denominator is not too small. The O-P-wave branch may be approximated by a similar expression

$$\omega^2 \simeq \omega_{pe}^2 \left(1 + \frac{\omega_{ce}^2 \sin^2\theta}{\omega_{pe}^2 - \omega_{ce}^2 + k^2c^2\sin^2\theta} \right) + k^2c^2\sin^2\theta + k^2 c_e^2 \cos^2\theta \gamma_\parallel \,. \tag{6.70}$$

These are sketched for an overdense plasma in Figure 6.3 and for an underdense plasma in Figure 6.4.

6.4.2 Electrostatic dispersion relation

When $k\lambda_{De} \geq 1$, equation (6.56) is inaccurate because of the neglect of space charge effects through the neglect of the $\boldsymbol{k}\cdot\boldsymbol{j}$ term. We could include this term through the charge continuity equation, whereby $\boldsymbol{k}\cdot\boldsymbol{j} = \omega\rho = \omega e(n_{1i} - n_{1e})$, but it is easier to use the electrostatic dispersion relation (ESDR) where the electric field is derived from a scalar potential. The large k limit, or low phase velocity limit, is a sufficient condition for the electrostatic approximation.

For this analysis, we may write the momentum equations for each species as

$$-\mathrm{i}\omega m_j \boldsymbol{v}_j = q_j\left[-\mathrm{i}\boldsymbol{k}\left(\frac{p_j}{n_0 q_j} + \varphi\right) + \boldsymbol{v}_j \times \boldsymbol{B}_0\right] \tag{6.71}$$

where φ is the scalar potential. This may be solved for the velocity components with $\boldsymbol{B}_0 = B_0\hat{e}_z$ such that

$$\boldsymbol{k}\cdot\boldsymbol{v}_j = \frac{1}{\omega m_j}\left(\frac{p_j}{n_0} + q_j\varphi\right)\left[\frac{k_x^2 + k_y^2}{1 - \omega_{cj}^2/\omega^2} + k_z^2\right]. \tag{6.72}$$

Then using the continuity equation and gas law ($p_j = c_j^2\rho_{1j}$) for each species, this may be written as

$$\boldsymbol{k}\cdot\boldsymbol{v}_j\left[1 - c_j^2\left(\frac{k_\perp^2}{\omega^2 - \omega_{cj}^2} + \frac{k_z^2}{\omega^2}\right)\right] = \frac{\omega q_j\varphi}{m_j}\left(\frac{k_\perp^2}{\omega^2 - \omega_{cj}^2} + \frac{k_z^2}{\omega^2}\right). \tag{6.73}$$

We may combine these expressions for each species through Poisson's equation,

$$k^2\varphi = \frac{n_0 e}{\omega\epsilon_0}(\boldsymbol{k}\cdot\boldsymbol{v}_i - \boldsymbol{k}\cdot\boldsymbol{v}_e),$$

and the result may be written as

$$1 + \frac{1}{k^2\lambda_{Di}^2} + \frac{1}{k^2\lambda_{De}^2} = \frac{1}{k^2\lambda_{Di}^2\left[1 - k^2c_i^2\left(\frac{\sin^2\theta}{\omega^2 - \omega_{ci}^2} + \frac{\cos^2\theta}{\omega^2}\right)\right]} + \frac{1}{k^2\lambda_{De}^2\left[1 - k^2c_e^2\left(\frac{\sin^2\theta}{\omega^2 - \omega_{ce}^2} + \frac{\cos^2\theta}{\omega^2}\right)\right]}, \tag{6.74}$$

where $\lambda_{Dj} = c_j/\omega_{pj}$. This dispersion relation is quartic in ω^2, but one root is the electron plasma wave and not a low frequency wave, so as long as $\omega \ll kc_e$, we may neglect the last term and reduce the dispersion relation to a quadratic in ω^2. This reduced dispersion relation may be used to plot the regions beyond D and H in both Figure 6.3 and Figure 6.4, but only the LFDR and the WPDR are actually plotted.

Problem 6.10 *Electrostatic dispersion relation.* Fill in the steps leading to equation (6.74).

6.4.2.1 Overdense case (high β)

Figure 6.3 shows the three roots of the LFDR (solid lines) in a low β hydrogen plasma ($c_s^2/V_A^2 = \beta = 0.01$, $m_i/m_e = 1836$) which is overdense ($\omega_{pe} > \omega_{ce}$). There are also roots of the warm plasma dispersion relation and the electrostatic dispersion relation.

We can identify the $R-X$ wave as the branch beginning at O_1 where it is a compressional Alfvén wave until A, after which it gradually changes slope until B where it enters the whistler wave region until it reaches C where it has a resonance. There is an apparent crossing of an ion-acoustic branch at D. In the original Stringer diagrams, these branches do not cross, since the ion-acoustic mode bends over and tends toward L while the first branch turns upward toward E, but they are not resolvable from equation (6.56).

The torsional Alfvén wave runs from O_2 to F where it approaches the ion cyclotron resonance at G, but for these parameters, it quickly couples to the ion-acoustic branch at H and is a simple ion-acoustic wave from I to D where it apparently crosses the first branch. The third branch that starts at O_3 does not occur in a cold plasma and is an ion-sound wave below M and approaches a resonance beyond P.

The warm plasma dispersion relation (WPDR), given by equation (6.65) with both electron and ion terms, is shown in figures 6.3 and 6.4 by the dot-dash lines. The corresponding roots of the WPDR generally follow those of the LFDR, but several additional roots are shown (the WPDR with both ions and electrons is 15th order in ω^2, but many roots are virtual double roots, so only 7 are distinct in this case). One additional root begins near O_3 as a magnetized ion-sound wave that experiences the resonance at the ion cyclotron resonance. Another begins at the ion cyclotron frequency and then becomes an unmagnetized ion sound wave. Yet another is a magnetized electron sound wave until it experiences the electron cyclotron resonance. The highest frequency branch shown begins at the electron cyclotron frequency and then follows the electron sound wave. Three more higher modes are beyond range of the figure.

One important difference between the LFDR and the WPDR is evident as one follows the LFDR (solid line) above I. As this root approaches J, the WPDR indicates a gradual transition from an ion-acoustic wave to an ion sound wave, so the LFDR is unreliable in describing this high frequency region.

6.4.2.2 Underdense case (low β)

For an underdense plasma ($\omega_{pe} < \omega_{ce}$), the roots of both the LFDR and the WPDR are shown in Figure 6.4 where now we have $c_s/V_A = 10^{-3}$ and $c/V_A = 10$. For this case, the first branch is little changed except that above B the LFDR indicates that the phase velocity exceeds c so the neglect of those

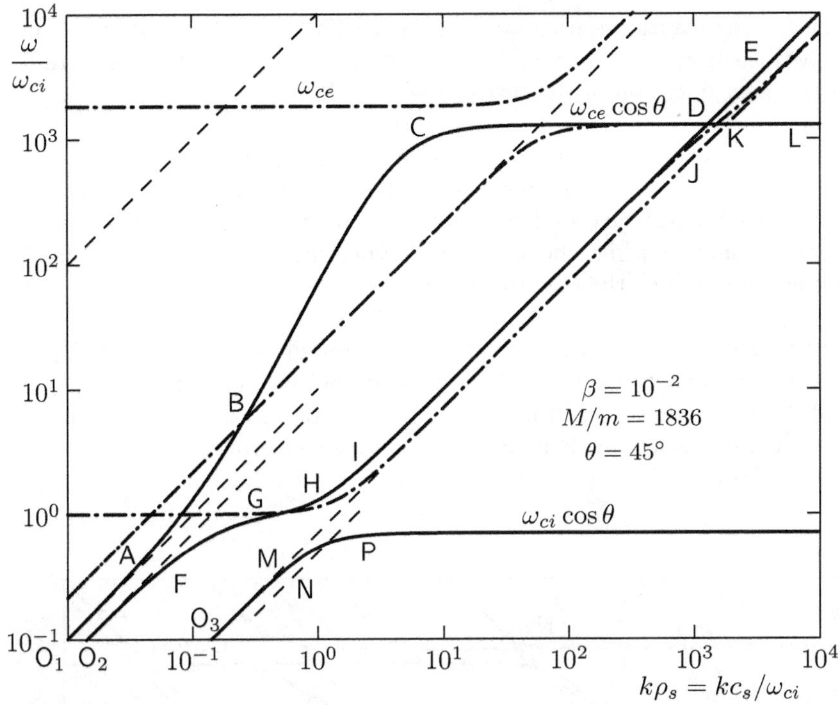

FIGURE 6.3
Dispersion curves for an overdense plasma. Solid lines are from the low frequency dispersion relation (LFDR), equation (6.56), while the dot-dash lines are from the WPDR, equation (6.65).

terms of order ω/kc is not valid in this region. Here the LFDR continues to the electron cyclotron resonance while the WPDR indicates that this root bends over near C to follow the electron plasma frequency until it makes an additional pair of transitions, first to the magnetized electron sound wave and finally to the electron cyclotron frequency. The other significant deviation from the LFDR is the transition from an ion-acoustic wave to an ion sound wave in the region between J and K which occurred at a much higher frequency in the overdense case of Figure 6.3. The figure is somewhat confusing near D, where it shows a crossing of two LFDR roots, but as noted above, the rising branch has already moved over to the ion sound root, so no crossing occurs near D. The crossing does occur just to the right of D, and even the WPDR cannot distinguish between a crossing and a transition (there are five separate roots in this vicinity, and some are barely distinct, but no transition is evident even at high precision).

The third branch of the LFDR is basically unchanged from the overdense case, since the acoustic waves are relatively insensitive to changes in the density. The three additional high frequency branches of the WPDR which were off scale for the overdense case are seen here. The highest frequency branch makes a transition to $v_p = c$ at N which is seen more clearly in Figure 6.5 which is a magnification of the box in Figure 6.4.

Two distinct roots are barely resolvable in the figure, the lower of which was also present in the overdense case. The two lower branches begin to asymptote to the velocity of light, but the upper one crosses the electron cyclotron resonance while the lower one experiences the resonance, finally making a transition to the electron sound wave.

Both figures 6.3 and 6.4 are similar to figures given by Stringer[42], but differ in some important ways. Some transitions which occur slowly in the original diagrams occur much more rapidly here or not at all, so the nature of the transitions is less evident. Several transitions in the original figures are not due to the equations given here or there.

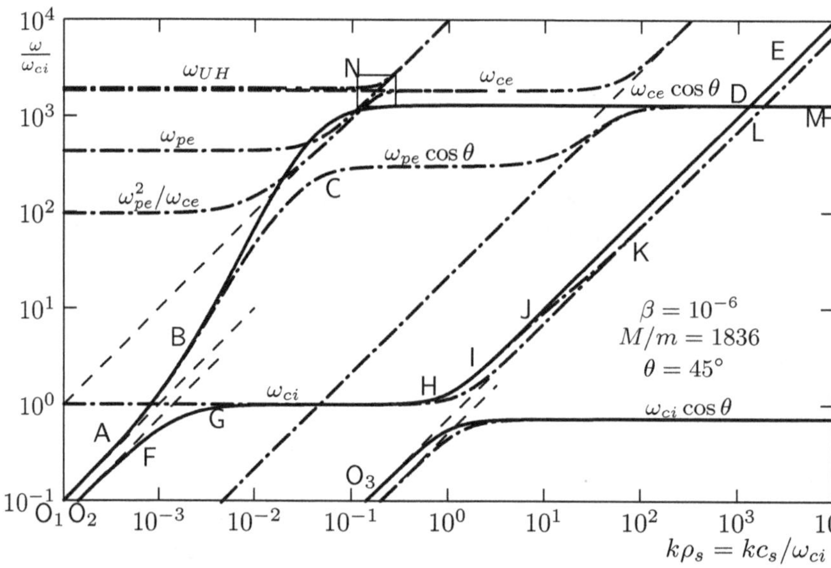

FIGURE 6.4
Dispersion curves for an underdense plasma. Solid lines are from the low frequency dispersion relation (LFDR), equation (6.56), while the dot-dash lines are from the WPDR, equation (6.65). The box near N is magnified in figure 6.5.

The region which has both high frequency and large k, which we may call the *electrostatic region*, we may neglect ω_{ci} compared to ω and also take

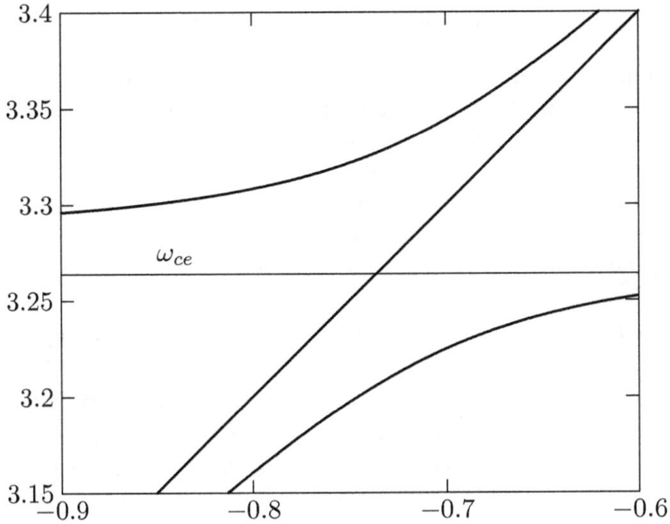

FIGURE 6.5
Magnified area near N from the warm plasma dispersion relation (WPDR) in figure 6.4.

$\omega \ll kc_e$ (except where $\omega \simeq \omega_{ce}\cos\theta$). In this case equation (6.74) reduces to

$$\frac{\omega^2}{k^2\lambda_{Di}^2(\omega^2 - k^2c_i^2)} = 1 + \frac{1}{k^2(\lambda_{Di}^2 + \lambda_{De}^2)} \tag{6.75}$$

with solution

$$\omega^2 = k^2c_i^2\left[1 + \frac{T_e}{T_i(1 + k^2\lambda_{De}^2)}\right]. \tag{6.76}$$

This gives $\omega = kc_s$ for $k\lambda_{De} \ll 1$ and $\omega = kc_i$ for $k\lambda_{De} \gg 1$. The transition occurs when $k\rho_s \sim c/V_A$ which is equivalent to $k\lambda_{De} = 1$. This transition is evident in both Figure 6.3 with $k\rho_s = c/V_A = 10^3$ and Figure 6.4 with $k\rho_s = c/V_A = 10$. This transition is missing from the LFDR, but evident from either the ESDR or the WPDR.

The other root behaves as $\omega = \omega_{ce}\cos\theta$ as $k \to \infty$, which is the K to L$\to \infty$ branch in Figure 6.3. This may be seen by multiplying equation (6.74) by $k^2\lambda_{De}^2$ and letting $k^2 \to \infty$ (holding ω fixed) whereupon the left hand side tends towards infinity as $k^2\lambda_{De}^2$, the first term on the right tends toward zero, and the second term also tends towards zero unless the denominator vanishes, so we have

$$\omega^2 \simeq \omega_{ce}^2\cos^2\theta - \frac{\omega^2(\omega_{ce}^2 - \omega^2)}{k^2c_e^2} \simeq \omega_{ce}^2\cos^2\theta\left[1 - \frac{\omega_{ce}^2\sin^2\theta}{k^2c_e^2}\right]. \tag{6.77}$$

6.4.3 Parallel and perpendicular propagation

The general character of the waves does not change greatly from the results above until $\theta \to 0$ or $\theta \to \pi/2$. As the angle of propagation approaches zero, the only significant change is that the transition regions become more localized and the transitions become sharper so that the coupling between the cold plasma waves and the thermal waves occurs only when the phase velocity is very close to one of the acoustic speeds. At $\theta = 0$, the R-wave and the L-wave are decoupled from thermal motions, but the plasma wave (a high frequency mode) remains coupled.

As the angle $\theta \to \pi/2$, the lowest frequency branch is essentially unchanged except that the resonant frequency $\omega = \omega_{ci} \cos\theta \to 0$ as $\cos\theta \to 0$.

The intermediate frequency branch, however, which is the shear Alfvén wave at the low frequency end with $\omega/k = V_A \cos\theta$, never experiences the ion cyclotron resonance when $V_A \cos\theta < c_s$ and propagates relatively smoothly up to its resonance at $\omega = \omega_{ce} \cos\theta \ll \omega_{ce}$ where it makes a transition to resonance.

The higher frequency branch, which is the magnetoacoustic branch at low frequencies, propagates at the Alfvén speed until it reaches the greater of $\omega_{ce} \cos\theta$ or the lower hybrid frequency where it makes a transition over to the ion-acoustic wave with $v_p = c_s$, or in the extreme case, to c_i.

6.4.4 Summary of fluid waves

The differences between cold plasma waves and the fluid plasma waves may now be seen to be due to the coupling of acoustic branches with the cold plasma waves. The L, R, and X waves simply tend towards infinity as they approach resonance, and the new feature is an acoustic wave with much lower phase velocity than the Alfvén wave which would cross these resonances if there were no coupling. The coupling, however, prevents a simple crossing and instead takes each resonant cold wave and converts it to an acoustic wave, and the original acoustic wave then converts to the resonant wave. Whether the acoustic wave propagates at the ion-acoustic speed c_s or the ion thermal speed (or for the high frequency branches, the electron thermal speed) is less obvious, and depends on the details of the dispersion relation.

In a homogeneous plasma, these waves are all linearly independent and energy does not couple from one branch to another in the transition regions. In inhomogeneous plasmas, however, where a transition region is approached in space, the waves are no longer independent and energy is coupled from one branch to another. For example, wave energy originating on branch O_3, the low frequency ion-acoustic branch, may couple some energy across from L to I by tunneling across the transition zone, and the remaining energy proceeds along the normal branch towards M. This kind of coupling is called linear mode conversion.

We also note that each of the transition regions violate the conditions for the validity of the fluid equations, except for the ion-acoustic wave transitions with $T_e \gg T_i$, since the moment expansion depended on the phase velocity being large compared to the thermal speed. Thus we would expect significant modifications of the dispersion relation near these regions from a full kinetic analysis, and we saw in Chapter 5 that both collisionless damping and modification of the dispersion do indeed occur when thermal effects are kept to higher order. The value of the fluid equations thus lies more in their ability to indicate the kinds of cold plasma wave – acoustic wave couplings that do occur than in an accurate description of all the thermal effects.

7

TRANSPORT IN A NONUNIFORM GAS

In the next three chapters, we will calculate the various transport coefficients that depend on collisions. These will include the pressure tensor, viscosity, thermal diffusion, and in electron-ion plasmas, the electrical conductivity. In the process, we will discuss entropy and show that maximizing the entropy will give the recipe we need to evaluate these various coefficients. We will also prove that the Maxwellian distribution is the equilibrium distribution. In each successive chapter, we will treat progressively more difficult plasma models and will apply the general methods introduced in this chapter to each case, ending with a plasma in a magnetic field.

In this chapter, we consider a simple nonuniform gas with only one charged species and no magnetic field. We assume a charge density of infinitely massive particles of opposite sign so that the gas is charge neutral. In the next chapter, we will extend the treatment to two species, electrons and one ion species, still with no magnetic field. In the final chapter, we will add the effects of the magnetic field in a one-fluid model, developing the equations of magnetohydrodynamics (MHD). The general development follows that of Marshall[44].

7.1 Boltzmann equation

Our basic equation for a simple gas is the Boltzmann equation

$$\left(\frac{\partial}{\partial t} + \boldsymbol{v}\cdot\nabla + \boldsymbol{a}\cdot\nabla_v\right)F = \Delta F \tag{7.1}$$

where $\boldsymbol{a}$ is the acceleration. The ΔF represents collisional effects. If Ψ is *any* function of $\boldsymbol{r}$, $\boldsymbol{v}$, and t, we define the average of Ψ by

$$\langle\Psi\rangle \equiv \frac{1}{n}\int \mathrm{d}\boldsymbol{v}\, F\Psi \tag{7.2}$$

where we use the shorthand expressions $\mathrm{d}\boldsymbol{v} = \mathrm{d}v_x\,\mathrm{d}v_y\,\mathrm{d}v_z$ and $\mathrm{d}\boldsymbol{r} = \mathrm{d}x\,\mathrm{d}y\,\mathrm{d}z$ and where

$$n \equiv \int \mathrm{d}\boldsymbol{v}\, F\,. \tag{7.3}$$

The average or drift velocity is

$$\boldsymbol{u} \equiv \langle \boldsymbol{v} \rangle = \frac{1}{n} \int \mathrm{d}\boldsymbol{v}\, F \boldsymbol{v}\,. \tag{7.4}$$

The random velocity is

$$\boldsymbol{w} \equiv \boldsymbol{v} - \boldsymbol{u}\,. \tag{7.5}$$

The kinetic temperature is defined by

$$\tfrac{3}{2} k_B T \equiv \tfrac{1}{2} m \langle w^2 \rangle = \frac{1}{n} \int \mathrm{d}\boldsymbol{v}\, F \tfrac{1}{2} m w^2\,. \tag{7.6}$$

The pressure tensor is given by

$$\mathsf{P} \equiv p_{\alpha\beta} = nm\langle w_\alpha w_\beta \rangle = m \int \mathrm{d}\boldsymbol{v}\, F\, w_\alpha w_\beta \tag{7.7}$$

where α and β run through x, y, z. We use the summation convention that if α or β are repeated, they are to be summed. There are only six independent components of the pressure tensor since $p_{\alpha\beta} = p_{\beta\alpha}$. The heat flux is given by

$$\boldsymbol{q} \equiv \frac{1}{2} nm \langle w^2 \boldsymbol{w} \rangle = \int \mathrm{d}\boldsymbol{v}\, F\, \frac{1}{2} nmw^2 \boldsymbol{w}\,. \tag{7.8}$$

Integrating over $\boldsymbol{v}$, we find

$$\frac{\partial n}{\partial t} + \nabla \cdot (n\boldsymbol{u}) = 0\,, \qquad \text{Continuity equation.} \tag{7.9}$$

Since particles are conserved, the collision term vanishes. Multiplying by $m\boldsymbol{v}$ and integrating, we find

$$\rho \frac{\partial \boldsymbol{u}}{\partial t} + \rho \boldsymbol{u} \cdot \nabla \boldsymbol{u} = -\nabla \cdot \mathsf{P} + \rho \langle \boldsymbol{a} \rangle\,, \qquad \text{Momentum equation} \tag{7.10}$$

and again the collision term vanishes because momentum is conserved in collisions. If we now multiply by $\frac{1}{2}mv^2$ and integrate over velocity, we find

$$\int \mathrm{d}\boldsymbol{v} \tfrac{1}{2} m v^2 \frac{\partial F}{\partial t} + \int \mathrm{d}\boldsymbol{v}\, \tfrac{1}{2} m v^2 \boldsymbol{v} \cdot \nabla F + \int \mathrm{d}\boldsymbol{v} \tfrac{1}{2} m v^2 \boldsymbol{a} \cdot \nabla_v F = \int \mathrm{d}\boldsymbol{v} \tfrac{1}{2} m v^2 \Delta F\,. \tag{7.11}$$

The first two terms may be written

$$\begin{aligned}
\frac{\partial}{\partial t}\left(\tfrac{1}{2} nm\langle v^2 \rangle\right) &= \frac{\partial}{\partial t}\left(\tfrac{1}{2} nmu^2\right) + \frac{\partial}{\partial t}\left(\tfrac{1}{2} nm \langle w^2 \rangle\right) \\
\nabla \cdot \left(\tfrac{1}{2} nm \langle v^2 \boldsymbol{v} \rangle\right) &= \nabla \cdot \left(\tfrac{1}{2} nmu^2 \boldsymbol{u}\right) + \nabla \cdot \left(\tfrac{1}{2} nm\boldsymbol{u} \langle w^2 \rangle\right) \\
&\quad + \nabla \cdot \left(nm\boldsymbol{u} \cdot \langle \boldsymbol{w}\boldsymbol{w} \rangle\right) + \nabla \cdot \left(\tfrac{1}{2} nm \langle w^2 \boldsymbol{w} \rangle\right)
\end{aligned}$$

while the third term may be written as

$$\begin{aligned}\int d\boldsymbol{v}\tfrac{1}{2}mv^2\boldsymbol{a}\cdot\nabla_v F &= \tfrac{1}{2}m\int d\boldsymbol{v}\nabla\cdot(v^2\boldsymbol{a}F) - \tfrac{1}{2}m\int d\boldsymbol{v}F\nabla_v\cdot(v^2\boldsymbol{a}) \\ &= \tfrac{1}{2}m\oint Fv^2\boldsymbol{a}\cdot d\boldsymbol{S}_v - \tfrac{1}{2}m\int d\boldsymbol{v}Fv^2\nabla_v\cdot\boldsymbol{a} \\ &\quad -m\int d\boldsymbol{v}F\boldsymbol{v}\cdot\boldsymbol{a} \\ &= -\rho\boldsymbol{u}\cdot\boldsymbol{a} - \rho\langle\boldsymbol{w}\cdot\boldsymbol{a}\rangle\end{aligned}$$

where the surface integral in velocity space vanishes as $|v| \to \infty$ because $F \to 0$, and we assume $\boldsymbol{a}$ is velocity independent in the absence of a magnetic field. These result in

$$\frac{D}{Dt}\left(\frac{3}{2}nk_BT\right) = \frac{3}{2}k_BT\frac{Dn}{Dt} - \nabla\cdot\boldsymbol{q} - \mathsf{P}\cdot\nabla\boldsymbol{u} + \rho\langle\boldsymbol{w}\cdot\boldsymbol{a}\rangle \tag{7.12}$$

or

$$\frac{3}{2}nk_B\frac{DT}{Dt} = -\nabla\cdot\boldsymbol{q} - \mathsf{P}\cdot\nabla\boldsymbol{u} + \rho\langle\boldsymbol{w}\cdot\boldsymbol{a}\rangle \tag{7.13}$$

where

$$\frac{D}{Dt} \equiv \frac{\partial}{\partial t} + \boldsymbol{u}\cdot\nabla\,.$$

Problem 7.1 Fill in the missing steps between equations (7.11) and (7.13).

7.2 Collision symmetries

Without specifying the interaction force law, there are certain symmetries characteristic of any collision model. We begin by specifying one particle with velocity $\boldsymbol{v}$ before the collision and velocity $\boldsymbol{v}'$ after the collision, and the other particle has velocity $\boldsymbol{s}$ before the collision and velocity $\boldsymbol{s}'$ after the collision. The velocity of the center of mass is invariant in the collision, such that

$$\boldsymbol{v}_{\rm cm} = \tfrac{1}{2}(\boldsymbol{v}+\boldsymbol{s}) = \tfrac{1}{2}(\boldsymbol{v}'+\boldsymbol{s}') \tag{7.14}$$

and the relative velocities are

$$\boldsymbol{g} = \boldsymbol{s} - \boldsymbol{v} \qquad\qquad \boldsymbol{g}' = \boldsymbol{s}' - \boldsymbol{v}'$$

so that

$$\boldsymbol{s} = \boldsymbol{v}_{\rm cm} + \tfrac{1}{2}\boldsymbol{g} \qquad\qquad \boldsymbol{s}' = \boldsymbol{v}_{\rm cm} + \tfrac{1}{2}\boldsymbol{g}' \tag{7.15}$$

and

$$\boldsymbol{v} = \boldsymbol{v}_{\rm cm} - \tfrac{1}{2}\boldsymbol{g} \qquad\qquad \boldsymbol{v}' = \boldsymbol{v}_{\rm cm} - \tfrac{1}{2}\boldsymbol{g}'\,. \tag{7.16}$$

Since energy is conserved, we have

$$\tfrac{1}{2}m(v^2+s^2)=\tfrac{1}{2}m(v'^2+s'^2) \tag{7.17}$$

and using equations (7.15) and (7.16), we find that

$$g=g' \tag{7.18}$$

or that the magnitude of the relative velocities is equivalent before and after. The geometry of a collision is illustrated in Figure 7.1 which shows the relative

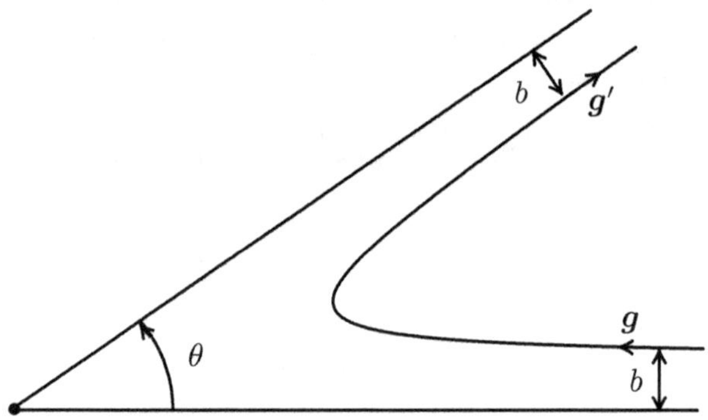

FIGURE 7.1
Collision trajectory.

velocity $\boldsymbol{g}$ before the collision and the relative velocity $\boldsymbol{g}'$ after the collision, having been deflected through an angle θ. The asymptotic distance from the action center is b before and after the collision from the conservation of angular momentum and is called the *impact parameter*. The deflection angle θ will be determined from the law of interaction as a function of b, but this law will be discussed later. The plane of interaction (the plane of the paper) is taken to be at an angle ϕ relative to a fixed plane.

We consider collisions that occur in a volume $\mathrm{d}\boldsymbol{r}$ in a time interval between t and $t+\mathrm{d}t$ between particles with velocities between $\boldsymbol{v}$ and $\boldsymbol{v}+\mathrm{d}\boldsymbol{v}$ and $\boldsymbol{s}$ and $\boldsymbol{s}+\mathrm{d}\boldsymbol{s}$, respectively and with impact parameters between b and $b+\mathrm{d}b$ and the azimuthal angle between ϕ and $\phi+\mathrm{d}\phi$. Assuming that we can neglect correlations with other particles, the number of collisions is proportional to the number of particles in the range $\mathrm{d}\boldsymbol{r}\mathrm{d}\boldsymbol{v}$, i.e., to $F(\boldsymbol{r},\boldsymbol{v},t)\mathrm{d}\boldsymbol{r}\mathrm{d}\boldsymbol{v}$ and to the number of particles in the range $\mathrm{d}\boldsymbol{s}$ that are in a cylinder in space of length $g\mathrm{d}t$ and of base area $b\,\mathrm{d}b\,\mathrm{d}\phi$. The number of such collisions is therefore

$$F(\boldsymbol{r},\boldsymbol{v},t)F(\boldsymbol{r},\boldsymbol{s},t)g\,b\,\mathrm{d}b\,\mathrm{d}\phi\,\mathrm{d}\boldsymbol{r}\,\mathrm{d}\boldsymbol{v}\,\mathrm{d}\boldsymbol{s}\,\mathrm{d}t\,. \tag{7.19}$$

For these collisions, the final velocities lie between $\boldsymbol{v}'$ and $\boldsymbol{v}'+\mathrm{d}\boldsymbol{v}'$ and between $\boldsymbol{s}'$ and $\boldsymbol{s}'+\mathrm{d}\boldsymbol{s}'$ where $\boldsymbol{v}'$ and $\boldsymbol{s}'$ depend on $\boldsymbol{v}$, $\boldsymbol{s}$, b, and ϕ through the law of interaction. It may be shown quite generally that

$$\mathrm{d}\boldsymbol{v}'\,\mathrm{d}\boldsymbol{s}' = \mathrm{d}\boldsymbol{v}\,\mathrm{d}\boldsymbol{s}\,. \tag{7.20}$$

There is a close relationship between inverse collisions, illustrated in Figure 7.2, where the initial velocities are $\boldsymbol{v}'$ and $\boldsymbol{s}'$ and the final velocities are $\boldsymbol{v}$ and $\boldsymbol{s}$ with the same b and ϕ. The number of these inverse collisions is

$$F(\boldsymbol{r},\boldsymbol{v}',t)F(\boldsymbol{r},\boldsymbol{s}',t)g'\,b\,\mathrm{d}b\,\mathrm{d}\phi\,\mathrm{d}\boldsymbol{r}\,\mathrm{d}\boldsymbol{v}'\,\mathrm{d}\boldsymbol{s}'\,\mathrm{d}t$$

which is equivalent to

$$F(\boldsymbol{r},\boldsymbol{v}',t)F(\boldsymbol{r},\boldsymbol{s}',t)g\,b\,\mathrm{d}b\,\mathrm{d}\phi\,\mathrm{d}\boldsymbol{r}\,\mathrm{d}\boldsymbol{v}\,\mathrm{d}\boldsymbol{s}\,\mathrm{d}t\,. \tag{7.21}$$

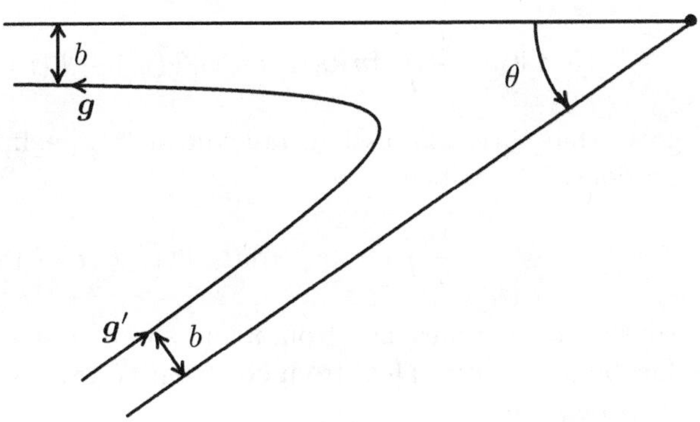

FIGURE 7.2
Inverse Collision trajectory.

The collisions of equation (7.19) scatter particles *out* of the range $\mathrm{d}\boldsymbol{v}$ while the collisions of equation (7.21) scatter particles *into* the range $\mathrm{d}\boldsymbol{v}$, so the net number of particles scattered into the range $\mathrm{d}\boldsymbol{v}$ is

$$\mathrm{d}\boldsymbol{v}\,\mathrm{d}\boldsymbol{r}\,\mathrm{d}t\int\mathrm{d}\boldsymbol{s}\int_0^{2\pi}\mathrm{d}\phi\int_0^d\mathrm{d}b\,bg[F(\boldsymbol{v}')F(\boldsymbol{s}')-F(\boldsymbol{v})F(\boldsymbol{s})]$$

where d is the maximum value of b which we take to be the Debye length, λ_D. Hence, the expression for ΔF in equation (7.1) is

$$\Delta F(\boldsymbol{r},\boldsymbol{v},t) = \int\mathrm{d}\boldsymbol{s}\int_0^{2\pi}\mathrm{d}\phi\int_0^d\mathrm{d}b\,bg[F(\boldsymbol{v}')F(\boldsymbol{s}')-F(\boldsymbol{v})F(\boldsymbol{s})] \tag{7.22}$$

where $\boldsymbol{v}'$ and $\boldsymbol{s}'$ are functions of $\boldsymbol{v}$, $\boldsymbol{s}$, b, and ϕ through the law of interaction.

7.3 Collision theorems

If Ψ is any property of the particles that depends on the position, velocity, and time, then the mean value of Ψ is $\langle\Psi\rangle$, and the rate of change of this mean value is from equation (7.2),

$$\langle\Delta\Psi\rangle = \frac{1}{n}\int \mathrm{d}\boldsymbol{v}\, \Psi\Delta F(\boldsymbol{v}) \tag{7.23}$$

$$= \frac{1}{n}\int \mathrm{d}\boldsymbol{v}\mathrm{d}\boldsymbol{s}\mathrm{d}\phi\mathrm{d}b\, bg\Psi(\boldsymbol{v})[F(\boldsymbol{v}')F(\boldsymbol{s}') - F(\boldsymbol{v})F(\boldsymbol{s})]\,. \tag{7.24}$$

We could also derive an expression for $\langle\Delta\Psi\rangle$ from equation (7.19). Equation (7.19) is the number of collisions of a specified type in which $\boldsymbol{v}$ changes to $\boldsymbol{v}'$. For each collision, Ψ changes by an amount $\Psi(\boldsymbol{v}) - \Psi(\boldsymbol{v}')$. Therefore, the total rate of change of $\langle\Psi\rangle$ due to collisions can also be written as

$$\langle\Delta\Psi\rangle = \frac{1}{n}\int \mathrm{d}\boldsymbol{v}\,\mathrm{d}\boldsymbol{s}\,\mathrm{d}\phi\,\mathrm{d}b\, bg[\Psi(\boldsymbol{v}') - \Psi(\boldsymbol{v})]F(\boldsymbol{v})F(\boldsymbol{s})]\,. \tag{7.25}$$

We prove that these expressions are equivalent, beginning with the first term in equation (7.24) which is

$$\frac{1}{n}\int \mathrm{d}\boldsymbol{v}\,\mathrm{d}\boldsymbol{s}\,\mathrm{d}\phi\,\mathrm{d}b\, bg\Psi(\boldsymbol{v}')F(\boldsymbol{v}')F(\boldsymbol{s}')\,. \tag{7.26}$$

We can then change variables from $\boldsymbol{v}$ and $\boldsymbol{s}$ to $\boldsymbol{v}'$ and $\boldsymbol{s}'$ and by equation (7.20), the Jacobian is unity. Then from equation (7.18), we may replace g by g' so that this becomes

$$\frac{1}{n}\int \mathrm{d}\boldsymbol{v}'\,\mathrm{d}\boldsymbol{s}'\,\mathrm{d}\phi\,\mathrm{d}b\, bg'\Psi(\boldsymbol{v})F(\boldsymbol{v}')F(\boldsymbol{s}')\,.$$

Now by symmetry, $\boldsymbol{v}$ is the same function of $\boldsymbol{v}'$, $\boldsymbol{s}'$, b, and ϕ as $\boldsymbol{v}'$ is of $\boldsymbol{v}$, $\boldsymbol{s}$, b, and ϕ. Hence, this last expression may be written as

$$\frac{1}{n}\int \mathrm{d}\boldsymbol{v}\,\mathrm{d}\boldsymbol{s}\,\mathrm{d}\phi\,\mathrm{d}b\, bg\Psi(\boldsymbol{v}')F(\boldsymbol{v})F(\boldsymbol{s})\,. \tag{7.27}$$

This result along with the second term of equation (7.24) is equation (7.25).

Because all particles are equivalent equation (7.25) may also be written as

$$\langle\Delta\Psi\rangle = \frac{1}{2n}\int \mathrm{d}\boldsymbol{v}\,\mathrm{d}\boldsymbol{s}\,\mathrm{d}\phi\,\mathrm{d}b\, bg[\Psi(\boldsymbol{v}') + \Psi(\boldsymbol{s}') - \Psi(\boldsymbol{v}) - \Psi(\boldsymbol{s})]F(\boldsymbol{v})F(\boldsymbol{s})\,,$$

which, in the same way that equation (7.26) is equivalent to equation (7.27), can be shown to be

$$\langle\Delta\Psi\rangle = -\frac{1}{2n}\int \mathrm{d}\boldsymbol{v}\,\mathrm{d}\boldsymbol{s}\,\mathrm{d}\phi\,\mathrm{d}b\, bg[\Psi(\boldsymbol{v}') + \Psi(\boldsymbol{s}') - \Psi(\boldsymbol{v}) - \Psi(\boldsymbol{s})]F(\boldsymbol{v}')F(\boldsymbol{s}')\,.$$

By adding these last two equations and dividing by 2, we have yet another formula,

$$\langle \Delta\Psi \rangle = \frac{1}{4n}\int d\boldsymbol{v}\, d\boldsymbol{s}\, d\phi\, db\, bg[\Psi(\boldsymbol{v}') + \Psi(\boldsymbol{s}') - \Psi(\boldsymbol{v}) - \Psi(\boldsymbol{s})] \times [F(\boldsymbol{v})F(\boldsymbol{s}) - F(\boldsymbol{v}')F(\boldsymbol{s}')].$$

These various formulas may be summarized as

$$\langle \Delta\Psi \rangle = \frac{1}{n}\int d\boldsymbol{v}\Psi(\boldsymbol{v})\Delta F(\boldsymbol{v}) \tag{7.28}$$

$$= \frac{1}{n}\int d\boldsymbol{v}\, d\boldsymbol{s}\, d\phi\, db\, bg\Psi(\boldsymbol{v})[F(\boldsymbol{v}')F(\boldsymbol{s}') - F(\boldsymbol{v})F(\boldsymbol{s})] \tag{7.29}$$

$$= \frac{1}{n}\int d\boldsymbol{v}\, d\boldsymbol{s}\, d\phi\, db\, bg[\Psi(\boldsymbol{v}') - \Psi(\boldsymbol{v})]F(\boldsymbol{v})F(\boldsymbol{s}) \tag{7.30}$$

$$= \frac{1}{2n}\int d\boldsymbol{v}\, d\boldsymbol{s}\, d\phi\, db\, bg[\Psi(\boldsymbol{v}') + \Psi(\boldsymbol{s}') - \Psi(\boldsymbol{v}) - \Psi(\boldsymbol{s})]F(\boldsymbol{v})F(\boldsymbol{s}) \tag{7.31}$$

$$= -\frac{1}{4n}\int d\boldsymbol{v}\, d\boldsymbol{s}\, d\phi\, db\, bg[\Psi(\boldsymbol{v}') + \Psi(\boldsymbol{s}') - \Psi(\boldsymbol{v}) - \Psi(\boldsymbol{s})] \times [F(\boldsymbol{v}')F(\boldsymbol{s}') - F(\boldsymbol{v})F(\boldsymbol{s})]. \tag{7.32}$$

The derivation of these expressions has depended only on the symmetries of the problem. It is evident that if Ψ is 1, $m\boldsymbol{v}$, or $\frac{1}{2}mv^2$, then $\langle \Delta\Psi \rangle$ vanishes from the conservation laws. Because these laws uniquely determine the final velocities in terms of the initial velocities and b and ϕ, it follows that only for these choices of Ψ does $\langle \Delta\Psi \rangle$ vanish. It also follows that if $\langle \Delta\Psi \rangle$ vanishes, then Ψ must be a linear combination of the "collision invariants," 1, $m\boldsymbol{v}$, and $\frac{1}{2}mv^2$. Since the particles also have spin, there are other invariants, but because the Coulomb force is so much stronger than the spin interactions, we neglect them. If the analysis were extended to include ionized molecules instead of particles, then some modifications would be necessary because molecules may convert kinetic energy into internal rotational and vibrational energy, so the conservation of energy would need to be changed.

7.4 The equilibrium state

It is the purpose of this section to show that the system, if left to itself, will approach an equilibrium state in which the distribution function will be Maxwellian. Further properties, such as the uniformity of the temperature and some restrictions on the drift velocity, will be included.

We begin with the quantity

$$S(t) = -k_B \int d\boldsymbol{r} \int d\boldsymbol{v} F(\boldsymbol{r}, \boldsymbol{v}, t) \ln F(\boldsymbol{r}, \boldsymbol{v}, t)\,. \tag{7.33}$$

We will show that in equilibrium, S will be the entropy of the system within a constant. Differentiating with respect to time, we find

$$\frac{\mathrm{d}S}{\mathrm{d}t} = -k_B \int \mathrm{d}\boldsymbol{r} \int \mathrm{d}\boldsymbol{v}(1 + \ln F)\frac{\partial F}{\partial t}\,. \tag{7.34}$$

Substituting $\partial F/\partial t$ from equation (7.1), we find

$$\frac{1}{k_B}\frac{\mathrm{d}S}{\mathrm{d}t} = \int \mathrm{d}\boldsymbol{r} \int \mathrm{d}\boldsymbol{v}(1 + \ln F)\boldsymbol{v}\cdot\nabla F + \int \mathrm{d}\boldsymbol{r} \int \mathrm{d}\boldsymbol{v}(1 + \ln F)\boldsymbol{a}\cdot\nabla_v F \tag{7.35}$$

$$- \int \mathrm{d}\boldsymbol{r} \int \mathrm{d}\boldsymbol{v}(1 + \ln F)\Delta F\,. \tag{7.36}$$

Taking the terms on the right one by one, the first may be written as

$$\int \mathrm{d}\boldsymbol{r}\nabla\cdot \int \mathrm{d}\boldsymbol{v}F\ln F$$

which can be transformed into a surface integral over the boundary of the system

$$\oint \mathrm{d}\boldsymbol{A}\cdot \int \mathrm{d}\boldsymbol{v}\,\boldsymbol{v}_N F\ln F \tag{7.37}$$

where $\mathrm{d}\boldsymbol{A}$ is a surface element and $\boldsymbol{v}_N$ is the normal component of the velocity at the surface. Assuming the wall of the container is either infinitely large so the integral vanishes since $F \to 0$ as $v \to \infty$ or smooth so that every particle bounces back elastically in which case F is an even function of $\boldsymbol{v}_N$ and equation (7.37) is an odd function of $\boldsymbol{v}_N$ so the integral vanishes by symmetry. Provided equation (7.37) does vanish, then the system must tend to an equilibrium state with a Maxwellian distribution function, but if the wall is not smooth or the wall leads to inelastic collisions, then the system will not approach a Maxwellian. If, for example, the wall is continuously absorbing energy, then the system cannot approach an equilibrium state.

Assuming equation (7.37) does vanish, then we are left with the second and third terms in equation (7.36). The second may be written as

$$\int \mathrm{d}\boldsymbol{r} \int \mathrm{d}\boldsymbol{v}\,\boldsymbol{a}\cdot\nabla_v F\ln F$$

which may be integrated by parts to obtain

$$\sum_\alpha \int \mathrm{d}\boldsymbol{r}\mathrm{d}v_\beta \mathrm{d}v_\gamma [a_\alpha F\ln F]_{v_\alpha=-\infty}^{v_\alpha=\infty}$$

which vanishes because F tends to zero more rapidly than $\ln F$ tends to infinity as $|v_\alpha| \to \infty$. The last term is

$$-\int \mathrm{d}\boldsymbol{r}\, n\nabla\langle 1 + \ln F\rangle\,.$$

Then using equation (7.32), we have

$$\frac{\mathrm{d}S}{\mathrm{d}t} = \frac{k_B}{4}\int \mathrm{d}\boldsymbol{r}\int \mathrm{d}\boldsymbol{v}\,\mathrm{d}\boldsymbol{s}\,\mathrm{d}\phi\, b\,\mathrm{d}b\, g[\ln F(\boldsymbol{v}') + \ln F(\boldsymbol{s}') - \ln F(\boldsymbol{v}) - \ln F(\boldsymbol{s})] \times [F(\boldsymbol{v}')F(\boldsymbol{s}') - F(\boldsymbol{v})F(\boldsymbol{s})]$$

which may be rewritten as

$$\frac{\mathrm{d}S}{\mathrm{d}t} = \frac{k_B}{4}\int \mathrm{d}\boldsymbol{r}\int \mathrm{d}\boldsymbol{v}\,\mathrm{d}\boldsymbol{s}\,\mathrm{d}\phi\, b\,\mathrm{d}b\, gF(\boldsymbol{v})F(\boldsymbol{s})\left[\frac{F(\boldsymbol{v}')F(\boldsymbol{s}')}{F(\boldsymbol{v})F(\boldsymbol{s})} - 1\right]\ln\frac{F(\boldsymbol{v}')F(\boldsymbol{s}')}{F(\boldsymbol{v})F(\boldsymbol{s})}\,.$$

The two factors, $[F(\boldsymbol{v}')F(\boldsymbol{s}')/F(\boldsymbol{v})F(\boldsymbol{s}) - 1]$, and $\ln[F(\boldsymbol{v}')F(\boldsymbol{s}')/F(\boldsymbol{v})F(\boldsymbol{s})]$, are both positive or both negative depending on whether $F(\boldsymbol{v}')\,F(\boldsymbol{s}')$ is greater than or less than $F(\boldsymbol{v})F(\boldsymbol{s})$. The product is therefore positive definite or zero, so we may conclude that

$$\frac{\mathrm{d}S}{\mathrm{d}t} \geq 0\,. \tag{7.38}$$

This requires that as the distribution function varies, it does so in such a fashion that entropy increases. We will show later that it changes in such a manner as to maximize the rate of change of entropy, subject to certain restraints. It cannot increase indefinitely, since it can be shown that this would require the energy content of the gas or plasma to increase indefinitely. Therefore, it must eventually reach a state where S no longer increases, which is the equilibrium state and the equality in equation (7.38) holds. In this state, we have

$$F(\boldsymbol{v}')F(\boldsymbol{s}') = F(\boldsymbol{v})F(\boldsymbol{s})$$

for all collisions, and furthermore,

$$\ln F(\boldsymbol{v}') + \ln F(\boldsymbol{s}') - \ln F(\boldsymbol{v}) - \ln F(\boldsymbol{s}) = 0\,. \tag{7.39}$$

This requires that in the equilibrium state, $F(\boldsymbol{v})$ must be a linear combination of the collision invariants, so that

$$\ln F(\boldsymbol{v}) = a_1 + a_2 m\boldsymbol{v} + \tfrac{1}{2}a_3 m v^2 \tag{7.40}$$

where a_1, a_2, and a_3 are independent of $\boldsymbol{v}$. Equation (7.39) leads directly to the solution

$$F(\boldsymbol{v}) = a_0 \mathrm{e}^{-\frac{1}{2}ma_3(\boldsymbol{v}-\boldsymbol{a}_4)^2} \tag{7.41}$$

where a_0 and a_4 are new quantities independent of $\boldsymbol{v}$ and are related to the quantities a_1, a_2, and a_3.

The significance of these quantities may be seen first by substituting $F(\boldsymbol{v})$ from equation (7.41) into equation (7.3) and integrating to find the number density as

$$n = a_0\left(\frac{2\pi}{ma_3}\right)^{3/2}\,.$$

The drift velocity from equation (7.4) is

$$\boldsymbol{u} = \boldsymbol{a}_4$$

and the temperature as defined in equation (7.6) is

$$k_B T = \frac{1}{a_3}.$$

Hence the equilibrium state is characterized by

$$F(\boldsymbol{v}) = f(\boldsymbol{v}),$$

where

$$f(\boldsymbol{v}) = \frac{n}{v_t^3 \pi^{3/2}} \mathrm{e}^{-(\boldsymbol{v}-\boldsymbol{u})^2/v_t^2}, \tag{7.42}$$

where v_t is a thermal speed given by

$$v_t \equiv \left(\frac{2k_B T}{m}\right)^{1/2}.$$

When we consider the spatial dependence of n, $\boldsymbol{u}$, and T in this equilibrium state, we find the possibilities by inserting equation (7.42) into the Boltzmann equation, equation (7.1), and assuming no variation in time, we find

$$(\boldsymbol{v} \cdot \nabla + \boldsymbol{a} \cdot \nabla_v) f(\boldsymbol{v}) = 0$$

since the right-hand side is zero. This gives, after dividing by $f(\boldsymbol{v})$,

$$(\boldsymbol{w} + \boldsymbol{u}) \cdot \left(\frac{\nabla n}{n} - \frac{3}{2}\frac{\nabla T}{T} + \frac{w^2}{v_t^2}\frac{\nabla T}{T} + 2\frac{\boldsymbol{w}\nabla\boldsymbol{u}}{v_t^2}\right) - 2\frac{\boldsymbol{a} \cdot \boldsymbol{w}}{v_t^2} = 0, \tag{7.43}$$

where

$$\boldsymbol{w} \cdot \boldsymbol{w}\nabla\boldsymbol{u} = w_\alpha w_\beta \nabla_\alpha u_\beta.$$

Equation (7.43) must hold over all space and for all velocities $\boldsymbol{w}$. This means the coefficient of each power of $\boldsymbol{w}$ must vanish independently. We assume for the remainder of this section that the acceleration, $\boldsymbol{a}$, is independent of $\boldsymbol{w}$. The only term of order $w^2\boldsymbol{w}$ is

$$\boldsymbol{w} \cdot \frac{w^2}{v_t^2}\frac{\nabla T}{T}$$

so

$$\nabla T = 0,$$

which means that the temperature is uniform in the equilibrium state (providing, of course, that collisions with the walls are elastic). The only term of order w^2 is

$$\frac{m}{k_B T} w_\alpha w_\beta \nabla_\alpha w_\beta$$

which requires

$$\nabla_\alpha u_\beta + \nabla_\beta u_\alpha = 0\,, \tag{7.44}$$

for all α and β, and the only nontrivial solution of this is

$$\boldsymbol{u} = \boldsymbol{u}_0 + \boldsymbol{\omega} \times \boldsymbol{r}\,,$$

where $\boldsymbol{u}_0$ and $\boldsymbol{\omega}$ are constants. This means that the only drift motion is a uniform drift plus a rotation with constant angular velocity $\boldsymbol{\omega}$. The term of order w is

$$w_\alpha \left[\nabla_\alpha \ln n + 2\frac{(u_\beta \nabla_\beta u_\alpha - a_\alpha)}{v_t^2}\right]$$

and from equation (7.44),

$$\nabla_\beta u_\alpha = -\nabla_\alpha u_\beta$$

so we have from setting the linear term to zero the result

$$\boldsymbol{a} = \tfrac{1}{2}\nabla(v_t^2 \ln n - u^2)\,. \tag{7.45}$$

It follows that the curl of $\boldsymbol{a}$ vanishes so that it can be derived from the gradient of a scalar

$$\boldsymbol{a} = -\nabla V\,, \tag{7.46}$$

so the force is conservative. Solving equations (7.45) and (7.46) for n, we find

$$n = n_0 \exp\left(-\frac{mV}{k_B T} + \frac{u^2}{v_t^2}\right), \tag{7.47}$$

where the unexpected positive sign is correct. The term in u^2 is due to the centripetal acceleration and is largest where the velocity is greatest, or at the outside of a rotating system.

The constant term in $\boldsymbol{w}$ is

$$\frac{\boldsymbol{u} \cdot \nabla n}{n} = \boldsymbol{u} \cdot \nabla \ln n$$

which requires

$$\boldsymbol{u} \cdot \nabla(V - \tfrac{1}{2}u^2) = 0\,.$$

But from equation (7.44),

$$\boldsymbol{u} \cdot \nabla u^2 = 2u_\alpha u_\beta \nabla_\alpha u_\beta = 0\,.$$

This implies

$$\boldsymbol{u} \cdot \nabla V = 0\,,$$

which leads to the conclusion that flows can only occur along equipotential surfaces.

Now that we know the equilibrium distribution function, we can calculate the pressure tensor and the heat flow from equations (7.7) and (7.8) with the result

$$p_{\alpha\beta} = p\,\delta_{\alpha\beta} = nk_BT\delta_{\alpha\beta}\,, \tag{7.48}$$

$$q = 0\,. \tag{7.49}$$

Equation (7.48) demonstrates that at equilibrium, only hydrostatic forces are present in the pressure tensor and equation (7.49) tells us that there is no heat flow in equilibrium.

In the equilibrium state, the quantity S of equation (7.33) becomes

$$S = -k_B \int \mathrm{d}\boldsymbol{r}\, n \left[\ln\left(\frac{n}{v_t^3\pi^{3/2}}\right) - \frac{3}{2}\right] \tag{7.50}$$

which is the entropy within an additive constant.

7.5 The mean free time theory

We discuss in this section the mean free path theory of kinetic processes such as viscosity and thermal conduction. The theory is useful for insight, but is not entirely consistent. For a simple plasma, it gives the same qualitative results as the more nearly complete method described in the next section. The theory is commonly called a "mean free path" theory, but we will treat it more as a "mean free time" theory since this approach can more easily be extended to the case with magnetic fields. We will begin with the exact theory before going over to the common theory in order to make the approximations more evident.

The basic idea of the theory is to follow the motion of each particle from the time it last suffered a collision. This will enable us to relate the distribution function to its form prior to the last collision. It is simply another way to solve the Boltzmann equation.

Between collisions, the motion of a single particle is especially simple and given by Newton's Law with a simple constant acceleration such that the solution of the equation of motion is simply

$$\boldsymbol{v} = \boldsymbol{v}_0 + \boldsymbol{a}(t - t_0) \tag{7.51}$$

$$\boldsymbol{r} = \boldsymbol{r}_0 + \boldsymbol{v}_0(t - t_0) + \tfrac{1}{2}\boldsymbol{a}(t - t_0)^2 \tag{7.52}$$

where $\boldsymbol{r}_0$ and $\boldsymbol{v}_0$ are the coordinate and velocity at time t_0.

Let $\tau(\boldsymbol{r}, \boldsymbol{v}, t)$ be the mean time between collisions for a particle at $\boldsymbol{r}$ with velocity $\boldsymbol{v}$ at time t. The number of collisions in a volume $\mathrm{d}\boldsymbol{r}$ in the interval between t and $t + \mathrm{d}t$ with velocities between $\boldsymbol{v}$ and $\boldsymbol{v} + \mathrm{d}\boldsymbol{v}$ is

$$\mathrm{d}\boldsymbol{r}\,\mathrm{d}\boldsymbol{v}\,\mathrm{d}t\frac{F(\boldsymbol{r},\boldsymbol{v},t)}{\tau(\boldsymbol{r},\boldsymbol{v},t)}\,.$$

Comparing this with equation (7.19) we may observe that

$$\frac{1}{\tau(\boldsymbol{r},\boldsymbol{v},t)} = \int \mathrm{d}\boldsymbol{s}\,\mathrm{d}\phi\, b\,\mathrm{d}b\, gF(\boldsymbol{r},\boldsymbol{s},t)\,.$$

We now define $p(t)$ to be the probability that a particle survives without making a collision up to time t given that it made a collision at time zero. To find $p(t)$, we consider this probability at a later time $t + \mathrm{d}t$. The probability that a particle survives up to time $t + \mathrm{d}t$ is the probability it survives up to time t minus the probability that it suffers a collision in the time interval between t and $t + \mathrm{d}t$. This latter is the probability that it survives up to time t times the probability that it suffers a collision in the interval $\mathrm{d}t$. Therefore,

$$p(t + \mathrm{d}t) = p(t) - p(t)\frac{\mathrm{d}t}{\tau}\,, \tag{7.53}$$

which, if τ were constant, would have the solution

$$p(t) = \mathrm{e}^{-t/\tau}\,.$$

Actually, however, τ in equation (7.53), τ is the collision time for a particle with position $\boldsymbol{r}$, velocity $\boldsymbol{v}$, at time t, where $\boldsymbol{r}$ and $\boldsymbol{v}$ are given in terms of the position and velocity at time zero by equations (7.51) and (7.52). A more nearly precise solution is therefore given by

$$p(t) = \exp\left(-\int_0^t \frac{\mathrm{d}t'}{\tau(t')}\right)\,. \tag{7.54}$$

Now in order to calculate the coefficients of thermal conductivity and viscosity, it is necessary to evaluate the number of particles crossing an element of area in time $\mathrm{d}t$ with velocities in a certain range. In general, the theory is formulated to give this directly, but since this number is related to F, we may alternatively regard the theory as an attempt to calculate F, i.e., to solve Boltzmann's equation. We now endeavor to do just this.

The number of particles in a given volume between $\boldsymbol{r}_0$ and $\boldsymbol{r}_0 + \mathrm{d}\boldsymbol{r}_0$ with velocities in the range between $\boldsymbol{v}_0$ and $\boldsymbol{v}_0 + \mathrm{d}\boldsymbol{v}_0$ at time t_0 is, by definition,

$$\mathrm{d}\boldsymbol{r}_0\,\mathrm{d}\boldsymbol{v}_0\, F(\boldsymbol{r}_0,\boldsymbol{v}_0,t_0)\,. \tag{7.55}$$

Consider now those particles of this set that suffered their last collision in the time interval between t and $t+\mathrm{d}t$, ($t < t_0$). These are the particles that suffer a collision in the time interval between t and $t + \mathrm{d}t$ between $\boldsymbol{r}_0$ and $\boldsymbol{r}_0 + \mathrm{d}\boldsymbol{r}_0$ such that their final velocities after the collision lie in the range between $\boldsymbol{v}$

and $\boldsymbol{v} + \mathrm{d}\boldsymbol{v}$ where $\boldsymbol{r}$ and $\boldsymbol{v}$ are given in terms of $\boldsymbol{r}_0$ and $\boldsymbol{v}_0$ from equations (7.51) and (7.52) with

$$\mathrm{d}\boldsymbol{r} = \mathrm{d}\boldsymbol{r}_0 \qquad \mathrm{d}\boldsymbol{v} = \mathrm{d}\boldsymbol{v}_0 \,.$$

The number of such particles is given by integrating equation (7.21) over $\boldsymbol{s}$, ϕ, and b to give

$$\mathrm{d}\boldsymbol{r}\,\mathrm{d}\boldsymbol{v}\,\mathrm{d}t \int \mathrm{d}\boldsymbol{s}\,\mathrm{d}\phi\, b\,\mathrm{d}b\, gF(\boldsymbol{r},\boldsymbol{v}',t)F(\boldsymbol{r},\boldsymbol{s}',t) \tag{7.56}$$

which by using equation (7.22) in Boltzmann's equation can also be written as

$$\mathrm{d}\boldsymbol{r}\,\mathrm{d}\boldsymbol{v}\,\mathrm{d}t \left[\frac{F(\boldsymbol{r},\boldsymbol{v},t)}{\tau(\boldsymbol{r},\boldsymbol{v},t)} + \left(\frac{\partial}{\partial t} + \boldsymbol{v}\cdot\nabla + \boldsymbol{a}\cdot\nabla_v\right)F(\boldsymbol{r},\boldsymbol{v},t)\right]. \tag{7.57}$$

Not every one of these particles contribute to the set given by equation (7.55), however, since some may suffer more than one collision in the interval $t_0 - t$. The probability they do survive is given by

$$\exp\left(-\int_t^{t_0} \frac{\mathrm{d}t_1}{\tau(\boldsymbol{r}_1,\boldsymbol{v}_1,t_1)}\right) \tag{7.58}$$

where

$$\begin{aligned} \boldsymbol{v}_1 &= \boldsymbol{v}_0 + \boldsymbol{a}(t_1 - t_0)\,, \\ \boldsymbol{r}_1 &= \boldsymbol{r}_0 + \boldsymbol{v}_0(t_1 - t_0) + \tfrac{1}{2}\boldsymbol{a}(t_1 - t_0)^2\,. \end{aligned}$$

We now multiply equation (7.57) by equation (7.58), integrate over all t and equate to equation (7.55) to find

$$\begin{aligned} F(\boldsymbol{r}_0,\boldsymbol{v}_0,t_0) = \int_{-\infty}^{t_0} \mathrm{d}t \exp\left(-\int_t^{t_0} \frac{\mathrm{d}t_1}{\tau(\boldsymbol{r}_1,\boldsymbol{v}_1,t_1)}\right) &\left[\frac{F(\boldsymbol{r},\boldsymbol{v},t)}{\tau(\boldsymbol{r},\boldsymbol{v},t)}\right. \\ &\left. + \left(\frac{\partial}{\partial t} + \boldsymbol{v}\cdot\nabla + \boldsymbol{a}\cdot\nabla_v\right)F(\boldsymbol{r},\boldsymbol{v},t)\right]. \end{aligned} \tag{7.59}$$

This formula relates the distribution function at time t_0 to the distribution function at an earlier time t. No approximations have been made so that equation (7.59) is exact.

In the mean free time theory, we now use an iteration method to solve a formula that is equivalent to equation (7.59),

$$F(\boldsymbol{r}_0,\boldsymbol{v}_0,t_0) = \int_{-\infty}^{t_0} \mathrm{d}t \exp\left(-\int_t^{t_0} \frac{\mathrm{d}t_1}{\tau(\boldsymbol{r}_1,\boldsymbol{v}_1,t_1)}\right) \int \mathrm{d}\boldsymbol{s}\,\mathrm{d}\phi\, b\,\mathrm{d}b\, gF(\boldsymbol{r},\boldsymbol{v}',t)F(\boldsymbol{r},\boldsymbol{s}',t) \tag{7.60}$$

where we used equation (7.56) in place of equation (7.57). We cannot use equation (7.59) directly because any distribution would satisfy the equation.

The procedure we use is to insert an approximate expression for F on the right-hand side of equation (7.60) and integrate to get a better approximation.

The first step is to insert a Maxwellian distribution into the right-hand side of equation (7.60). This approximation is good if the collision time is short compared to any macroscopic relaxation time for the plasma. This means we let

$$F(\boldsymbol{r},\boldsymbol{v}',t)F(\boldsymbol{r},\boldsymbol{s}',t) = f(\boldsymbol{r},\boldsymbol{v},t)f(\boldsymbol{r},\boldsymbol{s},t)\,,$$

where $f(\boldsymbol{r},\boldsymbol{v},t)$ is a Maxwellian at $\boldsymbol{r}$ at time t. If we now ignore the change of the collision time over the path of the particle, then equation (7.60) becomes

$$F(\boldsymbol{r}_0,\boldsymbol{v}_0,t) = \int_{-\infty}^{t_0} \mathrm{d}t\,\exp\left[\frac{t-t_0}{\tau(\boldsymbol{r}_0,\boldsymbol{v}_0,t_0)}\right]\int \mathrm{d}\boldsymbol{s}\mathrm{d}\phi\, b\,\mathrm{d}b\, g f(\boldsymbol{r},\boldsymbol{v},t)f(\boldsymbol{r},\boldsymbol{s},t) \tag{7.61}$$

$$= \int_{-\infty}^{t_0} \mathrm{d}t\,\exp\left(\frac{t-t_0}{\tau_0}\right)\frac{f(\boldsymbol{r},\boldsymbol{v},t)}{\tau_0}\,, \tag{7.62}$$

where τ_0 is shorthand for $\tau(\boldsymbol{r}_0,\boldsymbol{v}_0,t_0)$.

The next step is to expand $f(\boldsymbol{r},\boldsymbol{v},t)$ in powers of $t-t_0$ such that

$$f(\boldsymbol{r},\boldsymbol{v},t) = f(\boldsymbol{r}_0,\boldsymbol{v}_0,t_0) + (t-t_0)\left(\frac{\partial}{\partial t_0} + \boldsymbol{v}_0\cdot\nabla_0 + \boldsymbol{a}\cdot\nabla_{v0}\right) f(\boldsymbol{r}_0,\boldsymbol{v}_0,t_0) + \mathcal{O}(t-t_0)^2 \tag{7.63}$$

where

$$\nabla_0 = \hat{e}_x\frac{\partial}{\partial x_0} + \hat{e}_y\frac{\partial}{\partial y_0} + \hat{e}_z\frac{\partial}{\partial z_0}$$

$$\nabla_{v0} = \hat{e}_x\frac{\partial}{\partial v_{x0}} + \hat{e}_y\frac{\partial}{\partial v_{y0}} + \hat{e}_z\frac{\partial}{\partial v_{z0}}\,.$$

Ignoring the terms of order $(t-t_0)^2$, which is equivalent to ignoring terms of order τ^2, and then integrating over t yields

$$F(\boldsymbol{r},\boldsymbol{v},t) = f(\boldsymbol{r},\boldsymbol{v},t) - \tau\left(\frac{\partial}{\partial t} + \boldsymbol{v}\cdot\nabla + \boldsymbol{a}\cdot\nabla_v\right) f(\boldsymbol{r},\boldsymbol{v},t)\,, \tag{7.64}$$

where we have dropped the subscript 0. We now, without any justification at this point, write the result of the normal mean free time theory as

$$\begin{aligned}
F(\boldsymbol{r},\boldsymbol{v},t) &= f(\boldsymbol{r},\boldsymbol{v},t) - \tau n\boldsymbol{w}\cdot\nabla\left(\frac{m}{2\pi k_B T}\right)^{3/2}\mathrm{e}^{-(\boldsymbol{v}-\boldsymbol{u})^2/v_t^2} \\
&= f(\boldsymbol{r},\boldsymbol{v},t) - \tau n\left[\boldsymbol{w}\cdot\nabla T\frac{\partial}{\partial T} + w_\alpha(\nabla_\alpha u_\beta)\frac{\partial}{\partial u_\beta}\right]\frac{1}{v_t^3\pi^{3/2}}\mathrm{e}^{-(\boldsymbol{v}-\boldsymbol{u})^2/v_t^2} \\
&= f(\boldsymbol{r},\boldsymbol{v},t)\left[1 + \tau\left(\frac{3}{2} - \frac{w^2}{v_t^2}\right)\frac{\boldsymbol{w}\cdot\nabla T}{T} - \frac{2\tau}{v_t^2}w_\alpha w_\beta\nabla_\alpha u_\beta\right]\,.
\end{aligned} \tag{7.65}$$

This expression resembles the result we would get by using the Maxwellian $f(\boldsymbol{r},\boldsymbol{v},t)$ in equation (7.64) and presuming the density to be uniform so that $\nabla n = 0$. In fact, equation (7.65) is qualitatively correct to this order and can be made quantitatively correct by modifying only a few terms so that it becomes

$$F(\boldsymbol{r},\boldsymbol{v},t)=f(\boldsymbol{r},\boldsymbol{v},t)\left[1+\tau\left(\frac{5}{2}-\frac{w^2}{v_t^2}\right)\frac{\boldsymbol{w}\cdot\nabla T}{T}-\frac{2\tau}{v_t^2}(w_\alpha w_\beta-\tfrac{1}{2}w^2\delta_{\alpha\beta})\nabla_\alpha u_\beta\right]. \tag{7.66}$$

The details of this expression are given in the beginning of the next section. The essential idea is that the term $\partial f/\partial t$ in equation (7.64) involves $\partial n/\partial t$, $\partial \boldsymbol{u}/\partial t$, and $\partial T/\partial t$, and these three quantities are related to the spatial derivatives ∇n, $\nabla_\alpha u_\beta$, and ∇T by the equations of motion. When these relations are used to eliminate the time derivatives in equation (7.64), one finds that many terms cancel leaving one with equation (7.66).

Now we have no justification for this last step from normal theory, but even if we had, it has inherent difficulties. If we were to insert the non-Maxwellian parts of F into the right-hand side of equation (7.60), it would lead to terms comparable to the non-Maxwellian terms on the right-hand side of equation (7.66). Furthermore, in going from equation (7.61) to equation (7.62), it is inconsistent to take into account the differences between $f(\boldsymbol{r},\boldsymbol{v},t)$ and $f(\boldsymbol{r}_0,\boldsymbol{v}_0,t_0)$ and ignore the variation of τ over the trajectory of the particle. Thus we expect equation (7.66) to be only qualitatively correct.

A better procedure is to assume a small departure from Maxwellian and solve for this departure self-consistently. For this, we assume

$$F(\boldsymbol{r},\boldsymbol{v},t) = f(\boldsymbol{r},\boldsymbol{v},t)[1+\varphi(\boldsymbol{r},\boldsymbol{v},t)] \tag{7.67}$$

where φ is small, so that using this in equation (7.60), we obtain to first order in φ the result

$$\begin{aligned}F(\boldsymbol{r}_0,\boldsymbol{v}_0,t_0)[1+\varphi(\boldsymbol{r}_0,\boldsymbol{v}_0,t_0)] &= \int_{-\infty}^{t_0}\mathrm{d}t\,\mathrm{e}^{(t-t_0)/\tau}\int \mathrm{d}\boldsymbol{s}\,\mathrm{d}\phi\, b\,\mathrm{d}b\, g\, f(\boldsymbol{r},\boldsymbol{v},t)f(\boldsymbol{r},\boldsymbol{s},t)\\ &\quad+\tau\int \mathrm{d}\boldsymbol{s}\,\mathrm{d}\phi\, b\,\mathrm{d}b\, g\, f(\boldsymbol{r}_0,\boldsymbol{v}_0,t_0)f(\boldsymbol{r}_0,\boldsymbol{s}_0,t_0)\\ &\quad\times[\varphi(\boldsymbol{v}_0')+\varphi(\boldsymbol{s}_0')]\,.\end{aligned} \tag{7.68}$$

Using the expansion of equation (7.63) and the arguments leading to equation (7.66), we find (dropping the subscript 0),

$$\begin{aligned}f(\boldsymbol{w})\varphi(\boldsymbol{w}) &= \tau\, f(\boldsymbol{w})\left(\frac{5}{2}-\frac{w^2}{v_t^2}\right)\frac{\boldsymbol{w}\cdot\nabla T}{T} - f(\boldsymbol{w})\frac{2\tau}{v_t^2}w_\alpha w_\beta\nabla_\alpha u_\beta\\ &\quad+\tau^2\int \mathrm{d}\boldsymbol{s}\mathrm{d}\phi\, b\,\mathrm{d}b\, g\, f(\boldsymbol{w})f(\boldsymbol{s})\left[\left(\frac{5}{2}-\frac{s^2}{v_t^2}\right)\frac{\boldsymbol{s}\cdot\nabla T}{T}-\frac{2}{v_t^2}s_\alpha s_\beta\nabla_\alpha u_\beta\right]\\ &\quad+\tau\int \mathrm{d}\boldsymbol{s}\mathrm{d}\phi\, b\,\mathrm{d}b\, g\, f(\boldsymbol{w})f(\boldsymbol{s})[\varphi(\boldsymbol{w}')+\varphi(\boldsymbol{s}')]\,.\end{aligned} \tag{7.69}$$

This equation, however, is more difficult to solve for φ than the equation obtained by substituting equation (7.67) directly into the Boltzmann equation. This is the approach that was used by Chapman and Cowling[49].

Qualitatively, equation (7.66) is correct, and substituting it into the definition of the pressure tensor of equation (7.7), we obtain the formula,

$$p_{\alpha\beta} = \left(p + \frac{2}{3}\mu\nabla\cdot\boldsymbol{u}\right)\delta_{\alpha\beta} - \mu(\nabla_\alpha u_\beta + \nabla_\beta u_\alpha)\,, \tag{7.70}$$

so, for example,

$$p_{xx} = p + \frac{2}{3}\mu\nabla\cdot\boldsymbol{u} - 2\mu\frac{\partial u_x}{\partial x}\,,$$

$$p_{xy} = -\mu\left(\frac{\partial u_y}{\partial x} + \frac{\partial u_x}{\partial y}\right) = p_{yx}\,,$$

and similarly for the other terms where p is the hydrostatic pressure and μ is the coefficient of viscosity.

Similarly, if we substitute equation (7.66) into the definition of the heat flux vector, equation (7.8), we find

$$\boldsymbol{q} = -\lambda\nabla T$$

where

$$\lambda = \frac{5k_B}{2m}nk_BT\tau\,.$$

7.6 The formal theory of kinetic processes

In this section we describe the method of Chapman and Cowling[49] for solving the Boltzmann equation by successive approximations. This method will not provide an accurate representation of the distribution function, but it will give accurate expressions for the pressure tensor and the heat flux vector. For many applications, this is sufficient.

We begin by assuming that collisions are most important in determining the distribution function and that to a first approximation, we have a Maxwellian distribution at each point in space. We then calculate corrections to this first approximation. We indicate this process mathematically by writing equation (7.1) as

$$\left(\frac{\partial}{\partial t} + \boldsymbol{v}\cdot\nabla + \boldsymbol{a}\cdot\nabla_v\right)F = \frac{1}{\eta}\Delta F$$

$$= \frac{1}{\eta}\int \mathrm{d}\boldsymbol{s}\mathrm{d}\phi\, b\,\mathrm{d}b\, g\left[F(\boldsymbol{v}')F(\boldsymbol{s}') - F(\boldsymbol{v})F(\boldsymbol{s})\right]. \tag{7.71}$$

We will eventually set η to unity since it is only a formal expansion parameter, but it will help us to group the terms in the expansion. We write

$$F = F^{(0)}(1 + \eta\varphi + \eta^2\psi + \cdots), \tag{7.72}$$

where by keeping more and more terms, we obtain the successive approximations.

Substituting equation (7.72) into equation (7.71) and setting the coefficients of each power of η to zero, we obtain a set of equations which, in principle, give a complete solution of equation (7.71). We will only keep the first two equations because it gets too difficult very rapidly.

The first equation, of order $1/\eta$, is

$$\frac{1}{\eta}\int \mathrm{d}\boldsymbol{s}\mathrm{d}\phi\, b\, \mathrm{d}b\, g\, [F^{(0)}(\boldsymbol{v}')F^{(0)}(\boldsymbol{s}') - F^{(0)}(\boldsymbol{v})F^{(0)}(\boldsymbol{s})],$$

whose solution is

$$F^{(0)}(\boldsymbol{r}, \boldsymbol{v}, t) = f \equiv \frac{n}{v_t^3 \pi^{3/2}} \mathrm{e}^{-(\boldsymbol{v}-\boldsymbol{u})^2/v_t^2}, \tag{7.73}$$

which is the Maxwellian distribution.

The terms on the right-hand side of equation (7.71) that are independent of η are

$$\int \mathrm{d}\boldsymbol{s}\mathrm{d}\phi\, b\, \mathrm{d}b\, g\, \{f(\boldsymbol{v}')f(\boldsymbol{s}')[\varphi(\boldsymbol{v}') + \varphi(\boldsymbol{s}')] - f(\boldsymbol{v})f(\boldsymbol{s})[\varphi(\boldsymbol{v}) + \varphi(\boldsymbol{s})]\}, \tag{7.74}$$

and since

$$f(\boldsymbol{v}')f(\boldsymbol{s}') = f(\boldsymbol{v})f(\boldsymbol{s}),$$

the result may be written as

$$\int \mathrm{d}\boldsymbol{s}\mathrm{d}\phi\, b\, \mathrm{d}b\, g\, f(\boldsymbol{v})f(\boldsymbol{s})[\varphi(\boldsymbol{v}') + \varphi(\boldsymbol{s}') - \varphi(\boldsymbol{v}) - \varphi(\boldsymbol{s})]. \tag{7.75}$$

The terms that are independent of η coming from the last two terms on the left-hand side of equation (7.71) are

$$(\boldsymbol{v}\cdot\nabla + \boldsymbol{a}\cdot\nabla_v)f = -\frac{m}{k_B T} f \boldsymbol{a}\cdot\boldsymbol{w} + \frac{\partial f}{\partial n}\boldsymbol{v}\cdot\nabla n + \frac{\partial f}{\partial u_\alpha} v_\beta \nabla_\beta u_\alpha + \frac{\partial f}{\partial T}\boldsymbol{v}\cdot\nabla T \tag{7.76}$$

where it is apparent that

$$\begin{aligned} \frac{\partial f}{\partial n} &= \frac{f}{n} \\ \frac{\partial f}{\partial u_\alpha} &= \frac{2f}{v_t^2} w_\alpha \\ \frac{\partial f}{\partial T} &= -\frac{f}{T}\left(\frac{3}{2} - \frac{w^2}{v_t^2}\right). \end{aligned} \tag{7.77}$$

The terms that are independent of η coming from the first term on the left-hand side of equation (7.71) require more care, such that

$$\frac{\partial f}{\partial t} = \frac{\partial f}{\partial n}\frac{\partial n}{\partial t} + \frac{\partial f}{\partial u_\alpha}\frac{\partial u_\alpha}{\partial t} + \frac{\partial f}{\partial T}\frac{\partial T}{\partial t}, \tag{7.78}$$

and we can substitute for $\partial n/\partial t$, $\partial u_\alpha/\partial t$, and $\partial T/\partial t$ from equations (7.9), (7.10), and (7.13). We must be aware, however, that in our expansion approach, some of these terms are higher order in η. The pressure tensor in equations (7.10) and (7.13) and the heat flux tensor in equation (7.13) are given to zero order in η by equations (7.48) and (7.49). Therefore, to determine $\partial n/\partial t$, $\partial u_\alpha/\partial t$, and $\partial T/\partial t$ to zero order in η, we use

$$\begin{aligned}\frac{\partial n}{\partial t} + \nabla\cdot(n\boldsymbol{u}) &= 0\\ \frac{\partial \boldsymbol{u}}{\partial t} + (\boldsymbol{u}\cdot\nabla)\boldsymbol{u} &= -\frac{1}{\rho}\nabla p + \boldsymbol{a}\\ \frac{3}{2}nk_B\frac{\partial T}{\partial t} + \frac{3}{2}nk_B\boldsymbol{u}\cdot\nabla T &= -p\nabla\cdot\boldsymbol{u}\,.\end{aligned} \tag{7.79}$$

Substituting into equation (7.78) using equation (7.77) and adding to equation (7.76), it is found that many terms cancel and after some tedious algebra, we obtain an equation for φ,

$$\begin{aligned}&f(\boldsymbol{v})\left[2(\varpi_\alpha\varpi_\beta - \tfrac{1}{3}\varpi^2\delta_{\alpha\beta})\nabla_\alpha u_\beta - (\tfrac{5}{2} - \varpi^2)\boldsymbol{w}\cdot\nabla\ln T\right]\\ &\quad = \int \mathrm{d}\boldsymbol{s}\mathrm{d}\phi\, b\, \mathrm{d}b\, g\, f(\boldsymbol{v})f(\boldsymbol{s})[\varphi(\boldsymbol{v}') + \varphi(\boldsymbol{s}') - \varphi(\boldsymbol{v}) - \varphi(\boldsymbol{s})]\,,\end{aligned} \tag{7.80}$$

where $\boldsymbol{\varpi}$ is the normalized random velocity such that

$$\boldsymbol{\varpi} = \boldsymbol{w}/v_t\,. \tag{7.81}$$

Problem 7.2 Work out the left-hand side of equation (7.71) by using equations (7.76) and (7.78) along with equations (7.77) and (7.79) to show that equation (7.80) has the proper form.

If we now turn our attention to the terms of order η in equation (7.71), we can derive the next approximation to the distribution, but we shall not carry the process any further, since all of the higher order terms involve second derivatives of the drift velocity and temperature or products of first derivatives, so for slowly varying quantities, we may expect these terms to be small. For shock waves, they could be important, but no evidence has yet been found in experiments for them.

Before we attempt to solve equation (7.80), it will be convenient to introduce a notation for the tensor that appears on the left-hand side of equation (7.80). Let

$$(\boldsymbol{\varpi}^0\boldsymbol{\varpi})_{\alpha\beta} \equiv \varpi_\alpha\varpi_\beta - \tfrac{1}{3}\varpi^2\delta_{\alpha\beta}\,. \tag{7.82}$$

Since equation (7.80) is linear in φ, it follows that the solution is of the form

$$\varphi(\boldsymbol{v}) = -B_{\alpha\beta}\nabla_\alpha u_\beta - A_\alpha \nabla_\alpha \ln T\,, \tag{7.83}$$

where the tensor B satisfies

$$-2f(\boldsymbol{v})\boldsymbol{\varpi}^0\boldsymbol{\varpi} = \int \mathrm{d}\boldsymbol{s}\mathrm{d}\phi\, b\,\mathrm{d}b\, g\, f(\boldsymbol{v})f(\boldsymbol{s})[\mathsf{B}(\boldsymbol{v}') + \mathsf{B}(\boldsymbol{s}') - \mathsf{B}(\boldsymbol{v}) - \mathsf{B}(\boldsymbol{s})] \tag{7.84}$$

and the vector $\boldsymbol{A}$ satisfies

$$f(\boldsymbol{v})(\tfrac{5}{2} - \varpi^2)\boldsymbol{w} = \int \mathrm{d}\boldsymbol{s}\mathrm{d}\phi\, b\,\mathrm{d}b\, g\, f(\boldsymbol{v})f(\boldsymbol{s})[\boldsymbol{A}(\boldsymbol{v}') + \boldsymbol{A}(\boldsymbol{s}') - \boldsymbol{A}(\boldsymbol{v}) - \boldsymbol{A}(\boldsymbol{s})]\,. \tag{7.85}$$

From the form of these equations, it is apparent that the tensor B is of the form

$$\mathsf{B} = \boldsymbol{\varpi}^0\boldsymbol{\varpi}\mathcal{B}(\varpi^2) \tag{7.86}$$

where $\mathcal{B}(\varpi^2)$ is some scalar function of the magnitude ϖ^2. Similarly, it is clear that

$$\boldsymbol{A} = \boldsymbol{\varpi}\mathcal{A}(\varpi^2) \tag{7.87}$$

where $\mathcal{A}(\varpi^2)$ is some scalar function of the magnitude ϖ^2.

We have thus reduced the problem to solving for the scalar quantities $\mathcal{A}$ and $\mathcal{B}$. In addition to being solutions of equations (7.84) and (7.85) we must impose the condition on $\boldsymbol{A}$ and B that

$$\langle \boldsymbol{w} \rangle = 0$$

or that

$$\frac{1}{n}\int \mathrm{d}\boldsymbol{v}\, f(\boldsymbol{v})[1 + \varphi(\boldsymbol{v})]\boldsymbol{w} = 0\,.$$

Substituting for $\varphi(\boldsymbol{v})$ from equation (7.83) and for $\boldsymbol{A}$ and B from equations (7.87) and (7.86), this becomes

$$\frac{1}{\pi^{3/2}}\int \mathrm{d}\boldsymbol{\varpi}\, \mathrm{e}^{-\varpi^2}\varpi^2\mathcal{A}(\varpi^2) = 0\,, \tag{7.88}$$

and there is no corresponding condition on $\mathsf{B}(\varpi^2)$.

We now have an exact solution for $\varphi(\boldsymbol{v})$ if we can find $\mathcal{A}$ and $\mathcal{B}$ exactly. One reasonable method is to take them to be power series in ϖ^2 with undetermined coefficients such that, for example,

$$\mathcal{A} = c_0 + c_1\varpi^2 + c_2\varpi^4 + \cdots\,.$$

It is much more convenient to rearrange the terms of this expansion and redefine the coefficients so that we may write

$$\begin{aligned}\mathcal{A} &= a_0 + a_1(\tfrac{5}{2} - \varpi^2) + a_2(\tfrac{35}{8} - \tfrac{7}{2}\varpi^2 + \tfrac{1}{2}\varpi^4) + \cdots \\ &= \sum_{m=0}^{\infty} a_m S_{\frac{3}{2}}^m(\varpi^2)\end{aligned} \tag{7.89}$$

where $S_n^m(\varpi^2)$ is a Sonine polynomial defined by

$$S_n^m(\varpi^2) \equiv \sum_{j=0}^{m} \frac{(-1)^j (m+n)!}{(n+j)!(m-j)!} \frac{\varpi^{2j}}{j!} . \tag{7.90}$$

For these polynomials, if n is a half integer, we must take

$$n! = \Gamma(n+1) = n(n-1)(n-2)\cdots\frac{5}{2}\frac{3}{2}\frac{\sqrt{\pi}}{2} .$$

For this case, the Sonine polynomials are equivalent to the Generalized Laguerre polynomials which are tabulated in Abramowitz and Stegun[55] except that the notation reverses the superscripts and subscripts such that

$$L_m^{(\alpha)}(x) = \sum_{j=0}^{m} \frac{\Gamma(m+\alpha+1)(-1)^j}{\Gamma(\alpha+j+1)(m-j)!} \frac{x^j}{j!} . \tag{7.91}$$

For $x = \varpi^2$, they are equivalent. The normalization and orthogonality relations are (letting $x \to \varpi^2$)

$$\int_0^\infty \mathrm{e}^{-\varpi^2} \varpi^{2\alpha+1} L_m^{(\alpha)}(\varpi^2) L_n^{(\alpha)}(\varpi^2)\, \mathrm{d}\varpi = \begin{cases} 0\,, & m \neq n\,, \\ \dfrac{\Gamma(m+\alpha+1)}{2m!}\,, & m = n\,. \end{cases} \tag{7.92}$$

In equation (7.89), we used $n = \alpha = \frac{3}{2}$ for convenience, but for $\mathcal{B}$, it turns out to be more convenient to use $n = \alpha = \frac{5}{2}$, so that

$$\begin{aligned} \mathcal{B} &= b_0 + b_1(\tfrac{7}{2} - \varpi^2) + b_2(\tfrac{63}{8} - \tfrac{9}{2}\varpi^2 + \tfrac{1}{2}\varpi^4) + \cdots \\ &= \sum_{m=0}^{\infty} b_m S_{\frac{5}{2}}^m(\varpi^2) = \sum_{m=0}^{\infty} b_m L_m^{\frac{5}{2}}(\varpi^2) . \end{aligned} \tag{7.93}$$

For the complete solution, we need to know all of the a_m and b_m, but before we solve for the coefficients, we examine how the pressure tensor and heat flux vector depend on them.

Substituting equation (7.89) into equation (7.88), switching from the Sonine to the associated Laguerre polynomials, we find

$$\sum_m a_m \frac{1}{\pi^{3/2}} \int_0^\infty \mathrm{d}\boldsymbol{\varpi}\, \mathrm{e}^{-\varpi^2} \varpi^2 L_m^{\frac{3}{2}}(\varpi^2) = 0 .$$

Performing the integrals over the angles which introduces a factor of $4\pi\varpi^2$ and recalling that $L_0^{\frac{3}{2}}(\varpi^2) = 1$, this may be written as

$$\sum_m a_m \frac{4\pi}{\pi^{3/2}} \int_0^\infty \mathrm{d}\varpi\, \mathrm{e}^{-\varpi^2} \varpi^4 L_m^{\frac{3}{2}}(\varpi^2) L_0^{\frac{3}{2}}(\varpi^2) = 0 ,$$

so that the orthogonality integral leads to the simple result,

$$a_0 = 0\,. \tag{7.94}$$

The heat flux vector is defined by equation (7.8). Then using F from equation (7.72), φ from equation (7.83), and $\mathcal{A}$ and $\mathcal{B}$ from equations (7.87) and (7.86), this becomes

$$q_\gamma = \int \mathrm{d}\boldsymbol{v}\, f\left[1 - \nabla_\alpha u_\beta(\boldsymbol{\varpi}^0\boldsymbol{\varpi})_{\alpha\beta}\mathcal{B}(\varpi^2) - \nabla_\alpha \ln T\varpi_\alpha\mathcal{A}(\varpi^2)\right]\frac{1}{2}mw^2w_\gamma\,.$$

The term in $\mathcal{B}$ clearly vanishes because it is odd in the random velocity $\boldsymbol{w}$. Then substituting $\mathcal{A}$ from equation (7.89) using equation (7.81) and remembering equation (7.94), the other term becomes

$$q_\gamma = -\frac{1}{T}\nabla_\alpha T\, nk_BTv_t\sum_{m=1}^{\infty} a_m\frac{1}{\pi^{3/2}}\int \mathrm{d}\boldsymbol{\varpi}\,\mathrm{e}^{-\varpi^2}\varpi^2\varpi_\alpha\varpi_\gamma L_m^{\frac{3}{2}}(\varpi^2)\,.$$

The integral vanishes unless $\alpha = \gamma$, and we may replace ϖ_γ^2 by $\frac{1}{3}\varpi^2$ to give

$$q_\gamma = -\nabla_\gamma T\, nk_Bv_t\sum_{m=1}^{\infty} a_m\frac{4\pi}{3\pi^{3/2}}\int_0^\infty \mathrm{d}\varpi\,\mathrm{e}^{-\varpi^2}\varpi^6 L_m^{\frac{3}{2}}(\varpi^2) \tag{7.95}$$

since

$$\tfrac{5}{2}L_0^{\frac{3}{2}}(\varpi^2) - L_1^{\frac{3}{2}}(\varpi^2) = \varpi^2\,,$$

and the final integral can be written as

$$\int_0^\infty \mathrm{d}\varpi\,\mathrm{e}^{-\varpi^2}\varpi^4\left[\tfrac{5}{2}L_0^{\frac{3}{2}}(\varpi^2) - L_1^{\frac{3}{2}}(\varpi^2)\right]L_m^{\frac{3}{2}}(\varpi^2) = \frac{15\sqrt{\pi}}{16}(\delta_{m,0} - \delta_{m,1})\,.$$

Hence, equation (7.95) becomes

$$q_\gamma = \tfrac{5}{4}nk_Bv_ta_1\nabla_\gamma T\,. \tag{7.96}$$

If the coefficient of thermal conduction, λ, is defined by

$$q_\gamma \equiv -\lambda\nabla_\gamma T\,, \tag{7.97}$$

then

$$\lambda = -\tfrac{5}{4}nk_Bv_ta_1\,. \tag{7.98}$$

The important point to notice is that in equations (7.96) and (7.98), only the coefficient a_1 is involved. While we need all the a_m and b_m to describe the distribution function, only a_1 is involved in the thermal conductivity.

Turning our attention to the pressure tensor defined by equation (7.7), we use F from equation (7.72), φ from equation (7.83), and $\mathcal{A}$ and $\mathcal{B}$ from equations (7.87) and (7.86), to find

$$p_{\alpha\beta} = m\int \mathrm{d}\boldsymbol{v}\, w_\alpha w_\beta f\left[1 - \nabla_\gamma u_\epsilon(\boldsymbol{\varpi}^0\boldsymbol{\varpi})_{\gamma\epsilon}\mathcal{B}(\varpi^2) - \nabla_\gamma \ln T\varpi_\gamma\mathcal{A}(\varpi^2)\right]\,.$$

This time, the term involving $\mathcal{A}$ vanishes because it is odd in the random velocities. The first term also vanishes unless $\alpha = \beta$ and is

$$\delta_{\alpha\beta} nmv_t^2 \frac{1}{\pi^{3/2}} \int d\boldsymbol{\varpi}\, \varpi_\alpha^2 e^{-\varpi^2} = \delta_{\alpha\beta} nk_B T = \delta_{\alpha\beta} p \,.$$

The second term is

$$-nmv_t^2 \nabla_\gamma u_\epsilon \frac{1}{\pi^{3/2}} \int d\boldsymbol{\varpi}\, e^{-\varpi^2} \varpi_\alpha \varpi_\beta \{\varpi_\gamma \varpi_\epsilon - \tfrac{1}{3}\varpi^2 \delta_{\gamma\epsilon}\} \mathcal{B}(\varpi^2) \,, \tag{7.99}$$

which requires special attention. If $\gamma \neq \epsilon$, then the second term in the bracket contributes nothing while the first term contributes only if $\gamma = \alpha$ and $\epsilon = \beta$ or if $\gamma = \beta$ and $\epsilon = \alpha$. Therefore, if $\alpha \neq \beta$, equation (7.99) is

$$-2p(\nabla_\alpha u_\beta + \nabla_\beta u_\alpha) \frac{1}{\pi^{3/2}} \int d\boldsymbol{\varpi}\, \varpi_\alpha^2 \varpi_\beta^2 e^{-\varpi^2} \mathcal{B}(\varpi^2) = -\frac{pb_0}{2} (\nabla_\alpha u_\beta + \nabla_\beta u_\alpha) \,.$$

On the other hand, if $\alpha = \beta = \gamma = \epsilon$, then the second term in the bracket of equation (7.99) gives

$$2p(\nabla \cdot \boldsymbol{u}) \frac{1}{9\pi^{3/2}} \int d\boldsymbol{\varpi}\, e^{-\varpi^2} \varpi^4 \mathcal{B}(\varpi^2) = \frac{5}{6} pb_0 (\nabla \cdot \boldsymbol{u}) \,,$$

while the first term in the bracket gives

$$\begin{aligned}
&-2p\nabla_\alpha u_\alpha \frac{1}{\pi^{3/2}} \int d\boldsymbol{\varpi}\, e^{-\varpi^2} \varpi_\alpha^4 \mathcal{B}(\varpi^2) \\
&-2p \sum_{\gamma \neq \alpha} (\nabla_\gamma u_\gamma) \frac{1}{\pi^{3/2}} \int d\boldsymbol{\varpi}\, e^{-\varpi^2} \varpi_\alpha^2 \varpi_\gamma^2 \mathcal{B}(\varpi^2) = -\frac{b_0 p}{2} \left(3\nabla_\alpha u_\alpha + \sum_{\gamma \neq \alpha} \nabla_\gamma u_\gamma \right) \\
&\qquad = -b_0 p \left(\tfrac{1}{2} \nabla \cdot \boldsymbol{u} + \nabla_\alpha u_\alpha \right) \,,
\end{aligned}$$

where $\nabla_\alpha u_\alpha$ does not indicate a sum in this case.

Collecting these results, we find that

$$\begin{aligned}
p_{\alpha\beta} &= -\tfrac{1}{2} pb_0 (\nabla_\alpha u_\beta + \nabla_\beta u_\alpha) \,, \qquad \alpha \neq \beta \,, \\
p_{\alpha\alpha} &= p \left[1 + \tfrac{1}{3} b_0 (\nabla \cdot \boldsymbol{u}) - b_0 \nabla_\alpha u_\alpha \right] \,, \alpha = \beta \,.
\end{aligned} \tag{7.100}$$

In general, therefore, we have

$$p_{\alpha\beta} = [p + \tfrac{2}{3}\mu(\nabla \cdot \boldsymbol{u})]\delta_{\alpha\beta} - \mu(\nabla_\alpha u_\beta + \nabla_\beta u_\alpha) \,, \tag{7.101}$$

where

$$\mu = \tfrac{1}{2} nk_B T b_0 \tag{7.102}$$

is the coefficient of viscosity. We note that only the coefficient b_0 is needed for the coefficient of viscosity. This means that we are interested in only b_0 and a_1.

Problem 7.3 Evaluate the various velocity integrals involved in the pressure tensor.

To complete this section, we describe the formal variation procedure used by Hirschfelder et al.[50] which in principle enables us to find all of the a_m and b_m. In practice, it gives excellent approximations to b_0 and a_1, which are the only ones we really want.

If G and H are any properties of the particles, we define

$$G : H \equiv \begin{cases} \sum G_\alpha H_\alpha = \boldsymbol{G} \cdot \boldsymbol{H} \,, & \text{if } G \text{ and } H \text{ are both vectors,} \\ \sum_{\alpha,\beta} G_{\alpha\beta} H_{\beta\alpha} = G_{\alpha\beta} H_{\beta\alpha} \,, & \text{if } G \text{ and } H \text{ are both tensors.} \end{cases} \tag{7.103}$$

We then define what we call "collision integrals" as

$$[G, H] \equiv -\frac{1}{n^2} \int \mathrm{d}\boldsymbol{v} \mathrm{d}\boldsymbol{s} \mathrm{d}\phi \, b \, \mathrm{d}b \, g \, f(\boldsymbol{v}) f(\boldsymbol{s}) G(\boldsymbol{v}) : [H(\boldsymbol{v}') + H(\boldsymbol{s}') - H(\boldsymbol{v}) - H(\boldsymbol{s})] \,. \tag{7.104}$$

By symmetry, this is also

$$\begin{aligned} [G, H] = -\frac{1}{2n^2} \int \mathrm{d}\boldsymbol{v} \mathrm{d}\boldsymbol{s} \mathrm{d}\phi \, b \, \mathrm{d}b \, g \, f(\boldsymbol{v}) f(\boldsymbol{s}) [G(\boldsymbol{v}) + G(\boldsymbol{s})] : \\ [H(\boldsymbol{v}') + H(\boldsymbol{s}') - H(\boldsymbol{v}) - H(\boldsymbol{s})] \end{aligned}$$

or

$$\begin{aligned} [G, H] = -\frac{1}{4n^2} \int \mathrm{d}\boldsymbol{v} \mathrm{d}\boldsymbol{s} \mathrm{d}\phi \, b \, \mathrm{d}b \, g \, f(\boldsymbol{v}) f(\boldsymbol{s}) [G(\boldsymbol{v}') + G(\boldsymbol{s}') - G(\boldsymbol{v}) - G(\boldsymbol{s})] : \\ [H(\boldsymbol{v}') + H(\boldsymbol{s}') - H(\boldsymbol{v}) - H(\boldsymbol{s})] \,. \end{aligned}$$

From this last form, one may notice that

$$[G, G] \geq 0 \,. \tag{7.105}$$

Now the equations that we need to solve, equations (7.84) and (7.85), are of the form

$$R(\boldsymbol{v}) = \int \mathrm{d}\boldsymbol{s} \mathrm{d}\phi \, b \, \mathrm{d}b \, g \, f(\boldsymbol{v}) f(\boldsymbol{s}) [T(\boldsymbol{v}') + T(\boldsymbol{s}') - T(\boldsymbol{v}) - T(\boldsymbol{s})] \,, \tag{7.106}$$

where R and T are either vectors or tensors, and R is known and T is to be found. Let $t(\boldsymbol{v})$ be any trial solution of this equation subject only to the condition that

$$\begin{aligned} \int \mathrm{d}\boldsymbol{v} \, t(\boldsymbol{v}) : R(\boldsymbol{v}) &= \int \mathrm{d}\boldsymbol{v} \mathrm{d}\boldsymbol{s} \mathrm{d}\phi \, b \, \mathrm{d}b \, g \, f(\boldsymbol{v}) f(\boldsymbol{s}) t(\boldsymbol{v}) : [t(\boldsymbol{v}') + t(\boldsymbol{s}') - t(\boldsymbol{v}) - s(\boldsymbol{s})] \\ &= -n^2 [t, t] \,. \end{aligned} \tag{7.107}$$

But from equation (7.106),

$$\int \mathrm{d}\boldsymbol{v} \, t(\boldsymbol{v}) : R(\boldsymbol{v}) = -n^2 [t, T] \,,$$

so provided t is chosen to satisfy equation (7.107),

$$[t,t] = [t,T]\,. \tag{7.108}$$

Now consider

$$[t-T,t-T] = [t,t] - 2[t,T] + [T,T]\,.$$

By equation (7.105), this is positive or zero, so that using equation (7.108),

$$[t,t] \le [T,T]\,,$$

or using equation (7.107)

$$-\frac{1}{n^2}\int d\boldsymbol{v}\, t(\boldsymbol{v}) : R(\boldsymbol{v}) = [t,t] \le [T,T]\,. \tag{7.109}$$

This is the relation upon which the variation procedure is based. We first choose a trial function with as many undetermined parameters as is convenient. In our case, we take the trial functions to be

$$\boldsymbol{A} = \boldsymbol{\varpi}\mathcal{A}(\varpi^2) = \boldsymbol{\varpi}\sum_{m=1}^{\infty} a_m L_m^{\frac{3}{2}}(\varpi^2) \tag{7.110}$$

or

$$\mathsf{B} = \boldsymbol{\varpi}^0\boldsymbol{\varpi}\mathcal{B}(\varpi^2) = \boldsymbol{\varpi}^0\boldsymbol{\varpi}\sum_{m=0}^{\infty} b_m L_m^{\frac{5}{2}}(\varpi^2) \tag{7.111}$$

depending upon whether we are considering equation (7.85) or equation (7.84), respectively. We then take all but a few of the a_m and b_m to be zero. For example, in equation (7.110), the first trial function is obtained by putting all the coefficients a_m to zero except a_1. A second trial function giving a better result is obtained by putting all except a_1 and a_2 equal to zero. The third trial function takes a_1, a_2, and a_3 to be nonzero. Having picked our trial function, we then ensure that the coefficients are such that equation (7.109) is satisfied and then we maximize either the left or right hand side of this equation. Those values of the coefficients which give the maximum value give the best approximation to the correct answer. As we shall see in the next section when we apply this variation method, this procedure gives excellent values for a_1 and b_0 very rapidly.

7.7 Results of the variational procedure

We now endeavor to use the variational procedure to finally obtain the coefficients of viscosity and thermal conduction.

Considering equation (7.84) first, we see by comparing this equation with the general form of equation (7.106) that

$$R(\boldsymbol{v}) = -2f(\boldsymbol{v})\boldsymbol{\varpi}^0\boldsymbol{\varpi}\,.$$

If we use equation (7.111) for $t(\boldsymbol{v})$, then we find the left-hand side of equation (7.109) to be

$$\begin{aligned}
-\frac{1}{n^2}\int \mathrm{d}\boldsymbol{v}\, t(\boldsymbol{v}) : R(\boldsymbol{v}) &= \frac{2}{n^2}\sum_{m=0}^{\infty} b_m \int \mathrm{d}\boldsymbol{v}\, f(\boldsymbol{v})(\boldsymbol{\varpi}^0\boldsymbol{\varpi}) : (\boldsymbol{\varpi}^0\boldsymbol{\varpi}) L_m^{\frac{5}{2}}(\varpi^2) \\
&= \frac{2}{n}\sum_{m=0}^{\infty} b_m \frac{1}{\pi^{3/2}} \int \mathrm{d}\boldsymbol{\varpi}\, (\varpi_\alpha\varpi_\beta - \tfrac{1}{3}\varpi^2\delta_{\alpha\beta}) \\
&\qquad \times(\varpi_\alpha\varpi_\beta - \tfrac{1}{3}\varpi^2\delta_{\alpha\beta}) L_m^{\frac{5}{2}}(\varpi^2)\mathrm{e}^{-\varpi^2} \\
&= \frac{3}{4n}\sum_{m=0}^{\infty} b_m \frac{1}{\pi^{3/2}} \int \mathrm{d}\boldsymbol{\varpi}\, \varpi^4 L_m^{\frac{5}{2}}(\varpi^2)\mathrm{e}^{-\varpi^2} \\
&= \frac{5b_0}{n}.
\end{aligned} \tag{7.112}$$

This leads to the result that equation (7.109) becomes

$$\frac{5b_0}{n} = \sum_{m,m'} b_m b_{m'} \left[\boldsymbol{\varpi}^0\boldsymbol{\varpi} L_m^{\frac{5}{2}}(\varpi^2), \boldsymbol{\varpi}^0\boldsymbol{\varpi} L_{m'}^{\frac{5}{2}}(\varpi^2)\right]_1 < [T, T] \tag{7.113}$$

where the notation $[f_1(\varpi), f_2(\varpi)]_1$ indicates a collision integral between like particles for any functions $f_1(\varpi)$ and $f_2(\varpi)$. These integrals are evaluated in Appendix B.

The variational problem is now relatively easy because of the simple form on the left-hand side of equation (7.113). We only need to find the largest value of b_0 that equation (7.113) will allow. The simple form of the left-hand side of equation (7.113) is another consequence of the use of the Sonine or Generalized Laguerre polynomials. From the general principles of variational methods, we may expect a relatively good value for b_0 with quite poor trial functions, which in this case means with only a few nonzero coefficients. From equation (7.102), we see that μ is directly proportional to b_0, so we expect a relatively good value for μ.

By including more and more nonzero coefficients, we will obtain successive approximations to b_0 which we will denote as

$$[b_0]_1\,,\quad [b_0]_2\,,\quad [b_0]_3\,,\cdots \quad \text{etc.}$$

with corresponding approximations to μ,

$$[\mu]_1\,,\quad [\mu]_2\,,\quad [\mu]_3\,,\cdots \quad \text{etc.}$$

Each time we improve the trial function, we will obtain a larger value for b_0, so that

$$[b_0]_1 < [b_0]_2 < [b_0]_3 , \cdots \quad \text{etc.}$$

and correspondingly,

$$[\mu]_1 < [\mu]_2 < [\mu]_3 , \cdots \quad \text{etc.} \tag{7.114}$$

The first trial function is obtained by setting all of the coefficients except b_0 to zero. Equation (7.113) then becomes

$$\frac{5b_0}{n} = b_0^2[\boldsymbol{\varpi}^0\boldsymbol{\varpi}, \boldsymbol{\varpi}^0\boldsymbol{\varpi}]_1 < [T, T] \,. \tag{7.115}$$

This has only two possible solutions, either $b_0 = 0$, or

$$[b_0]_1 = \frac{5}{n}[\boldsymbol{\varpi}^0\boldsymbol{\varpi}, \boldsymbol{\varpi}^0\boldsymbol{\varpi}]_1^{-1},$$

and this second solution is positive by equation (7.105). Therefore it is the value that maximizes equation (7.115) in our first approximation. From equation (7.102), the corresponding value of μ is

$$[\mu]_1 = \frac{5}{2}k_BT[\boldsymbol{\varpi}^0\boldsymbol{\varpi}, \boldsymbol{\varpi}^0\boldsymbol{\varpi}]_1^{-1} \,. \tag{7.116}$$

The evaluation of integrals like the one that appears in equation (7.116) is discussed in Appendix B. There it is shown that

$$\left[\boldsymbol{\varpi}^0\boldsymbol{\varpi}, \boldsymbol{\varpi}^0\boldsymbol{\varpi}\right]_1 = \sqrt{2}\varphi = \frac{e^4\psi}{8\epsilon_0^2\sqrt{m}(\pi k_BT)^{3/2}} \,, \tag{7.117}$$

where m is the particle mass, e the particle charge, $\psi = 2\ln\Lambda$, and $\Lambda \simeq 9N_D$ where N_D is given by

$$N_D = \tfrac{4}{3}\pi n\lambda_D^3 \tag{7.118}$$

and is the number of particles in a Debye sphere where λ_D is the Debye length. Because λ_D appears only in the logarithmic term, equation (7.118) is insensitive to the precise value of the Debye length and adds minimum uncertainty to the value of μ, whose first approximation is

$$[\mu]_1 = \frac{20\epsilon_0^2\pi^{3/2}\sqrt{m}(k_BT)^{5/2}}{e^4\psi} \,. \tag{7.119}$$

We now consider the next approximation to μ so that we can get an estimate of how accurate equation (7.119) might be. For this case, we set all of the coefficients in equation (7.113) to zero except b_0 and b_1 in the trial function. This leads to

$$\begin{aligned}\frac{5b_0}{n} = b_0^2\left[\boldsymbol{\varpi}^0\boldsymbol{\varpi}, \boldsymbol{\varpi}^0\boldsymbol{\varpi}\right]_1 + 2b_0b_1\left[\boldsymbol{\varpi}^0\boldsymbol{\varpi}, \boldsymbol{\varpi}^0\boldsymbol{\varpi}L_1^{\frac{5}{2}}(\varpi^2)\right]_1 \\ +b_1^2\left[\boldsymbol{\varpi}^0\boldsymbol{\varpi}L_1^{\frac{5}{2}}(\varpi^2), \boldsymbol{\varpi}^0\boldsymbol{\varpi}L_1^{\frac{5}{2}}(\varpi^2)\right]_1 \,,\end{aligned} \tag{7.120}$$

and the maximum value of b_0 is the one where equation (7.120) has a real solution for b_1. This maximum value of b_0 is therefore obtained from maximizing equation (7.120) as a function of b_1 by setting the derivative of equation (7.120) with respect to b_1 to zero with the result that

$$b_1 = -\frac{b_0 b^{(0,1)}}{b^{(1,1)}},$$

where we have defined

$$b^{(i,j)} \equiv \left[\varpi^0 \varpi L_i^{\frac{5}{2}}(\varpi^2), \varpi^0 \varpi L_j^{\frac{5}{2}}(\varpi^2)\right]_1, \tag{7.121}$$

remembering that $L_0^\alpha(x) = 1$ for any α. Inserting this value of b_1 into equation (7.120), we find

$$\frac{5b_0}{n} = b_0^2 \left[b^{(0,0)} - \frac{(b^{(0,1)})^2}{b^{(1,1)}}\right],$$

so that the only nonzero value for b_0 is

$$[b_0]_2 = \frac{5}{nb^{(0,0)}(1 - \Delta_b)}, \tag{7.122}$$

where

$$\Delta_b = \frac{\left[b^{(0,1)}\right]^2}{b^{(0,0)} b^{(1,1)}}.$$

Evaluating these integrals from Appendix B, we find

$$b^{(0,0)} = \sqrt{2}\varphi, \quad b^{(0,1)} = \frac{3\sqrt{2}}{4}\varphi, \quad b^{(1,1)} = \frac{205\sqrt{2}}{48}\varphi,$$

where

$$\varphi = \frac{e^4 \psi}{4\epsilon_0^2 \sqrt{m}(2\pi k_B T)^{3/2}},$$

so $\Delta_b = 27/205$ and

$$[b_0]_2 = 1.15\,[b_0]_1, \tag{7.123}$$

so it follows that

$$[\mu]_2 = 1.15\,[\mu]_1. \tag{7.124}$$

Since this second approximation makes a 15% correction, we may want to add additional terms in the trial function.*

*W. Marshall finds $[b_0]_2 = 1.025[b_0]_1$ with $\Delta_b = 5/205$, but lists no values for $b^{(0,1)}$ or $b^{(1,1)}$.

Problem 7.4 *Third order correction for b_0.* Show that $[b_0]_3$ is given by

$$[b_0]_3 = [b_0]_1 \left[1 - \frac{\left[b^{(0,1)}\right]^2 b^{(2,2)} - 2b^{(0,1)}b^{(0,2)}b^{(1,2)} + b^{(1,1)}\left[b^{(0,2)}\right]^2}{b^{(0,0)}\left\{b^{(1,1)}b^{(2,2)} - \left[b^{(1,2)}\right]^2\right\}} \right]^{-1},$$

and show that $[b_0]_3 = 1.1583\,[b_0]_1$. Show also that if $b^{(0,2)} \to 0$ and $b^{(1,2)} \to 0$ that $b_0^{(3)} \to b_0^{(2)}$.

For the thermal conductivity of equation (7.85), we note the general form of equation (7.106) is

$$R(\boldsymbol{v}) = (\tfrac{5}{2} - \varpi^2)\boldsymbol{w} = L_1^{\frac{3}{2}}(\varpi^2)\boldsymbol{w}\,.$$

Using the expression in equation (7.110) as t, we find

$$\begin{aligned} -\frac{1}{n^2}\int d\boldsymbol{v}\, t : R &= -\frac{1}{n^2}\sum_{m=1}^{\infty} a_m \int d\boldsymbol{v}\, f(\boldsymbol{v}) L_1^{\frac{3}{2}}(\varpi^2)\boldsymbol{w}\cdot\boldsymbol{\varpi} L_m^{\frac{3}{2}}(\varpi^2) \\ &= -\frac{15}{4n} v_t a_1 \end{aligned}$$

so that equation (7.109) becomes

$$-\frac{15}{4n} v_t a_1 = \sum_{m,m'} a_m a_{m'} \left[\boldsymbol{\varpi} L_m^{\frac{3}{2}}(\varpi^2), \boldsymbol{\varpi} L_{m'}^{\frac{3}{2}}(\varpi^2)\right]_1 < [T,T]\,, \qquad (7.125)$$

so once again the variation method is very simple and consists of finding the most negative value of a_1 possible. We also have here a relationship similar to equation (7.114) where

$$[\lambda]_1 < [\lambda]_2 < [\lambda]_3\,, \cdots \quad \text{etc.}$$

The first trial function is obtained by setting all of the coefficients to zero except a_1 so that equation (7.125) becomes

$$-\frac{15}{4n} v_t a_1 = a_1^2 \left[\boldsymbol{\varpi} L_1^{\frac{3}{2}}(\varpi^2), \boldsymbol{\varpi} L_1^{\frac{3}{2}}(\varpi^2)\right]_1\,, \qquad (7.126)$$

which again has only two solutions. The first is $a_1 = 0$, the second is

$$[a_1]_1 = -\frac{15}{4n} v_t \left[\boldsymbol{\varpi} L_1^{\frac{3}{2}}(\varpi^2), \boldsymbol{\varpi} L_1^{\frac{3}{2}}(\varpi^2)\right]_1^{-1},$$

and this is the value that maximizes equation (7.126). The corresponding value of the thermal conductivity is

$$[\lambda]_1 = \frac{75 k_B}{16} v_t^2 \left[\boldsymbol{\varpi} L_1^{\frac{3}{2}}(\varpi^2), \boldsymbol{\varpi} L_1^{\frac{3}{2}}(\varpi^2)\right]_1^{-1}. \qquad (7.127)$$

This collision integral has exactly the same value as that in equation (7.117). Therefore,

$$[\lambda]_1 = \frac{75\pi^{3/2}\epsilon_0^2 k_B (k_B T)^{5/2}}{\sqrt{m} e^4 \ln\Lambda} . \tag{7.128}$$

The second approximation is obtained by setting all of the coefficients to zero except for a_1 and a_2, so that equation (7.125) becomes

$$-\frac{15}{4n} v_t a_1 = a_1^2 \left[\boldsymbol{\varpi} L_1^{\frac{3}{2}}(\varpi^2), \boldsymbol{\varpi} L_1^{\frac{3}{2}}(\varpi^2)\right]_1 + 2a_1 a_2 \left[\boldsymbol{\varpi} L_1^{\frac{3}{2}}(\varpi^2), \boldsymbol{\varpi} L_2^{\frac{3}{2}}(\varpi^2)\right]_1$$
$$+ a_2^2 \left[\boldsymbol{\varpi} L_2^{\frac{3}{2}}(\varpi^2), \boldsymbol{\varpi} L_2^{\frac{3}{2}}(\varpi^2)\right]_1 , \tag{7.129}$$

which, by analogy with equation (7.120), is maximized by

$$[a_1]_2 = [a_1]_1 (1 - \Delta_a)^{-1} ,$$

where

$$\Delta_a = \frac{\left[a^{(1,2)}\right]^2}{a^{(1,1)} a^{(2,2)}} .$$

Defining these integrals by

$$a^{(i,j)} \equiv \left[\boldsymbol{\varpi} L_i^{\frac{3}{2}}(\varpi^2), \boldsymbol{\varpi} L_j^{\frac{3}{2}}(\varpi^2)\right]_1 , \tag{7.130}$$

and evaluating these collision integrals from Appendix B we find

$$a^{(1,1)} = \sqrt{2}\varphi , \qquad a^{(1,2)} = \frac{3\sqrt{2}}{4}\varphi , \qquad a^{(2,2)} = \frac{45}{16}\sqrt{2}\varphi ,$$

so $\Delta_a = 1/5$ and

$$[a_1]_2 = 1.25 \, [a_1]_1 . \tag{7.131}$$

Therefore our second approximation gives

$$[\lambda]_2 = 1.25 \, [\lambda]_1 . \tag{7.132}$$

In this case the second order correction leads to a 25% correction which is almost twice the correction for the viscosity, but we still expect this result is within a few percent, and we do not carry it further except as a problem.†

Problem 7.5 *Third order correction for a_1.* Show that $[a_1]_3$ is given by

$$[a_1]_3 = [a_1]_1 \left[1 - \frac{\left[a^{(1,2)}\right]^2 a^{(3,3)} - 2a^{(1,2)} a^{(1,3)} a^{(2,3)} + a^{(2,2)} \left[a^{(1,3)}\right]^2}{a^{(1,1)} \left\{a^{(2,2)} a^{(3,3)} - \left[a^{(2,3)}\right]^2\right\}}\right]^{-1} ,$$

and show that $[a_1]_3 = 1.264 \, [a_1]_1$.

†W. Marshall finds $[a_1]_2 = 1.08[a_1]_1$ from $\Delta_a = 3/41$, but lists no values for $a^{(1,2)}$ or $a^{(2,2)}$.

It may be noted that

$$\frac{[\lambda]_1}{[\mu]_1} = \frac{15k_B}{4m} ,$$

which is a well known result, but it is true only to first order. Using the results of the problems above, an improved value is

$$\frac{[\lambda]_3}{[\mu]_3} = 1.091\frac{15k_B}{4m} . \tag{7.133}$$

8

TRANSPORT IN A NONUNIFORM BINARY GAS

In this chapter we extend the treatment in the previous chapter to two species[45], electrons and one ion species, still with no magnetic field. In addition to recalculating the transport coefficients of the previous chapter, we add the calculation of the electrical conductivity for the two-component plasma.

8.1 The Boltzmann equations

With both electrons and ions, we need to define two distribution functions, $F_1(\boldsymbol{r}, \boldsymbol{v}, t)$ and $F_2(\boldsymbol{r}, \boldsymbol{v}, t)$, referring to electrons and ions respectively. We can derive Boltzmann equations giving the rate of change of these functions in the same way as in Chapter 7. These are

$$\left[\frac{\partial}{\partial t} + \boldsymbol{v}\cdot\nabla + \left(\frac{e_1}{m_1}\boldsymbol{E} + \boldsymbol{X}\right)\cdot\nabla_v\right] F_1(\boldsymbol{r}, \boldsymbol{v}, t) = \Delta F_1(\boldsymbol{r}, \boldsymbol{v}, t) \qquad (8.1)$$

$$\left[\frac{\partial}{\partial t} + \boldsymbol{v}\cdot\nabla + \left(\frac{e_2}{m_2}\boldsymbol{E} + \boldsymbol{X}\right)\cdot\nabla_v\right] F_2(\boldsymbol{r}, \boldsymbol{v}, t) = \Delta F_2(\boldsymbol{r}, \boldsymbol{v}, t)\,. \qquad (8.2)$$

The force due to the electric field $\boldsymbol{E}$ is written explicitly, and $\boldsymbol{X}$ indicates the acceleration due to any other force that might be present, but there is no magnetic field considered in this chapter. Since the gas is presumed to be made of electrons and singly charged ions, it follows that $-e_1 = e_2 = e$.

The right hand sides of these equations, representing the rates of change due to collisions, are the sum of two parts, such that

$$\Delta F_i(\boldsymbol{r}, \boldsymbol{v}, t) = \sum_{j=1}^{2} \Delta_j F_j(\boldsymbol{r}, \boldsymbol{v}, t)\,. \qquad (8.3)$$

The term with $j = 1$ represents the rate of change of F_i due to collisions with electrons, and the term with $j = 2$ represents that due to ions.

8.2 The equations of hydrodynamics

The basic definitions and equations are the same as in Chapter 7 except that now we add subscripts to designate which species and in some cases we need the sum over species.

If Ψ_i is *any* property of particles i, depending in general on position, velocity, and time, then the mean value of Ψ_i at the position $\boldsymbol{r}$ and time t is

$$\langle \Psi_i \rangle \equiv \frac{1}{n_i} \int \mathrm{d}\boldsymbol{v}\, \Psi_i(\boldsymbol{r}, \boldsymbol{v}, t) F_i(\boldsymbol{r}, \boldsymbol{v}, t)$$

where

$$n_i \equiv \int \mathrm{d}\boldsymbol{v}\, F_i(\boldsymbol{r}, \boldsymbol{v}, t) .$$

The partial mass densities are

$$\rho_i = n_i m_i ,$$

and the total number density is

$$n = \sum_i n_i = n_1 + n_2 ,$$

and the total mass density is

$$\rho = \sum_i \rho_i = n_1 m_1 + n_2 m_2 .$$

The mean velocities are

$$\langle \boldsymbol{v}_i \rangle = \frac{1}{n_i} \int \mathrm{d}\boldsymbol{v}\, \boldsymbol{v} F_i(\boldsymbol{r}, \boldsymbol{v}, t)$$

and the drift velocity is

$$\boldsymbol{u} = \frac{1}{\rho} \sum_i \rho_i \langle \boldsymbol{v}_i \rangle .$$

The random velocity is

$$\boldsymbol{w} \equiv \boldsymbol{v} - \boldsymbol{u} \tag{8.4}$$

and the mean random velocities are

$$\langle \boldsymbol{w}_i \rangle = \frac{1}{n_i} \int \mathrm{d}\boldsymbol{v}\, \boldsymbol{w} F_i(\boldsymbol{r}, \boldsymbol{v}, t) . \tag{8.5}$$

Because of equation (8.4),

$$\sum_i \rho_i \langle \boldsymbol{w}_i \rangle = n_1 m_1 \langle \boldsymbol{w}_1 \rangle + n_2 m_2 \langle \boldsymbol{w}_2 \rangle = 0 . \tag{8.6}$$

The kinetic temperature is defined by

$$\tfrac{3}{2}nk_BT \equiv \sum_i n_i \tfrac{1}{2}m_i\langle w_i^2\rangle = \sum_i \tfrac{1}{2}m_i \int \mathrm{d}\boldsymbol{v} w^2 F_i\,. \tag{8.7}$$

The pressure tensor is

$$\mathsf{P} \equiv p_{\alpha\beta} = \sum_i \rho_i\langle w_{i\alpha}w_{i\beta}\rangle = \sum_i \int \mathrm{d}\boldsymbol{v}\, F_i\, w_\alpha w_\beta\,. \tag{8.8}$$

The heat flux vector is

$$\boldsymbol{q} \equiv \sum_i \tfrac{1}{2}\langle w_i^2\boldsymbol{w}_i\rangle = \sum_i \tfrac{1}{2}m_i \int \mathrm{d}\boldsymbol{v}\, F_i\, w_i^2\boldsymbol{w}_i\,. \tag{8.9}$$

The charge density is

$$Q \equiv \sum_i n_i e_i = (n_2 - n_1)e\,. \tag{8.10}$$

The total current is

$$\boldsymbol{J} = \sum_i n_i e_i\langle \boldsymbol{v}_i\rangle = Q\boldsymbol{u} + \boldsymbol{j}$$

where

$$\boldsymbol{j} = \sum_i n_i e_i\langle \boldsymbol{w}_i\rangle = (n_2\langle \boldsymbol{w}_2\rangle - n_1\langle \boldsymbol{w}_1\rangle)e \tag{8.11}$$

is the conduction current and $Q\boldsymbol{u}$ is the convection current. From equation (8.6), this relationship for $\boldsymbol{j}$ can be rearranged as

$$\begin{aligned}\boldsymbol{j} &= \frac{n_1n_2}{\rho}(e_1m_2 - e_2m_1)(\langle \boldsymbol{w}_1\rangle - \langle \boldsymbol{w}_2\rangle)\\ &= -\frac{n_1n_2}{\rho}e(m_2 + m_1)(\langle \boldsymbol{w}_1\rangle - \langle \boldsymbol{w}_2\rangle)\,.\end{aligned} \tag{8.12}$$

The continuity equations are obtained by integrating equations (8.1) or (8.2) to give

$$\int \mathrm{d}\boldsymbol{v}\frac{\partial F_i}{\partial t} + \int \mathrm{d}\boldsymbol{v}\, \boldsymbol{v}\cdot\nabla F_i + \int \mathrm{d}\boldsymbol{v}\left(\frac{e_i}{m_i}\boldsymbol{E} + \boldsymbol{X}_i\right)\cdot\nabla_v F_i = \int \mathrm{d}\boldsymbol{v}\,\Delta F_i\,.$$

These terms may be treated as in Chapter 7 to yield

$$\frac{\partial n_i}{\partial t} + \nabla\cdot(n_i\boldsymbol{u} + n_i\langle \boldsymbol{w}_i\rangle) = 0\,. \tag{8.13}$$

Multiplying this equation by m_i and summing over i, we notice that the terms involving $\langle \boldsymbol{w}_1\rangle$ and $\langle \boldsymbol{w}_2\rangle$ vanish from equation (8.6) and we are left with

$$\frac{\partial \rho}{\partial t} + \nabla\cdot(\rho\boldsymbol{u}) = 0\,, \tag{8.14}$$

or

$$\frac{\mathrm{D}\rho}{\mathrm{D}t} + \rho\nabla\cdot\boldsymbol{u} = 0 \tag{8.15}$$

where

$$\frac{\mathrm{D}}{\mathrm{D}t} \equiv \frac{\partial}{\partial t} + \boldsymbol{u}\cdot\nabla\,.$$

These are alternative forms of the continuity equation.

The equation of motion is obtained by multiplying equations (8.1) or (8.2) by $m_1 v_\alpha$ or $m_2 v_\alpha$, respectively, integrating and adding to give

$$\sum_i \int \mathrm{d}\boldsymbol{v} m_i v_\alpha \frac{\partial F_i}{\partial t} + \sum_i \int \mathrm{d}\boldsymbol{v}\, m_i v_\alpha \boldsymbol{v}\cdot\nabla F_i + \sum_i \int \mathrm{d}\boldsymbol{v} m_i v_\alpha \left(\frac{e_i}{m_i}\boldsymbol{E} + \boldsymbol{X}_i\right)\cdot\nabla_v F_i$$
$$= \sum_i m_i v_\alpha \int \mathrm{d}\boldsymbol{v}\,\Delta F_i\,.$$

The right-hand side of this equation is the rate of change of momentum due to collisions and is therefore equal to zero. The first term on the left-hand side may be rewritten as

$$\frac{\partial}{\partial t}\sum_i n_i m_i \langle \boldsymbol{v}_{i\alpha}\rangle = \frac{\partial}{\partial t}(\rho u_\alpha)\,.$$

The second term may be written

$$\nabla_\beta \sum_i n_i m_i \langle v_{i\alpha} v_{i\beta}\rangle = \nabla_\beta \sum_i n_i m_i (u_\beta u_\alpha + u_\beta\langle w_{i\alpha}\rangle + u_\alpha\langle w_{i\beta}\rangle + \langle w_{i\alpha} w_{i\beta}\rangle)$$
$$= \nabla_\beta(\rho u_\beta u_\alpha) + \nabla_\beta p_{\alpha\beta}$$

where equation (8.6) has been used again. The third term is

$$-\sum_i n_i m m_i \left(\frac{e_i}{m_i}E_\alpha + \langle X_{i\alpha}\rangle\right) = -QE_\alpha - \sum_i \rho_i\langle X_{i\alpha}\rangle\,.$$

Collecting terms, the equation of motion becomes

$$\frac{\partial}{\partial t}(\rho u_\alpha) + \nabla_\beta(\rho u_\beta u_\alpha) = -\nabla_\beta p_{\alpha\beta} + QE_\alpha + \sum_i \rho_i\langle X_{i\alpha}\rangle\,,$$

which, using equation (8.14), is

$$\rho\frac{\mathrm{D}u_\alpha}{\mathrm{D}t} = \rho\left(\frac{\partial u_\alpha}{\partial t} + \boldsymbol{u}\cdot\nabla u_\alpha\right) = -\nabla_\beta p_{\alpha\beta} + QE_\alpha + \sum_i \rho_i\langle X_{i\alpha}\rangle\,. \tag{8.16}$$

$Q\boldsymbol{E}$ is the net electric force acting per unit volume of gas.

The energy equation is obtained by multiplying equations (8.1) or (8.2) by $\frac{1}{2}m_1v^2$ or $\frac{1}{2}m_2v^2$, respectively, integrating and adding to give

$$\sum_i \int \mathrm{d}\boldsymbol{v} \tfrac{1}{2}m_i v^2 \left[\frac{\partial F_i}{\partial t} + \boldsymbol{v}\cdot\nabla F_i + \left(\frac{e_i}{m_i}\boldsymbol{E} + \boldsymbol{X}_i\right)\cdot\nabla_v F_i\right] = \sum_i \int \mathrm{d}\boldsymbol{v} \tfrac{1}{2}m_i v^2 \Delta F_i. \tag{8.17}$$

The right-hand side is the rate of change of energy due to collisions and is therefore zero. The first term on the left-hand side is

$$\frac{\partial}{\partial t}\sum_i \tfrac{1}{2}n_i m_i(u^2 + 2\boldsymbol{u}\cdot\langle\boldsymbol{w}_i\rangle + \langle w_i^2\rangle) = \frac{\partial}{\partial t}\left(\tfrac{1}{2}\rho u^2\right) + \frac{\partial}{\partial t}\left(\tfrac{3}{2}nk_BT\right).$$

The second term is

$$\begin{aligned}\nabla\cdot\sum_i \tfrac{1}{2}n_i m_i\langle(u^2 + 2\boldsymbol{u}\cdot\boldsymbol{w}_i + w_i^2)(\boldsymbol{u}+\boldsymbol{w}_i)\rangle \\ = \nabla\cdot\left(\tfrac{1}{2}\rho u^2\boldsymbol{u}\right) + \nabla_\alpha(u_\beta p_{\beta\alpha}) + \nabla\cdot\left(\tfrac{3}{2}nk_BT\boldsymbol{u}\right) + \nabla\cdot\boldsymbol{q},\end{aligned}$$

and the third term is

$$-\sum_i n_i m_i \left\langle(\boldsymbol{u}+\boldsymbol{w})\cdot\left(\frac{e_i}{m_i}\boldsymbol{E}+\boldsymbol{X}_i\right)\right\rangle = -\boldsymbol{J}\cdot\boldsymbol{E} - \boldsymbol{u}\cdot\sum_i \rho_i\langle\boldsymbol{X}_i\rangle - \sum_i \rho_i\langle\boldsymbol{w}_i\cdot\boldsymbol{X}_i\rangle.$$

Collecting terms, equation (8.17) becomes

$$\begin{aligned}\frac{\mathrm{D}}{\mathrm{D}t}\left(\tfrac{1}{2}\rho u^2 + \tfrac{3}{2}nk_BT\right) = &-\left(\tfrac{1}{2}\rho u^2 + + \tfrac{3}{2}nk_BT\right)\nabla\cdot\boldsymbol{u} - \nabla_\alpha(u_\beta p_\beta\alpha) \\ &-\nabla\cdot\boldsymbol{q} + \boldsymbol{J}\cdot\boldsymbol{E} + \boldsymbol{u}\cdot\sum_i \rho_i\langle\boldsymbol{X}_i\rangle + \sum_i \rho_i\langle\boldsymbol{w}_i\cdot\boldsymbol{X}_i\rangle.\end{aligned}$$

The physical significance of each term in this equation is the following:

1. The left-hand side is the rate of change of the total energy (ordered energy plus heat) per unit volume of an element of fluid moving with the gas drift velocity.

2. The first term on the right-hand side represents the contribution to this rate of change due to the increase of density [this may be seen by substituting for $\nabla\cdot\boldsymbol{u}$ from equation (8.15)].

3. The second term is the work done by the surface forces acting on the element and is made up of two parts; $-u_\beta\nabla_\alpha p_{\alpha\beta}$ is the work done in moving the element as a whole and $p_{\beta\alpha}\nabla_\alpha u_\beta$ is the work done in changing the shape and size of the element.

4. The third term is the heat flow into the element due to thermal conduction.

5. The fourth term is the Joule heating.

6. The fifth term is the work done by the force $\boldsymbol{X}$ in moving the element as a whole.

7. The sixth term is the work done by the force $\boldsymbol{X}$ on the random motions of the particles.

Using equations (8.15) and (8.16), this equation can also be written as

$$\frac{\mathrm{D}}{\mathrm{D}t}\left(\tfrac{3}{2}nk_BT\right) = -\tfrac{3}{2}nk_BT(\nabla\cdot\boldsymbol{u}) - p_{\beta\alpha}\nabla_\alpha u_\beta - \nabla\cdot\boldsymbol{q} + \boldsymbol{j}\cdot\boldsymbol{E} + \sum_i \rho_i\langle\boldsymbol{X}_i\cdot\boldsymbol{w}i\rangle\,. \tag{8.18}$$

Just as for a simple gas, for most problems we would be quite content to obtain a solution to these equations of continuity, motion, and energy without knowing the distribution functions $F_i(\boldsymbol{r},\boldsymbol{v},t)$ accurately. However, in order to do this we must first obtain expressions for the pressure tensor, heat flux vector, and current and these can only be obtained from a solution of Boltzmann's equation for the distribution function. Later we shall show how the procedure we used in Chapter 7 can be extended to consider this binary gas and so give us good approximations with only a poor approximation to the F_i.

Once we have these expressions for $p_{\alpha\beta}$, $\boldsymbol{q}$, and $\boldsymbol{j}$, then in principle these three conservation equations, together with the relevant Maxwell equations, can be solved.

8.3 The collision terms

Collisions between like particles have been considered in Chapter 7. We shall now consider collisions between unlike particles. We need only give a short discussion because formally these collisions can be described in much the same way as those in a simple gas.

Consider the collision between a particle of mass m_1 and velocity $\boldsymbol{v}$ with a particle of mass m_2 and velocity $\boldsymbol{s}$. Suppose after the collision the velocities are $\boldsymbol{v}'$ and $\boldsymbol{s}'$ respectively. The velocity of the center of mass, $\boldsymbol{G}$, must remain constant, so

$$\boldsymbol{G} = \frac{m_1\boldsymbol{v} + m_2\boldsymbol{s}}{m_1+m_2} = \frac{m_1\boldsymbol{v}' + m_2\boldsymbol{s}'}{m_1+m_2}\,.$$

We define the relative velocities before and after the collision to be

$$\boldsymbol{g} \equiv \boldsymbol{s} - \boldsymbol{v}\,, \qquad \boldsymbol{g}' \equiv \boldsymbol{s}' - \boldsymbol{v}'\,.$$

Therefore,

$$\begin{aligned} \boldsymbol{s} &= \boldsymbol{G} + \frac{m_1}{m_1+m_2}\boldsymbol{g}\,, \quad & \boldsymbol{s}' &= \boldsymbol{G} + \frac{m_1}{m_1+m_2}\boldsymbol{g}'\,, \\ \boldsymbol{v} &= \boldsymbol{G} - \frac{m_1}{m_1+m_2}\boldsymbol{g}\,, \quad & \boldsymbol{v}' &= \boldsymbol{G} - \frac{m_1}{m_1+m_2}\boldsymbol{g}'\,. \end{aligned} \tag{8.19}$$

Then the conservation of energy gives

$$g = g' . \tag{8.20}$$

To complete the specification of the collision we must also give the geometry of the collision and we do this, as in Chapter 7, by specifying the asymptotic impact parameter, b, and the azimuthal angle, ϕ. Then the number of collisions in a volume $\mathrm{d}\boldsymbol{r}$ and time $\mathrm{d}t$ between particles 1 in the range $\boldsymbol{v}$ to $\boldsymbol{v}+\mathrm{d}\boldsymbol{v}$ and particles 2 in the range $\boldsymbol{s}$ to $\boldsymbol{s}+\mathrm{d}\boldsymbol{s}$ such that the asymptotic distance of approach lies between b and $b+\mathrm{d}b$ and the azimuthal angle ϕ to $\phi+\mathrm{d}\phi$ is

$$F_1(\boldsymbol{r},\boldsymbol{v},t)F_2(\boldsymbol{r},\boldsymbol{s},t)g\,b\,\mathrm{d}b\,\mathrm{d}\phi\,\mathrm{d}\boldsymbol{v}\,\mathrm{d}\boldsymbol{s}\,\mathrm{d}\boldsymbol{r}\,\mathrm{d}t .$$

In these collisions, the final velocities lie in the range $\boldsymbol{v}'$ to $\boldsymbol{v}'+\mathrm{d}\boldsymbol{v}'$ and $\boldsymbol{s}'$ to $\boldsymbol{s}'+\mathrm{d}\boldsymbol{s}'$ where $\boldsymbol{v}'$ and $\boldsymbol{s}'$ depend upon $\boldsymbol{v}$, $\boldsymbol{s}$, b, and ϕ and the law of interaction between the particles. It can be shown quite generally that

$$\mathrm{d}\boldsymbol{v}'\,\mathrm{d}\boldsymbol{s}' = \mathrm{d}\boldsymbol{v}\,\mathrm{d}\boldsymbol{s} . \tag{8.21}$$

The number of inverse collisions, where the initial velocities are $\boldsymbol{v}'$ and $\boldsymbol{s}'$ and the other specifications, b and ϕ, are the same, is

$$F_1(\boldsymbol{r},\boldsymbol{v}',t)F_2(\boldsymbol{r},\boldsymbol{s}',t)g'b\,\mathrm{d}b\,\mathrm{d}\phi\,\mathrm{d}\boldsymbol{v}'\,\mathrm{d}\boldsymbol{s}'\,\mathrm{d}\boldsymbol{r}\,\mathrm{d}t ,$$

and from equations (8.20) and (8.21), this is equal to

$$F_1(\boldsymbol{r},\boldsymbol{v}',t)F_2(\boldsymbol{r},\boldsymbol{s}',t)g\,b\,\mathrm{d}b\,\mathrm{d}\phi\,\mathrm{d}\boldsymbol{v}\,\mathrm{d}\boldsymbol{s}\,\mathrm{d}\boldsymbol{r}\,\mathrm{d}t .$$

For these collisions the final velocities are $\boldsymbol{v}$ and $\boldsymbol{s}$. Hence the net number of particles 1 scattered into the range $\boldsymbol{v}$ to $\boldsymbol{v}+\mathrm{d}\boldsymbol{v}$ by particles 2 is

$$\mathrm{d}\boldsymbol{r}\,\mathrm{d}\boldsymbol{v}\,\mathrm{d}t \int \mathrm{d}\boldsymbol{s} \int_0^{2\pi} \mathrm{d}\phi \int_0^{\lambda_D} b\,\mathrm{d}b\,g[F_1(\boldsymbol{v}')F_2(\boldsymbol{s}') - F_1(\boldsymbol{v})F_2(\boldsymbol{s})] ,$$

where we have set the maximum interaction distance to be the Debye length, λ_D. Hence the rate of change of the distribution function $F_i(\boldsymbol{v})$ due to collisions with particles j is

$$\Delta_j F_i(\boldsymbol{v}) = \int \mathrm{d}\boldsymbol{s} \int_0^{2\pi} \mathrm{d}\phi \int_0^{\lambda_D} b\,\mathrm{d}b\,g[F_i(\boldsymbol{v}')F_j(\boldsymbol{s}') - F_i(\boldsymbol{v})F_j(\boldsymbol{s})] ,$$

where $\boldsymbol{v}'$ and $\boldsymbol{s}'$ are functions of $\boldsymbol{v}$, $\boldsymbol{s}$, b, and ϕ and the law of interaction between the particles i, j. The total rate of change of $F_i(\boldsymbol{v})$ due to collisions is

$$\Delta F_i(\boldsymbol{v}) = \sum_{j=1}^{2} \Delta_j F_i(\boldsymbol{v}) . \tag{8.22}$$

If Ψ_i is any property of the particles i depending in general on velocity, position, and time, then the rate of change due to collisions of the mean value of Ψ_i is

$$\Delta\langle\Psi_i\rangle = \frac{1}{n_i}\int \mathrm{d}\boldsymbol{v}\,\Psi_i(\boldsymbol{v})\Delta F_i(\boldsymbol{v}) = \sum_j \Delta_j\langle\Psi_i\rangle\,, \tag{8.23}$$

where

$$\begin{aligned}
\Delta_j\langle\Psi_i\rangle &= \frac{1}{n_i}\int \mathrm{d}\boldsymbol{v}\,\Psi_i(\boldsymbol{v})\Delta_j F_i(\boldsymbol{v}) \\
&= \frac{1}{n_i}\int \mathrm{d}\boldsymbol{v}\,\mathrm{d}\boldsymbol{s}\,\mathrm{d}\phi\, b\,\mathrm{d}b\, g\Psi_i(\boldsymbol{v})[F_i(\boldsymbol{v}')F_j(\boldsymbol{s}') - F_i(\boldsymbol{v})F_j(\boldsymbol{s})] \\
&= \frac{1}{n_i}\int \mathrm{d}\boldsymbol{v}\,\mathrm{d}\boldsymbol{s}\,\mathrm{d}\phi\, b\,\mathrm{d}b\, g[\Psi_i(\boldsymbol{v}') - \Psi_i(\boldsymbol{v})]F_i(\boldsymbol{v})F_j(\boldsymbol{s}) \\
&= -\frac{1}{2n_i}\int \mathrm{d}\boldsymbol{v}\,\mathrm{d}\boldsymbol{s}\,\mathrm{d}\phi\, b\,\mathrm{d}b\, g[\Psi_i(\boldsymbol{v}') - \Psi_i(\boldsymbol{v})] \\
&\qquad\qquad \times[F_i(\boldsymbol{v}')F_j(\boldsymbol{s}') - F_i(\boldsymbol{v})F_j(\boldsymbol{s})]\,.
\end{aligned} \tag{8.24}$$

These formulas for $\Delta_j\langle\Psi_i\rangle$ can be proved equivalent by simple algebraic manipulation. In the special case when $i = j$, another formula is, by symmetry,

$$\begin{aligned}
\Delta_i\langle\Psi_i\rangle = -\frac{1}{4n_i}\int \mathrm{d}\boldsymbol{v}\,\mathrm{d}\boldsymbol{s}\,\mathrm{d}\phi\, b\,\mathrm{d}b\, g[\Psi_i(\boldsymbol{v}') + \Psi_i(\boldsymbol{s}') - \Psi_i(\boldsymbol{v}) - \Psi_i(\boldsymbol{s})] \\
\times[F_i(\boldsymbol{v}')F_i(\boldsymbol{s}') - F_i(\boldsymbol{v})F_i(\boldsymbol{s})]\,.
\end{aligned} \tag{8.25}$$

Using equation (8.25) and the last form of equation (8.24), we find that

$$\begin{aligned}
\sum_{i=1}^{2} n_i\Delta\langle\Psi_i\rangle &= n_1\Delta_1\langle\Psi_1\rangle + n_1\Delta_2\langle\Psi_1\rangle + n_2\Delta_1\langle\Psi_2\rangle + n_2\Delta_2\langle\Psi_2\rangle \qquad (8.26)\\
&= -\frac{1}{4}\int \mathrm{d}\boldsymbol{v}\,\mathrm{d}\boldsymbol{s}\,\mathrm{d}\phi\, b\,\mathrm{d}b\, g[\Psi_1(\boldsymbol{v}') + \Psi_1(\boldsymbol{s}') - \Psi_1(\boldsymbol{v}) - \Psi_1(\boldsymbol{s})] \\
&\qquad \times[F_1(\boldsymbol{v}')F_1(\boldsymbol{s}') - F_1(\boldsymbol{v})F_1(\boldsymbol{s})] \\
&= -\frac{1}{2}\int \mathrm{d}\boldsymbol{v}\,\mathrm{d}\boldsymbol{s}\,\mathrm{d}\phi\, b\,\mathrm{d}b\, g[\Psi_1(\boldsymbol{v}') + \Psi_2(\boldsymbol{s}') - \Psi_1(\boldsymbol{v}) - \Psi_2(\boldsymbol{s})] \\
&\qquad \times[F_1(\boldsymbol{v}')F_2(\boldsymbol{s}') - F_1(\boldsymbol{v})F_2(\boldsymbol{s})] \\
&= -\frac{1}{4}\int \mathrm{d}\boldsymbol{v}\,\mathrm{d}\boldsymbol{s}\,\mathrm{d}\phi\, b\,\mathrm{d}b\, g[\Psi_2(\boldsymbol{v}') + \Psi_2(\boldsymbol{s}') - \Psi_2(\boldsymbol{v}) - \Psi_2(\boldsymbol{s})] \\
&\qquad \times[F_2(\boldsymbol{v}')F_2(\boldsymbol{s}') - F_2(\boldsymbol{v})F_2(\boldsymbol{s})] \\
&= -\frac{1}{4}\sum_{i,j}\int \mathrm{d}\boldsymbol{v}\,\mathrm{d}\boldsymbol{s}\,\mathrm{d}\phi\, b\,\mathrm{d}b\, g[\Psi_i(\boldsymbol{v}') + \Psi_j(\boldsymbol{s}') - \Psi_i(\boldsymbol{v}) - \Psi_j(\boldsymbol{s})] \\
&\qquad \times[F_i(\boldsymbol{v}')F_j(\boldsymbol{s}') - F_i(\boldsymbol{v})F_j(\boldsymbol{s})]\,.
\end{aligned}$$

We shall use this last formula in the next section.

8.4 The equilibrium state

In this section we shall prove that the system if left to itself will approach an equilibrium state in which each of the distribution functions are of Maxwell form. The proof is very similar to that given in Chapter 7 for a simple gas.

Consider the "entropy" of the gas defined as

$$S(t) \equiv -k_B \int \mathrm{d}\boldsymbol{r} \sum_j \int \mathrm{d}\boldsymbol{v}\, F_j(\boldsymbol{r},\boldsymbol{v},t) \ln F_j(\boldsymbol{r},\boldsymbol{v},t)\,.$$

Then

$$\frac{\mathrm{d}S}{\mathrm{d}t} = -k_B \int \mathrm{d}\boldsymbol{r} \sum_j \int \mathrm{d}\boldsymbol{v}\, [1+\ln F_j(\boldsymbol{r},\boldsymbol{v},t)] \frac{\partial}{\partial t} F_j(\boldsymbol{r},\boldsymbol{v},t)\,.$$

Substituting for $\partial F_j/\partial t$ from equation (8.1), this becomes

$$\begin{aligned}\frac{1}{k_B}\frac{\mathrm{d}S}{\mathrm{d}t} = &-\int \mathrm{d}\boldsymbol{r} \sum_j \int \mathrm{d}\boldsymbol{v}\, \boldsymbol{v}\cdot\nabla[F_j \ln F_j] \\ &+\int \mathrm{d}\boldsymbol{r} \sum_j \int \mathrm{d}\boldsymbol{v} \left(\frac{e_j}{m_j}\boldsymbol{E}+\boldsymbol{X}_j\right)\cdot\nabla_v[F_j \ln F_j] \\ &-\int \mathrm{d}\boldsymbol{r} \sum_j \int \mathrm{d}\boldsymbol{v}\, [1+\ln F_j(\boldsymbol{r},\boldsymbol{v},t)]\Delta F_j(\boldsymbol{r},\boldsymbol{v},t)\,.\end{aligned}$$

As in Chapter 7, the first term can be transformed into an integral over the surface of the container and vanishes if we assume elastic collisions with the walls. The second term can be integrated by parts and is certainly zero provided $\nabla\cdot\boldsymbol{X}$ is zero. The third term by equation (8.23) is

$$-\int \mathrm{d}\boldsymbol{r} \sum_j n_j \Delta\langle 1+\ln F_j\rangle\,.$$

Hence, by equation (8.26),

$$\frac{\mathrm{d}S}{\mathrm{d}t} = \frac{k_B}{4}\int \mathrm{d}\boldsymbol{r} \sum_{i,j} \int \mathrm{d}\boldsymbol{v}\,\mathrm{d}\boldsymbol{s}\,\mathrm{d}\phi\, b\,\mathrm{d}b\, g \ln\left[\frac{F_i(\boldsymbol{v}')F_j(\boldsymbol{s}')}{F_i(\boldsymbol{v})F_j(\boldsymbol{s})}\right][F_i(\boldsymbol{v}')F_j(\boldsymbol{s}')-F_i(\boldsymbol{v})F_j(\boldsymbol{s})].$$

Now the two principal factors in the integrand, $[F_i(\boldsymbol{v}')F_j(\boldsymbol{s}')-F_i(\boldsymbol{v})F_j(\boldsymbol{s})]$ and $\ln[F_i(\boldsymbol{v}')F_j(\boldsymbol{s}')\,/F_i(\boldsymbol{v})F_j(\boldsymbol{s})]$, are both positive or both negative according as $F_i(\boldsymbol{v}')F_j(\boldsymbol{s}')$ is greater or less than $F_i(\boldsymbol{v})F_j(\boldsymbol{s})$. Thus the product is always positive or zero, so that

$$\frac{\mathrm{d}S}{\mathrm{d}t} \geq 0\,.$$

Thus the entropy increases until eventually a steady state is reached where $\mathrm{d}S/\mathrm{d}t$ is zero (since it cannot increase indefinitely). In this equilibrium state,

$$F_i(\boldsymbol{v}')F_j(\boldsymbol{s}') = F_i(\boldsymbol{v})F_j(\boldsymbol{s})$$
$$\ln F_i(\boldsymbol{v}') + \ln F_j(\boldsymbol{s}') - \ln F_i(\boldsymbol{v}) - \ln F_j(\boldsymbol{s}) = 0 \tag{8.27}$$

for all collisions. It therefore follows that $\ln F$ must be a collision invariant.

From the fact that $\ln F_1$ is a collision invariant for collisions between particles 1 it follows, as in Chapter 7, that

$$F_1(\boldsymbol{r},\boldsymbol{v},t) = \frac{n_1}{v_{t1}^3\pi^{3/2}}\mathrm{e}^{-(\boldsymbol{v}-\boldsymbol{u}_1)^2/v_{t1}^2},$$

where $v_{ti}^2 = 2k_BT_i/m_i$. Similarly, because $\ln F_2$ is invariant for collisions between particles 2,

$$F_2(\boldsymbol{r},\boldsymbol{v},t) = \frac{n_2}{v_{t2}^3\pi^{3/2}}\mathrm{e}^{-(\boldsymbol{v}-\boldsymbol{u}_2)^2/v_{t2}^2}.$$

Then putting $i = 1$ and $j = 2$ in equation (8.27) gives

$$-\frac{(\boldsymbol{v}'-\boldsymbol{u}_1)^2}{v_{t1}^2} - \frac{(\boldsymbol{s}'-\boldsymbol{u}_2)^2}{v_{t2}^2} + \frac{(\boldsymbol{v}-\boldsymbol{u}_1)^2}{v_{t1}^2} + \frac{(\boldsymbol{s}-\boldsymbol{u}_2)^2}{v_{t2}^2} = 0$$

for all collisions. Thus,

$$T_1 = T_2 = T, \quad \text{and} \quad \boldsymbol{u}_1 = \boldsymbol{u}_2 = \boldsymbol{u}.$$

Therefore, in the equilibrium state,

$$F_i(\boldsymbol{r},\boldsymbol{v},t) = f_i(\boldsymbol{r},\boldsymbol{v},t),$$

where $f_i(\boldsymbol{r},\boldsymbol{v},t)$ is the Maxwellian distribution

$$f_i(\boldsymbol{r},\boldsymbol{v},t) = \frac{n_i}{v_{ti}^3\pi^{3/2}}\mathrm{e}^{-(\boldsymbol{v}-\boldsymbol{u})^2/v_{ti}^2}.$$

It is easy to show that, as for the simple gas, in this equilibrium state the temperature is uniform and the most general drift velocity is

$$\boldsymbol{u} = \boldsymbol{u}_0 + \boldsymbol{\omega}\times\boldsymbol{r}$$

where $\boldsymbol{u}_0$ and $\boldsymbol{\omega}$ are constant vectors, and the number densities are given by

$$n_i = n_{i0}\exp[-(m_iU_i + e_i\Phi)/k_BT + u^2/v_{ti}^2] \tag{8.28}$$

where Φ is the electrostatic potential and U_i is the potential from which $\boldsymbol{X}_i$ is derived, i.e.,

$$\boldsymbol{X}_i = -\nabla U_i.$$

For this equilibrium state, it is clear that

$$p_{\alpha\beta} = p\delta_{\alpha\beta} = nk_BT\delta_{\alpha\beta} \tag{8.29}$$
$$\boldsymbol{q} = 0. \tag{8.30}$$

8.5 The formal theory of kinetic processes

In this section we again describe the method of Chapman and Cowling[49] for solving the Boltzmann equations in a binary gas by successive approximations. Just as in the simple gas we find that we will not obtain an accurate representation of the distribution function, but we will get quite accurate expressions for the pressure tensor, the heat flux vector, and the electric current, i.e., for the coefficients of viscosity, thermal conduction, and electrical conductivity. This analysis also predicts the thermal diffusion effect which turns out to be quite large for an ionized gas.

Just as for the simple gas, we begin by assuming that collisions are important in determining the distribution function and that to a first approximation, we have a Maxwellian distribution at each point in space. We then calculate corrections to this first approximation. We indicate this process mathematically by rewriting the Boltzmann equations as

$$\left[\frac{\partial}{\partial t} + \boldsymbol{v}\cdot\nabla + \left(\frac{e_i}{m_i}\boldsymbol{E} + \boldsymbol{X}_i\right)\cdot\nabla_v\right] F_i$$
$$= \frac{1}{\eta}\Delta F_i = \frac{1}{\eta}\sum_j \int \mathrm{d}\boldsymbol{s}\mathrm{d}\phi\, b\, \mathrm{d}b\, g\,[F_i(\boldsymbol{v}')F_j(\boldsymbol{s}') - F_i(\boldsymbol{v})F_j(\boldsymbol{s})]\,. \tag{8.31}$$

We will eventually set η to unity since it is only a formal expansion parameter, but it will help us to group the terms in the expansion. We look for solutions of the type

$$F_i = F_i^{(0)}(1 + \eta\varphi_i + \eta^2\xi_i + \cdots) \tag{8.32}$$

where by keeping more and more terms, we obtain the successive approximations. Substituting equation (8.32) into equation (8.31), the terms of order $1/\eta$ are

$$\sum_j \int \mathrm{d}\boldsymbol{s}\mathrm{d}\phi\, b\, \mathrm{d}b\, g\,[F_i^{(0)}(\boldsymbol{v}')F_j^{(0)}(\boldsymbol{s}') - F_i^{(0)}(\boldsymbol{v})F_j^{(0)}(\boldsymbol{s})] = 0\,.$$

Since this is true for all $\boldsymbol{v}$ and for $i = 1, 2$, the only solution is the Maxwell solution,

$$F_i^{(0)}(\boldsymbol{r}, \boldsymbol{v}, t) = f_i(\boldsymbol{v}) = \frac{n_i}{v_{ti}^3\pi^{3/2}}\mathrm{e}^{-(\boldsymbol{v}-\boldsymbol{u})^2/v_{ti}^2}\,,$$

where now $F_i^{(0)}$ depends on $\boldsymbol{r}$ and t only through the dependence on n_i, $\boldsymbol{u}$, and T.

We now consider the terms on the right-hand side that are independent of η. These are

$$\sum_j \int \mathrm{d}\boldsymbol{s}\mathrm{d}\phi\, b\, \mathrm{d}b\, g\,\{f_i(\boldsymbol{v}')f_j(\boldsymbol{s}')[\varphi_i(\boldsymbol{v}') + \varphi_j(\boldsymbol{s}')] - f_i(\boldsymbol{v})f_j(\boldsymbol{s})[\varphi_i(\boldsymbol{v}) + \varphi_j(\boldsymbol{s})]\}\,,$$

which, because

$$f_i(\boldsymbol{v}')f_j(\boldsymbol{s}') = f_i(\boldsymbol{v})f_j(\boldsymbol{s}),$$

may be written as

$$\sum_j \int \mathrm{d}\boldsymbol{s}\mathrm{d}\phi\, b\, \mathrm{d}b\, g\, f_i(\boldsymbol{v})f_j(\boldsymbol{s})[\varphi_i(\boldsymbol{v}') + \varphi_j(\boldsymbol{s}') - \varphi_i(\boldsymbol{v}) - \varphi_j(\boldsymbol{s})].$$

The last two terms on the left-hand side of equation (8.31) give

$$\left[\boldsymbol{v}\cdot\nabla + \left(\frac{e_i}{m_i}\boldsymbol{E} + \boldsymbol{X}_i\right)\cdot\nabla_v\right] f_i(\boldsymbol{v}) = \frac{\partial f_i}{\partial n_i}\boldsymbol{v}\cdot\nabla n_i + \frac{\partial f_i}{\partial u_\alpha}\boldsymbol{v}\cdot\nabla u_\alpha + \frac{\partial f_i}{\partial T}\boldsymbol{v}\cdot\nabla T - \frac{m_i}{k_B T} f_i \boldsymbol{w}\cdot\left(\frac{e_i}{m_i}\boldsymbol{E} + \boldsymbol{X}_i\right), \tag{8.33}$$

where clearly

$$\frac{\partial f_i}{\partial n_i} = \frac{f_i}{n_i},$$
$$\frac{\partial f_i}{\partial u_\alpha} = 2 f_i \frac{w_\alpha}{v_{ti}^2},$$
$$\frac{\partial f_i}{\partial T} = -\frac{f_i}{T}\left(\frac{3}{2} - \frac{w^2}{v_{ti}^2}\right).$$

The terms that are independent of η coming from the first term on the left-hand side of equation (8.31) require more care, such that

$$\frac{\partial f_i}{\partial t} = \frac{\partial f_i}{\partial n_i}\frac{\partial n_i}{\partial t} + \frac{\partial f_i}{\partial u_\alpha}\frac{\partial u_\alpha}{\partial t} + \frac{\partial f_i}{\partial T}\frac{\partial T}{\partial t}, \tag{8.34}$$

and we can substitute for $\partial n_i/\partial t$, $\partial u_\alpha/\partial t$, and $\partial T/\partial t$ from the continuity equations of equation (8.13), the momentum equation of equation (8.16) and the energy equation of equation (8.18). We must be reminded again, however, that in our expansion approach, some of these terms are higher order in η and should be ignored. Thus in equation (8.13) to the order we are working, $\langle \boldsymbol{w}_i \rangle$ is zero so in equation (8.34) we replace $\partial n_i/\partial t$ by

$$\frac{\partial n_i^{(0)}}{\partial t} = -\nabla\cdot(n_i\boldsymbol{u}).$$

In equation (8.16), to zero order in η, the pressure tensor, $p_{\alpha\beta}$, is $p\delta_{\alpha\beta}$ and so in equation (8.34) we replace $\partial u_\alpha/\partial t$ by

$$\frac{\partial u_\alpha^{(0)}}{\partial t} = -(\boldsymbol{u}\cdot\nabla)u_\alpha - \frac{1}{\rho}\nabla_\alpha p + \frac{Q}{\rho}E_\alpha + \frac{1}{\rho}\sum_i \rho_i X_{i\alpha}.$$

Finally, in the energy equation (8.18), to zero order in η, $p_{\alpha\beta}$ is $p\delta_{\alpha\beta}$ and $\boldsymbol{q}$, $\boldsymbol{j}$, and $\langle \boldsymbol{w} \rangle$ are zero, so in equation (8.34), we replace $\partial T/\partial t$ by

$$\frac{\partial T^{(0)}}{\partial t} = -(\boldsymbol{u} \cdot \nabla)T - \frac{2T}{3}\nabla \cdot \boldsymbol{u}\,.$$

Collecting together all these terms of zero order in η we find that many terms cancel and finally we obtain

$$f_i(\boldsymbol{v})\left[2(\boldsymbol{\varpi}_i^0\boldsymbol{\varpi}_i)_{\alpha\beta}\nabla_\alpha u_\beta - \left(\tfrac{5}{2} - \varpi_i^2\right)\boldsymbol{w} \cdot \nabla \ln T + \frac{n}{n_i}\boldsymbol{w} \cdot \boldsymbol{d}_i\right]$$
$$= \sum_j \int \mathrm{d}\boldsymbol{s}\mathrm{d}\phi\, b\,\mathrm{d}b\, g\, f_i(\boldsymbol{v})f_j(\boldsymbol{s})[\varphi_i(\boldsymbol{v}') + \varphi_j(\boldsymbol{s}') - \varphi_i(\boldsymbol{v}) - \varphi_j(\boldsymbol{s})]\,, \quad (8.35)$$

where $\boldsymbol{\varpi}_i$ is the dimensionless random velocity defined by equation (7.81), and $\boldsymbol{\varpi}_i^0\boldsymbol{\varpi}_i$ is a tensor defined by

$$(\boldsymbol{\varpi}_i^0\boldsymbol{\varpi}_i)_{\alpha\beta} \equiv \varpi_{i\alpha}\varpi_{i\beta} - \tfrac{1}{3}\varpi_i^2\delta_{\alpha\beta}\,. \quad (8.36)$$

The vector $\boldsymbol{d}_i$ is given by

$$\boldsymbol{d}_i \equiv \nabla\left(\frac{n_i}{n}\right) + \frac{1}{p}\left[\frac{n_i}{n} - \frac{n_i m_i}{\rho}\right]\nabla p - \frac{1}{p}\left[\rho_i \boldsymbol{X}_i - \frac{\rho_i}{\rho}\left(\sum_j \rho_j \langle \boldsymbol{X}_i \rangle\right)\right]$$
$$-\frac{1}{p}\left(n_i e_i - \frac{\rho_i Q}{\rho}\right)\boldsymbol{E}\,. \quad (8.37)$$

Notice that if $\boldsymbol{X}$ is independent of velocity so that $\langle \boldsymbol{X}_i \rangle = \boldsymbol{X}_i$, then

$$\boldsymbol{d}_1 + \boldsymbol{d}_2 = 0\,. \quad (8.38)$$

The analysis simplifies considerably if equation (8.38) is true so from now on we shall assume $\boldsymbol{X}_i$ is velocity independent. This restriction is not a serious one for probably the only force $\boldsymbol{X}$ need stand for is gravity and even this is often unimportant.

Problem 8.1 Show that $\boldsymbol{d}_i$ has the form indicated in equation (8.37).

Equation (8.35) is a difficult integral equation to be solved for $\varphi(\boldsymbol{v})$, the first order correction to the distribution function. In principle we could now go on to consider terms proportional to η in equation (8.31) and so get an equation for the second order correction to F_i. But in fact this would be very difficult and has never been done for a binary gas. These higher order terms would give contributions to the distribution function proportional to the second derivatives of the temperature and drift velocity and to products of the first derivatives. Hence if all gradients are small it is reasonable to neglect these higher order terms.

We now discuss the solution of equation (8.35). Because it is linear in $\varphi(\boldsymbol{v})$ we can write immediately

$$\varphi_1 = \mathsf{B}_1^{\alpha\beta}\nabla_\alpha u_\beta - \boldsymbol{A}_1 \cdot \nabla \ln T - n\boldsymbol{E}_1 \cdot \boldsymbol{d}_1 \tag{8.39}$$
$$\varphi_2 = \mathsf{B}_2^{\alpha\beta}\nabla_\alpha u_\beta - \boldsymbol{A}_2 \cdot \nabla \ln T - n\boldsymbol{E}_2 \cdot \boldsymbol{d}_1 \tag{8.40}$$

where $\mathsf{B}_i^{\alpha\beta}$ is a tensor function and $\boldsymbol{A}_i$ and $\boldsymbol{E}_i$ are vector functions of $\boldsymbol{w}_i$ satisfying the equations

$$-2f_i\boldsymbol{\varpi}_i^0\boldsymbol{\varpi}_i = \sum_j \int \mathrm{d}\boldsymbol{s}\mathrm{d}\phi\, b\, \mathrm{d}b\, g\, f_i(\boldsymbol{v})f_j(\boldsymbol{s}) \times[\mathsf{B}_i(\boldsymbol{v}') + \mathsf{B}_j(\boldsymbol{s}') - \mathsf{B}_i(\boldsymbol{v}) - \mathsf{B}_j(\boldsymbol{s})]\,, \tag{8.41}$$

$$f_i\left(\tfrac{5}{2} - \varpi_i^2\right)\boldsymbol{w} = \sum_j \int \mathrm{d}\boldsymbol{s}\mathrm{d}\phi\, b\, \mathrm{d}b\, g\, f_i(\boldsymbol{v})f_j(\boldsymbol{s}) \times[\boldsymbol{A}_i(\boldsymbol{v}') + \boldsymbol{A}_j(\boldsymbol{s}') - \boldsymbol{A}_i(\boldsymbol{v}) - \boldsymbol{A}_j(\boldsymbol{s})]\,, \tag{8.42}$$

$$-\frac{1}{n_1}f_1\boldsymbol{w} = \sum_j \int \mathrm{d}\boldsymbol{s}\mathrm{d}\phi\, b\, \mathrm{d}b\, g\, f_1(\boldsymbol{v})f_j(\boldsymbol{s}) \times[\boldsymbol{E}_1(\boldsymbol{v}') + \boldsymbol{E}_j(\boldsymbol{s}') - \boldsymbol{E}_1(\boldsymbol{v}) - \boldsymbol{E}_j(\boldsymbol{s})]\,, \tag{8.43}$$

$$-\frac{1}{n_2}f_2\boldsymbol{w} = \sum_j \int \mathrm{d}\boldsymbol{s}\mathrm{d}\phi\, b\, \mathrm{d}b\, g\, f_2(\boldsymbol{v})f_j(\boldsymbol{s}) \times[\boldsymbol{E}_2(\boldsymbol{v}') + \boldsymbol{E}_j(\boldsymbol{s}') - \boldsymbol{E}_2(\boldsymbol{v}) - \boldsymbol{E}_j(\boldsymbol{s})]\,. \tag{8.44}$$

From the form of these equations, B, $\boldsymbol{A}$, and $\boldsymbol{E}$ must be of the form

$$\begin{aligned} \mathsf{B}_i &= \boldsymbol{\varpi}_i^0\boldsymbol{\varpi}_i\mathcal{B}(\varpi_i^2) \\ \boldsymbol{A}_i &= \boldsymbol{\varpi}_i\mathcal{A}(\varpi_i^2) \\ \boldsymbol{E}_i &= \boldsymbol{\varpi}_i\mathcal{E}(\varpi_i^2) \end{aligned} \tag{8.45}$$

where $\mathcal{B}_i$, $\mathcal{A}_i$, and $\mathcal{E}_i$ are scalar functions of ϖ_i^2 and the problem has been reduced to calculating these functions.

There are some subsidiary conditions which these functions must satisfy to ensure that n_i, $\boldsymbol{u}$, and T really are the number density, drift velocity, and temperature, respectively. The number density is

$$n_i = \int \mathrm{d}\boldsymbol{v}\, f_i(1+\varphi_i) = n_i + \int \mathrm{d}\boldsymbol{v}\, f_i\varphi_i\,,$$

so we must have

$$\int \mathrm{d}\boldsymbol{v}\, f_i\varphi_i = 0\,.$$

Substituting for φ_i from equations (8.39) and (8.40) and using equation (8.45) we find that this equation is automatically satisfied for any $\mathcal{A}$, $\mathcal{B}$, and $\mathcal{E}$, so this gives no subsidiary condition.

The temperature is given by

$$\tfrac{3}{2}nk_BT = \sum_i \tfrac{1}{2}m_i \int \mathrm{d}\boldsymbol{v}\, w^2 F_i$$

so we must have

$$\tfrac{1}{2}m_i \int \mathrm{d}\boldsymbol{v}\, w^2 f_i\varphi_i = 0\,,$$

which is also automatically satisfied.

Finally, the drift velocity must be given by

$$\boldsymbol{u} = \frac{1}{\rho}\sum_i m_i \int \mathrm{d}\boldsymbol{v}\, \boldsymbol{v}F_i$$

so we must have

$$\sum_i m_i \int \mathrm{d}\boldsymbol{v}\, \boldsymbol{v}f_i\varphi_i = 0\,.$$

This equation does give a condition on $\mathcal{A}$ and $\mathcal{E}$. Substituting for φ_i and using equation (8.45) and changing variables from $\boldsymbol{w}_i$ to $\boldsymbol{\varpi}_i$ by means of equation (7.81) gives

$$\sum_i n_i\sqrt{m_i}\frac{1}{\pi^{3/2}}\int \mathrm{d}\boldsymbol{\varpi}_i\, \varpi_i^2\mathcal{A}_i(\varpi_i^2)\mathrm{e}^{-\varpi_i^2} = 0 \tag{8.46}$$

and

$$\sum_i n_i\sqrt{m_i}\frac{1}{\pi^{3/2}}\int \mathrm{d}\boldsymbol{\varpi}_i\, \varpi_i^2\mathcal{E}_i(\varpi_i^2)\mathrm{e}^{-\varpi_i^2} = 0\,. \tag{8.47}$$

There is no subsidiary condition to be applied to $\mathcal{B}_i(\varpi_i^2)$.

Just as for the simple gas considered in Chapter 7, it is convenient at this stage to express $\mathcal{A}_i$, $\mathcal{B}_i$, and $\mathcal{E}_i$ as an infinite series of Sonine, S_n^m, or Generalized Laguerre, L_m^n, polynomials. These were defined in Chapter 7, and some polynomials we will need explicitly are

$$L_0^{\frac{3}{2}}(y^2) = 1\,, \quad L_1^{\frac{3}{2}}(y^2) = \tfrac{5}{2} - y^2\,,$$

$$L_0^{\frac{5}{2}}(y^2) = 1\,, \quad L_1^{\frac{5}{2}}(y^2) = \tfrac{7}{2} - y^2\,.$$

Expanding in terms of these polynomials, we write

$$\mathcal{A}_i(\varpi_i^2) = \sum_{m=0}^{\infty} a_i^m L_m^{\frac{3}{2}}(\varpi_i^2)\,, \tag{8.48}$$

$$\mathcal{E}_i(\varpi_i^2) = \sum_{m=0}^{\infty} e_i^m L_m^{\frac{3}{2}}(\varpi_i^2)\,, \tag{8.49}$$

$$\mathcal{B}_i(\varpi_i^2) = \sum_{m=0}^{\infty} b_i^m L_m^{\frac{5}{2}}(\varpi_i^2)\,. \tag{8.50}$$

To obtain an exact solution for φ_i we need to know all the coefficients a_i^m, b_i^m, and e_i^m, but just as in the case for a simple gas, to obtain the coefficients of viscosity, thermal conduction, etc., we need only a few of them. Later we shall describe a variation method which is different from the method used in Chapter 7 that can be used to give these few coefficients quite accurately without too much work. Before describing this new method, we shall consider what conditions equations (8.46) and (8.47) place on the coefficients and how the physical quantities we are interested in are related to them.

Substituting equation (8.48) into the subsidiary condition of equation (8.46) gives

$$\sum_i n_i \sqrt{m_i} \sum_m a_i^m \frac{1}{\pi^{3/2}} 4\pi \int_0^\infty \mathrm{d}\varpi_i \, \varpi_i^4 \mathrm{e}^{-\varpi_i^2} L_0^{\frac{3}{2}}(\varpi_i^2) L_m^{\frac{3}{2}}(\varpi_i^2) = 0 \,,$$

since $L_0^{\frac{3}{2}}(\varpi_i^2) = 1$. Evaluating the integral by equation (7.92) gives

$$\sum_i n_i \sqrt{m_i} a_i^0 = n_1 \sqrt{m_1} a_1^0 + n_2 \sqrt{m_2} a_2^0 = 0$$

so

$$a_2^0 = -\frac{n_1}{n_2} \left(\frac{m_1}{m_2} \right)^{1/2} a_1^0 \,. \tag{8.51}$$

Similarly, equations (8.49) and (8.47) give the condition,

$$e_2^0 = -\frac{n_1}{n_2} \left(\frac{m_1}{m_2} \right)^{1/2} e_1^0 \,. \tag{8.52}$$

Now consider the pressure tensor defined by equation (8.8). Substituting for F_i from equation (8.32) and for φ_i from equation (8.39) gives

$$p_{\gamma\epsilon} = p\delta_{\gamma\epsilon} - \sum_i \int \mathrm{d}\boldsymbol{w}_i \, w_{i\gamma} w_{i\epsilon} f_i \left[\nabla_\alpha u_\beta \mathsf{B}_i^{\alpha\beta} + \boldsymbol{A}_i \cdot \nabla \ln T + n \boldsymbol{E}_i \cdot \boldsymbol{d}_i \right] .$$

On using equation (8.45) we find that the terms involving $\boldsymbol{A}_i$ and $\boldsymbol{E}_i$ vanish so

$$p_{\gamma\epsilon} = \left(p + \tfrac{2}{3} \mu \nabla \cdot \boldsymbol{u} \right) \delta_{\gamma\epsilon} - \mu (\nabla_\gamma u_\epsilon + \nabla_\epsilon u_\gamma) \tag{8.53}$$

where

$$\mu = \tfrac{1}{2} k_B T (n_1 b_1^0 + n_2 b_2^0) \,. \tag{8.54}$$

Thus to calculate the coefficient of viscosity we need to know only b_1^0 and b_2^0 accurately, while the other b-coefficients are relatively unimportant.

Now consider the mean random velocity, $\langle \boldsymbol{w}_i \rangle$, defined by equation (8.5). Substituting for F_i from equation (8.32) and for φ_i from equation (8.39) gives

$$\langle \boldsymbol{w}_i \rangle = -\frac{1}{n_i} \int \mathrm{d}\boldsymbol{w}_i \, \boldsymbol{w}_i \, f_i \left[\nabla_\alpha u_\beta \mathsf{B}_i^{\alpha\beta} + \boldsymbol{A}_i \cdot \nabla \ln T + n \boldsymbol{E}_i \cdot \boldsymbol{d}_i \right] .$$

On using equation (8.45) we find that the first term vanishes. The other terms are

$$\langle \boldsymbol{w}_i \rangle = -\frac{1}{n_i}\sum_m [a_i^m \nabla_\alpha \ln T + e_i^m d_{1\alpha}] \int \mathrm{d}\boldsymbol{w}_i\, \boldsymbol{w}_i\, f_i \varpi_{i\alpha} L_m^{\frac{3}{2}}(\varpi_i^2)$$
$$= -\frac{1}{3n_i}\sum_m [a_i^m \nabla \ln T + e_i^m \boldsymbol{d}_1] \int \mathrm{d}\boldsymbol{w}_i\, f_i \boldsymbol{w}_i \cdot \boldsymbol{\varpi}_i L_m^{\frac{3}{2}}(\varpi_i^2)\,. \quad (8.55)$$

The integral in equation (8.55) can be evaluated by changing variables from $\boldsymbol{w}_i$ to $\boldsymbol{\varpi}_i$ and using equation (7.92) with $m' = 0$ to give $I = \frac{3}{2} n_i v_{ti} \delta_{m,0}$. Therefore,

$$\langle \boldsymbol{w}_i \rangle = -\tfrac{1}{2} v_{ti} (a_i^0 \nabla \ln T + n e_i^0 \boldsymbol{d}_1)\,, \quad (8.56)$$

and so, using equations (8.51) and (8.52)

$$\langle \boldsymbol{w}_1 \rangle - \langle \boldsymbol{w}_2 \rangle = -\frac{v_{t1}\rho}{2n_2 m_2} a_i^0 \nabla \ln T - \frac{n v_{t1}\rho}{2n_2 m_2} e_i^0 \boldsymbol{d}_1\,.$$

This equation is frequently written as

$$\langle \boldsymbol{w}_1 \rangle - \langle \boldsymbol{w}_2 \rangle = -\frac{n^2}{n_1 n_2}[D_{12}\boldsymbol{d}_1 + D_T \nabla \ln T]\,, \quad (8.57)$$

where

$$D_{12} \equiv \frac{n_1 \rho v_{t1}}{2n m_2} e_1^0 \quad (8.58)$$

is called the diffusion coefficient, and

$$D_T \equiv \frac{n_1 \rho v_{t1}}{2n^2 m_2} a_1^0 \quad (8.59)$$

is the thermal diffusion coefficient. It is also convenient to define the ratio

$$k_T \equiv \frac{D_T}{D_{12}} = \frac{1}{n}\frac{a_1^0}{e_1^0}\,. \quad (8.60)$$

From equations (8.12) and (8.37) we find

$$\boldsymbol{j} = \sigma \Bigg[\boldsymbol{E} + \frac{p\rho}{n_1 n_2 e (m_2 + m_1)} \nabla \left(\frac{n_1}{n}\right) + \frac{(m_2 - m_1)}{n e (m_2 + m_1)} \nabla p$$
$$- \frac{m_1 m_2}{e(m_1 + m_2)} (\boldsymbol{X}_1 - \boldsymbol{X}_2) + \frac{p\rho}{n_1 n_2 e (m_2 + m_1)} k_T \frac{1}{T} \nabla T \Bigg]\,, \quad (8.61)$$

where

$$\sigma = \frac{n_1 n_2 n^2}{p\rho^2} (e_1 m_2 - e_2 m_1)^2 D_{12} = \frac{n_1^2 n_2 n e^2 (m_1 + m_2)^2 v_{t1}}{2p\rho m_2} e_1^0 \quad (8.62)$$

is the electrical conductivity as usually defined. These expressions simplify considerably if we ignore small terms $\simeq m_1/m_2$ and if we use the quasi-neutrality approximation $n_1 - n_2 \ll n$. Then we find approximately,

$$\boldsymbol{j} = \sigma \left[\boldsymbol{E} + \frac{1}{2n_1 e} \nabla p - \frac{m_1}{e} (\boldsymbol{X}_1 - \boldsymbol{X}_2) + \frac{2k_B}{e} k_T \nabla T \right] , \tag{8.63}$$

with

$$\sigma \simeq \frac{n_1^2 e^2 v_{t1}}{2k_B T} e_1^0 .$$

From equation (8.63) we see that a conduction current flows because of the electric field, the pressure gradient, the difference between the effects of the nonelectromagnetic force, $\boldsymbol{X}_1 - \boldsymbol{X}_2$, and the temperature gradient. This last effect is known as the thermal diffusion effect and was first predicted by Chapman from this analysis and later discovered experimentally. In normal gases it is a small effect but in ionized gases, because of the great difference in mass between ions and electrons, it is quite important and comparable to the other effects.

From these expressions we see that to calculate σ we need only to know the coefficient e_1^0 and to calculate the thermal effects we only need to know a_1^0.

The last quantity we want to know is the heat flux vector $\boldsymbol{q}$ defined by equation (8.9). Substituting for F_i from equation (8.32) and for φ_i from equation (8.39) we get

$$\boldsymbol{q} = -\sum_i \tfrac{1}{2} m_i \int \mathrm{d}\boldsymbol{v}\, w^2 \boldsymbol{w} f_i [\mathsf{B}_i^{\alpha\beta} \nabla_\alpha u_\beta + \boldsymbol{A}_i \cdot \nabla \ln T + n\boldsymbol{E}_i \cdot \boldsymbol{d}_1] .$$

Using equation (8.45) we find that the first term vanishes and we get

$$\boldsymbol{q} = -\sum_i \frac{1}{6} m_i \sum_m [a_i^m \nabla \ln T + n e_i^m \boldsymbol{d}_1] \int \mathrm{d}\boldsymbol{v}\, w^2 \boldsymbol{w} \cdot \boldsymbol{\varpi}_i f_i L_m^{\frac{3}{2}}(\varpi^2) . \tag{8.64}$$

The integral in equation (8.64) can be evaluated by changing variables from $\boldsymbol{w}$ to $\boldsymbol{\varpi}_i$ to give

$$I \equiv \frac{n_i v_{ti}^3}{\pi^{3/2}} \int \mathrm{d}\boldsymbol{\varpi} \varpi^4 \mathrm{e}^{-\varpi^2} L_m^{\frac{3}{2}}(\varpi^2) = \frac{4\pi n_i v_{ti}^3}{\pi^{3/2}} \int \mathrm{d}\varpi \varpi^6 \mathrm{e}^{-\varpi^2} L_m^{\frac{3}{2}}(\varpi^2) .$$

We now write

$$\varpi^6 = -\varpi^4 \left(\tfrac{5}{2} - \varpi^2 - \tfrac{5}{2} \right) = -\varpi^4 \left[L_1^{\frac{3}{2}}(\varpi^2) - \tfrac{5}{2} L_0^{\frac{3}{2}}(\varpi^2) \right]$$

and the integrals can be evaluated using equation (7.92) to give

$$I = -\frac{15 n_i v_{ti}^3}{4} (\delta_{m,1} - \delta_{m,0}) .$$

Hence, equation (8.64) becomes

$$q = -\lambda' \nabla T + \frac{5nk_BT}{4} \sum_i n_i v_{ti} e_i^1 d_1 - \frac{5k_B n_1 v_{t1}}{4} a_1^0 \frac{m_2 - m_1}{m_2} \nabla T$$
$$- \frac{5nk_BTn_1v_{t1}}{4} e_1^0 \frac{m_2 - m_1}{m_2} d_1 \,, \qquad (8.65)$$

where

$$\lambda' = -\frac{5k_B}{4} \sum_i n_i v_{ti} a_i^1 \,. \qquad (8.66)$$

The form of this expression is interesting. From equations (8.56), (8.51), and (8.52), we notice that the last two terms are

$$\tfrac{5}{2} k_B T (n_1 \langle w_1 \rangle + n_2 \langle w_2 \rangle) \,,$$

so equation (8.65) can be rewritten as

$$q = -\lambda' \nabla T + \tfrac{5}{2} k_B T (n_1 \langle w_1 \rangle + n_2 \langle w_2 \rangle) + \frac{5nk_BT}{4} \sum_i n_i v_{ti} e_i^1 d_1 \,. \qquad (8.67)$$

Problem 8.2 Show

1. where the factor of $\frac{1}{3}$ came from in obtaining equation (8.64), and
2. that equation (8.64) can be reduced to the form of equation (8.67).

There are several points of interest to note in this equation. The first term represents the conduction of heat due to a temperature gradient. The second term is the heat carried by the flux of particles $(n_1 \langle w_1 \rangle + n_2 \langle w_2 \rangle)$ relative to the drift velocity u, each particle carrying on average a heat energy $\frac{5}{2} k_B T$ (the reason why this is $\frac{5}{2} k_B T$ and not $\frac{3}{2} k_B T$ is explained below). The third term is the heat flow produced by the force d_1 and we shall see later that it is related to the thermal diffusion effect. From the form of this equation it appears that λ' is the coefficient of thermal conduction but this is not so. The reason for this is that, because of the thermal diffusion effect, a temperature gradient tends to set up a current j as we see from equation (8.57) or equation (8.63). Now the coefficient of thermal conduction is usually defined relative to a state in which no current j flows, i.e., in such a state that a force d_1 is set up which just balances off the thermal diffusion effect. From equation (8.57) we see that the force required is

$$d_1 = -\frac{D_T}{D_{12}} \frac{1}{T} \nabla T = -k_T \frac{1}{T} \nabla T$$

and this force produces additional heat flow through the last term of equation (8.58). We therefore define the coefficient of thermal conduction as

$$\lambda = \lambda' + \frac{5k_B}{4} k_T n \sum_i n_i v_{ti} e_i^1 \,. \qquad (8.68)$$

Notice that the second term of equation (8.58) can, using equation (8.6), be written in the alternative form

$$\tfrac{5}{2}k_BT(n_1\langle \boldsymbol{w}_1\rangle + n_2\langle \boldsymbol{w}_2\rangle) = \frac{5k_BT}{2\rho}n_1n_2(m_2 - m_1)(\langle \boldsymbol{w}_1\rangle - \langle \boldsymbol{w}_2\rangle)$$

so it vanishes when $\langle \boldsymbol{w}_1\rangle - \langle \boldsymbol{w}_2\rangle$ vanishes, i.e., when $\boldsymbol{j}$ vanishes. In general, from equation (8.57),

$$\boldsymbol{d}_1 = -k_T\frac{1}{T}\nabla T - \frac{n_1n_2}{n^2D_{12}}(\langle \boldsymbol{w}_1\rangle - \langle \boldsymbol{w}_2\rangle)\,. \tag{8.69}$$

When we use this expression for $\boldsymbol{d}_1$ in equation (8.67) and use equation (8.68), we obtain

$$\boldsymbol{q} = -\lambda\nabla T + \tfrac{5}{2}k_BT(n_1\langle \boldsymbol{w}_1\rangle + n_2\langle \boldsymbol{w}_2\rangle) - \frac{5k_BTn_1n_2}{4nD_{12}}\sum_i n_iv_{ti}e_i^1(\langle \boldsymbol{w}_1\rangle - \langle \boldsymbol{w}_2\rangle)\,. \tag{8.70}$$

This is an alternate formula for $\boldsymbol{q}$ which is convenient to use when the current $\boldsymbol{j} = 0$, for then the second and third terms vanish. Later we shall see how the coefficients e_i^1 are related to the thermal diffusion coefficient D_T and then we shall give another form for $\boldsymbol{q}$.

At first sight it seems a little surprising that the second term of equations (8.67) and (8.70) should turn up with a coefficient $\frac{5}{2}k_BT$ instead of $\frac{3}{2}k_BT$ as we might naively have expected, but we can understand this from the following argument. $(n_1\langle \boldsymbol{w}_1\rangle + n_2\langle \boldsymbol{w}_2\rangle)$ is the flux of particles crossing a surface per unit area. Now each particle of the gas has on average an energy $\frac{3}{2}k_BT$ but each particle contributing to this flux effectively contributes more than $\frac{3}{2}k_BT$ energy on average because the faster particles with the greater energy contribute more to the flux. In other words, the higher energy particles contribute more to the heat flow across some surface in the fluid not only because of their higher energy but also because of the greater rate that they cross the surface. We can verify that this term really does represent the energy flow due to the flux of particles $n_1\langle \boldsymbol{w}_1\rangle + n_2\langle \boldsymbol{w}_2\rangle$ by calculating the energy flow relative to the mean velocity of the particles defined by

$$\langle \boldsymbol{w}\rangle = \frac{(n_1\langle \boldsymbol{w}_1\rangle + n_2\langle \boldsymbol{w}_2\rangle)}{n}\,.$$

The flux of particles relative to this velocity is

$$n_1\langle \boldsymbol{w}_1 - \langle \boldsymbol{w}\rangle\rangle + n_2\langle \boldsymbol{w}_2 - \langle \boldsymbol{w}\rangle\rangle = 0$$

and so, relative to this velocity, there should be no term like the second term of equation (8.67) contributing to the heat flux. The α component of heat flux relative to this velocity is

$$\tfrac{1}{2}n_1m_1\langle(\boldsymbol{w}_1 - \langle \boldsymbol{w}\rangle)^2(\boldsymbol{w}_1 - \langle \boldsymbol{w}\rangle)_\alpha\rangle + \tfrac{1}{2}n_2m_2\langle(\boldsymbol{w}_2 - \langle \boldsymbol{w}\rangle)^2(\boldsymbol{w}_2 - \langle \boldsymbol{w}\rangle)_\alpha\rangle\,. \tag{8.71}$$

Writing this out and remembering that

$$n_1 m_1 \langle \boldsymbol{w}_1 \rangle + n_2 m_2 \langle \boldsymbol{w}_2 \rangle = 0$$

gives

$$\begin{aligned} &\tfrac{1}{2} n_1 m_1 [\langle w_1^2 w_{1\alpha} \rangle - \langle w_1^2 \rangle \langle w_\alpha \rangle - 2 \langle w_\beta \rangle \langle w_{1\beta} w_{1\alpha} \rangle - \langle w^2 \rangle \langle w_\alpha \rangle] \\ &\quad + \tfrac{1}{2} n_2 m_2 [\langle w_2^2 w_{2\alpha} \rangle - \langle w_2^2 \rangle \langle w_\alpha \rangle - 2 \langle w_\beta \rangle \langle w_{2\beta} w_{1\alpha} \rangle - \langle w^2 \rangle \langle w_\alpha \rangle] . \end{aligned}$$

The first terms of these brackets just give q_α. The second terms, on replacing $\frac{1}{2} m_i \langle w_i^2 \rangle$ by $\frac{3}{2} k_B T$, give $-\frac{3}{2} k_B T n \langle w_\alpha \rangle$. The third terms give a contribution when $\beta = \alpha$ of $-k_B T n \langle w_\alpha \rangle$, and the fourth terms give $\frac{1}{2} \rho \langle w^2 \rangle \langle w_\alpha \rangle$. Hence the heat flux relative to the velocity $\langle \boldsymbol{w} \rangle$ is

$$\boldsymbol{q} - \tfrac{5}{2} k_B T n \langle \boldsymbol{w} \rangle - \tfrac{1}{2} \rho \langle w^2 \rangle \langle \boldsymbol{w} \rangle .$$

The last term of this expression is a small term representing the difference between what is called random and what is called ordered energy relative to $\langle \boldsymbol{w} \rangle$ or to the drift velocity $\boldsymbol{u}$. The important point to note is that, relative to this velocity, the heat flux is given by equation (8.70) without the second term. Hence this second term does represent the heat carried by the flux $(n_1 \langle \boldsymbol{w}_1 \rangle + n_2 \langle \boldsymbol{w}_2 \rangle)$.

To sum up, we have seen that for many purposes we do not need to know all the coefficients a_i^m, b_i^m, or e_i^m of the expansions of equation (8.45) but we would like to have accurate expressions for the particular coefficients which are related to the quantities of physical interest. These are a_1^0, a_2^0, a_1^1, a_2^1, b_1^0, b_2^0, e_1^0, e_2^0, e_1^1, and e_2^1. We shall see that this is just what the variation method gives us.

We have already seen that a_2^0 and e_2^0 are given in terms of a_1^0 and e_1^0 respectively by equations (8.51) and (8.52). Before describing the variation method we shall prove that e_1^1 and e_2^1 are simply related to a_1^0. This shortens the list of unknowns we have to calculate. The scalar product of equation (8.43) with $\boldsymbol{A}_1$ integrated over velocities plus the scalar product of equation (8.44) with $\boldsymbol{A}_2$ integrated is

$$\begin{aligned} &-\frac{1}{n_1} \int \mathrm{d}\boldsymbol{w}\, f_1 \boldsymbol{w} \cdot \boldsymbol{A}_1 \\ &+\frac{1}{n_2} \int \mathrm{d}\boldsymbol{w}\, f_2 \boldsymbol{w} \cdot \boldsymbol{A}_2 = \sum_j \int \mathrm{d}\boldsymbol{v} \mathrm{d}\boldsymbol{s} \mathrm{d}\phi\, b\, \mathrm{d}b\, g\, f_1(\boldsymbol{v}) f_j(\boldsymbol{s}) \boldsymbol{A}_1(\boldsymbol{v}) \\ &\qquad\qquad \cdot [\boldsymbol{E}_1(\boldsymbol{v}') + \boldsymbol{E}_j(\boldsymbol{s}') - \boldsymbol{E}_1(\boldsymbol{v}) - \boldsymbol{E}_j(\boldsymbol{s})] \\ &\qquad + \sum_j \int \mathrm{d}\boldsymbol{v} \mathrm{d}\boldsymbol{s} \mathrm{d}\phi\, b\, \mathrm{d}b\, g\, f_2(\boldsymbol{v}) f_j(\boldsymbol{s}) \boldsymbol{A}_2(\boldsymbol{v}) \\ &\qquad\qquad \cdot [\boldsymbol{E}_2(\boldsymbol{v}') + \boldsymbol{E}_j(\boldsymbol{s}') - \boldsymbol{E}_2(\boldsymbol{v}) - \boldsymbol{E}_j(\boldsymbol{s})] . \end{aligned}$$

The right-hand side can be rewritten as

$$\int d\boldsymbol{v}d\boldsymbol{s}d\phi\, b\, db\, g\, f_1(\boldsymbol{v})f_2(\boldsymbol{s})\boldsymbol{A}_1(\boldsymbol{v})\cdot[\boldsymbol{E}_1(\boldsymbol{v}')+\boldsymbol{E}_1(\boldsymbol{s}')-\boldsymbol{E}_1(\boldsymbol{v})-\boldsymbol{E}_1(\boldsymbol{s})]$$
$$+\int d\boldsymbol{v}d\boldsymbol{s}d\phi\, b\, db\, g\, f_1(\boldsymbol{v})f_2(\boldsymbol{s})[\boldsymbol{A}_1(\boldsymbol{v})+\boldsymbol{A}_2(\boldsymbol{s})]$$
$$\cdot[\boldsymbol{E}_1(\boldsymbol{v}')+\boldsymbol{E}_2(\boldsymbol{s}')-\boldsymbol{E}_1(\boldsymbol{v})-\boldsymbol{E}_2(\boldsymbol{s})]$$
$$+\int d\boldsymbol{v}d\boldsymbol{s}d\phi\, b\, db\, g\, f_2(\boldsymbol{v})f_2(\boldsymbol{s})\boldsymbol{A}_2(\boldsymbol{v})\cdot[\boldsymbol{E}_2(\boldsymbol{v}')+\boldsymbol{E}_2(\boldsymbol{s}')-\boldsymbol{E}_2(\boldsymbol{v})-\boldsymbol{E}_2(\boldsymbol{s})]$$

which is easily shown to be equal to

$$-\frac{1}{4}\sum_{i,j=1}^{2}\int d\boldsymbol{v}d\boldsymbol{s}d\phi\, b\, db\, g\, f_i(\boldsymbol{v})f_j(\boldsymbol{s})[\boldsymbol{A}_i(\boldsymbol{v}')+\boldsymbol{A}_j(\boldsymbol{s}')-\boldsymbol{A}_i(\boldsymbol{v})-\boldsymbol{A}_j(\boldsymbol{s})]$$
$$\cdot[\boldsymbol{E}_i(\boldsymbol{v}')+\boldsymbol{E}_j(\boldsymbol{s}')-\boldsymbol{E}_i(\boldsymbol{v})-\boldsymbol{E}_j(\boldsymbol{s})]\,. \tag{8.72}$$

The scalar product of equation (8.42) with $\boldsymbol{E}_i$, integrated, and summed over i is

$$\sum_i\int d\boldsymbol{w}\, f_i\left(\tfrac{5}{2}-w_i^2\right)\boldsymbol{w}\cdot\boldsymbol{E}_i=\sum_{i,j}\int d\boldsymbol{v}d\boldsymbol{s}d\phi\, b\, db\, g\, f_i(\boldsymbol{v})f_j(\boldsymbol{s})\boldsymbol{E}_i(\boldsymbol{v})$$
$$\cdot[\boldsymbol{A}_i(\boldsymbol{v}')+\boldsymbol{A}_j(\boldsymbol{s}')-\boldsymbol{A}_i(\boldsymbol{v})-\boldsymbol{A}_j(\boldsymbol{s})]\,.$$

The right-hand side of this expression can also be shown to be equal to equation (8.72), hence we have the relation

$$-\frac{1}{n_1}\int d\boldsymbol{w}\, f_1\boldsymbol{w}\cdot\boldsymbol{A}_1+\frac{1}{n_2}\int d\boldsymbol{w}\, f_2\boldsymbol{w}\cdot\boldsymbol{A}_2=\sum_i\int d\boldsymbol{w}\, f_i\left(\tfrac{5}{2}-\varpi_i^2\right)\boldsymbol{w}\cdot\boldsymbol{E}_i\,. \tag{8.73}$$

Substituting from equation (8.45) for $\boldsymbol{A}_i$ and $\boldsymbol{E}_i$ and from equations (8.48) and (8.49) for the expansions, these integrals can be evaluated by noting that

$$L_1^{\frac{3}{2}}(\varpi^2)=\tfrac{5}{2}-\varpi^2\,,\qquad L_0^{\frac{3}{2}}(\varpi^2)=1$$

and using equation (7.92). This gives

$$\tfrac{15}{4}\sum_i n_i v_{ti}e_i^1=-\tfrac{3}{2}\left[v_{t1}a_1^0-v_{t2}a_2^0\right]$$
$$=-\tfrac{3}{2}v_{t1}\frac{\rho}{n_2m_2}a_1^0\,.$$

The expression on the left is just that which we need to evaluate the terms involving e_1^1 in equations (8.68), (8.67), and (8.70). Substituting we find these can be rewritten as

$$\lambda=\lambda'-\frac{nk_B\rho v_{t1}}{2n_2m_2}k_Ta_1^0$$
$$=\lambda'-\frac{k_B\rho v_{t1}}{2n_2m_2}\frac{(a_1^0)^2}{e_1^0}\,, \tag{8.74}$$

where we have used equation (8.60) to give k_T,

$$\boldsymbol{q} = -\lambda'\nabla T + \tfrac{5}{2}k_BT(n_1\langle\boldsymbol{w}_1\rangle + n_2\langle\boldsymbol{w}_2\rangle) - \frac{nk_BT\rho v_{t1}}{2n_2m_2}a_1^0\boldsymbol{d}_1 \tag{8.75}$$

and

$$\boldsymbol{q} = -\lambda\nabla T + \tfrac{5}{2}k_BT(n_1\langle\boldsymbol{w}_1\rangle + n_2\langle\boldsymbol{w}_2\rangle) + \frac{n_1k_BT\rho v_{t1}}{2nm_2D_{12}}a_1^0(\langle\boldsymbol{w}_1\rangle - \langle\boldsymbol{w}_2\rangle)\,. \tag{8.76}$$

8.6 The variation method

In this section we shall describe the variation method of Hirschfelder et al.[50] as it applies to a binary gas. First, however, we shall prove some important integral theorems.

If G, H, K, and L be any four functions of the same character (i.e., all scalars, all vectors, or all tensors and all of the same dimensions) representing some property of the particles, depending in general on position, velocity, and time, and defined for both types of particles, then we define

$$\begin{aligned}[G_i, H_j; K_i, L_j]_{ij} \equiv \frac{1}{2n_in_j}\int & \mathrm{d}\boldsymbol{v}\mathrm{d}\boldsymbol{s}\mathrm{d}\phi\, b\,\mathrm{d}b\, g\, f_i(\boldsymbol{v})f_j(\boldsymbol{s}) \\ & \times[G_i(\boldsymbol{v}') + H_j(\boldsymbol{s}') - G_i(\boldsymbol{v}) - H_j(\boldsymbol{s})] \\ & : [K_i(\boldsymbol{v}') + L_j(\boldsymbol{s}') - K_i(\boldsymbol{v}) - L_j(\boldsymbol{s})] \end{aligned} \tag{8.77}$$

where the symbol : indicates that the scalar product must be taken if the quantities are vectors and the full scalar product must be taken if the quantities are tensors. Equation (8.77) can also be written

$$\begin{aligned}[G_i, H_j; K_i, L_j]_{ij} = -\frac{1}{n_in_j}\int & \mathrm{d}\boldsymbol{v}\mathrm{d}\boldsymbol{s}\mathrm{d}\phi\, b\,\mathrm{d}b\, g\, f_i(\boldsymbol{v})f_j(\boldsymbol{s})[G_i(\boldsymbol{v}) + H_j(\boldsymbol{s})] \\ & : [K_i(\boldsymbol{v}') + L_j(\boldsymbol{s}') - K_i(\boldsymbol{v}) - L_j(\boldsymbol{s})] \end{aligned} \tag{8.78}$$

$$\begin{aligned} = -\frac{1}{n_in_j}\int & \mathrm{d}\boldsymbol{v}\mathrm{d}\boldsymbol{s}\mathrm{d}\phi\, b\,\mathrm{d}b\, g\, f_i(\boldsymbol{v})f_j(\boldsymbol{s}) \\ & [G_i(\boldsymbol{v}') + H_j(\boldsymbol{s}') - G_i(\boldsymbol{v}) - H_j(\boldsymbol{s})] : [K_i(\boldsymbol{v}) + L_j(\boldsymbol{s})]\,. \end{aligned} \tag{8.79}$$

To prove this we change variables in equation (8.78) from $\boldsymbol{v}$ and $\boldsymbol{s}$ to $\boldsymbol{v}'$ and $\boldsymbol{s}'$. Then by equation (8.21),

$$\mathrm{d}\boldsymbol{v}\,\mathrm{d}\boldsymbol{s} = \mathrm{d}\boldsymbol{v}\,'\mathrm{d}\boldsymbol{s}\,,'$$

and by equation (8.20), $g = g'$, and by the form of $f_i(\boldsymbol{v})$,

$$f_i(\boldsymbol{v})f_j(\boldsymbol{s}) = f_i(\boldsymbol{v}')f_j(\boldsymbol{s}')\,,$$

and hence equation (8.78) becomes

$$[G_i, H_j; K_i, L_j]_{ij} = \frac{1}{2n_in_j}\int \mathrm{d}\boldsymbol{v}'\mathrm{d}\boldsymbol{s}'\mathrm{d}\phi\, b\,\mathrm{d}b\, g\, f_i(\boldsymbol{v}')f_j(\boldsymbol{s}')[G_i(\boldsymbol{v}) + H_j(\boldsymbol{s})] \\ : [K_i(\boldsymbol{v}') + L_j(\boldsymbol{s}') - K_i(\boldsymbol{v}) - L_j(\boldsymbol{s})] . \quad (8.80)$$

This may be regarded as an integral over inverse collisions. Now in an inverse collision, the final velocities $\boldsymbol{v}$ and $\boldsymbol{s}$ are the same functions of $\boldsymbol{v}'$, $\boldsymbol{s}'$, b, and ϕ as $\boldsymbol{v}'$ and $\boldsymbol{s}'$ are of $\boldsymbol{v}$, $\boldsymbol{s}$, b, and ϕ in direct collisions. Hence, equation (8.80) is equal to

$$[G_i, H_j; K_i, L_j]_{ij} = \frac{1}{2n_in_j}\int \mathrm{d}\boldsymbol{v}\mathrm{d}\boldsymbol{s}\mathrm{d}\phi\, b\,\mathrm{d}b\, g\, f_i(\boldsymbol{v})f_j(\boldsymbol{s})[G_i(\boldsymbol{v}') + H_j(\boldsymbol{s}')] \\ : [K_i(\boldsymbol{v}) + L_j(\boldsymbol{s}) - K_i(\boldsymbol{v}') - L_j(\boldsymbol{s}')] . \quad (8.81)$$

Adding equations (8.78) and (8.81) and dividing by 2 gives equation (8.77). The equalities (8.78) and (8.79) can be proved in a similar fashion.

Clearly,

$$[G_i, H_j; K_i, L_j]_{ij} = [K_i, L_j; G_i, H_j]_{ij} .$$

Define

$$\begin{aligned}[G_1; K_1]_{12} \equiv [K_1; G_1]_{12} &= [G_1, 0; K_1, 0]_{12} \\ &= \frac{1}{2n_1n_2}\int \mathrm{d}\boldsymbol{v}\mathrm{d}\boldsymbol{s}\mathrm{d}\phi\, b\,\mathrm{d}b\, g\, f_1(\boldsymbol{v})f_2(\boldsymbol{s}) \\ &\qquad \times[G_1(\boldsymbol{v}') - G_1(\boldsymbol{v})] : [K_1(\boldsymbol{v}') - K_1(\boldsymbol{v})] \\ &= -\frac{1}{n_1n_2}\int \mathrm{d}\boldsymbol{v}\mathrm{d}\boldsymbol{s}\mathrm{d}\phi\, b\,\mathrm{d}b\, g\, f_1(\boldsymbol{v})f_2(\boldsymbol{s}) \\ &\qquad \times G_1(\boldsymbol{v}) : [K_1(\boldsymbol{v}') - K_1(\boldsymbol{v})] . \end{aligned} \quad (8.82)$$

$$\begin{aligned}[H_2; L_2]_{12} \equiv [L_2; H_2]_{12} &= [0, H_2; 0, L_2]_{12} \\ &= \frac{1}{2n_1n_2}\int \mathrm{d}\boldsymbol{v}\mathrm{d}\boldsymbol{s}\mathrm{d}\phi\, b\,\mathrm{d}b\, g\, f_1(\boldsymbol{v})f_2(\boldsymbol{s}) \\ &\qquad \times[H_2(\boldsymbol{s}') - H_2(\boldsymbol{s})] : [L_2(\boldsymbol{s}') - L_2(\boldsymbol{s})] \\ &= -\frac{1}{n_1n_2}\int \mathrm{d}\boldsymbol{v}\mathrm{d}\boldsymbol{s}\mathrm{d}\phi\, b\,\mathrm{d}b\, g\, f_1(\boldsymbol{v})f_2(\boldsymbol{s}) \\ &\qquad \times H_2(\boldsymbol{s}) : [L_2(\boldsymbol{s}') - L_2(\boldsymbol{s})] . \end{aligned}$$

$$\begin{aligned}[G_1; L_2]_{12} \equiv [L_2; G_1]_{12} &= [G_1, 0; 0, L_2]_{12} \\ &= \frac{1}{2n_1n_2}\int \mathrm{d}\boldsymbol{v}\mathrm{d}\boldsymbol{s}\mathrm{d}\phi\, b\,\mathrm{d}b\, g\, f_1(\boldsymbol{v})f_2(\boldsymbol{s}) \\ &\qquad \times[G_1(\boldsymbol{v}') - G_1(\boldsymbol{v})] : [L_2(\boldsymbol{s}') - L_2(\boldsymbol{s})] \end{aligned}$$

$$= -\frac{1}{n_1 n_2}\int \mathrm{d}\boldsymbol{v}\mathrm{d}\boldsymbol{s}\mathrm{d}\phi\, b\,\mathrm{d}b\, g\, f_1(\boldsymbol{v})f_2(\boldsymbol{s}) \times G_1(\boldsymbol{v}) : [L_2(\boldsymbol{s}') - L_2(\boldsymbol{s})]$$
$$= -\frac{1}{n_1 n_2}\int \mathrm{d}\boldsymbol{v}\mathrm{d}\boldsymbol{s}\mathrm{d}\phi\, b\,\mathrm{d}b\, g\, f_1(\boldsymbol{v})f_2(\boldsymbol{s}) \times L_2(\boldsymbol{s}) : [G_1(\boldsymbol{v}') - G_1(\boldsymbol{v})]\,.$$

$$\begin{aligned}[G_i; K_i]_i &\equiv [G_i, 0; K_i, K_i]_{ii} = [0, G_i; K_i, K_i]_{ii} \\ &= \frac{1}{2n_i^2}\int \mathrm{d}\boldsymbol{v}\mathrm{d}\boldsymbol{s}\mathrm{d}\phi\, b\,\mathrm{d}b\, g\, f_i(\boldsymbol{v})f_i(\boldsymbol{s}) \\ &\quad \times[G_i(\boldsymbol{v}') - G_i(\boldsymbol{v})] : [K_i(\boldsymbol{v}') + K_i(\boldsymbol{s}') - K_i(\boldsymbol{v}) - K_i(\boldsymbol{s})] \\ &= -\frac{1}{n_i^2}\int \mathrm{d}\boldsymbol{v}\mathrm{d}\boldsymbol{s}\mathrm{d}\phi\, b\,\mathrm{d}b\, g\, f_i(\boldsymbol{v})f_i(\boldsymbol{s}) \\ &\quad \times G_i(\boldsymbol{v}) : [K_i(\boldsymbol{v}') + K_i(\boldsymbol{s}') - K_i(\boldsymbol{v}) - K_i(\boldsymbol{s})]\,. \end{aligned} \tag{8.83}$$

This last expression is precisely the integral $[G, K]$ defined in Chapter 7 for collisions in a simple gas, except the subscripts i are required now, however, to indicate whether the collision is between two electrons or between two ions.

Setting $G = H = K = L$ in equation (8.77), the integrand on the right-hand side becomes positive definite, hence

$$[G_i, G_j; G_i, G_j]_{ij} \geq 0\,. \tag{8.84}$$

8.6.1 Definition of $\{G; H\}$

We define the bracketed quantity

$$\begin{aligned}\{G; H\} &\equiv \sum_{ij} n_i n_j [G_i, G_j; H_i, H_j]_{ij} \\ &= n_1^2[G_1, G_1; H_1, H_1]_{11} + n_2^2[G_2, G_2; H_2, H_2]_{22} \\ &\quad + 2n_1 n_2[G_1, G_2; H_1, H_2]_{12}\,. \end{aligned} \tag{8.85}$$

Then, using equations (8.82) to (8.83)

$$\begin{aligned}\{G; H\} &= 2n_i^2[G_1, H_1]_1 + 2n_2^2[G_2, H_2]_2 + 2n_1 n_2\{[G_1, H_1]_{12} \\ &\quad + [G_1, H_2]_{12} + [G_2, H_1]_{12} + [G_2, H_2]_{12}\}\,. \end{aligned} \tag{8.86}$$

From equations (8.84) and (8.85) it follows that

$$\{G; G\} \geq 0\,. \tag{8.87}$$

This is a very important result.

Now consider the integral equations we want to solve, equations (8.41) to (8.44). They are of the form

$$R_i(\boldsymbol{v}) = \sum_j \int \mathrm{d}\boldsymbol{s}\mathrm{d}\phi\, b\,\mathrm{d}b\, g\, f_i(\boldsymbol{v})f_j(\boldsymbol{s})[T_i(\boldsymbol{v}') + T_j(\boldsymbol{s}') - T_i(\boldsymbol{v}) - T_j(\boldsymbol{s})]\,. \tag{8.88}$$

Here R_i is known, T_i is to be found, and both are either vectors or tensors. Let $t_i(\boldsymbol{v})$ be a trial function, containing as many parameters as is convenient, such that

$$\sum_i \int \mathrm{d}\boldsymbol{v}\, R_i(\boldsymbol{v}) : t_i(\boldsymbol{v}) \equiv \sum_{ij} \int \mathrm{d}\boldsymbol{v}\mathrm{d}\boldsymbol{s}\mathrm{d}\phi\, b\, \mathrm{d}b\, g\, f_i(\boldsymbol{v}) f_j(\boldsymbol{s})$$
$$\times t_i(\boldsymbol{v}) : [t_i(\boldsymbol{v}') + t_j(\boldsymbol{s}') - t_i(\boldsymbol{v}) - t_j(\boldsymbol{v})]$$
$$= -n_1^2[t_1 : t_1]_1 - n_1 n_2 [t_1, t_2; t_1, t_2]_{12} - n_1^2 [t_2; t_2]_2$$
$$= -\tfrac{1}{2}\{t; t\}\,. \tag{8.89}$$

Multiplying equation (8.88) by t_1, integrating, and summing over i gives

$$\sum_i \int \mathrm{d}\boldsymbol{v}\, R_i(\boldsymbol{v}) : t_i(\boldsymbol{v}) = -\tfrac{1}{2}\{t; T\}\,.$$

Hence, provided t satisfies equation (8.89) we also have

$$\{t; t\} = \{t; T\}\,. \tag{8.90}$$

Now consider

$$\{t - T; t - T\} = \{t; t\} - 2\{t; T\} + \{T; T\}\,.$$

By equation (8.87), this expression is positive or zero and hence substituting for $\{t; T\}$ from equation (8.90) we get

$$-2\sum_i \int t_i : R_i\, \mathrm{d}\boldsymbol{v} = \{t; t\} \leq \{T; T\}\,. \tag{8.91}$$

This equation is the basis of the variation method for the solution of the equations of the form of equation (8.89). For the trial function t we use the appropriate expansion in Sonine or Associated Laguerre polynomials given by equation (8.45) and equations (8.48) through (8.50) with as many nonzero coefficients as is convenient. We then ensure that the coefficients are such that the first part of equation (8.91) is satisfied and maximize either side of the equation. In practice, we always maximize the left-hand side because this turns out to be easiest. In principle this variation method could give us all the coefficients in the expansions, i.e., solve the equations exactly. In practice, we use it to give quite accurate values for those few coefficients that we are particularly interested in because it turns out that the method gives these particular coefficients very easily.

8.6.2 The Davison function

We will sometimes use a variational method described by B. Davison[53]. We first compare this method with the direct method of Chapter 7 which we will use also. In general terms, we may write a function of several variables as

$$ax = r_1 x^2 + 2r_2 xy + 2r_3 xz + r_4 y^2 + r_5 z^2 + 2r_6 yz\,. \tag{8.92}$$

As a first approximation, we may set $y = z = 0$ and find that $[x]_1 = a/r_1$ where the subscript on x indicates the first approximation. For the next approximation, we choose y and z nonzero, and then we maximize the term on the left by differentiating the right-hand side with respect to y and z and solving the two resulting equations for y and z in terms of x. We then substitute these expressions into the original expression and solve it for $[x]_2$. From this second approximation to x, we can then solve for $[y]_2$ and $[z]_2$ to complete the problem.

Davison suggested writing a general expression of the form

$$\mathcal{D}(x,y,z) = ax - \tfrac{1}{2}[r_1x^2 + 2r_2xy + 2r_3xz + r_4y^2 + r_5z^2 + 2r_6yz]\,, \qquad (8.93)$$

and then maximizing $\mathcal{D}(x, y, z)$ with respect to x, y, and z, resulting in three equations which are simultaneously solved for the three variables in terms of a. The factor of $\frac{1}{2}$ appears because the expression in terms of the r_i is a quadratic form. The need for this factor is evident from considering equation (8.93) with y and z set to zero so that if we now differentiate with respect to x, we find

$$ax = r_1x^2\,,$$

with solution $[x]_1 = a/r_1$. If we had constructed a Davison-type function as

$$\mathcal{D} = ax - r_1x^2\,,$$

then the derivative would have led to $x = a/2r_1$ which is inconsistent. It is the quadratic form that requires the factor of $\frac{1}{2}$ to be consistent. Because this type of quadratic form appears frequently, we will need this form.

Problem 8.3 Show that the direct method of solving equation (8.92) by differentiating only with respect to y and z gives the same results for $[x]_2$, $[y]_2$, and $[z]_2$ as constructing the Davison function of equation (8.93) and differentiating with respect to all three variables.

8.7 Results

8.7.1 Viscosity

In this section we shall apply the variation methods we have just described to calculate the various quantities we want to know in terms of certain collision integrals that are evaluated in Appendix B. Consider equation (8.41) first. This will give us the coefficient of viscosity. We take as our trial function, t, the expansion given by equations (8.45) and (8.50), i.e.,

$$t_i = \mathsf{B}_i = \boldsymbol{\varpi}_i^0\boldsymbol{\varpi}_i\mathcal{B}_i(\varpi_i^2) = \boldsymbol{\varpi}_i^0\boldsymbol{\varpi}_i\sum_{m=0}^{\infty} b_i^m L_m^{\frac{5}{2}}(\varpi_i^2)\,.$$

Comparing equation (8.41) to the general form of equation (8.88) we see that for this case,

$$R_i = -2 f_i \boldsymbol{\varpi}_i^0 \boldsymbol{\varpi}_i \,.$$

The left-hand side of equation (8.91) is therefore

$$-2\sum_i \int t_i : R_i \, \mathrm{d}\boldsymbol{v} = 4\sum_i \int \mathrm{d}\boldsymbol{v}\, f_i (\boldsymbol{\varpi}_i^0 \boldsymbol{\varpi}_i)_{\alpha\beta} (\boldsymbol{\varpi}_i^0 \boldsymbol{\varpi}_i)_{\alpha\beta} \sum_{m=0}^{\infty} b_i^m L_m^{\frac{5}{2}}(\varpi_i^2) \,. \tag{8.94}$$

The integral on the right-hand side of equation (8.94) is of the same form as that which we evaluated in Chapter 7 and gives

$$-2\sum_i \int t_i : R_i \, \mathrm{d}\boldsymbol{v} = 10 \sum_i n_i b_i^0 = 10(n_1 b_1^0 + n_2 b_2^0) \,. \tag{8.95}$$

Comparing with the formula of equation (8.54), we see that this is simply related to the coefficient of viscosity μ. So, just as for the simple gas case, we can expect to get good values for μ with quite poor trial functions. Equation (8.91) now becomes

$$10(n_1 b_1^0 + n_2 b_2^0) = \left\{ \boldsymbol{\varpi}^0 \boldsymbol{\varpi} \sum_m b^m L_m^{\frac{5}{2}}(\varpi^2); \boldsymbol{\varpi}^0 \boldsymbol{\varpi} \sum_m b^m L_m^{\frac{5}{2}}(\varpi^2) \right\} \tag{8.96}$$

and we must ensure that the coefficients satisfy this equation and find that choice which maximizes this same expression. The simplest trial function is obtained by setting all the coefficients except b_1^0 and b_2^0 to zero. Then equation (8.96) becomes

$$\begin{aligned} 10(n_1 b_1^0 + n_2 b_2^0) = {} & 2n_1^2 (b_1^0)^2 [\boldsymbol{\varpi}_1^0 \boldsymbol{\varpi}_1, \boldsymbol{\varpi}_1^0 \boldsymbol{\varpi}_1]_1 + 2n_2^2 (b_2^0)^2 [\boldsymbol{\varpi}_2^0 \boldsymbol{\varpi}_2, \boldsymbol{\varpi}_2^0 \boldsymbol{\varpi}_2]_2 \\ & + 2n_1 n_2 \left\{ (b_1^0)^2 [\boldsymbol{\varpi}_1^0 \boldsymbol{\varpi}_1, \boldsymbol{\varpi}_1^0 \boldsymbol{\varpi}_1]_{12} + 2 b_1^0 b_2^0 [\boldsymbol{\varpi}_1^0 \boldsymbol{\varpi}_1, \boldsymbol{\varpi}_2^0 \boldsymbol{\varpi}_2]_{12} \right. \\ & \left. + (b_2^0)^2 [\boldsymbol{\varpi}_2^0 \boldsymbol{\varpi}_2, \boldsymbol{\varpi}_2^0 \boldsymbol{\varpi}_2]_{12} \right\} , \end{aligned} \tag{8.97}$$

where we have used equation (8.86) to expand the right-hand side.

Simplifying, we write this as

$$5(n_1 b_1^0 + n_2 b_2^0) = (b_1^0)^2 (n_1^2 \ell_1 + n_1 n_2 \ell_2) + (b_2^0)^2 (n_2^2 \ell_3 + n_1 n_2 \ell_4) + 2 b_1^0 b_2^0 n_1 n_2 \ell_5 \,, \tag{8.98}$$

where ℓ_1, ℓ_2, ℓ_3, ℓ_4, and ℓ_5 stand for the various collision integrals occurring in equation (8.97). Constructing the Davison function,

$$\begin{aligned} \mathcal{D}(b_1^0, b_2^0) = 5(n_1 b_1^0 + n_2 b_2^0) - \tfrac{1}{2} [& (b_1^0)^2 (n_1^2 \ell_1 + n_1 n_2 \ell_2) + (b_2^0)^2 (n_2^2 \ell_3 + n_1 n_2 \ell_4) \\ & + 2 b_1^0 b_2^0 n_1 n_2 \ell_5] \,, \end{aligned}$$

the maximum of the left-hand side of equation (8.98) is given by

$$n_1 b_1^0 + n_2 b_2^0 = \frac{5(\ell_1 + \ell_3 - 2\ell_5 + \frac{n_2}{n_1}\ell_2 + \frac{n_1}{n_2}\ell_4)}{(\ell_1 + \frac{n_2}{n_1}\ell_2)(\ell_3 + \frac{n_1}{n_2}\ell_4) - \ell_5^2} \,. \tag{8.99}$$

From equation (8.54) the corresponding value for the coefficient of viscosity, written $[\mu]_1$ to signify it is the first approximation, is

$$[\mu]_1 = \frac{5}{2}k_B T \frac{(\ell_1 + \ell_3 - 2\ell_5 + \frac{n_2}{n_1}\ell_2 + \frac{n_1}{n_2}\ell_4)}{(\ell_1 + \frac{n_2}{n_1}\ell_2)(\ell_3 + \frac{n_1}{n_2}\ell_4) - \ell_5^2}.$$

When evaluating this expression we can use the quasineutrality condition $n_1 \simeq n_2$. Then comparing equation (8.98) with equation (8.97) to see what integrals $\ell_1 \ldots \ell_5$ stand for, and writing down the values of the integrals from Appendix B, we get

$$[\mu]_1 = \frac{5}{2}k_B T \frac{\sqrt{2}\varphi + \sqrt{2M_1}\varphi + \frac{8}{3}M_1\varphi + 2\varphi + \frac{10}{3}M_1\varphi}{(\sqrt{2}\varphi + 2\varphi)(\sqrt{2M_1}\varphi + \frac{10}{3}M_1\varphi) - (\frac{4}{3}M_1\varphi)^2},$$

where

$$M_1 = \frac{m_1}{m_1 + m_2},$$

and φ is given by equation (B.113). To an excellent approximation, because $M_1 \ll 1$,

$$[\mu]_1 = \frac{5k_B T}{2\sqrt{2M_1}\varphi}. \tag{8.100}$$

Problem 8.4 List the expressions for the various ℓ_i and then show that equation (8.99) does produce the extremum of equation (8.98). Then find the expressions for the various ℓ_i from Appendix B to verify equation (8.100).

This last expression is just the result we derived in Chapter 7 for an ion gas alone. Thus we see that the electrons do not contribute to the viscosity of the gas. This is only to be expected because the viscosity of an electron gas alone would have $\sqrt{m_1}$ in place of $\sqrt{m_2}$ in equation (8.100). Thus the viscosity of an ion gas alone is $\sqrt{m_2/m_1}$ times, i.e., at least 43 times, as large as the viscosity of an electron gas alone and therefore, in the mixture of ions and electrons, we expect the viscosity to be due almost entirely to the ions.

We could improve the solution we have obtained for B by including more nonzero coefficients in equation (8.96). But, just as we found for the simple gas, this would probably not increase our value for μ by more than a few percent, so we will not trouble to do this.

8.7.2 Diffusion and electrical conductivity

Now consider the variation method applied to equations (8.43) and (8.44). These will give the coefficient of diffusion, D_{12}, and the electrical conductivity, σ. Comparing these equations with the general form of equation (8.88) we see that, in this case,

$$R_1 = -\frac{1}{n_1}f_1\boldsymbol{w}_1 \quad \text{and} \quad R_2 = \frac{1}{n_2}f_2\boldsymbol{w}_2.$$

We take the trial function to be given by equation (8.45) and the expression of equation (8.49), so

$$t = \boldsymbol{E}(\boldsymbol{\varpi}) = \boldsymbol{\varpi} \sum_{m=0}^{\infty} e^m L_m^{\frac{3}{2}}(\varpi^2)$$

and e_2^0 is related to e_1^0 by the subsidiary condition of equation (8.52). The left-hand side of equation (8.91) is

$$\begin{aligned} -2\sum_i \int t_i : R_i \, d\boldsymbol{v} &= 2\int d\boldsymbol{v}\, \frac{1}{n_1} f_1 \boldsymbol{w}_1 \cdot \boldsymbol{\varpi}_1 \sum_{m=0}^{\infty} e_1^m L_m^{\frac{3}{2}}(\varpi_1^2) \\ &\quad -2\int d\boldsymbol{v}\, \frac{1}{n_2} f_2 \boldsymbol{w}_2 \cdot \boldsymbol{\varpi}_2 \sum_{m=0}^{\infty} e_2^m L_m^{\frac{3}{2}}(\varpi_2^2) \\ &= 3v_{t1}e_1^0 - 3v_{t2}e_2^0 \\ &\simeq 3v_{t1}e_1^0 \end{aligned} \tag{8.101}$$

where we have used equation (8.52) to eliminate e_2^0.

Equation (8.91) then becomes

$$3v_{t1}e_1^0 = \left\{ \boldsymbol{\varpi} \sum_m e^m L_m^{\frac{3}{2}}(\varpi^2) ; \boldsymbol{\varpi} \sum_m e^m L_m^{\frac{3}{2}}(\varpi^2) \right\} . \tag{8.102}$$

The simplest trial function is obtained by setting all the coefficients e_i^m equal to zero except those with $m = 0$. This gives, using equation (8.86),

$$3\frac{v_{t1}\rho}{n_2 m_2} e_1^0 = 2n_1 n_2 \left\{ (e_1^0)^2 [\boldsymbol{\varpi}_1, \boldsymbol{\varpi}_1]_{12} + 2e_1^0 e_2^0 [\boldsymbol{\varpi}_1, \boldsymbol{\varpi}_2]_{12} + (e_2^0)^2 [\boldsymbol{\varpi}_2, \boldsymbol{\varpi}_2]_{12} \right\} . \tag{8.103}$$

The other terms of equation (8.86) do not contribute in this case because $[\boldsymbol{\varpi}_i, \boldsymbol{\varpi}_i]_i$ is proportional to the rate of change of momentum of particles i due to collisions with particles i and this is zero. Eliminating e_2^0 from equation (8.103) by using equation (8.52) gives, as a simple equation for e_1^0

$$\frac{3v_{t1}}{2} e_1^0 = (e_1^0)^2 n_1 n_2 \left\{ [\boldsymbol{\varpi}_1, \boldsymbol{\varpi}_1]_{12} - 2\frac{n_1\sqrt{m_1}}{n_2\sqrt{m_2}} [\boldsymbol{\varpi}_1, \boldsymbol{\varpi}_2]_{12} + \frac{n_1^2 m_1}{n_2^2 m_2} [\boldsymbol{\varpi}_2, \boldsymbol{\varpi}_2]_{12} \right\} . \tag{8.104}$$

Now from Appendix B we see that $[\boldsymbol{\varpi}_1, \boldsymbol{\varpi}_1]_{12} = \varphi$ where φ is our basic collision integral given by equation (B.113) and

$$[\boldsymbol{\varpi}_1, \boldsymbol{\varpi}_2]_{12} = -\sqrt{M_1}[\boldsymbol{\varpi}_1, \boldsymbol{\varpi}_1]_{12} = -\sqrt{M_1}\varphi$$

and

$$[\boldsymbol{\varpi}_2, \boldsymbol{\varpi}_2]_{12} = M_1[\boldsymbol{\varpi}_1, \boldsymbol{\varpi}_1]_{12} = M_1\varphi$$

so equation (8.104) is more simply written as

$$3v_{t1}e_1^0 \simeq (e_1^0)^2 2n_1 n_2 \varphi . \tag{8.105}$$

This equation has one solution with $e_1^0 = 0$ which clearly does not give a maximum for the left-hand side of equation (8.105). The other solution is therefore the one we want and is

$$[e_1^0]_1 = \frac{3}{2}\frac{v_{t1}}{n_1 n_2 \varphi} \simeq \frac{v_{t1}\tau_e}{n_2},$$

where τ_e is the mean collision time for electrons given by

$$\tau_e = \frac{3}{2n_1\varphi},$$

and we have written the result as $[e_1^0]_1$ to indicate that it is the first approximation to e_1^0. Using equation (8.58), this gives us a first approximation to the diffusion coefficient,

$$[D_{12}]_1 = \frac{3}{4n}\frac{2k_BT}{m_1\varphi} = \tfrac{1}{4}v_{t1}^2\tau_e\,. \tag{8.106}$$

Problem 8.5 Use the indicated substitutions to prove equation (8.105). Then use the expressions from Appendix B to verify equation (8.106).

From equation (8.62) we also get a first approximation to the electrical conductivity as

$$[\sigma]_1 = \frac{3}{2\sqrt{2\pi}}\frac{n_1 n_2 (m_1+m_2)^2}{\rho^2}\frac{(4\pi\epsilon_0)^2 (k_BT)^{3/2}}{\sqrt{m_1}e^2 \ln\Lambda},$$

which, since $m_2 \gg m_1$ and n_1 and n_2 are approximately equal, becomes

$$[\sigma]_1 = \frac{3e^2}{2m_1\varphi} = \frac{n_1 e^2\tau_e}{m_1}\,. \tag{8.107}$$

Thus we see that the electrical conductivity increases with temperature like $T^{3/2}$ and is practically independent of density since $\ln\Lambda$ varies only slowly with n.

It is quite easy to calculate the second approximation to these quantities. This second approximation is obtained by setting all the coefficients except e_1^0, e_2^0, e_1^1, and e_2^1 equal to zero in equation (8.102). Instead of equation (8.103) we then get, after eliminating e_2^0,

$$\begin{aligned}\frac{3v_{t1}}{2n_1n_2}e_1^0 = (e_1^0)^2 &\left([\varpi_1,\varpi_1]_{12} - 2\frac{n_1\sqrt{m_1}}{n_2\sqrt{m_2}}[\varpi_1,\varpi_2]_{12} + \frac{n_1^2 m_1}{n_2^2 m_2}[\varpi_2,\varpi_2]_{12}\right)\\ +2e_1^0e_1^1&\left([\varpi_1,\varpi_1 L_1^{\frac{3}{2}}]_{12} - \frac{n_1\sqrt{m_1}}{n_2\sqrt{m_2}}[\varpi_2,\varpi_1 L_1^{\frac{3}{2}}]_{12}\right)\\ +2e_1^0e_2^1&\left([\varpi_1,\varpi_2 L_1^{\frac{3}{2}}]_{12} - \frac{n_1\sqrt{m_1}}{n_2\sqrt{m_2}}[\varpi_2,\varpi_2 L_1^{\frac{3}{2}}]_{12}\right)\end{aligned}$$

$$+(e_1^1)^2\left([\varpi_1 L_1^{\frac{3}{2}},\varpi_1 L_1^{\frac{3}{2}}]_{12}+\frac{n_1}{n_2}[\varpi L_1^{\frac{3}{2}},\varpi L_1^{\frac{3}{2}}]_1\right)$$
$$+(e_2^1)^2\left([\varpi_2 L_1^{\frac{3}{2}},\varpi_2 L_1^{\frac{3}{2}}]_{12}+\frac{n_2}{n_1}[\varpi L_1^{\frac{3}{2}},\varpi L_1^{\frac{3}{2}}]_2\right)$$
$$+2e_1^1e_2^1[\varpi_1 L_1^{\frac{3}{2}},\varpi_2 L_1^{\frac{3}{2}}]_{12} \tag{8.108}$$

which we write in shorthand as

$$\frac{3v_{t1}}{2n_1n_2}e_1^0=r_1(e_1^0)^2+2r_2e_1^0e_1^1+2r_3e_1^0e_2^1+r_4(e_1^1)^2+r_5(e_2^1)^2+2r_6e_1^1e_2^1\,, \tag{8.109}$$

where $r_1\ldots r_6$ stand for the various combinations of the collision integrals which occur in equation (8.108).

Using the method of Lagrange's multipliers we find the maximum of this expression is given by

$$e_1^1=\frac{-r_2r_5+r_3r_6}{r_4r_5-r_6^2}e_1^0\,,\qquad e_2^1=\frac{-r_3r_4+r_2r_6}{r_4r_5-r_6^2}e_1^0\,, \tag{8.110}$$

and

$$[e_1^0]_2=[e_1^0]_1\frac{1}{1-\Delta}\,, \tag{8.111}$$

where

$$\Delta=\frac{r_2^2r_5+r_3^2r_4-2r_2r_3r_6}{r_1(r_4r_5-r_6^2)}\,.$$

In equation (8.111) we have written $[e_1^0]_2$ to indicate that this is the second approximation to e_1^0. Looking up the values of $r_1\ldots r_6$ from Appendix B we find that, neglecting terms of order m_1/m_2,

$$\Delta=\frac{\frac{9}{4}}{\frac{13}{4}+\sqrt{2}}=0.4824\,.$$

Therefore, from equation (8.111)*,

$$[e_1^0]_2=\frac{9\sqrt{2}-5}{4}[e_1^0]_1=1.93198[e_1^0]_1=\frac{v_{t1}\tau_e}{n_1}1.932 \tag{8.112}$$

and

$$[e_1^1]_2=-\frac{v_{t1}\tau_e}{n_1}.621\,,\qquad [e_2^1]_2=-\frac{v_{t1}\tau_eM_1}{n_1}.916\,,$$

and thus the second approximations to D_{12} and σ are

$$[D_{12}]_2=1.932[D_{12}]_1\qquad\text{and}\qquad[\sigma]_2=1.932[\sigma]_1\,. \tag{8.113}$$

*W. Marshall gives $\Delta=.483$ and $(1-\Delta)^{-1}=1.94$ instead of 1.932.

Problem 8.6 Verify equation (8.108) and then fill in the steps that lead to equation (8.113).

In view of this large factor of 1.93 between the first and second approximations we might think our result to be an order of magnitude calculation only for there is no a priori reason why the third, fourth, fifth, and higher approximations should not each introduce factors of the same order of magnitude. However, this is not so, for Landshoff[51] has calculated the third and fourth approximation to σ and he finds that the method converges much more rapidly than the first factor of 1.93 leads us to expect. He finds that

$$[\sigma]_3 = 1.95[\sigma]_1 \qquad \text{and} \qquad [\sigma]_4 = 1.96[\sigma]_1 \tag{8.114}$$

so we may be confident that this last value is correct to a few percent, especially as it agrees closely with the value given by Spitzer[52], namely,

$$\sigma = 1.97[\sigma]_1\,,$$

which he obtained by an entirely different method.

8.7.3 Thermal conduction

Finally, let us consider the variation method applied to equation (8.42). This will give the coefficient of thermal conduction, λ, the thermal diffusion coefficient, D_T, and hence the ratio, k_T, defined by equation (8.60). Comparing equation (8.42) to the general form of equation (8.88) we see that in this case

$$R_i = f_i \left(\tfrac{5}{2} - \varpi_i^2\right) \boldsymbol{\varpi}_i\,.$$

We use as our trial function the expressions given by equations (8.45) and (8.48), i.e.,

$$t_i = \boldsymbol{\varpi}_i \sum_{m=0}^{\infty} a_i^m L_1^{\frac{3}{2}}(\varpi_i^2)$$

where a_2^0 is related to a_1^0 by the subsidiary condition equation (8.51). Equation (8.91) now becomes

$$-\tfrac{15}{2} \sum_i n_i v_{ti} a_i^1 = \left\{ \boldsymbol{\varpi} \sum_m a_i^m L_1^{\frac{3}{2}}(\varpi^2); \boldsymbol{\varpi} \sum_m a_i^m L_1^{\frac{3}{2}}(\varpi^2) \right\}. \tag{8.115}$$

Once again we note that the variation method maximizes the very expression we are interested in, for the left-hand side of equation (8.115) is simply related to the coefficient λ', defined by equation (8.66), which appears in the definition of λ, equation (8.76).

We can obtain a first approximation by setting all the coefficients except a_1^1 and a_2^1 equal to zero. This is slightly different from the first approximation with the e_i^m because only a_1^1 and a_2^1 appear on the left so the simplest form is to keep only these two terms nonzero. We include the a_1^0 and a_2^0 at the next level of approximation. Equation (8.115) then becomes

$$-\tfrac{15}{2}\sum_i n_i v_{ti} a_i^1 = 2n_1 n_2 [r_4 (a_1^1)^2 + r_5 (a_2^1)^2 + 2r_6 a_1^1 a_2^1] \,, \tag{8.116}$$

where r_4, r_5, and r_6 are the same combination of collision integrals as occur in equation (8.109). Constructing the appropriate Davison function,

$$\mathcal{D}(a_1^1, a_2^1) = -\tfrac{15}{2}(n_1 v_{t1} a_1^1 + n_2 v_{t2} a_2^1) - n_1 n_2 [r_4 (a_1^1)^2 + r_5 (a_2^1)^2 + 2r_6 a_1^1 a_2^1] \,,$$

the first approximation yields

$$[a_1^1]_1 \simeq -\frac{15 v_{t1}}{4 n_1 r_4} \quad \text{and} \quad [a_2^1]_1 \simeq -\frac{15 v_{t2}}{4 n_1 r_5} .$$

Evaluating the integrals from Appendix B gives

$$r_4 = \left(\frac{13}{4} + \sqrt{2}\right)\varphi \,,$$
$$r_5 \simeq \sqrt{2M_1}\varphi \,,$$

where $M_1 = m_1/(m_1 + m_2) \simeq m_1/m_2$ so

$$\left[a_1^1\right]_1 \simeq -\frac{15 v_{t1}}{4 n_1 \left(\frac{13}{4} + \sqrt{2}\right)\varphi} = -\frac{5 v_{t1} \tau_e}{2(\frac{13}{4} + \sqrt{2})} \,,$$
$$\left[a_2^1\right]_1 \simeq -\frac{15 v_{t1}}{4 n_1 \sqrt{2}\varphi} = -\frac{5 v_{t1} \tau_e}{2\sqrt{2}} \,,$$

and equation (8.66) gives

$$[\lambda']_1 \simeq \frac{75 k_B v_{t1}^2}{16\varphi}\left(\frac{1}{\frac{13}{4} + \sqrt{2}}\right) \simeq \frac{25 k_B v_{t1}^2 n_1 \tau_e}{2(13 + 4\sqrt{2})} .$$

Now in Chapter 7 we derived an expression for the thermal conductivity of a simple gas of particles of mass m_1 as

$$[\lambda_1]_1 = \frac{75 k_B v_{t1}^2}{16\sqrt{2}\varphi} \,, \tag{8.117}$$

so we see that

$$[\lambda']_1 = \frac{\sqrt{2}}{\frac{13}{4} + \sqrt{2}}[\lambda_1]_1 = 0.3032[\lambda_1]_1 \,, \tag{8.118}$$

where $[\lambda_1]_1$ as defined by equation (8.117) is the first approximation to the thermal conductivity of a simple gas composed only of electrons and $[\lambda']_1$ is the first approximation to the coefficient λ' of the gas mixture. To obtain this first approximation we set a_1^0 and a_2^0 equal to zero. Hence from equations (8.59), (8.60), and (8.76) to this approximation,

$$[D_T]_1 = 0\,, \qquad [k_T]_1 = 0\,, \qquad [\lambda']_1 = 0.3032[\lambda_1]_1\,. \tag{8.119}$$

A second approximation is obtained by setting all the coefficients except a_1^1, a_2^1, a_1^0, and a_2^0 equal to zero. Then the Davison function of equation (8.115) becomes, within a constant,

$$\begin{aligned}\mathcal{D}(a_1^0, a_1^1, a_2^1) = -\frac{15v_{t1}}{2n_2}\left[a_1^1 + \frac{n_2\sqrt{m_1}}{n_1\sqrt{m_2}}a_2^1\right] - \big[r_1(a_1^0)^2 + 2r_2a_1^0a_1^1 + 2r_3a_1^0a_2^1 \\ + r_4(a_1^1)^2 + r_5(a_2^1)^2 + 2r_6a_1^1a_2^1\big]\,,\end{aligned}$$

which is maximized by (a_2^0 is of order $\sqrt{M_1}$ compared to a_1^0 and neglected)

$$\begin{aligned}\left[a_1^0\right]_2 &= \frac{45v_{t1}}{8n_2(1+\sqrt{2})\varphi} \simeq \frac{15v_{t1}\tau_e}{4(1+\sqrt{2})}\,,\\ \left[a_1^1\right]_2 &= -\frac{15v_{t1}}{4n_2(1+\sqrt{2})\varphi} \simeq -\frac{5v_{t1}\tau_e}{2(1+\sqrt{2})}\,,\\ \left[a_2^1\right]_2 &= -\frac{15v_{t1}}{4n_1\sqrt{2}\varphi} \simeq -\frac{5v_{t1}\tau_e}{2\sqrt{2}}\,.\end{aligned} \tag{8.120}$$

Using equation (8.66) to give λ', we get

$$\begin{aligned}[\lambda']_2 &= \frac{75k_Bv_{t1}^2}{16\varphi}\left(\frac{1}{1+\sqrt{2}}\right)\\ &= \frac{\sqrt{2}}{1+\sqrt{2}}\left[\lambda_1\right]_1 = 0.5858\left[\lambda_1\right]_1\,.\end{aligned} \tag{8.121}$$

Then equation (8.59) gives

$$[D_T]_2 = \frac{45v_{t1}^2}{64n_1(1+\sqrt{2})\varphi}\,, \tag{8.122}$$

so that

$$[k_T]_2 = \frac{15}{2(9\sqrt{2}-5)(1+\sqrt{2})} = 0.402\,.$$

The coefficient of thermal conductivity is given by equation (8.74), or

$$[\lambda]_2 = \frac{\sqrt{2}}{1+\sqrt{2}}\left[1 - \frac{6}{5}\left[k_T\right]_2\right]\left[\lambda_1\right]_1 \tag{8.123}$$

$$= 0.3032[\lambda_1]_1\,. \tag{8.124}$$

Problem 8.7 Solve for the second approximations for $\left[a_1^0\right]_2$, $\left[a_1^1\right]_2$, and $\left[a_2^1\right]_2$ to verify these last results.

We see from equation (8.123) that the thermal diffusion term is quite large and comparable to the "normal" thermal diffusion term. From equation (8.124) we see that the thermal conductivity of the ionized gas is of the same order of magnitude as that of an electron gas alone. This is just what we expect because a simple gas of ions alone would have a thermal conductivity given by equation (8.117) with m_2 in place of m_1. Hence the thermal conductivity of a simple electron gas is $\sqrt{m_2/m_1} = 43$ times as large as that of a simple ion gas and so we would expect the electrons to dominate the thermal conduction effect.

A useful check on the accuracy of these results is given by equation (8.5) which is a relation between the values of a_1^0, e_1^1, and e_2^1. In quite independent calculations we have obtained approximate values of e_1^1 and e_2^1 in equation (8.110) and of a_1^0 in equation (8.120). Comparing these results we find that the agreement is within a few percent.

On the other hand, the convergence of the process using higher order calculations is not as rapid as with the conductivity. Comparing λ' of equation (8.121) with that of equation (8.118) gives

$$\frac{[\lambda']_2}{[\lambda']_1} = \frac{13 + 4\sqrt{2}}{4(1+\sqrt{2})} = 1.932$$

which is the same change as in the conductivity between the second and first approximations. For the thermal conductivity, however, the successive approximations give [47]

$$\frac{[\lambda']_3}{[\lambda']_2} = 1.601$$
$$\frac{[\lambda']_4}{[\lambda']_3} = 1.017$$

so

$$\frac{[\lambda']_4}{[\lambda']_1} \simeq 3.15$$

so the convergence is relatively slower in this case.

Problem 8.8 Find the percentage difference from the left-hand and right-hand sides of equation (8.5) using the values of e_1^1 and e_2^1 from equation (8.110) and of a_1^0 from equation (8.120).

9

TRANSPORT WITH A FINITE MAGNETIC FIELD

In this last chapter, we introduce a magnetic field. In addition to introducing a preferred direction so that some of the transport coefficients become tensors, they also become functions of $\omega_{ce}\tau_e$ or $\omega_{ci}\tau_i$, the product of the electron or ion cyclotron frequency times the electron or ion collision frequency. These coefficients are first calculated, then summarized at the end of the chapter.

9.1 Boltzmann equations

In this chapter, we consider a gas of electrons and one species of ions but in a magnetic field, again following Marshall[46]. We thus require the Boltzmann equation for each species such that

$$\left\{\frac{\partial}{\partial t} + \boldsymbol{v}\cdot\nabla + \left[\boldsymbol{X} + \frac{e_i}{m_i}(\boldsymbol{E} + \boldsymbol{v}\times\boldsymbol{B})\right]\cdot\nabla_v\right\} F_i(\boldsymbol{r},\boldsymbol{v},t) = \Delta F_i = \sum_j \Delta_j F_i \tag{9.1}$$

where $\boldsymbol{E}(\boldsymbol{r},t)$ is the electric field, $\boldsymbol{X}$ is any nonelectromagnetic acceleration, and $\boldsymbol{B}(\boldsymbol{r},t)$ is the magnetic field. The collisions are the same as in Chapter 8, namely,

$$\Delta_j F_i = \int \mathrm{d}\boldsymbol{s} \int_0^{2\pi} \mathrm{d}\phi \int_0^{\lambda_D} \mathrm{d}b\, bg[F_i(\boldsymbol{v}')F_j(\boldsymbol{s}') - F_i(\boldsymbol{v})F_j(\boldsymbol{s})]\,, \tag{9.2}$$

or in terms of the differential cross section σ into a solid angle $\mathrm{d}\Omega = \mathrm{d}\phi\,\sin\theta\mathrm{d}\theta$,

$$\Delta_j F_i = \int \mathrm{d}\boldsymbol{s}\,\sigma\mathrm{d}\Omega\, g[F_i(\boldsymbol{v}')F_j(\boldsymbol{s}') - F_i(\boldsymbol{v})F_j(\boldsymbol{s})]\,. \tag{9.3}$$

Actually, Equations (9.2) and (9.3) are correct only if the Debye length, λ_D, is much smaller than a Larmor radius. They are usually valid except in extreme cases with very low densities.

9.2 The magnetohydrodynamic (MHD) equations

The equations of magnetohydrodynamics (MHD) shall be deduced in this section from the Boltzmann equations of equation (9.1). Reviewing some of the quantities from Chapter 8, we define the mean value of Ψ, where Ψ is any quantity that depends on the position, velocity, and time for a particle, as

$$\langle \Psi_i \rangle = \frac{1}{n_i} \int \mathrm{d}\boldsymbol{v}\, \Psi_i(\boldsymbol{r}, \boldsymbol{v}, t) F_i(\boldsymbol{r}, \boldsymbol{v}, t)$$

where n_i is the particle density of particles of species i, or

$$n_i = \frac{1}{n_i} \int \mathrm{d}\boldsymbol{v}\, F_i(\boldsymbol{r}, \boldsymbol{v}, t)\,.$$

The partial mass densities are

$$\rho_i = n_i m_i\,,$$

and the total number density is

$$n = \sum_i n_i = n_1 + n_2\,,$$

and the mass density is

$$\rho = \sum_i \rho_i = n_1 m_1 + n_2 m_2\,.$$

The mean velocities are

$$\langle \boldsymbol{v}_i \rangle = \frac{1}{n_i} \int \mathrm{d}\boldsymbol{v}\, F_i \boldsymbol{v}\,,$$

and the drift velocity is

$$\boldsymbol{u} = \frac{1}{\rho} \sum_i \rho_i \langle \boldsymbol{v}_i \rangle\,. \tag{9.4}$$

The random velocity is

$$\boldsymbol{w} = \boldsymbol{v} - \boldsymbol{u}\,,$$

and the mean random velocities are

$$\langle \boldsymbol{w}_i \rangle = \frac{1}{n_i} \int \mathrm{d}\boldsymbol{v}\, \boldsymbol{w} F_i\,. \tag{9.5}$$

Hence by definition,

$$\sum_i \rho_i \langle \boldsymbol{w}_i \rangle \equiv 0 = n_1 m_1 \langle \boldsymbol{w}_1 \rangle + n_2 m_2 \langle \boldsymbol{w}_2 \rangle\,. \tag{9.6}$$

It may seem surprising that the mean random velocities do not vanish, since with a single species they do. If we were to define mean velocities for each species as

$$\boldsymbol{u}_i = \langle \boldsymbol{v}_i \rangle \,, \tag{9.7}$$

so that there are two separate mean velocities, as is frequently done in two-fluid theories, then $\langle \boldsymbol{w}_i \rangle = 0$, but MHD theory is a one-fluid theory with the drift velocity given by equation (9.4), so by equation (9.5), they do not, in general, vanish.

The temperature T is defined by

$$\tfrac{3}{2} n k_B T \equiv \tfrac{1}{2} \sum_i n_i m_1 \langle w_i^2 \rangle \,,$$

and the pressure tensor is

$$p_{\alpha\beta} = \sum_i \rho_i \langle w_{i\alpha} w_{i\beta} \rangle \,. \tag{9.8}$$

The heat flux vector is

$$\boldsymbol{q} = \sum_i \tfrac{1}{2} \rho \langle w_i^2 \boldsymbol{w}_i \rangle \,. \tag{9.9}$$

The charge density is

$$Q = \sum_i n_i e_i = (n_2 - n_1) e \,, \tag{9.10}$$

and the total current is

$$\boldsymbol{J} = \sum_i n_i e_i \langle \boldsymbol{v}_i \rangle = Q \boldsymbol{u} + \boldsymbol{j} \,, \tag{9.11}$$

where $\boldsymbol{j}$ is the conduction current given by

$$\begin{aligned} \boldsymbol{j} &= \sum_i n_i e_i \langle \boldsymbol{w}_i \rangle = (n_2 \langle \boldsymbol{w}_2 \rangle - n_1 \langle \boldsymbol{w}_1 \rangle) e \\ &= -\frac{n_1 n_2}{\rho} e (m_2 + m_1)(\langle \boldsymbol{w}_1 \rangle - \langle \boldsymbol{w}_2 \rangle) \,. \end{aligned} \tag{9.12}$$

The MHD equations can be obtained from equation (9.1) by simply integrating the various moments. The new term in the equation of continuity is given by

$$\frac{e_i}{m_i} \int \mathrm{d}\boldsymbol{v} \, (\boldsymbol{v} \times \boldsymbol{B}) \cdot \nabla_v F_i \,,$$

which can be integrated by parts, and since $(\boldsymbol{v} \times \boldsymbol{B})_\beta$ is independent of v_β, gives zero. Thus we obtain the continuity equation, equation (8.13) from Chapter 8, namely,

$$\frac{\partial n_i}{\partial t} + \nabla \cdot (n_i \boldsymbol{u} + n_i \langle \boldsymbol{w}_i \rangle) = 0 \,.$$

Multiplying by m_i and summing over i gives

$$\frac{\partial \rho}{\partial t} + \nabla \cdot (\rho \boldsymbol{u}) = 0 .$$

The equation of motion is found by multiplying equation (9.1) by $m_i v_\alpha$ and then integrating and summing over the species. The new term is

$$\sum_i e_i \int \mathrm{d}\boldsymbol{v}\, v_\alpha (\boldsymbol{v} \times \boldsymbol{B}) \cdot \nabla F_i(\boldsymbol{r}, \boldsymbol{v}, t) ,$$

which can be integrated by parts to give

$$-\sum_i e_i \int \mathrm{d}\boldsymbol{v}\, (\boldsymbol{v} \times \boldsymbol{B})_\alpha F_i(\boldsymbol{r}, \boldsymbol{v}, t) = -(\boldsymbol{J} \times \boldsymbol{B})_\alpha .$$

The equation of motion therefore becomes

$$\rho \frac{D}{Dt} u_\alpha = -\nabla_\beta p_{\alpha\beta} + Q E_\alpha + (\boldsymbol{J} \times \boldsymbol{B})_\alpha + \sum_i \rho_i X_{i\alpha} ,$$

or using equation (7.10),

$$\rho \frac{D}{Dt} u_\alpha = -\nabla_\beta p_{\alpha\beta} + Q[E_\alpha + (\boldsymbol{u} \times \boldsymbol{B})_\alpha] + (\boldsymbol{j} \times \boldsymbol{B})_\alpha + \sum_i \rho_i X_{i\alpha} . \quad (9.13)$$

Multiplying equation (9.1) by $\frac{1}{2} m_i v^2$ and integrating and summing over i gives the energy equation. The new term is

$$\sum_i \tfrac{1}{2} e_i \int \mathrm{d}\boldsymbol{v}\, v^2 (\boldsymbol{v} \times \boldsymbol{B}) \cdot \nabla_v F_i ,$$

which can again be integrated by parts with the result that this term vanishes. The equation of motion is therefore

$$\frac{\mathrm{D}}{\mathrm{D}t}\left(\tfrac{3}{2} n k_B T\right) = -\tfrac{3}{2} n k_B T \nabla \cdot \boldsymbol{u} - p_{\alpha\beta} \nabla_\alpha u_\beta - \nabla \cdot \boldsymbol{q} + \boldsymbol{j} \cdot (\boldsymbol{E} + \boldsymbol{u} \times \boldsymbol{B}) + \sum_i \rho_i \langle \boldsymbol{w}_i \cdot \boldsymbol{X}_i \rangle . \quad (9.14)$$

These equations, along with the Maxwell equations, give, at least in principle, the complete solution of the problem once we have expressions for the pressure tensor, $p_{\alpha\beta}$, the heat flux vector, $\boldsymbol{q}$, and the conduction current, $\boldsymbol{j}$. The remaining sections endeavor to provide these quantities.

9.3 The formal theory of kinetic processes

The expansion procedure used without a magnetic field is quite different from the procedure we use for this case. Previously we assumed that the collision terms on the right-hand side of the Boltzmann equation were the only large terms (of course in equilibrium they all cancel but each term is large). With a large magnetic field, however, the magnetic force term on the left-hand side of the Boltzmann equation can be very large, especially for high velocity particles, so it is not adequate to treat this term the way we have treated the other terms on the left-hand side. In order to illustrate the new expansion procedure, we write equation (9.1) as

$$\left\{\frac{\partial}{\partial t}+\boldsymbol{v}\cdot\nabla+\left[\boldsymbol{X}_i+\frac{e_i}{m_i}(\boldsymbol{E}+\boldsymbol{u}\times\boldsymbol{B})\right]\cdot\nabla_v\right\}F_i(\boldsymbol{r},\boldsymbol{v},t)$$
$$+\frac{e_i}{\eta m_i}(\boldsymbol{w}\times\boldsymbol{B})\cdot\nabla_v F_i(\boldsymbol{r},\boldsymbol{v},t)=\frac{1}{\eta}\Delta F_i(\boldsymbol{r},\boldsymbol{v},t)\,, \tag{9.15}$$

and look for solutions of the form

$$F_i = F_i^{(0)}(1+\eta\varphi+\eta^2\psi+\cdots)\,. \tag{9.16}$$

We will eventually set η to unity.

Because we have included a term of order $1/\eta$ on the left-hand side of equation (9.15), we no longer need to insist on collisions being the dominant process determining the distribution function. In fact, we will see later that we may let the collision time approach infinity and still get sensible results (except for some elements of the pressure tensor). This is important, since as the temperature increases, the collision time increases as $T^{3/2}$ so that for high T, collisions are rare events.

The manner we use to break up the magnetic force term is very important. We do not treat the complete term,

$$\frac{e}{m}\boldsymbol{v}\times\boldsymbol{B}\cdot\nabla_v F\,,$$

as being of order $1/\eta$, but only that part due to the random velocity $\boldsymbol{w}$. The part involving the drift velocity $\boldsymbol{u}$ is grouped with the electric field $\boldsymbol{E}$. For the remainder of this chapter, we will define

$$\boldsymbol{E}'\equiv\boldsymbol{E}+\boldsymbol{u}\times\boldsymbol{B}\,.$$

It is clear that $\boldsymbol{E}'$ is the electric field in the system of coordinates moving with the fluid element. The expansion procedure treats this term as small as we would expect for a medium of high electrical conductivity.

Substituting equation (9.16) into equation (9.15), the terms of order $1/\eta$ are

$$\frac{e_i}{m_i}\boldsymbol{w}\times\boldsymbol{B}\nabla_v F_i^{(0)}=\Delta F_i^{(0)}\,,$$

which has the solution

$$F_i^{(0)}=f_i=\frac{n_i}{v_{ti}^3\pi^{3/2}}\mathrm{e}^{-(\boldsymbol{v}-\boldsymbol{u})^2/v_{ti}^2}\,.$$

The terms of order $\eta^{(0)}$ are

$$\left(\frac{\partial f_i}{\partial t}\right)^{(0)} + \left[\boldsymbol{v}\cdot\nabla + \left(\boldsymbol{X}_i + \frac{e_i}{m_i}\boldsymbol{E}'\right)\cdot\nabla_v\right] f_i + \frac{e_i}{m_i}\boldsymbol{w}\times\boldsymbol{B}\cdot\nabla_v(f_i\varphi_i)$$
$$= \int \mathrm{d}\boldsymbol{s}\,\sigma\mathrm{d}\Omega\, g f_i(\boldsymbol{v}) f_j(\boldsymbol{s})[\varphi_i(\boldsymbol{v}') + \varphi_j(\boldsymbol{s}') - \varphi_i(\boldsymbol{v}) - \varphi_j(\boldsymbol{s})]. \quad (9.17)$$

The last term on the left-hand side of equation (9.17) is of a new type and comes from the new expansion procedure. We also need to take extra care in dealing with the first term of equation (9.17). We may write

$$\left(\frac{\partial f_i}{\partial t}\right)^{(0)} = f_i\left[\frac{1}{n_i}\left(\frac{\partial n_i}{\partial t}\right)^{(0)} - \frac{1}{T}\left(\frac{3}{2} - \varpi^2\right)\frac{\partial T}{\partial t}^{(0)} + \frac{m_i}{k_B T} w_\alpha \left(\frac{\partial u_\alpha}{\partial t}\right)^{(0)}\right] \quad (9.18)$$

and now we can substitute for the time derivatives in this expression from the equations of motion. When doing this, however, we must remember that these equations should be derived from the Boltzmann equation with the factor $1/\eta$ in the magnetic force term. The equations then become

$$\begin{aligned}
\frac{\partial n_i}{\partial t} &= -\nabla_\alpha(n_i u_\alpha + n_i\langle w_{i\alpha}\rangle) \\
\frac{\partial \rho}{\partial t} &= -\nabla\cdot(\rho\boldsymbol{u}) \\
\rho\frac{\mathrm{D}}{\mathrm{D}t}u_\alpha &= -\nabla_\beta p_{\alpha\beta} + QE'_\alpha + \frac{1}{\eta}(\boldsymbol{j}\times\boldsymbol{B})_\alpha + \sum_i \rho_i X_{i\alpha} \\
\frac{\mathrm{D}}{\mathrm{D}t}\left(\frac{3}{2}nk_B T\right) &= -\frac{3}{2}nk_B T\nabla\cdot\boldsymbol{u} - p_{\alpha\beta}\nabla_\alpha u_\beta - \nabla\cdot\boldsymbol{q} + \boldsymbol{j}\cdot\boldsymbol{E}' \\
&\quad + \sum_i \rho_i\langle\boldsymbol{w}_i\cdot\boldsymbol{X}_i\rangle\,.
\end{aligned} \quad (9.19)$$

In these equations,

$$\begin{aligned}
\boldsymbol{j} &= \sum_i n_i e_i\langle\boldsymbol{w}_i\rangle = \sum_i e_i\int \mathrm{d}\boldsymbol{v}\,\boldsymbol{w}_i F_i \\
&= \eta\sum_i e_i\int\mathrm{d}\boldsymbol{v}\,\boldsymbol{w}_i f_i\varphi_i\,.
\end{aligned} \quad (9.20)$$

From equation (9.20) we see that $\boldsymbol{j}$ is of order η so that in the energy equation the Joule heating term, $\boldsymbol{j}\cdot\boldsymbol{E}'$, is also of order η. In the equation of motion, however, the magnetic force term, $\boldsymbol{j}\times\boldsymbol{B}$, appears multiplied by $1/\eta$, so this term is of order zero in η. Hence, to zero order in η, we can neglect $\langle\boldsymbol{w}_i\rangle$ in the equation of continuity to give

$$\left(\frac{\partial n_i}{\partial t}\right)^{(0)} = -\nabla_\alpha(n_i u_\alpha)\,, \quad (9.21)$$

and we can replace $p_{\alpha\beta}$ by $p\delta_{\alpha\beta}$ in the equation of motion to give

$$\left(\frac{\partial u_\alpha}{\partial t}\right)^{(0)} = -(\boldsymbol{u}\cdot\nabla)u_\alpha - \frac{1}{\rho}\nabla_\alpha p + \frac{Q}{\rho}E'_\alpha + \frac{1}{\rho}\sum_i e_i \int \mathrm{d}\boldsymbol{v}\, f_i\varphi(\boldsymbol{w}_i\times\boldsymbol{B})_\alpha + \frac{1}{\rho}\sum_i \rho_i X_{i\alpha}\,. \tag{9.22}$$

In the energy equation, we can replace $p_{\alpha\beta}$ by $p\delta_{\alpha\beta}$, $\boldsymbol{q}$ by zero, and $\boldsymbol{j}\cdot\boldsymbol{E}'$ by zero to give

$$\left(\frac{\partial T}{\partial t}\right)^{(0)} = -(\boldsymbol{u}\cdot\nabla)T - \frac{2p}{3nk_B}\nabla\cdot\boldsymbol{u}\,. \tag{9.23}$$

Substituting Equations (9.21), (9.22), and (9.23) into equation (9.18) and (9.18) into (9.17), we find that many terms cancel with the result

$$\begin{aligned} f_i\left[2(\boldsymbol{\varpi}_i^0\boldsymbol{\varpi}_i)_{\alpha\beta}\nabla_\alpha u_\beta - \left(\tfrac{5}{2}-\varpi_i^2\right)\boldsymbol{w}_i\cdot\nabla\ln T + \frac{n}{n_i}\boldsymbol{w}_i\cdot\boldsymbol{d}_i\right] \\ = -f_i\frac{m_i}{\rho k_B T}\boldsymbol{w}_i\cdot\sum_j e_j\int \mathrm{d}\boldsymbol{v}\, f_j\varphi_j\boldsymbol{w}_j\times\boldsymbol{B} - f_i\frac{e_i}{m_i}(\boldsymbol{\varpi}\times\boldsymbol{B})\cdot\nabla_{\boldsymbol{\varpi}}\varphi \\ +\sum_j\int \mathrm{d}\boldsymbol{s}\,\sigma \mathrm{d}\Omega\, g f_i(\boldsymbol{v})f_j(\boldsymbol{s})[\varphi_i(\boldsymbol{v}')+\varphi_j(\boldsymbol{s}')-\varphi_i(\boldsymbol{v})-\varphi_j(\boldsymbol{s})], \end{aligned} \tag{9.24}$$

where

$$\begin{aligned} -d_2 &= d_1 \\ &= \nabla\left(\frac{n_1}{n}\right) + \frac{n_1 n_2(m_2-m_1)}{pn\rho}\nabla p - \frac{\rho_1\rho_2}{p\rho}(\boldsymbol{X}_1-\boldsymbol{X}_2) \\ &\quad -\frac{n_1n_2}{p\rho}(e_1m_2-e_2m_1)\boldsymbol{E}'\,. \end{aligned}$$

Again, ϖ_i is a dimensionless velocity given by

$$\boldsymbol{\varpi}_i = \frac{\boldsymbol{w}_i}{v_{ti}}\,,$$

and $\boldsymbol{\varpi}_i^0\boldsymbol{\varpi}_i$ is a tensor given by

$$\boldsymbol{\varpi}_i^0\boldsymbol{\varpi}_i = \varpi_{i\alpha}\varpi_{i\beta} - \tfrac{1}{3}\varpi_i^2\delta_{\alpha\beta}\,.$$

In the expansion procedure we have just described, the current $\boldsymbol{j}$ is formally a quantity of first order in smallness, i.e., $\boldsymbol{j}$ is proportional to η. But at high temperatures, the electrical conductivity becomes very large, so large currents can flow and would therefore seem to appear that we should take $\boldsymbol{j}$ to be of zero order in η. Looking at the problem this way is misleading, however, since when we say that the current is large, we do not mean that all terms depending on the current are large. We know, for example, that as the temperature and electrical resistivity increase, the Joule heating term, $\boldsymbol{j} \cdot \boldsymbol{E}'$, in the energy equation becomes less important because $\boldsymbol{E}'$ becomes smaller. In the momentum equation, on the other hand, the magnetic force term, $\boldsymbol{j} \times \boldsymbol{B}$, can be very large and comparable if not greater than the pressure term, $-\nabla p$. Examining the procedure we have used, we see that it gives exactly such a description so that the magnetic force term appears as a term of zero order in η while the Joule heating term appears as a term of first order in η.

Because equation (9.24) is linear in φ_i, the solution can be written immediately in the form,

$$\varphi_i = -\mathsf{B}_i^{\alpha\beta} \nabla_\alpha u_\beta - \boldsymbol{A}_i \cdot \nabla \ln T - n\boldsymbol{E}_i \cdot \boldsymbol{d}_1 \,, \tag{9.25}$$

where

$$\begin{aligned} -2f_i \boldsymbol{\varpi}_i^0 \boldsymbol{\varpi}_i = &-\frac{1}{\rho k_B T} m_i f_i \boldsymbol{w}_i \cdot \sum_j e_j \int \mathrm{d}\boldsymbol{v}\, f_j(\boldsymbol{w}_j \times \boldsymbol{B}) \mathsf{B}_j \\ &-\frac{e_i}{m_i} f_i(\boldsymbol{\varpi}_i \times \boldsymbol{B}) \cdot \nabla_{\boldsymbol{\varpi}_i} \mathsf{B}_i + \sum_j \int \mathrm{d}\boldsymbol{s}\, \sigma \mathrm{d}\Omega\, g f_i(\boldsymbol{v}) f_j(\boldsymbol{s}) \\ &\qquad \times [\mathsf{B}_i(\boldsymbol{v}') + \mathsf{B}_j(\boldsymbol{s}') - \mathsf{B}_i(\boldsymbol{v}) - \mathsf{B}_j(\boldsymbol{s})] \end{aligned} \tag{9.26}$$

$$\begin{aligned} f_i\left(\tfrac{5}{2} - \varpi_i^2\right) \boldsymbol{w}_i = &-\frac{1}{\rho k_B T} m_i f_i \boldsymbol{w}_i \cdot \sum_j e_j \int \mathrm{d}\boldsymbol{v}\, f_j(\boldsymbol{w}_j \times \boldsymbol{B}) \boldsymbol{A}_j \\ &-\frac{e_i}{m_i} f_i(\boldsymbol{\varpi}_i \times \boldsymbol{B}) \cdot \nabla_{\boldsymbol{\varpi}_i} \boldsymbol{A}_i \\ &+\sum_j \int \mathrm{d}\boldsymbol{s}\, \sigma \mathrm{d}\Omega\, g f_i(\boldsymbol{v}) f_j(\boldsymbol{s}) \\ &\qquad \times [\boldsymbol{A}_i(\boldsymbol{v}') + \boldsymbol{A}_j(\boldsymbol{s}') - \boldsymbol{A}_i(\boldsymbol{v}) - \boldsymbol{A}_j(\boldsymbol{s})] \end{aligned} \tag{9.27}$$

$$\begin{aligned} \frac{(-1)^i}{n_i} f_i \boldsymbol{w}_i = &-\frac{1}{\rho k_B T} m_i f_i \boldsymbol{w}_i \cdot \sum_j e_j \int \mathrm{d}\boldsymbol{v}\, f_j(\boldsymbol{w}_i \times \boldsymbol{B}) \boldsymbol{E}_j \\ &-\frac{e_i}{m_i} f_i(\boldsymbol{\varpi}_i \times \boldsymbol{B}) \cdot \nabla_{\boldsymbol{\varpi}_i} \boldsymbol{E}_i \\ &+\sum_j \int \mathrm{d}\boldsymbol{s}\, \sigma \mathrm{d}\Omega\, g f_i(\boldsymbol{v}) f_j(\boldsymbol{s}) \\ &\qquad \times [\boldsymbol{E}_i(\boldsymbol{v}') + \boldsymbol{E}_j(\boldsymbol{s}') - \boldsymbol{E}_i(\boldsymbol{v}) - \boldsymbol{E}_j(\boldsymbol{s})] \,. \end{aligned} \tag{9.28}$$

The analysis of equation (9.26) is lengthy and complicated, and will be deferred until Section 9.6. Because the form of equations (9.27) and (9.28) is similar in form (although they are completely independent of one another), we will treat them together.

The vectors $\boldsymbol{A}_i$ and $\boldsymbol{E}_i$ are vector functions of the vector $\boldsymbol{\varpi}$ and the pseudo-vector $\boldsymbol{B}$. The only vectors that can be formed from these are

$$\boldsymbol{\varpi}_i\,; \quad \boldsymbol{\varpi}_i \times \boldsymbol{B}\,; \quad (\boldsymbol{\varpi}_i \times \boldsymbol{B}) \times \boldsymbol{B}\,; \quad [(\boldsymbol{\varpi}_i \times \boldsymbol{B}) \times \boldsymbol{B}] \times \boldsymbol{B}\,; \quad \text{etc.} \tag{9.29}$$

(We cannot include $\boldsymbol{B}$ itself because it is a pseudo-vector.) Now

$$[(\boldsymbol{\varpi}_i \times \boldsymbol{B}) \times \boldsymbol{B}] \times \boldsymbol{B} = -B^2 \boldsymbol{\varpi}_i \times \boldsymbol{B} ,$$

so we only need to consider the first three of the vectors in equation (9.29). Furthermore,

$$(\boldsymbol{\varpi}_i \times \boldsymbol{B}) \times \boldsymbol{B} = -B^2 \boldsymbol{\varpi}_i + \boldsymbol{B}(\boldsymbol{\varpi}_i \cdot \boldsymbol{B}) ,$$

so we may consider the three independent vectors to be

$$\boldsymbol{\varpi}_i ; \quad \boldsymbol{\varpi}_i \times \boldsymbol{B} ; \quad \boldsymbol{B}(\boldsymbol{\varpi}_i \cdot \boldsymbol{B}) .$$

The last of these is the product of a pseudo-vector and a pseudo-scalar, so it is a vector. The only scalars we can construct from $\boldsymbol{\varpi}_i$ and $\boldsymbol{B}$ must be functions of

$$\varpi_i^2 ; \quad B^2 ; \quad (\boldsymbol{\varpi}_i \cdot \boldsymbol{B})^2 .$$

(We cannot include $\boldsymbol{\varpi}_i \cdot \boldsymbol{B}$ itself because it is a pseudo-scalar). We therefore write

$$\boldsymbol{A}_i = \boldsymbol{\varpi}_i \mathcal{A}_i^I + \boldsymbol{\varpi}_i \times \boldsymbol{B} \mathcal{A}_i^{II} + (\boldsymbol{\varpi}_i \times \boldsymbol{B}) \times \boldsymbol{B} \mathcal{A}_i^{III} , \tag{9.30}$$

$$\boldsymbol{E}_i = \boldsymbol{\varpi}_i \mathcal{E}_i^I + \boldsymbol{\varpi}_i \times \boldsymbol{B} \mathcal{E}_i^{II} + (\boldsymbol{\varpi}_i \times \boldsymbol{B}) \times \boldsymbol{B} \mathcal{E}_i^{III} , \tag{9.31}$$

where $\mathcal{A}_i^I$, $\mathcal{A}_i^{II}$, $\mathcal{A}_i^{III}$, $\mathcal{E}_i^I$, $\mathcal{E}_i^{II}$, and $\mathcal{E}_i^{III}$ are scalar functions of ϖ_i^2, B^2, and $(\boldsymbol{\varpi}_i \cdot \boldsymbol{B})^2$.

We now consider the operator, $(\boldsymbol{\varpi}_i \times \boldsymbol{B}) \cdot \nabla_{\boldsymbol{\varpi}_i}$, acting on $\boldsymbol{A}_i$. We note that the operator acting on either ϖ_i^2 or $(\boldsymbol{\varpi}_i \cdot \boldsymbol{B})^2$ gives zero. Therefore,

$$\begin{aligned}(\boldsymbol{\varpi}_i \times \boldsymbol{B}) \cdot \nabla_{\boldsymbol{\varpi}_i} \boldsymbol{A}_i &= \boldsymbol{\varpi}_i \times \boldsymbol{B} \mathcal{A}_i^I + (\boldsymbol{\varpi}_i \times \boldsymbol{B}) \times \boldsymbol{B} \mathcal{A}_i^{II} \\ &= -\boldsymbol{\varpi}_i B^2 \mathcal{A}_i^{II} + \boldsymbol{\varpi}_i \times \boldsymbol{B} \mathcal{A}_i^I + \boldsymbol{B}(\boldsymbol{\varpi}_i \times \boldsymbol{B}) \mathcal{E}_i^{II} .\end{aligned} \tag{9.32}$$

Now consider the vector

$$\begin{aligned}&\boldsymbol{w}_i \cdot \sum_j e_j \int \mathrm{d}\boldsymbol{v}\, f_j (\boldsymbol{w}_j \times \boldsymbol{B}) \boldsymbol{A}_j \\ &= (\boldsymbol{B} \times \boldsymbol{w}_i) \cdot \sum_j e_j \int \mathrm{d}\boldsymbol{v}\, f_j \boldsymbol{w}_j \boldsymbol{A}_j \\ &= \frac{2k_B T}{\sqrt{m_i}} (\boldsymbol{B} \times \boldsymbol{\varpi}_i)_\gamma \sum_j \frac{n_j e_j}{\sqrt{m_j}} \frac{1}{\pi^{3/2}} \int \frac{\mathrm{d}\boldsymbol{\varpi}_j}{j} \mathrm{e}^{-\varpi_j^2} \varpi_{j\gamma} \boldsymbol{A}_j ,\end{aligned} \tag{9.33}$$

where

$$\int \mathrm{d}\boldsymbol{\varpi}_j \mathrm{e}^{-\varpi_j^2} \varpi_{j\ell} A_{jm} = \int \mathrm{d}\boldsymbol{\varpi}_j \mathrm{e}^{-\varpi_j^2} \varpi_{j\ell} \times [\varpi_{jm} \mathcal{A}_j^I + (\boldsymbol{\varpi}_j \times \boldsymbol{B})_m \mathcal{A}_j^{II} + B_m (\boldsymbol{\varpi}_i \cdot \boldsymbol{B}) \mathcal{A}_j^{III}] \tag{9.34}$$

and we use the subscripts ℓ, m instead of Greek subscripts because no summation is implied. If $\mathcal{A}_j^I$ were independent of $(\boldsymbol{\varpi}_i \cdot \boldsymbol{B})^2$, then it would be obvious that the first of these integrals would be zero unless $\ell = m$. It actually is easy to prove that this is still true even if $\mathcal{A}_j^I$ does depend on $(\boldsymbol{\varpi}_i \cdot \boldsymbol{B})^2$. Hence, equation (9.34) becomes

$$\int \mathrm{d}\boldsymbol{\varpi}_j \mathrm{e}^{-\varpi_j^2} \varpi_{j\ell}^2 \left[\delta_{\ell m}\mathcal{A}_j^I + \varepsilon_{m\ell\alpha} B_\alpha \mathcal{A}_j^{II} + B_m B_\ell \mathcal{A}_j^{III}\right],$$

where

$$\varepsilon_{m\ell\alpha} = \begin{cases} +1 \text{ if } m\ell\alpha \text{ are in cyclic order,} \\ -1 \text{ if } m\ell\alpha \text{ are not in cyclic order,} \\ \;\;0 \text{ if } m,\ \ell, \text{ and } \alpha \text{ are not all different.} \end{cases} \tag{9.35}$$

Therefore,

$$\begin{aligned} &\sum_\ell (\boldsymbol{B} \times \boldsymbol{\varpi}_i)_\ell \int \mathrm{d}\boldsymbol{\varpi}_j \mathrm{e}^{-\varpi_j^2} \varpi_{j\ell} A_{jm} \\ &\quad = (\boldsymbol{B} \times \boldsymbol{\varpi}_i)_m \int \mathrm{d}\boldsymbol{\varpi}_j \mathrm{e}^{-\varpi_j^2} \varpi_{jm} \mathcal{A}_j^I \\ &\qquad + \sum_\ell \varepsilon_{m\ell\alpha} (\boldsymbol{B} \times \boldsymbol{\varpi}_i)_\ell B_\alpha \int \mathrm{d}\boldsymbol{\varpi}_j \mathrm{e}^{-\varpi_j^2} \varpi_{j\ell}^2 \mathcal{A}_j^{II}. \end{aligned} \tag{9.36}$$

We note that both of these integrals involve the square of a velocity component that must be perpendicular to $\boldsymbol{B}$. By symmetry the integrals are therefore equal so that the right-hand side of equation (9.36) may be rewritten as

$$\begin{aligned} &(\boldsymbol{B} \times \boldsymbol{\varpi}_i)_m \tfrac{1}{2} \int \mathrm{d}\boldsymbol{\varpi}_j \mathrm{e}^{-\varpi_j^2} \left[\varpi_j^2 - \frac{(\boldsymbol{w}_j \cdot \boldsymbol{B})^2}{B^2}\right] \mathcal{A}_j^I \\ &+ [\varpi_{im} B^2 - B_m (\boldsymbol{w}_i \times \boldsymbol{B})] \tfrac{1}{2} \int \mathrm{d}\boldsymbol{\varpi}_j \mathrm{e}^{-\varpi_j^2} \left[\varpi_j^2 - \frac{(\boldsymbol{w}_j \cdot \boldsymbol{B})^2}{B^2}\right] \mathcal{A}_j^{II}, \end{aligned}$$

so that the right-hand side of equation (9.33) becomes

$$\begin{aligned} &\frac{k_B T}{\sqrt{m_i}} (\boldsymbol{B} \times \boldsymbol{\varpi}_i) \sum_j \frac{n_j e_j}{\sqrt{m_j}} \frac{1}{\pi^{3/2}} \int \mathrm{d}\boldsymbol{\varpi}_j \, \mathrm{e}^{-\varpi_j^2} \left[\varpi_j^2 - \frac{(\boldsymbol{w}_j \cdot \boldsymbol{B})^2}{B^2}\right] \mathcal{A}_j^I \\ &+ \frac{k_B T}{\sqrt{m_i}} [\boldsymbol{\varpi}_i B^2 - \boldsymbol{B}(\boldsymbol{w}_i \cdot \boldsymbol{B})] \sum_j \frac{n_j e_j}{\sqrt{m_j}} \frac{1}{\pi^{3/2}} \int \mathrm{d}\boldsymbol{\varpi}_j \, \mathrm{e}^{-\varpi_j^2} \left[\varpi_j^2 - \frac{(\boldsymbol{w}_j \cdot \boldsymbol{B})^2}{B^2}\right] \mathcal{A}_j^{II}. \end{aligned} \tag{9.37}$$

Substituting equations (9.30), (9.32), and (9.37) into equation (9.27), we obtain

$$f_i \left(\tfrac{5}{2} - \varpi_i^2\right) \boldsymbol{w}_i = \frac{e_i}{m_i} f_i \boldsymbol{\varpi}_i B^2 \mathcal{A}_j^{II} - \frac{1}{\rho} \sqrt{m_i} f_i \boldsymbol{\varpi}_i B^2 G^{II}(B^2)$$

$$
\begin{aligned}
&+\sum_j \int \mathrm{d}\boldsymbol{s}\,\sigma \mathrm{d}\Omega\, g f_i(\boldsymbol{v}) f_j(\boldsymbol{s}) \\
&\quad\times \left[(\boldsymbol{\varpi}_i \mathcal{A}_i^I)' + (\boldsymbol{\varpi}_j \mathcal{A}_j^I)' - (\boldsymbol{\varpi}_i \mathcal{A}_i^I) - (\boldsymbol{\varpi}_j \mathcal{A}_j^I)\right] \\
&-\frac{e_i}{m_i} f_i \boldsymbol{\varpi}_i \times \boldsymbol{B} \mathcal{A}_i^I + \frac{1}{\rho}\sqrt{m_i} f_i \boldsymbol{\varpi}_i \times \boldsymbol{B} G^I(B^2) \\
&+\sum_j \int \mathrm{d}\boldsymbol{s}\,\sigma \mathrm{d}\Omega\, g f_i(\boldsymbol{v}) f_j(\boldsymbol{s}) \\
&\quad\times \left[(\boldsymbol{\varpi}_i \mathcal{A}_i^{II})' + (\boldsymbol{\varpi}_j \mathcal{A}_j^{II})' - (\boldsymbol{\varpi}_i \mathcal{A}_i^{II}) - (\boldsymbol{\varpi}_j \mathcal{A}_j^{II})\right] \\
&-\frac{e_i}{m_i} f_i \boldsymbol{B}(\boldsymbol{\varpi}_i \cdot \boldsymbol{B}) \mathcal{A}_i^{II} + \frac{1}{\rho}\sqrt{m_i} f_i \boldsymbol{B}(\boldsymbol{\varpi}_i \cdot \boldsymbol{B}) G^{II}(B^2) \\
&+\boldsymbol{B}\boldsymbol{B} \cdot \sum_j \int \mathrm{d}\boldsymbol{s}\,\sigma \mathrm{d}\Omega\, g f_i(\boldsymbol{v}) f_j(\boldsymbol{s}) \\
&\quad\times \left[(\boldsymbol{\varpi}_i \mathcal{A}_i^{III})' + (\boldsymbol{\varpi}_j \mathcal{A}_j^{III})' - (\boldsymbol{\varpi}_i \mathcal{A}_i^{III}) - (\boldsymbol{\varpi}_j \mathcal{A}_j^{III})\right], \qquad (9.38)
\end{aligned}
$$

where $G^I(B^2)$ is shorthand for

$$
G^I(B^2) = \sum_j \frac{n_j e_j}{\sqrt{m_j}} \frac{1}{\pi^{3/2}} \int \mathrm{d}\boldsymbol{\varpi}_j \mathrm{e}^{-\varpi_j^2} \left[\varpi_j^2 - \frac{1}{B^2}(\boldsymbol{\varpi}_j \cdot \boldsymbol{B})^2\right] \mathcal{A}_j^I , \qquad (9.39)
$$

and $G^{II}(B^2)$ is defined similarly.

Now by examining the form of equation (9.38), it can be shown that it is plausible that the scalars $\mathcal{A}_i^I$, $\mathcal{A}_i^{II}$, and $\mathcal{A}_i^{III}$ do not in fact depend upon $(\boldsymbol{\varpi}_j \cdot \boldsymbol{B})^2$ but are functions only of ϖ_i^2 and B^2. It must be emphasized that this has not been proved. We now assume no dependence on $(\boldsymbol{\varpi}_j \cdot \boldsymbol{B})^2$ in these scalars and we shall then go on to prove that we can obtain a solution of the equation. Strictly speaking, because we do this, it is then necessary to prove a uniqueness theorem to show that this is the only solution to the equation, but we shall not do this.

With this assumption, it is clear that equation (9.38) breaks into three simultaneous vector equations, namely

$$
\begin{aligned}
f_i\left(\tfrac{5}{2} - \varpi_i^2\right) \boldsymbol{w}_i = {} & \frac{e_i}{m_i} f_i B^2 \boldsymbol{\varpi}_i \mathcal{A}_i^{II} - \frac{1}{\rho}\sqrt{m_i} f_i B^2 \boldsymbol{\varpi}_i G^{II}(B^2) \\
&+\sum_j \int \mathrm{d}\boldsymbol{s}\,\sigma \mathrm{d}\Omega\, g f_i(\boldsymbol{v}) f_j(\boldsymbol{s}) \\
&\quad\times \left[(\boldsymbol{\varpi}_i \mathcal{A}_i^I)' + (\boldsymbol{\varpi}_j \mathcal{A}_j^I)' - (\boldsymbol{\varpi}_i \mathcal{A}_i^I) - (\boldsymbol{\varpi}_j \mathcal{A}_j^I)\right] \qquad (9.40) \\
0 = {} & -\frac{e_i}{m_i} f_i \boldsymbol{\varpi}_i \mathcal{A}_i^I + \frac{\sqrt{m_i}}{\rho} f_i \boldsymbol{\varpi}_i G^I(B^2) \\
&+\sum_j \int \mathrm{d}\boldsymbol{s}\,\sigma \mathrm{d}\Omega\, g f_i(\boldsymbol{v}) f_j(\boldsymbol{s})
\end{aligned}
$$

$$\times\left[(\varpi_i\mathcal{A}_i^{II})'+(\varpi_j\mathcal{A}_j^{II})'-(\varpi_i\mathcal{A}_i^{II})-(\varpi_j\mathcal{A}_j^{II})\right] \quad (9.41)$$

$$0=-\frac{e_i}{m_i}f_i\varpi_i\mathcal{A}_i^{II}+\frac{\sqrt{m_i}}{\rho}f_i\varpi_iG^{II}(B^2)$$
$$+\sum_j\int \mathrm{d}s\,\sigma\mathrm{d}\Omega\,gf_i(\boldsymbol{v})f_j(\boldsymbol{s})$$
$$\times\left[(\varpi_i\mathcal{A}_i^{III})'+(\varpi_j\mathcal{A}_j^{III})'-(\varpi_i\mathcal{A}_i^{III})-(\varpi_j\mathcal{A}_j^{III})\right] \quad (9.42)$$

and equation (9.39) becomes

$$G^I(B^2)=\frac{2}{3}\sum_j\frac{n_je_j}{\sqrt{m_j}}\frac{1}{\pi^{3/2}}\int \mathrm{d}\varpi_j\mathrm{e}^{-\varpi_j^2}\varpi_j^2\mathcal{A}_j^I\,. \quad (9.43)$$

Multiplying equation (9.42) by B^2 and adding to equation (9.40) gives the simple equation,

$$f_i\left(\tfrac{5}{2}-\varpi_i^2\right)\boldsymbol{w}_i=\sum_j\int \mathrm{d}s\,\sigma\mathrm{d}\Omega\,gf_i(\boldsymbol{v})f_j(\boldsymbol{s})$$
$$\times\left[\left\{\varpi_i\left(\mathcal{A}_i^I+B^2\mathcal{A}_i^{III}\right)\right\}'+\left\{\varpi_j\left(\mathcal{A}_j^I+B^2\mathcal{A}_j^{III}\right)\right\}'\right.$$
$$\left.-\left\{\varpi_i\left(\mathcal{A}_i^I+B^2\mathcal{A}_i^{III}\right)\right\}-\left\{\varpi_j\left(\mathcal{A}_j^I+B^2\mathcal{A}_j^{III}\right)\right\}\right], \quad (9.44)$$

which is of the same form as that of the integral equations we solved in Chapter 8.

Multiplying equation (9.41) by iB and adding to equation (9.40) gives a single complex vector equation,

$$f_i\left(\tfrac{5}{2}-\varpi_i^2\right)\boldsymbol{w}_i=-\mathrm{i}B\frac{e_i}{m_i}f_i\varpi_i\mathcal{A}_i+\frac{\mathrm{i}B}{\rho}\sqrt{m_i}f_i\varpi_iG(B^2)$$
$$+\sum_j\int \mathrm{d}s\,\sigma\mathrm{d}\Omega\,gf_i(\boldsymbol{v})f_j(\boldsymbol{s})$$
$$\times\left[(\varpi_i\mathcal{A}_i)'+(\varpi_j\mathcal{A}_j)'-(\varpi_i\mathcal{A}_i)-(\varpi_j\mathcal{A}_j)\right], \quad (9.45)$$

where

$$\mathcal{A}_i=\mathcal{A}_i^I+\mathrm{i}B\mathcal{A}_i^{II}\,,\quad\text{and}\quad G=G^I+\mathrm{i}BG^{II}\,. \quad (9.46)$$

Equation (9.45) is of a different type from those we have met previously due to the presence of the first two terms on the right-hand side. We shall discuss how to solve this equation later.

In just the same way that we have discussed equation (9.27), we can discuss equation (9.28) and derive equations for the scalars $\mathcal{E}_i^I$, $\mathcal{E}_i^{II}$, and $\mathcal{E}_i^{III}$ introduced in equation (9.31). Clearly in place of equation (9.44), we get

$$(-1)^i\frac{1}{n_i}f_i\boldsymbol{w}_i=\sum_j\int \mathrm{d}s\,\sigma\mathrm{d}\Omega\,gf_i(\boldsymbol{v})f_j(\boldsymbol{s})$$

$$\times\Big[\{\boldsymbol{\varpi}_i\left(\mathcal{E}_i^I+B^2\mathcal{E}_i^{III}\right)\}'+\{\boldsymbol{\varpi}_j\left(\mathcal{E}_j^I+B^2\mathcal{E}_j^{III}\right)\}'$$
$$-\{\boldsymbol{\varpi}_i\left(\mathcal{E}_i^I+B^2\mathcal{E}_i^{III}\right)\}-\{\boldsymbol{\varpi}_j\left(\mathcal{E}_j^I+B^2\mathcal{E}_j^{III}\right)\}'\Big], \quad (9.47)$$

and in place of equation (9.45), we get

$$(-1)^i\frac{1}{n_i}f_i\boldsymbol{w}_i=-\mathrm{i}B\frac{e_i}{m_i}f_i\boldsymbol{\varpi}_i\mathcal{E}_i+\frac{\mathrm{i}B}{\rho}\sqrt{m_i}f_i\boldsymbol{\varpi}_iK(B^2)$$
$$+\sum_j\int \mathrm{d}\boldsymbol{s}\,\sigma\mathrm{d}\Omega\,gf_i(\boldsymbol{v})f_j(\boldsymbol{s})$$
$$\times\left[(\boldsymbol{\varpi}_i\mathcal{E}_i)'+(\boldsymbol{\varpi}_j\mathcal{E}_j)'-(\boldsymbol{\varpi}_i\mathcal{E}_i)-(\boldsymbol{\varpi}_j\mathcal{E}_j)\right], \quad (9.48)$$

where

$$\mathcal{E}_i=\mathcal{E}_i^I+\mathrm{i}B\mathcal{E}_i^{II}, \quad (9.49)$$

and $K(B^2)$ is a quantity similar to $G(B^2)$, defined by

$$K(B^2)\equiv K^I+\mathrm{i}BK^{II}, \quad (9.50)$$

where

$$K^I=\frac{2}{3}\sum_j\frac{n_je_j}{\sqrt{m_j}}\frac{1}{\pi^{3/2}}\int \mathrm{d}\boldsymbol{\varpi}_j\,\mathrm{e}^{-\varpi_j^2}\varpi_j^2\mathcal{E}_j^I\,, \quad (9.51)$$

and K^{II} is defined similarly.

Besides obeying these equations, $\boldsymbol{A}_i$ and $\boldsymbol{E}_i$ must satisfy subsidiary conditions obtained by requiring that n_i, T, and $\boldsymbol{u}$ really be the number density, temperature, and drift velocity, respectively. The first condition is

$$n_i=\int \mathrm{d}\boldsymbol{v}\,F_i=\int \mathrm{d}\boldsymbol{v}\,f_i(1+\varphi_i)\,.$$

Substituting for φ_i from equation (9.25) and for $\boldsymbol{A}_i$ and $\boldsymbol{E}_i$ from equations (9.30) and (9.31), we find that this condition is automatically satisfied because $\boldsymbol{A}_i$ and $\boldsymbol{E}_i$ are odd in the random velocity. Similarly, we find that the second condition,

$$\tfrac{3}{2}nk_BT=\sum_i\int \mathrm{d}\boldsymbol{v}\,F_i\tfrac{1}{2}m_iw_i^2\,,$$

is also automatically satisfied. The third condition is

$$\boldsymbol{u}=\frac{1}{\rho}\sum_i\int \mathrm{d}\boldsymbol{v}\,F_im_i\boldsymbol{v}_i\,,$$

or that

$$\sum_i\int \mathrm{d}\boldsymbol{v}\,f_i\varphi_im_i\boldsymbol{w}_i=0\,.$$

On substituting for φ_i and then for $\boldsymbol{A}_i$ and $\boldsymbol{E}_i$ this gives

$$0 = \nabla \ln T \sum_i \int \mathrm{d}\boldsymbol{v} f_i m_i \boldsymbol{w}_i \cdot \boldsymbol{\varpi}_i \mathcal{A}_i^I + \boldsymbol{B} \times \nabla \ln T \sum_i \int \mathrm{d}\boldsymbol{v} f_i m_i \boldsymbol{w}_i \cdot \boldsymbol{\varpi}_i \mathcal{A}_i^{II}$$
$$+\boldsymbol{B}(\boldsymbol{B} \cdot \nabla \ln T) \sum_i \int \mathrm{d}\boldsymbol{v} f_i m_i \boldsymbol{w}_i \cdot \boldsymbol{\varpi}_i \mathcal{A}_i^{III} \qquad (9.52)$$

and a similar equation with $\boldsymbol{d}_1$ in place of $\nabla \ln T$ and $\mathcal{E}$ in place of $\mathcal{A}$. Because we are supposing the scalars $\mathcal{A}_i^I$, etc. to be independent of $(\boldsymbol{\varpi}_i \cdot \boldsymbol{B})^2$, it follows that each of these three terms must be zero independently. Hence, we obtain, in terms of the unknowns appearing in Equations (9.44) and (9.45),

$$\sum_i n_i \sqrt{m_i} \int \mathrm{d}\boldsymbol{\varpi}_i \, \mathrm{e}^{-\varpi_i^2} \varpi_j^2 (\mathcal{A}_i^I + B^2 \mathcal{A}_i^{III}) = 0 \qquad (9.53)$$

$$\sum_i n_i \sqrt{m_i} \int \mathrm{d}\boldsymbol{\varpi}_i \, \mathrm{e}^{-\varpi_i^2} \varpi_j^2 \mathcal{A}_i = 0 \, .$$

Similarly,

$$\sum_i n_i \sqrt{m_i} \int \mathrm{d}\boldsymbol{\varpi}_i \, \mathrm{e}^{-\varpi_i^2} \varpi_j^2 (\mathcal{E}_i^I + B^2 \mathcal{E}_i^{III}) = 0$$

$$\sum_i n_i \sqrt{m_i} \int \mathrm{d}\boldsymbol{\varpi}_i \, \mathrm{e}^{-\varpi_i^2} \varpi_j^2 \mathcal{E}_i = 0 \, . \qquad (9.54)$$

It is convenient to expand $\mathcal{A}_i^I$ etc. as a power series of generalized Laguerre polynomials, as

$$\mathcal{A}_i^I(\varpi_i^2, B^2) = \sum_{m=0}^{\infty} a_i^{I,m} L_m^{\frac{3}{2}}(\varpi_i^2) \, ,$$

with exactly similar expansions for $\mathcal{A}_i^{II}$, $\mathcal{A}_i^{III}$, $\mathcal{E}_i^I$, $\mathcal{E}_i^{II}$, and $\mathcal{E}_i^{III}$. Then defining

$$a_i^m \equiv a_i^{I,m} + \mathrm{i} B a_i^{II,m} \quad \text{and} \quad e_i^m \equiv e_i^{I,m} + \mathrm{i} B e_i^{II,m} \, ,$$

we have by Equations (9.46) and (9.49),

$$\mathcal{A}_i = \sum_m a_i^m L_m^{\frac{3}{2}}(\varpi_i^2) \quad \text{and} \quad \mathcal{E}_i = \sum_m e_i^m L_m^{\frac{3}{2}}(\varpi_i^2) \, . \qquad (9.55)$$

Of course, in all these expressions, the coefficients a_i^I, e_i^I, etc. are functions only of B^2. A solution of the equations is equivalent to a complete knowledge of these coefficients.

In terms of these coefficients, the subsidiary conditions of Equations (9.53) to (9.54) become

$$\sum_i n_i \sqrt{m_i} \left(a_i^{I,0} + B^2 a_i^{III,0} \right) = 0$$

$$\sum_i n_i \sqrt{m_i} a_i^0 = 0$$

$$\sum_i n_i \sqrt{m_i} \left(e_i^{I,0} + B^2 e_i^{III,0} \right) = 0$$

$$\sum_i n_i \sqrt{m_i} e_i^0 = 0 . \tag{9.56}$$

In deriving these results we have used the useful orthogonality property of the generalized Laguerre polynomials of equation (7.92). The quantity G defined by equations (9.46) and (9.43) is

$$G \equiv \sum_j \frac{n_j e_j}{\sqrt{m_j}} a_j^0 ,$$

and K, defined by equation (9.50), is

$$K \equiv \sum_j \frac{n_j e_j}{\sqrt{m_j}} e_j^0 . \tag{9.57}$$

Before proceeding any further with the analysis, we shall examine what the heat flux and electric current are in terms of these coefficients. As in Chapter 8, we shall find that these quantities involve only a few coefficients and it is therefore only these few coefficients which we want to find.

The electric current is defined by equation (9.12)

$$\boldsymbol{j} \equiv \sum_i n_i e_i \boldsymbol{w}_i = \sum_i e_i \int \mathrm{d}\boldsymbol{v}\, f i \varphi_i \boldsymbol{w}_i .$$

Substituting for φ_i from equation (9.25) and for $\boldsymbol{A}_i$ and $\boldsymbol{E}_i$ from equations (9.30) and (9.31) gives

$$\begin{aligned} \boldsymbol{j} = &-(\nabla \ln T)\tfrac{1}{2} \sum_i n_i e_i v_{ti} a_i^{I,0} - (\boldsymbol{B} \times \nabla \ln T)\tfrac{1}{2} \sum_i n_i e_i v_{ti} a_i^{II,0} \\ &-\boldsymbol{B}(\boldsymbol{B} \cdot \nabla \ln T)\tfrac{1}{2} \sum_i n_i e_i v_{ti} a_i^{III,0} \\ &-(n\boldsymbol{d}_1)\tfrac{1}{2} \sum_i n_i e_i v_{ti} e_i^{I,0} - (\boldsymbol{B} \times n\boldsymbol{d}_1)\tfrac{1}{2} \sum_i n_i e_i v_{ti} e_i^{II,0} \\ &-\boldsymbol{B}(\boldsymbol{B} \cdot n\boldsymbol{d}_1)\tfrac{1}{2} \sum_i n_i e_i v_{ti} e_i^{III,0} . \end{aligned} \tag{9.58}$$

If we now resolve the vectors ∇T, $\boldsymbol{d}_1$, etc. into components parallel and perpendicular to the magnetic field, denoting these superscripts $\|$ and $\perp$ respectively, we can write this equation in the form

$$\boldsymbol{j} = \sigma^I \boldsymbol{D}^{\|} + \sigma^{II} \boldsymbol{D}^{\perp} + \sigma^{III} \hat{\boldsymbol{b}} \times \boldsymbol{D}^{\perp} + \varphi^I (\nabla T)^{\|} + \varphi^{II} (\nabla T)^{\perp} + \varphi^{III} \hat{\boldsymbol{b}} \times (\nabla T)^{\perp}, \tag{9.59}$$

where $\boldsymbol{D}$ is a "generalized" electric field given by

$$\boldsymbol{D} = \boldsymbol{E} + \boldsymbol{u} \times \boldsymbol{B} - \frac{m_2 - m_1}{n(e_1 m_2 - e_2 m_1)} \nabla p - \frac{m m_1 m_2}{e(m_1 + m_2)} (\boldsymbol{X}_1 - \boldsymbol{X}_2) - \frac{p\rho}{n_1 n_2 (e_1 m_2 - e_2 m_1)} \nabla \left(\frac{n_1}{n} \right), \tag{9.60}$$

and $\hat{\boldsymbol{b}}$ is a unit vector in the direction of $\boldsymbol{B}$ so

$$\boldsymbol{B} = B\hat{\boldsymbol{b}},$$

$$\sigma^I = \frac{n n_1^2 n_2 v_{t1}}{2p\rho m_2} (e_1 m_2 - e_2 m_1)^2 \left(e_1^{I,0} + B^2 e_1^{III,0} \right) \tag{9.61}$$

$$\sigma^{II} + \mathrm{i}\sigma^{III} = \frac{n n_1^2 n_2 v_{t1}}{2p\rho m_2} (e_1 m_2 - e_2 m_1)^2 e_1^0 \tag{9.62}$$

$$\varphi^I = -\frac{n_1 v_{t1}}{2 m_2 T} (e_1 m_2 - e_2 m_1) \left(a_1^{I,0} + B^2 a_1^{III,0} \right) \tag{9.63}$$

$$\varphi^{II} + \mathrm{i}\varphi^{III} = -\frac{n_1 v_{t1}}{2 m_2 T} (e_1 m_2 - e_2 m_1) a_1^0 . \tag{9.64}$$

Clearly σ^I, σ^{II}, and σ^{III} are coefficients of electrical conductivity and φ^I, φ^{II}, and φ^{III} are coefficients of thermal diffusion.

The heat flux vector is defined by equation (9.9) and is

$$\boldsymbol{q} \equiv \sum_i n_i \tfrac{1}{2} m_i \langle w_i^2 \boldsymbol{w}_i \rangle = \sum_i \tfrac{1}{2} m_i \int \mathrm{d}\boldsymbol{v}\, w_i^2 \boldsymbol{w}_i f_i \varphi_i .$$

Substituting for φ_i from equation (9.25) and for $\boldsymbol{A}_i$ and $\boldsymbol{E}_i$ from equations (9.30) and (9.31) gives

$$\begin{aligned}
\boldsymbol{q} = &-\tfrac{5}{4} k_B T \sum_i n_i v_{ti} \left(a_i^{I,0} - a_i^{I,1} \right) \nabla \ln T \\
&-\tfrac{5}{4} k_B T \sum_i n_i v_{ti} \left(a_i^{II,0} - a_i^{II,1} \right) \boldsymbol{B} \times \nabla \ln T \\
&-\tfrac{5}{4} k_B T \sum_i n_i v_{ti} \left(a_i^{III,0} - a_i^{III,1} \right) \boldsymbol{B} (\boldsymbol{B} \cdot \nabla \ln T) \\
&-\tfrac{5}{4} k_B T \sum_i n_i v_{ti} \left(e_i^{I,0} - e_i^{I,1} \right) \boldsymbol{d}_1 \\
&-\tfrac{5}{4} k_B T \sum_i n_i v_{ti} \left(e_i^{II,0} - e_i^{II,1} \right) \boldsymbol{B} \times \boldsymbol{d}_1 \\
&-\tfrac{5}{4} k_B T \sum_i n_i v_{ti} \left(e_i^{III,0} - e_i^{III,1} \right) \boldsymbol{B} (\boldsymbol{B} \cdot \boldsymbol{d}_1) ,
\end{aligned}$$

which can be rewritten as

$$\boldsymbol{q} = -\lambda'^{I}(\nabla T)^{\parallel} - \lambda'^{II}(\nabla T)^{\perp} - \lambda'^{III}\hat{\boldsymbol{b}} \times (\nabla T)^{\perp} + \frac{5}{2}k_B T \frac{m_2 - m_1}{e_1 m_2 - e_2 m_1}\boldsymbol{j}$$
$$-\frac{5n_1 n_2}{4\rho}(e_1 m_2 - e_2 m_1)\left[\boldsymbol{D}^{\parallel}\sum_i n_i v_{ti}\left(e_i^{I,1} + B^2 e_i^{III,1}\right)\right.$$
$$\left. + \boldsymbol{D}^{\perp}\sum_i n_i v_{ti} e_i^{I,1} + \boldsymbol{B} \times \boldsymbol{D}^{\perp}\sum_i n_i v_{ti} e_i^{II,1}\right], \qquad (9.65)$$

where

$$\lambda'^{I} = -\tfrac{5}{4}k_B \sum_i n_i v_{ti}\left(a_i^{I,1} + B^2 a_i^{III,1}\right) \qquad (9.66)$$
$$\lambda'^{II} + \mathrm{i}\lambda'^{III} = -\tfrac{5}{4}k_B \sum_i n_i v_{ti} a_i^1 . \qquad (9.67)$$

Clearly the first three terms of equation (9.65) represent "true thermal conduction." The next term is just another form for the expression,

$$\tfrac{5}{2}k_B T(n_1\langle \boldsymbol{w}_1\rangle + n_2\langle \boldsymbol{w}_2\rangle),$$

which turned up in Chapter 8. The last term is closely connected with the thermal diffusion terms in equation (9.59) and we shall now derive an alternative form for it.

By taking the scalar product of equation (9.48) with $\boldsymbol{\varpi}_i\mathcal{A}_i$, integrating and summing over i, it is possible to show that

$$\sum_i (-1)^i \frac{1}{n_i}\int \mathrm{d}\boldsymbol{v}\, f_i \boldsymbol{\varpi}\cdot\boldsymbol{w}_i\mathcal{A}_i = \sum_i \int \mathrm{d}\boldsymbol{v}\, f_i\left(\tfrac{5}{2} - \varpi_i^2\right)\boldsymbol{\varpi}\cdot\boldsymbol{w}_i\mathcal{E}_i ,$$

or

$$-\frac{v_{t1}\rho}{n_2 m_2}a_1^0 = \tfrac{5}{2}\sum_i n_i v_{ti} e_i^1 . \qquad (9.68)$$

Similarly, from Equations (9.44) and (9.47) we find

$$-\frac{v_{t1}\rho}{n_2 m_2}\left(a_1^{I,0} + B^2 a_1^{III,0}\right) = \tfrac{5}{2}\sum_i n_i v_{ti}\left(e_i^{I,0} + B^2 e_i^{III,0}\right),$$

so equation (9.65) becomes

$$\boldsymbol{q} = -\lambda'^{I}(\nabla T)^{\parallel} - \lambda'^{II}(\nabla T)^{\perp} - \lambda'^{III}\hat{\boldsymbol{b}} \times (\nabla T)^{\perp} + \frac{5}{2}k_B T \frac{m_2 - m_1}{e_1 m_2 - e_2 m_1}\boldsymbol{j}$$
$$-T\left[\varphi^I \boldsymbol{D}^{\parallel} + \varphi^{II}\boldsymbol{D}^{\perp} + \varphi^{III}\hat{\boldsymbol{b}} \times \boldsymbol{D}^{\perp}\right]. \qquad (9.69)$$

Now we can rewrite equation (9.65) as

$$\boldsymbol{D}^{\parallel} = \frac{\boldsymbol{j}^{\parallel} - \varphi^I(\nabla T)^{\parallel}}{\sigma^I}, \qquad (9.70)$$

where

$$\boldsymbol{D}^{\perp} = \frac{1}{(\sigma^{II})^2 + (\sigma^{III})^2}\Big[\sigma^{II}\boldsymbol{j}^{\perp} - \sigma^{III}\hat{\boldsymbol{b}}\times\boldsymbol{j}^{\perp} \tag{9.71}$$

$$- \left(\sigma^{II}\varphi^{II} + \sigma^{III}\varphi^{III}\right)(\nabla T)^{\perp} - \left(\sigma^{II}\varphi^{III} - \sigma^{III}\varphi^{II}\right)\hat{\boldsymbol{b}}\times(\nabla T)^{\perp}\Big]$$

$$\hat{\boldsymbol{b}}\times\boldsymbol{D}^{\perp} = \frac{1}{(\sigma^{II})^2 + (\sigma^{III})^2}\Big[\sigma^{III}\boldsymbol{j}^{\perp} + \sigma^{II}\hat{\boldsymbol{b}}\times\boldsymbol{j}^{\perp} \tag{9.72}$$

$$- \left(\sigma^{III}\varphi^{II} - \sigma^{II}\varphi^{III}\right)(\nabla T)^{\perp} - \left(\sigma^{III}\varphi^{III} + \sigma^{II}\varphi^{II}\right)\hat{\boldsymbol{b}}\times(\nabla T)^{\perp}\Big].$$

Substituting into equation (9.69) now gives

$$\boldsymbol{q} = -\lambda^{I}(\nabla T)^{\parallel} - \lambda^{II}(\nabla T)^{\perp} - \lambda^{III}\hat{\boldsymbol{b}}\times(\nabla T)^{\perp} + \psi^{I}\boldsymbol{j}^{\parallel} + \psi^{II}\boldsymbol{j}^{\perp} + \psi^{III}\hat{\boldsymbol{b}}\times\boldsymbol{j}^{\perp} \tag{9.73}$$

where

$$\lambda^{I} = \lambda'^{I} - T\frac{(\varphi^{I})^2}{\sigma^{I}} \tag{9.74}$$

$$\lambda^{II} = \lambda'^{II} - T\frac{\left[\sigma^{II}(\varphi^{II})^2 - \sigma^{II}(\varphi^{III})^2 + 2\sigma^{III}\varphi^{II}\varphi^{III}\right]}{(\sigma^{II})^2 + (\sigma^{III})^2} \tag{9.75}$$

$$\lambda^{III} = \lambda'^{III} - T\frac{\left[\sigma^{III}(\varphi^{III})^2 - \sigma^{III}(\varphi^{II})^2 + 2\sigma^{II}\varphi^{II}\varphi^{III}\right]}{(\sigma^{II})^2 + (\sigma^{III})^2} \tag{9.76}$$

$$\psi^{I} = \frac{5}{2}k_B T\frac{m_2 - m_1}{e_1 m_2 - e_2 m_1} - \frac{T\varphi^{I}}{\sigma^{I}} \tag{9.77}$$

$$\psi^{II} = \frac{5}{2}k_B T\frac{m_2 - m_1}{e_1 m_2 - e_2 m_1} - \frac{T(\sigma^{II}\varphi^{II} + \sigma^{III}\varphi^{III})}{(\sigma^{II})^2 + (\sigma^{III})^2} \tag{9.78}$$

$$\psi^{III} = -\frac{T\left(\sigma^{II}\varphi^{III} - \sigma^{III}\varphi^{II}\right)}{(\sigma^{II})^2 + (\sigma^{III})^2}. \tag{9.79}$$

Clearly λ^{I}, λ^{II}, and λ^{III} are the coefficients of thermal conductivity as usually defined, i.e., with respect to the heat flux when $\boldsymbol{j}$ is zero. The terms involving the current in equation (9.73) are those representing the heat flow which accompanies any electric current.

Alternatively we can write equation (9.69) as

$$\boldsymbol{q} = -\theta^{I}(\nabla T)^{\parallel} - \theta^{II}(\nabla T)^{\perp} - \theta^{III}\hat{\boldsymbol{b}}\times(\nabla T)^{\perp} + \xi^{I}\boldsymbol{D}^{\parallel} + \xi^{II}\boldsymbol{D}^{\perp} + \xi^{III}\hat{\boldsymbol{b}}\times\boldsymbol{D}^{\perp} \tag{9.80}$$

where

$$\begin{aligned} \theta^{I} &= \lambda'^{I} - \frac{5}{2}k_B T\frac{m_2 - m_1}{e_1 m_2 - e_2 m_1}\varphi^{I} \\ \theta^{II} &= \lambda'^{II} - \frac{5}{2}k_B T\frac{m_2 - m_1}{e_1 m_2 - e_2 m_1}\varphi^{II} \\ \theta^{III} &= \lambda'^{III} - \frac{5}{2}k_B T\frac{m_2 - m_1}{e_1 m_2 - e_2 m_1}\varphi^{III} \end{aligned} \tag{9.81}$$

$$\begin{aligned}\xi^{I} &= \frac{5}{2}k_BT\frac{m_2-m_1}{e_1m_2-e_2m_1}\sigma^{I} - T\varphi^{I}\\ \xi^{II} &= \frac{5}{2}k_BT\frac{m_2-m_1}{e_1m_2-e_2m_1}\sigma^{II} - T\varphi^{II}\\ \xi^{III} &= \frac{5}{2}k_BT\frac{m_2-m_1}{e_1m_2-e_2m_1}\sigma^{III} - T\varphi^{III}\,.\end{aligned} \tag{9.82}$$

9.4 Solutions for the electrical conductivity

In this section we shall examine the solutions of equations (9.47) and (9.48) to obtain values for the various coefficients we want.

We first of all note that equation (9.47), as an equation for $\mathcal{E}_i^I + B^2\mathcal{E}_i^{III}$, is of precisely the same form as the equations we considered in Chapter 8. We can therefore take over the results we derived in Chapter 8. Alternatively we can note, by comparing equations (9.47) and (9.48), that

$$\mathcal{E}_i^I + B^2\mathcal{E}_i^{III} = \lim_{B\to 0}\mathcal{E}_i\,. \tag{9.83}$$

Similarly,

$$\mathcal{A}_i^I + B^2\mathcal{A}_i^{III} = \lim_{B\to 0}\mathcal{A}_i\,. \tag{9.84}$$

Clearly the particular combinations of coefficients on the left-hand side of these equations do not involve the magnetic field although as written they appear to do so. We see from equations (9.59) and (9.69) that it is this particular combination of coefficients which occurs in the expressions for electrical and thermal conduction along the magnetic field. We are able to say immediately therefore that transport properties along the magnetic field are the same as if the field were absent.

Because of equation (9.83), we only have to consider the equations (9.48), which will give us values for the electrical conductivity. First we shall consider an important property of the equation and then we shall describe a variation procedure which gives successive approximations to the coefficients we want. Substitute the expansion of equation (9.55) into equation (9.48) and obtain an infinite set of simultaneous equations for the coefficients e_i^m by taking the scalar product with $\boldsymbol{\varpi}_i L_r(\varpi_i^2)$ and integrating over velocities. The set of equations is, if $r \neq 0$,

$$\begin{aligned}0 = -\mathrm{i}B\frac{n_ie_i}{m_i}\frac{2}{\sqrt{\pi}}\frac{\Gamma(r+\frac{5}{2})}{r!}e_i^r + \sum_j\int \mathrm{d}\boldsymbol{v}\,\mathrm{d}\boldsymbol{s}\,\sigma\mathrm{d}\Omega\, g f_i(\boldsymbol{v})f_j(\boldsymbol{s})\boldsymbol{\varpi}_i L_r(\varpi_i^2)\\ \cdot\left[(\boldsymbol{\varpi}_i\mathcal{E}_i)' + (\boldsymbol{\varpi}_j\mathcal{E}_j)' - \boldsymbol{\varpi}_i\mathcal{E}_i - \boldsymbol{\varpi}_j\mathcal{E}_j\right],\end{aligned} \tag{9.85}$$

and if $r = 0$,

$$(-1)^i \frac{3v_{ti}}{2} = -\frac{3\mathrm{i}Bn_ie_i}{2m_i}e_i^0 + \frac{3\mathrm{i}Bn_i\sqrt{m_i}}{2\rho}K(B^2) + \sum_j \int \mathrm{d}\boldsymbol{v}\,\mathrm{d}\boldsymbol{s}\,\sigma\mathrm{d}\Omega\, g f_i(\boldsymbol{v}) f_j(\boldsymbol{s})\boldsymbol{\varpi}_i \cdot [(\boldsymbol{\varpi}_i\mathcal{E}_i)' + (\boldsymbol{\varpi}_j\mathcal{E}_j)' - \boldsymbol{\varpi}_i\mathcal{E}_i - \boldsymbol{\varpi}_j\mathcal{E}_j]\,. \tag{9.86}$$

The infinite set of equations (9.85) we can regard as giving the infinite set of coefficients, e_i^1, $e_i^2, \dots$ in terms of e_1^0 and e_2^0 and then it appears at first sight that the two equations of (9.86) (with $i = 1, 2$) determines e_1^0 and e_2^0. If this were the case we would have no freedom left to apply the subsidiary condition of equation (9.56). But we shall now show that the two equations are identical so that the solution is not fixed completely by equation (9.48).

Multiply equation (9.86) by $\sqrt{m_i}$ and sum over i to obtain

$$0 = -\frac{3\mathrm{i}B}{2}\sum_i \frac{n_ie_i}{\sqrt{m_i}}e_i^0 + \frac{3\mathrm{i}B}{2}K(B^2) + \sum_{i,j}\int \mathrm{d}\boldsymbol{v}\,\mathrm{d}\boldsymbol{s}\,\sigma\mathrm{d}\Omega\, g f_i(\boldsymbol{v}) f_j(\boldsymbol{s})\sqrt{m_i}\boldsymbol{\varpi}_i \cdot [(\boldsymbol{\varpi}_i\mathcal{E}_i)' + (\boldsymbol{\varpi}_j\mathcal{E}_j)' - \boldsymbol{\varpi}_i\mathcal{E}_i - \boldsymbol{\varpi}_j\mathcal{E}_j]\,. \tag{9.87}$$

Recalling that $\sqrt{m_i}\boldsymbol{\varpi}_i$ is $\sqrt{m_i}\boldsymbol{w}_i/v_{ti}$, some manipulation of the last term of equation (9.87) shows that it vanishes identically whatever $\mathcal{E}_i$ may be because it is proportional to the total rate of change of momentum due to collisions. Also from the definition of $K(B^2)$ given by equation (9.57) we see that the first two terms cancel exactly so equation (9.87) is satisfied identically. Hence the two equations of equation (9.86) are identical and equations (9.85) and (9.86) do not fix all the coefficients e_i^m, so one coefficient is still arbitrary. This last coefficient is fixed by the subsidiary equation (9.56). Notice that this property of the equation is connected with the very special relation between the first and second term on the right-hand side of equation (9.48) and would not hold if, for example, the second term were multiplied by some factor which is not unity. In other words, we only have the freedom to satisfy the subsidiary condition if the second term is precisely as written in equation (9.48). The importance of this is that when we describe the variation method which solves this equation, we will find that the second term drops out of the analysis completely so, looking at the variation method alone, it appears that the second term could be multiplied by any factor, including zero, and yet the method would give the same solution. This, however, is only apparent; the foregoing analysis shows it is only consistent to apply a subsidiary condition if the second term is precisely as written.

9.4.1 Solving for the e_i^m

Suppose $\mathcal{J}_i$ is a trial function for the unknown $\mathcal{E}_i$ of equation (9.48). Then we construct the Davison function, defined by

$$\mathcal{D}(\mathcal{J}) \equiv \sum_i \int \mathrm{d}\boldsymbol{v} \mathcal{J}_i \boldsymbol{\varpi}_i \left\{ -2(-1)^i \frac{1}{n_i} f_i \boldsymbol{w}_i - \mathrm{i}B \frac{e_i}{m_i} f_i \boldsymbol{\varpi}_i \mathcal{J}_i \right.$$
$$+ \frac{\mathrm{i}B\sqrt{m_i}}{\rho} f_i \boldsymbol{\varpi}_i \frac{2}{3} \left(\sum_j \frac{n_j e_j}{\sqrt{m_j}} \frac{1}{\pi^{3/2}} \int \mathrm{d}\boldsymbol{\varpi}_j \, \mathrm{e}^{-\varpi_j^2} \varpi_j^2 \mathcal{J}_j \right) \tag{9.88}$$
$$\left. + \sum_j \int \mathrm{d}\boldsymbol{s} \, \sigma \mathrm{d}\Omega \, g f_i(\boldsymbol{v}) f_j(\boldsymbol{s}) [(\boldsymbol{\varpi}_i \mathcal{J}_i)' + (\boldsymbol{\varpi}_j \mathcal{J}_j)' - \boldsymbol{\varpi}_i \mathcal{J}_i - \boldsymbol{\varpi}_j \mathcal{J}_j] \right\}.$$

We also ensure that the trial function satisfies the subsidiary condition of equation (9.54), namely,

$$\sum_i n_i \sqrt{m_i} \int \mathrm{d}\boldsymbol{\varpi}_i \, \mathrm{e}^{-\varpi_i^2} \varpi_i^2 \mathcal{J}_i = 0 \,, \tag{9.89}$$

and we do not permit variations of $\mathcal{J}_i$ which violate this condition. But now we notice that if all our trial functions satisfy equation (9.89), then the third term in equation (9.88) disappears, so then

$$\mathcal{D}(\mathcal{J}) = -2 \sum_i \frac{(-1)^i}{n_i} \int \mathrm{d}\boldsymbol{v} \, f_i \boldsymbol{w}_i \cdot \boldsymbol{\varpi}_i \mathcal{J}_i - \mathrm{i}B \sum_i \frac{e_i}{m_i} \int \mathrm{d}\boldsymbol{v} \, f_i \varpi_i^2 \mathcal{J}_i^2$$
$$+ \sum_{i,j} \int \mathrm{d}\boldsymbol{v} \, \mathrm{d}\boldsymbol{s} \, \sigma \mathrm{d}\Omega \, g f_i(\boldsymbol{v}) f_j(\boldsymbol{s}) \mathcal{J}_i \boldsymbol{\varpi}_i$$
$$\cdot [(\boldsymbol{\varpi}_i \mathcal{J}_i)' + (\boldsymbol{\varpi}_j \mathcal{J}_j)' - \boldsymbol{\varpi}_i \mathcal{J}_i - \boldsymbol{\varpi}_j \mathcal{J}_j] \,. \tag{9.90}$$

Varying $\mathcal{J}$, we find after some manipulation of the last term,

$$\delta\mathcal{D}(\mathcal{J}) = -2 \sum_i \frac{(-1)^i}{n_i} \int \mathrm{d}\boldsymbol{v} \, f_i \boldsymbol{w}_i \cdot \boldsymbol{\varpi}_i (\delta \mathcal{J}_i) - 2\mathrm{i}B \sum_i \frac{e_i}{m_i} \int \mathrm{d}\boldsymbol{v} \, f_i \varpi_i^2 \mathcal{J}_i (\delta \mathcal{J}_i)$$
$$+ 2 \sum_i \int \mathrm{d}\boldsymbol{v} \, (\delta \mathcal{J}_i) \boldsymbol{\varpi}_i \cdot \sum_j \int \mathrm{d}\boldsymbol{v} \, \mathrm{d}\boldsymbol{s} \, \sigma \mathrm{d}\Omega \, g f_i(\boldsymbol{v}) f_j(\boldsymbol{s})$$
$$\times [(\boldsymbol{\varpi}_i \mathcal{J}_i)' + (\boldsymbol{\varpi}_j \mathcal{J}_j)' - \boldsymbol{\varpi}_i \mathcal{J}_i - \boldsymbol{\varpi}_j \mathcal{J}_j] \,. \tag{9.91}$$

If we therefore ask for that function which makes equation (9.90) stationary, i.e., makes equation (9.91) vanish, subject to the condition of equation (9.89), then this function is a solution of the equation,

$$-\frac{(-1)^i}{n_i} f_i \boldsymbol{w}_i = -\mathrm{i}B \frac{e_i}{m_i} f_i \boldsymbol{\varpi}_i \mathcal{J}_i + \alpha \sqrt{m_i} f_i \boldsymbol{\varpi}_i + \sum_j \int \mathrm{d}\boldsymbol{v} \, \mathrm{d}\boldsymbol{s} \, \sigma \mathrm{d}\Omega \, g f_i(\boldsymbol{v}) f_j(\boldsymbol{s})$$
$$\times [(\boldsymbol{\varpi}_i \mathcal{J}_i)' + (\boldsymbol{\varpi}_j \mathcal{J}_j)' - \boldsymbol{\varpi}_i \mathcal{J}_i - \boldsymbol{\varpi}_j \mathcal{J}_j] \,, \tag{9.92}$$

where α is any quantity independent of the subscript i. There would, of course, be no term like this if we permitted all variations $\delta\mathcal{J}_i$, but because in fact we do restrict $\delta\mathcal{J}_i$ to obey equation (9.89), we must in general include a term like this (which is orthogonal to the variations $\varpi_i\delta\mathcal{J}_i$). So far α can be arbitrary. But now we recall that, in fact, equation (9.92) will only have a solution satisfying the subsidiary equation if α has the value,

$$\alpha = \frac{\mathrm{i}B}{\rho}\frac{2}{3}\sum_j \frac{n_j e_j}{\sqrt{m_j}}\frac{1}{\pi^{3/2}}\int \mathrm{d}\boldsymbol{\varpi}_j\, \mathrm{e}^{-\varpi_j^2}\varpi_j^2\mathcal{J}_j\,,$$

and then equation (9.92) becomes exactly equation (9.48). This means that the trial function which makes $\mathcal{D}(\mathcal{J})$ stationary is the correct solution of equation (9.48). Furthermore if

$$\mathcal{J} = \mathcal{E} + \mathcal{O}(\delta)\,,$$

where δ is small, then

$$\mathcal{D}(\mathcal{J}) = \mathcal{D}(\mathcal{E}) + \mathcal{O}(\delta^2)\,,$$

so a fairly poor trial function will give a good value for $\mathcal{D}(\mathcal{E})$ which is precisely what we would like to know accurately. To demonstrate this, we substitute the expansion equation (9.55) for $\mathcal{J}_i$ in equation (9.88). Then since equation (9.55) is the exact solution, equation (9.88) reduces to

$$-\sum_i \frac{(-1)^i}{n_i}\int \mathrm{d}\boldsymbol{v}\, f_i \boldsymbol{w}_i \cdot \boldsymbol{\varpi}_i \mathcal{E}_i = \frac{3}{2}\frac{v_{ti}\rho}{n_2 m_2} e_1^0\,.$$

Hence the variation method gives e_1^0 correct to order δ^2 if δ is the error in the trial function. But e_1^0 is precisely what we want to know to evaluate the conductivities of equations (9.61) and (9.62). The method therefore gives σ^I, σ^{II}, and σ^{III} correct to order δ^2 and we can expect to get good values for these conductivities with quite poor trial functions.

It is now convenient to introduce a condensed notation for various kinds of collision integrals. This notation is precisely the same as we used in Chapter 8. If G_i, H_i are any properties connected with particles i, depending in general on position, velocity, and time, we define

$$
\begin{aligned}
[G_1, H_1]_{12} &\equiv -\frac{1}{n_1 n_2}\int \mathrm{d}\boldsymbol{v}\mathrm{d}\boldsymbol{s}\sigma\mathrm{d}\Omega\, g f_1(\boldsymbol{v}) f_2(\boldsymbol{s}) G_1(\boldsymbol{v}) : [H_1(\boldsymbol{s}') - H_1(\boldsymbol{s})]\\
[G_2, H_2]_{12} &\equiv \frac{1}{n_1 n_2}\int \mathrm{d}\boldsymbol{v}\mathrm{d}\boldsymbol{s}\sigma\mathrm{d}\Omega\, g f_1(\boldsymbol{v}) f_2(\boldsymbol{s}) G_2(\boldsymbol{s}) : [H_2(\boldsymbol{s}') - H_2(\boldsymbol{s})]\\
[G_1, H_2]_{12} &= [H_2, G_1]_{12}\\
&\equiv -\frac{1}{n_1 n_2}\int \mathrm{d}\boldsymbol{v}\mathrm{d}\boldsymbol{s}\sigma\mathrm{d}\Omega\, g f_1(\boldsymbol{v}) f_2(\boldsymbol{s}) G_1(\boldsymbol{v}) : [H_2(\boldsymbol{s}') - H_2(\boldsymbol{s})]\\
&= -\frac{1}{n_1 n_2}\int \mathrm{d}\boldsymbol{v}\mathrm{d}\boldsymbol{s}\sigma\mathrm{d}\Omega\, g f_1(\boldsymbol{v}) f_2(\boldsymbol{s}) H_2(\boldsymbol{s}) : [G_1(\boldsymbol{v}') - G_1(\boldsymbol{v})]
\end{aligned}
$$

$$[G_i, H_i]_i \equiv -\frac{1}{n_i^2}\int \mathrm{d}\boldsymbol{v}\mathrm{d}\boldsymbol{s}\sigma\mathrm{d}\Omega\, g f_i(\boldsymbol{v}) f_i(\boldsymbol{s}) G_i(\boldsymbol{v}) \\ : \left[H_i(\boldsymbol{v}') + H_i(\boldsymbol{s}') - H_i(\boldsymbol{v}) - H_i(\boldsymbol{s})\right]. \tag{9.93}$$

In all these expressions, $G_i(v) : H_j(s)$ stands for the full scalar product as indicated by equation (7.103). Finally, we define

$$\{G_1; H_1\} \equiv -2\sum_{i,j}\int \mathrm{d}\boldsymbol{v}\mathrm{d}\boldsymbol{s}\sigma\mathrm{d}\Omega\, g f_i(\boldsymbol{v}) f_j(\boldsymbol{s}) G_i(\boldsymbol{v}) \\ : \left[H_i(\boldsymbol{v}') + H_j(\boldsymbol{s}') - H_i(\boldsymbol{v}) - H_j(\boldsymbol{s})\right] \\ = 2n_1^2[G_1, H_1]_1 + 2n_2^2[G_2, H_2]_2 \\ +2n_1 n_2\left([G_1, H_1]_{12} + [G_1, H_2]_{12} + [G_2, H_1]_{12} + [G_2, H_2]_{12}\right). \tag{9.94}$$

As the trial function $\mathcal{J}_i$ to be used in equation (9.90), we use a *finite* expansion of the form of equation (9.55), i.e.,

$$\mathcal{J}_i = \sum_{m=0}^{M} e_i^m L_m^{\frac{3}{2}}(\varpi_i^2),$$

where successive approximations will be made by increasing M by unity. The first approximation will be given by $M = 0$, i.e., $\mathcal{J}_i = e_i^0$, etc.

The subsidiary condition of equation (9.89) is now just equation (9.56), i.e.,

$$e_2^0 = -\frac{n_1\sqrt{m_1}}{n_2\sqrt{m_2}} e_1^0, \tag{9.95}$$

and equation (9.90) becomes

$$\mathcal{D}(\mathcal{J}) = -3\sum_i (-1)^i v_{ti} e_i^0 - \frac{3\mathrm{i}B}{2}\sum_i \frac{n_i e_i}{m_i}\sum_{m=0}^{M}(e_i^m)^2\frac{4(m+\frac{3}{2})!}{3\sqrt{\pi}m!} \\ -\frac{1}{2}\left\{\boldsymbol{\varpi}\sum_{m=0}^{M} e^m L_m^{\frac{3}{2}}(\varpi^2); \boldsymbol{\varpi}\sum_{m=0}^{M} e^m L_m^{\frac{3}{2}}(\varpi^2)\right\}. \tag{9.96}$$

The first approximation is obtained by setting $M = 0$. Then using equation (9.95), $\mathcal{D}(\mathcal{J})$ becomes

$$\mathcal{D}(\mathcal{J}) = 3v_{t1}e_1^0 - \frac{3\mathrm{i}B}{2}\frac{n_1 e_1}{m_1}(e_1^0)^2 - n_1 n_2[\boldsymbol{\varpi}_1, \boldsymbol{\varpi}_1]_{12}(e_1^0)^2, \tag{9.97}$$

where we have used the tabulation of the collision integrals given in Appendix B to express $[\boldsymbol{\varpi}_2, \boldsymbol{\varpi}_2]_{12}$ and $[\boldsymbol{\varpi}_1, \boldsymbol{\varpi}_2]_{12}$ in terms of $[\boldsymbol{\varpi}_1, \boldsymbol{\varpi}_1]_{12}$, neglecting terms of order m_1/m_2.

From now on it is convenient to express all the collision integrals in terms of the collision time τ_e defined by

$$\tau_e \equiv \frac{3}{2n_2[\boldsymbol{\varpi}_1, \boldsymbol{\varpi}_1]_{12}}, \tag{9.98}$$

and, in place of the magnetic field B, to define

$$\omega_{ce} \equiv \left| \frac{eB}{m_1} \right| . \tag{9.99}$$

From the value given for $[\boldsymbol{\varpi}_1, \boldsymbol{\varpi}_1]_{12}$ in Appendix B we have

$$\tau_e \simeq \frac{3}{n\varphi} = \frac{3(4\pi\epsilon_0 k_B T)^2}{\sqrt{\pi} v_{t1} n e^4 \ln \Lambda} . \tag{9.100}$$

Here, and throughout this section, the sign $\simeq$ signifies that the usual plasma approximations have been made, i.e.,

$$m_2 \gg m_1 , \qquad n_1 \simeq n_2 \simeq \tfrac{1}{2} n , \qquad e_1 = -e_2 = -e . \tag{9.101}$$

These approximations are very slight ones. Hence, from equation (9.99),

$$\omega_{ce} = 1.759\,10^{11}\, B \sec^{-1}$$

where B is in Tesla.

In terms of these variables, equation (9.97) becomes

$$\mathcal{D}(\mathcal{J}) \simeq 3 v_{t1} e_1^0 - \frac{3n_1}{2\tau_e}(1 - \mathrm{i}\omega_{ce}\tau_e)(e_1^0)^2 , \tag{9.102}$$

which is stationary ($\mathrm{d}\mathcal{D}/\mathrm{d}e_1^0 = 0$) for

$$\left[e_1^0\right]_1 = \frac{\tau_e v_{t1}}{n_1(1 - \mathrm{i}\omega_{ce}\tau_e)} = \frac{\tau_e v_{t1}(1 + \mathrm{i}\omega_{ce}\tau_e)}{n_1(1 + \omega_{ce}^2\tau_e^2)} , \tag{9.103}$$

where we have written this as $\left[e_1^0\right]_1$ to indicate it is the first approximation to e_1^0.

Now in Chapter 8 we found that the first approximation to e_1^0 was not a very good one, but that the second approximation was very good. We shall therefore not discuss the first approximations to the electrical conductivity which equation (9.103) leads to, but will go immediately to consider the second approximation which is obtained by letting $M = 1$ in equation (9.96). Then the additional terms to be added to equation (9.97) are, using the expansion of $\{\mathcal{J};\mathcal{J}\}$ in equation (9.94),

$$\begin{aligned}
&-\frac{15\mathrm{i}B}{4}\sum_i \frac{n_i e_i}{m_i}(e_i^1)^2 - n_1^2[\boldsymbol{\varpi}_1 L_1^{\frac{3}{2}}, \boldsymbol{\varpi}_1 L_1^{\frac{3}{2}}]_1 (e_1^1)^2 - n_2^2[\boldsymbol{\varpi}_2 L_1^{\frac{3}{2}}, \boldsymbol{\varpi}_2 L_1^{\frac{3}{2}}]_2 (e_2^1)^2 \\
&-n_1 n_2 \Big\{ 2e_1^0 e_1^1[\boldsymbol{\varpi}_1, \boldsymbol{\varpi}_1 L_1^{\frac{3}{2}}]_{12} + (e_1^1)^2[\boldsymbol{\varpi}_1 L_1^{\frac{3}{2}}, \boldsymbol{\varpi}_1 L_1^{\frac{3}{2}}]_{12} + 2e_2^0 e_2^1[\boldsymbol{\varpi}_2, \boldsymbol{\varpi}_2 L_1^{\frac{3}{2}}]_{12} \\
&+(e_2^1)^2[\boldsymbol{\varpi}_2 L_1^{\frac{3}{2}}, \boldsymbol{\varpi}_2 L_1^{\frac{3}{2}}]_{12} + 2e_1^0 e_2^1[\boldsymbol{\varpi}_1, \boldsymbol{\varpi}_2 L_1^{\frac{3}{2}}]_{12} + 2e_1^1 e_2^0[\boldsymbol{\varpi}_1 L_1^{\frac{3}{2}}, \boldsymbol{\varpi}_2]_{12} \\
&\qquad +2e_1^1 e_2^1[\boldsymbol{\varpi}_1 L_1^{\frac{3}{2}}, \boldsymbol{\varpi}_2 L_1^{\frac{3}{2}}]_{12} \Big\} .
\end{aligned}$$

Adding these terms to equation (9.102) we get

$$\mathcal{D}(\mathcal{J}) \simeq 3v_{t1}e_1^0 - \frac{3n}{4\tau_e}\Big\{(1-\mathrm{i}\omega_{ce}\tau_e)(e_1^0)^2 + 3e_1^0e_1^1 - 3M_1^{3/2}e_1^0e_2^1$$
$$+(e_1^1)^2\left(\tfrac{13}{4}+\sqrt{2}-\tfrac{5}{2}\mathrm{i}\omega_{ce}\tau_e\right)$$
$$+(e_2^1)^2\left[\tfrac{5}{2}(3+\mathrm{i}\omega_{ce}\tau_e)M_1+\sqrt{2M_1}\right] - \tfrac{27}{2}e_1^1e_2^1M_1^{3/2}\Big\}, \quad (9.104)$$

where

$$M_1 = \frac{m_1}{m_1+m_2} \simeq \frac{m_1}{m_2}.$$

The values of e_1^0, e_1^1, and e_2^1 which make this stationary can be readily found. They are, using $n_1 = n_e \simeq n/2$,

$$\left[e_1^0\right]_2 \simeq \frac{\tau_e v_{t1}}{n_1}\left[\frac{\omega_{ce}^2\tau_e^2 + 1.8017 + \mathrm{i}\omega_{ce}\tau_e(\omega_{ce}^2\tau_e^2+4.381)}{\omega_{ce}^4\tau_e^4 + 6.281\omega_{ce}^2\tau_e^2 + .9325}\right] \quad (9.105)$$

$$\left[e_1^1\right]_2 \simeq \frac{3v_{t1}\tau_e}{5n_1}\left[\frac{\omega_{ce}^2\tau_e^2 - .9657 - 2.866\mathrm{i}\omega_{ce}\tau_e}{\omega_{ce}^4\tau_e^4 + 6.281\omega_{ce}^2\tau_e^2 + .9325}\right] \quad (9.106)$$

$$\left[e_2^1\right]_2 \simeq \frac{v_{t1}\tau_e}{n_1}M_1\left[\frac{3.924\omega_{ce}^2\tau_e^2 - .8546 - 3.560\mathrm{i}\omega_{ce}\tau_e + 1.0607\omega_{ce}^3\tau_e^3}{\omega_{ce}^4\tau_e^4 + 6.281\omega_{ce}^2\tau_e^2 + .9325}\right]. \quad (9.107)$$

From equations (9.62) and (9.105) we have, to this approximation,

$$\left[\sigma^{II}\right]_2 \simeq \frac{n_1e^2\tau_e}{m_1}\frac{\omega_{ce}^2\tau_e^2 + 1.8017}{\omega_{ce}^4\tau_e^4 + 6.281\omega_{ce}^2\tau_e^2 + .9325} \quad (9.108)$$

$$\left[\sigma^{III}\right]_2 \simeq \frac{n_1e^2\tau_e}{m_1}\frac{\omega_{ce}\tau_e(\omega_{ce}^2\tau_e^2 + 4.381)}{\omega_{ce}^4\tau_e^4 + 6.281\omega_{ce}^2\tau_e^2 + .9325}, \quad (9.109)$$

and from Equations (9.61) and (9.83)

$$\left[\sigma^{I}\right]_2 = \frac{n_1e^2\tau_e}{m_1}1.932 = 1.932\left[\sigma^I\right]_1. \quad (9.110)$$

Problem 9.1 Verify equation (9.104) and the expressions for $\left[e_1^0\right]_2$, $\left[e_1^1\right]_2$, and $\left[e_2^1\right]_2$.*

One combination of the conductivities σ^{II} and σ^{III} is particularly useful. Suppose that there are no temperature gradients so that we need consider only the first three terms of equation (9.59), and that an electric field $\boldsymbol{E}^{\perp}$ is applied perpendicular to the direction of the magnetic field. Then equation (9.59) tells us that $\boldsymbol{E}^{\perp}$ gives rise to a current $\sigma^{II}\boldsymbol{E}^{\perp}$ and also a current perpendicular

*The results for $\left[e_2^1\right]_2$ differ from those in Marshall's work because his expression for equation (B.131) has been corrected. Also, the odd order terms in $\omega_{ce}\tau_e$ change sign since he uses $\omega_{ce} = -|\omega_{ce}|$.

to this, $\sigma^{III}\hat{\boldsymbol{b}} \times \boldsymbol{E}^{\perp}$. Now very frequently the experimental arrangement is such that no current can flow in this latter direction. What happens then is that when $\boldsymbol{E}^{\perp}$ is switched on, an additional electric field $\boldsymbol{E}_0^{\perp}$ is set up in the direction of $\hat{\boldsymbol{b}} \times \boldsymbol{E}^{\perp}$ of just such strength to cancel off this current. In this situation the total electric field is $\boldsymbol{E}^{\perp} + \boldsymbol{E}_0^{\perp}$ and the current parallel to $\boldsymbol{E}^{\perp}$ is

$$\boldsymbol{j}^{\perp} = \sigma^{II}\boldsymbol{E}^{\perp} + \sigma^{III}\hat{\boldsymbol{b}} \times \boldsymbol{E}_0^{\perp} ,$$

and the current perpendicular to $\boldsymbol{E}^{\perp}$ and $\boldsymbol{B}$ is

$$\boldsymbol{j}_0^{\perp} = \sigma^{II}\boldsymbol{E}_0^{\perp} + \sigma^{III}\hat{\boldsymbol{b}} \times \boldsymbol{E}^{\perp} .$$

If the experimental arrangement is such that $\boldsymbol{j}_0^{\perp}$ must be zero we have

$$\boldsymbol{E}_0^{\perp} = -\frac{\sigma^{III}}{\sigma^{II}}\hat{\boldsymbol{b}} \times \boldsymbol{E}^{\perp} ,$$

and hence,

$$\boldsymbol{j}^{\perp} = \frac{(\sigma^{II})^2 + (\sigma^{III})^2}{\sigma^{II}}\boldsymbol{E}^{\perp} .$$

For such an experimental arrangement it is convenient to define a "perpendicular conductivity" by

$$\boldsymbol{j}^{\perp} = \sigma^{\perp}\boldsymbol{E}^{\perp}$$

where

$$\begin{aligned} \sigma^{\perp} &= \frac{(\sigma^{II})^2 + (\sigma^{III})^2}{\sigma^{II}} \\ &= \sigma^{II}\left[1 + \omega_{ce}^2\tau_e^2\left(\frac{\omega_{ce}^2\tau_e^2 + 4.381}{\omega_{ce}^2\tau_e^2 + 1.802}\right)^2\right] . \end{aligned} \tag{9.111}$$

Notice that in very strong magnetic fields such that $\omega_{ce}\tau_e \gg 1$,

$$\sigma^{\perp} \simeq \frac{n_1 e^2 \tau_e}{m_1} = \frac{\sigma^I}{1.932} ,$$

so in this situation $\sigma^{\perp}$ is approximately half of the longitudinal conductivity. It is even closer to half than appears here since higher order approximations [see equation (8.114)] are even closer to a factor of two.

We can be confident that these conductivities are correct to about 10%. The greatest error comes from the uncertainty in the cut-off distance.

9.5 Thermal conductivity and diffusion

9.5.1 Variational results

In this section we shall consider equation (9.45) and obtain values for the thermal conductivity and thermal diffusion. The analysis is clearly similar to that given in the previous section and we need not repeat it in detail. This time we take as our trial function the finite sum,

$$\mathcal{J}_i = \sum_{m=0}^{M} a_i^m \boldsymbol{\varpi}_i L_m^{\frac{3}{2}}(\varpi_i^2)\,,$$

and in place of equation (9.96) we get for the Davison function

$$\mathcal{D}(\mathcal{J}) = -\frac{15}{2}\sum_i n_i v_{ti} a_i^1 - \frac{3\mathrm{i}B}{2}\sum_i \frac{n_i e_i}{m_i}\sum_{m=0}^{M}(a_i^m)^2 \frac{4(m+\frac{3}{2})!}{3\sqrt{\pi}m!}$$
$$-\frac{1}{2}\left\{\sum_{m=0}^{M} a^m \boldsymbol{\varpi} L_m^{\frac{3}{2}}(\varpi^2); \sum_{m=0}^{M} a^m \boldsymbol{\varpi} L_m^{\frac{3}{2}}(\varpi^2)\right\}, \quad (9.112)$$

and in place of equation (9.95), the subsidiary condition is

$$a_2^0 = -\frac{n_1\sqrt{m_1}}{n_2\sqrt{m_2}}a_1^0\,.$$

We could now consider a first approximation to equation (9.112) by setting all the coefficients except a_1^1 and a_2^1 to zero. But this would ignore all thermal diffusion effects so we shall consider the second approximation immediately. Setting all the coefficients except a_i^0 and a_i^1 equal to zero we can write down $\mathcal{D}(\mathcal{J})$ immediately by comparing Equations (9.96), (9.104), and (9.112). This gives

$$\mathcal{D}(\mathcal{J}) \simeq -\frac{15}{2}\sum_i n_i v_{ti} a_i^1 - \frac{3n}{4\tau_e}\Big\{(1-\mathrm{i}\omega_{ce}\tau_e)(a_1^0)^2 + 3a_1^0 a_1^1 - 3M_1^{3/2}a_1^0 a_2^1$$
$$+(a_1^1)^2\left(\tfrac{13}{4}+\sqrt{2}-\tfrac{5}{2}\mathrm{i}\omega_{ce}\tau_e\right)$$
$$+(a_2^1)^2\left[\tfrac{5}{2}(3+\mathrm{i}\omega_{ce}\tau_e)M_1+\sqrt{2M_1}\right] - \tfrac{27}{2}a_1^1 a_2^1 M_1^{3/2}\Big\}. \quad (9.113)$$

The values of a_1^0, a_1^1, and a_2^1 which make this stationary are

$$\left[a_1^0\right]_2 \simeq -\frac{3v_{t1}\tau_e}{2}\left[\frac{\omega_{ce}^2\tau_e^2 - .9657 - 2.866\mathrm{i}\omega_{ce}\tau_e}{\omega_{ce}^4\tau_e^4 + 6.281\omega_{ce}^2\tau_e^2 + .9325}\right] \quad (9.114)$$

$$\left[a_1^1\right]_2 \simeq -v_{t1}\tau_e\left[\frac{1.866\omega_{ce}^2\tau_e^2 + .9657 + \mathrm{i}\omega_{ce}\tau_e(1.9+\omega_{ce}^2\tau_e^2)}{\omega_{ce}^4\tau_e^4 + 6.281\omega_{ce}^2\tau_e^2 + .9325}\right] \quad (9.115)$$

$$\left[a_2^1\right]_2 \simeq -\frac{5v_{t1}\tau_e}{2\sqrt{2}}\,. \quad (9.116)$$

There is one check we can apply to these solutions. Earlier we proved a relation, equation (9.68), between the exact values of a_1^0, e_1^1, and e_2^1. Now in the course of deriving e_1^0 and the electrical conductivity in the previous section we calculated approximations to e_1^1 and e_2^1 from equation (9.106), and now we have an approximation for a_1^0 from equation (9.114). It is therefore interesting to see how well these approximations satisfy equation (9.68) and it is easy to show that they satisfy equation (9.68) exactly apart from terms $\sim M_1$ which we have neglected throughout. This helps to give us confidence in the results we have derived.

Problem 9.2 Verify the expressions for $\left[a_1^0\right]_2$, $\left[a_1^1\right]_2$, and $\left[a_2^1\right]_2$. †

9.5.2 The transport coefficients

From equations (9.64) and (9.114) we have

$$\left[\varphi^{II}\right]_2 \simeq -\frac{3k_B n_1 e\tau_e}{2m_1}\frac{\omega_{ce}^2\tau_e^2 - .9657}{\omega_{ce}^4\tau_e^4 + 6.281\omega_{ce}^2\tau_e^2 + .9325} \tag{9.117}$$

$$\left[\varphi^{III}\right]_2 \simeq \frac{3k_B n_1 e\tau_e}{2m_1}\frac{2.8657\omega_{ce}\tau_e}{\omega_{ce}^4\tau_e^4 + 6.281\omega_{ce}^2\tau_e^2 + .9325}, \tag{9.118}$$

and therefore by equation (9.84),

$$\left[\varphi^{I}\right]_2 \simeq \frac{3k_B n_1 e\tau_e}{2m_1}\frac{.9657}{.9325} \simeq 1.553\frac{k_B n_1 e\tau_e}{m_1}. \tag{9.119}$$

By equations (9.67), (9.115), and (9.116),

$$\left[\lambda'^{II}\right]_2 \simeq \frac{5n_1\tau_e k_B^2 T}{2m_1}\left[\frac{1.8657\omega_{ce}^2\tau_e^2 + .9657}{\omega_{ce}^4\tau_e^4 + 6.281\omega_{ce}^2\tau_e^2 + .9325}\right] \tag{9.120}$$

$$\left[\lambda'^{III}\right]_2 \simeq \frac{5n_1\tau_e k_B^2 T}{2m_1}\left[\frac{\omega_{ce}\tau_e(\omega_{ce}^2\tau_e^2 + 1.9)}{\omega_{ce}^4\tau_e^4 + 6.281\omega_{ce}^2\tau_e^2 + .9325}\right], \tag{9.121}$$

and by equation (9.84),

$$\left[\lambda'^{I}\right]_2 \simeq \frac{5n_1\tau_e k_B^2 T}{2m_1}\left(\frac{.9657}{.9325}\right) \simeq \frac{n_1\tau_e k_B^2 T}{m_1}2.589. \tag{9.122}$$

Then from equation (9.81),

$$\left[\theta^{I}\right]_2 \simeq \frac{n_1\tau_e k_B^2 T}{m_1}6.472 \tag{9.123}$$

$$\left[\theta^{II}\right]_2 \simeq \frac{5n_1\tau_e k_B^2 T}{2m_1}\left[\frac{.3657\omega_{ce}^2\tau_e^2 + 2.415}{\omega_{ce}^4\tau_e^4 + 6.281\omega_{ce}^2\tau_e^2 + .9325}\right] \tag{9.124}$$

$$\left[\theta^{III}\right]_2 \simeq \frac{5n_1\tau_e k_B^2 T}{2m_1}\left[\frac{\omega_{ce}\tau_e(\omega_{ce}^2\tau_e^2 + 6.199)}{\omega_{ce}^4\tau_e^4 + 6.281\omega_{ce}^2\tau_e^2 + .9325}\right], \tag{9.125}$$

†These results differ from those in Marshall's work because his expression for equation (B.131) has been corrected.

and by equation (9.82),

$$\left[\xi^{I}\right]_2 \simeq -\frac{n_1 e\tau_e k_B T}{m_1} 6.383 \tag{9.126}$$

$$\left[\xi^{II}\right]_2 \simeq -\frac{n_1 e\tau_e k_B T}{m_1} \frac{\omega_{ce}^2\tau_e^2 + 5.953}{\omega_{ce}^4\tau_e^4 + 6.281\omega_{ce}^2\tau_e^2 + .9325} \tag{9.127}$$

$$\left[\xi^{III}\right]_2 \simeq -\frac{5n_1 e\tau_e k_B T}{2m_1} \frac{\omega_{ce}\tau_e(\omega_{ce}^2\tau_e^2 + 6.10)}{\omega_{ce}^4\tau_e^4 + 6.281\omega_{ce}^2\tau_e^2 + .9325}\,. \tag{9.128}$$

From equation (9.74)

$$\left[\lambda^{I}\right]_2 \simeq \frac{n_1 \tau_e k_B^2 T}{m_1} 1.340\,. \tag{9.129}$$

After some tedious algebra, equation (9.75) becomes

$$\left[\lambda^{II}\right]_2 \simeq \frac{n_1 \tau_e k_B^2 T}{m_1} \left[\frac{4.664}{\omega_{ce}^2\tau_e^2 + 3.481}\right], \tag{9.130}$$

and equation (9.76) is

$$\left[\lambda^{III}\right]_2 \simeq \frac{5n_1 \tau_e k_B^2 T}{2m_1} \left[\frac{\omega_{ce}\tau_e}{\omega_{ce}^2\tau_e^2 + 3.481}\right]. \tag{9.131}$$

Equation (9.77) gives

$$\left[\psi^{I}\right]_2 \simeq -3.304\frac{k_B T}{e}, \tag{9.132}$$

equation (9.78) gives

$$\left[\psi^{II}\right]_2 \simeq -\frac{5k_B T}{2e} \left[\frac{\omega_{ce}^2\tau_e^2 + 4.600}{\omega_{ce}^2\tau_e^2 + 3.481}\right], \tag{9.133}$$

and equation (9.79) gives

$$\left[\psi^{III}\right]_2 \simeq -\frac{3k_B T}{2e} \frac{\omega_{ce}\tau_e}{\omega_{ce}^2\tau_e^2 + 3.481}\,. \tag{9.134}$$

These expressions, though simplified, are good to a better than a part per thousand. The errors due to stopping at the second approximation are larger than those due to the accuracy of the various constants.

Problem 9.3 Show that equation (9.68) is satisfied exactly with the listed values of $\left[a_1^0\right]_2$ and $\left[e_1^1\right]_2$ except for terms of order $\sqrt{M_1} \ll 1$.

9.5.3 Convergence and accuracy

When considering convergence, it was noted in Chapter 7 that when evaluating the coefficients to first, second, and third order that the second approximation made a significant change but that the third approximation made a relatively small additional correction. We have generally stopped at this approximation, but we can note that for the electrical conductivity, even higher order calculations have been made. It was found that the correction factor was $1.93198 \sim 1.932$ using $[e_i^j]_2$ coefficients, but the next higher approximation is given by Balescu[47] as 1.950 and the next as 1.953 so the convergence is rather rapid. Even higher approximations have been given by Kaneko and Yamao[48] where they used 50 polynomials to achieve an accuracy of six significant figures. This limit of 1.96 is used by Spitzer[1] and Braginskii[43], but the full expressions for $[e_1^0]_n$ and $[e_1^1]_n$ are not given. The corresponding successive approximations by Balescu for φ^I are 1.55, 1.39, and 1.40 while the coefficients for λ'^I are 2.59, 4.15, and 4.22. These indicate that the expressions for φ are probably accurate enough but that the expressions for λ'^I should go at least one step further.

The accuracy of the listed expressions depends on two factors. The coefficients have all been calculated to at least six significant figures and then rounded to three or four figures, so the numerical values are accurate to the approximation level used. On the other hand, we have neglected terms of the order of $\sqrt{M_1}$ and smaller. If we were to keep terms to this order, the various coefficients $[a_1^0]_2$, $[a_1^1]_2$, and $[a_2^1]_2$ would vary only in the fourth significant figure, so the neglect is justifiable. For the $[e_1^0]_2$, $[e_1^1]_2$, and $[e_2^1]_2$ coefficients, however, the denominator changes so that

$$0.932548 + 6.28078x^2 + x^4 \rightarrow 1.273(0.924724 + 6.22933x^2 + x^4 + 0.001336x^6)$$

but σ^I is unchanged to four significant figures.

9.6 The pressure tensor

In this section we examine equation (9.26) which for convenience we repeat here as

$$\begin{aligned}
-2f_i\boldsymbol{\varpi}_i^0\boldsymbol{\varpi}_i = & -\frac{1}{\rho k_B T}m_i f_i \boldsymbol{w}_i \cdot \sum_j e_j \int \mathrm{d}\boldsymbol{v}\, f_j(\boldsymbol{w}_j \times \boldsymbol{B})\mathsf{B}_j \\
& -\frac{e_i}{m_i} f_i(\boldsymbol{\varpi}_i \times \boldsymbol{B}) \cdot \nabla_{\boldsymbol{\varpi}_i} \mathsf{B}_i \\
& +\sum_j \int \mathrm{d}\boldsymbol{s}\,\sigma \mathrm{d}\Omega\, g f_i(\boldsymbol{v}) f_j(\boldsymbol{s})
\end{aligned}$$

$$\times[\mathsf{B}_i(\boldsymbol{v}') + \mathsf{B}_j(\boldsymbol{s}') - \mathsf{B}_i(\boldsymbol{v}) - \mathsf{B}_j(\boldsymbol{s})] . \tag{9.135}$$

The solution B_i will lead to values for the elements in the pressure tensor.

We first consider the possible forms that the tensor B_i can take. B_i can only depend on the vector $\boldsymbol{\varpi}_i$ and the pseudo-vector $\boldsymbol{B}$. The scalars which can be formed from these are

$$\varpi_i^2 , \qquad B^2 , \qquad (\boldsymbol{\varpi}_i \cdot \boldsymbol{B})^2 .$$

The vectors which can be formed are

$$\boldsymbol{\varpi}_i , \qquad \boldsymbol{\varpi}_i \times \boldsymbol{B} , \qquad \boldsymbol{B}(\boldsymbol{\varpi}_i \cdot \boldsymbol{B}) .$$

From these vectors, we can form six traceless tensors which are

$$\begin{aligned}
(\mathsf{T}_i^1)_{\alpha\beta} &= (\boldsymbol{\varpi}_i^0 \boldsymbol{\varpi}_i)_{\alpha\beta} \\
(\mathsf{T}_i^2)_{\alpha\beta} &= \tfrac{1}{2}[\varpi_{i\alpha}(\boldsymbol{\varpi}_i \times \boldsymbol{B})_\beta + \varpi_{i\beta}(\boldsymbol{\varpi}_i \times \boldsymbol{B})_\alpha] \\
(\mathsf{T}_i^3)_{\alpha\beta} &= (\boldsymbol{\varpi}_i \times \boldsymbol{B})_\alpha(\boldsymbol{\varpi}_i \times \boldsymbol{B})_\beta - \tfrac{1}{3}[B^2\varpi_i^2 - (\boldsymbol{\varpi}_i \cdot \boldsymbol{B})^2]\delta_{\alpha\beta} \\
(\mathsf{T}_i^4)_{\alpha\beta} &= \tfrac{1}{2}(\varpi_{i\alpha}B_\beta + \varpi_{i\beta}B_\alpha)(\boldsymbol{\varpi}_i \cdot \boldsymbol{B}) - \tfrac{1}{3}(\boldsymbol{\varpi}_i \cdot \boldsymbol{B})^2\delta_{\alpha\beta} \\
(\mathsf{T}_i^5)_{\alpha\beta} &= \tfrac{1}{2}\left[B_\alpha(\boldsymbol{\varpi}_i \times \boldsymbol{B})_\beta + B_\beta(\boldsymbol{\varpi}_i \times \boldsymbol{B})_\alpha\right](\boldsymbol{\varpi}_i \cdot \boldsymbol{B}) \\
(\mathsf{T}_i^6)_{\alpha\beta} &= B_\alpha B_\beta(\boldsymbol{\varpi}_i \cdot \boldsymbol{B})^2 - \tfrac{1}{3}B^2(\boldsymbol{\varpi}_i \cdot \boldsymbol{B})^2\delta_{\alpha\beta} .
\end{aligned} \tag{9.136}$$

In addition, of course, we have the unit tensor,

$$(\mathsf{T}_i^0)_{\alpha\beta} = \delta_{\alpha\beta} .$$

We thus write

$$\mathsf{B}_i = \sum_{n=0}^{6} B_i^n \mathsf{T}_i^n , \tag{9.137}$$

where the B_i^n are scalars that are in general functions of ϖ_i^2, B^2, and $(\boldsymbol{\varpi}_i \cdot \boldsymbol{B})^2$.

We now notice that equation (9.137) is even in $\boldsymbol{\varpi}_i$ so the integral in the first term on the right-hand side of equation (9.135) must vanish. Equation (9.135) then becomes

$$-2f_i\boldsymbol{\varpi}_i^0\boldsymbol{\varpi}_i = -\frac{e_i}{m_i}f_i(\boldsymbol{\varpi}_i \times \boldsymbol{B}) \cdot \nabla_{\boldsymbol{\varpi}_i}\mathsf{B}_i + \mathcal{I}(\mathsf{B}_i) , \tag{9.138}$$

where throughout this section we shall use $\mathcal{I}$ to stand for the integral operator defined by

$$\mathcal{I}(\mathsf{B}_i) = \sum_j \int \mathrm{d}\boldsymbol{s}\,\sigma \mathrm{d}\Omega\, g f_i(\boldsymbol{v}) f_j(\boldsymbol{s})[\mathsf{B}_i(\boldsymbol{v}') + \mathsf{B}_j(\boldsymbol{s}') - \mathsf{B}_i(\boldsymbol{v}) - \mathsf{B}_j(\boldsymbol{s})] . \tag{9.139}$$

Now consider the effect of the operator $(\boldsymbol{\varpi}_i \times \boldsymbol{B}) \cdot \nabla_{\boldsymbol{\varpi}_i}$ acting on B_i. When this operator acts on the scalars B_i^n it gives zero because these are only functions of ϖ_i^2, B^2, and $(\boldsymbol{\varpi}_i \cdot \boldsymbol{B})^2$. Thus, for example,

$$
\begin{aligned}
(\boldsymbol{\varpi}_i \times \boldsymbol{B}) \cdot \nabla_{\boldsymbol{\varpi}_i} B_i^n(\varpi_i^2) &= \left[(\boldsymbol{\varpi}_i \times \boldsymbol{B}) \cdot \nabla_{\boldsymbol{\varpi}_i}(\varpi_i^2)\right] \frac{\partial}{\partial \varpi_i^2} B_i^n(\varpi_i^2) \\
&= 2[(\boldsymbol{\varpi}_i \times \boldsymbol{B}) \cdot \boldsymbol{\varpi}_i] \frac{\partial}{\partial \varpi_i^2} B_i^n(\varpi_i^2) = 0 \,.
\end{aligned}
$$

Effectively, the operator only acts on the tensors T_i^n. It is easy to derive the following results:

$$
\begin{aligned}
(\boldsymbol{\varpi}_i \times \boldsymbol{B}) \cdot \nabla_{\boldsymbol{\varpi}_i} \mathsf{T}_i^0 &= 0 \\
(\boldsymbol{\varpi}_i \times \boldsymbol{B}) \cdot \nabla_{\boldsymbol{\varpi}_i} \mathsf{T}_i^1 &= 2\mathsf{T}_i^2 \\
(\boldsymbol{\varpi}_i \times \boldsymbol{B}) \cdot \nabla_{\boldsymbol{\varpi}_i} \mathsf{T}_i^2 &= -B^2 \mathsf{T}_1^i + \mathsf{T}_i^3 + \mathsf{T}_i^4 \\
(\boldsymbol{\varpi}_i \times \boldsymbol{B}) \cdot \nabla_{\boldsymbol{\varpi}_i} \mathsf{T}_i^3 &= -2B^2 \mathsf{T}_i^2 + 2\mathsf{T}_i^5 \\
(\boldsymbol{\varpi}_i \times \boldsymbol{B}) \cdot \nabla_{\boldsymbol{\varpi}_i} \mathsf{T}_i^4 &= \mathsf{T}_i^5 \\
(\boldsymbol{\varpi}_i \times \boldsymbol{B}) \cdot \nabla_{\boldsymbol{\varpi}_i} \mathsf{T}_i^5 &= -B^2 \mathsf{T}_i^4 + \mathsf{T}_i^6 \\
(\boldsymbol{\varpi}_i \times \boldsymbol{B}) \cdot \nabla_{\boldsymbol{\varpi}_i} \mathsf{T}_i^6 &= 0 \,.
\end{aligned}
\tag{9.140}
$$

Notice that these tensors form a closed group, as they must if equation (9.137) is to be a solution, so this operator does not create any new tensors.

Problem 9.4 *The T_i^n tensors.*

1. Write out the tensors T_i^n for $n = 0$ through $n = 6$.
2. Prove that the relationships of equation (9.140) are satisfied. Assume that $\boldsymbol{B}$ is in the z-direction.
3. Show that T_i^n, $n = 1$ through $n = 6$ are traceless.

The tensors T_i^n are not easy to work with because they are constructed from components of $\boldsymbol{\varpi}_i$ and $\boldsymbol{B}$ in a complicated way. It is therefore convenient to define a new set of nine tensors, each of which has a simple dependence on the components of $\boldsymbol{\varpi}_i$. These are:

$$
\begin{aligned}
\left(\mathsf{Q}_i^0\right)_{\alpha\beta} &= \delta_{\alpha\beta} \\
\left(\mathsf{Q}_i^1\right)_{\alpha\beta} &= \delta_{\alpha\gamma}\delta_{\beta\lambda}(\boldsymbol{\varpi}_i^0 \boldsymbol{\varpi}_i)_{\gamma\lambda} \\
\left(\mathsf{Q}_i^2\right)_{\alpha\beta} &= \tfrac{1}{2}(\delta_{\alpha\gamma}\varepsilon_{\beta\lambda\varphi} + \delta_{\beta\gamma}\varepsilon_{\alpha\lambda\varphi}) B_\varphi (\boldsymbol{\varpi}_i^0 \boldsymbol{\varpi}_i)_{\gamma\lambda} \\
\left(\mathsf{Q}_i^3\right)_{\alpha\beta} &= \varepsilon_{\alpha\gamma\varphi}\varepsilon_{\beta\lambda\psi}(\boldsymbol{B} : \boldsymbol{B})_{\varphi\psi}(\boldsymbol{\varpi}_i^0 \boldsymbol{\varpi}_i)_{\gamma\lambda} \\
\left(\mathsf{Q}_i^4\right)_{\alpha\beta} &= \delta_{\alpha\beta}(\boldsymbol{B} : \boldsymbol{B})_{\gamma\lambda}(\boldsymbol{\varpi}_i^0 \boldsymbol{\varpi}_i)_{\gamma\lambda} \\
\left(\mathsf{Q}_i^5\right)_{\alpha\beta} &= \varpi_i^2 (\boldsymbol{B} : \boldsymbol{B})_{\alpha\beta}
\end{aligned}
$$

$$\begin{aligned}
(\mathsf{Q}_i^6)_{\alpha\beta} &= \tfrac{1}{2}[\delta_{\alpha\gamma}(\boldsymbol{B}:\boldsymbol{B})_{\beta\lambda} + \delta_{\beta\gamma}(\boldsymbol{B}:\boldsymbol{B})_{\alpha\lambda}](\boldsymbol{\varpi}_i^0\boldsymbol{\varpi}_i)_{\gamma\lambda} \\
(\mathsf{Q}_i^7)_{\alpha\beta} &= \tfrac{1}{2}[\varepsilon_{\beta\gamma\varphi}(\boldsymbol{B}:\boldsymbol{B})_{\alpha\varphi} + \varepsilon_{\alpha\gamma\varphi}(\boldsymbol{B}:\boldsymbol{B})_{\beta\varphi}]B_\lambda(\boldsymbol{\varpi}_i^0\boldsymbol{\varpi}_i)_{\gamma\lambda} \\
(\mathsf{Q}_i^8)_{\alpha\beta} &= (\boldsymbol{B}:\boldsymbol{B})_{\alpha\beta}(\boldsymbol{B}:\boldsymbol{B})_{\gamma\lambda}(\boldsymbol{\varpi}_i^0\boldsymbol{\varpi}_i)_{\gamma\lambda}\,, \qquad (9.141)
\end{aligned}$$

where $\varepsilon_{\alpha\beta\gamma}$ was defined in equation (9.35) and

$$(\boldsymbol{B}:\boldsymbol{B})_{\alpha\beta} \equiv B_\alpha B_\beta - \tfrac{1}{3}B^2\delta_{\alpha\beta}\,.$$

This may also be written as

$$\boldsymbol{B}:\boldsymbol{B} = \begin{pmatrix} -\frac{1}{3}B^2 & 0 & 0 \\ 0 & -\frac{1}{3}B^2 & 0 \\ 0 & 0 & \frac{2}{3}B^2 \end{pmatrix}.$$

Notice that all the tensors except Q_i^0 and Q_i^5 are expressed in terms of $\boldsymbol{\varpi}_i^0\boldsymbol{\varpi}_i$. In terms of these Q_i^n tensors, the T_i^n tensors may be written

$$\begin{aligned}
\mathsf{T}_i^0 &= \mathsf{Q}_i^0 \\
\mathsf{T}_i^1 &= \mathsf{Q}_i^1 \\
\mathsf{T}_i^2 &= \mathsf{Q}_i^2 \\
\mathsf{T}_i^3 &= \mathsf{Q}_i^3 + \tfrac{1}{3}\mathsf{Q}_i^4 - \tfrac{1}{3}B^2\mathsf{Q}_i^1 - \tfrac{1}{3}\mathsf{Q}_i^5 \\
\mathsf{T}_i^4 &= \mathsf{Q}_i^6 - \tfrac{1}{3}\mathsf{Q}_i^4 + \tfrac{1}{3}B^2\mathsf{Q}_i^1 + \tfrac{1}{3}\mathsf{Q}_i^5 \\
\mathsf{T}_i^5 &= \mathsf{Q}_i^7 \\
\mathsf{T}_i^6 &= \mathsf{Q}_i^8 + \tfrac{1}{3}B^2\mathsf{Q}_i^5\,. \qquad (9.142)
\end{aligned}$$

Problem 9.5 Prove the relationships of equation (9.142) for the Q_i^n.

The transformation properties of the Q_i^n are

$$\begin{aligned}
(\boldsymbol{\varpi}_i \times \boldsymbol{B})\cdot\nabla_{\boldsymbol{\varpi}_i}\mathsf{Q}_i^0 &= 0 \\
(\boldsymbol{\varpi}_i \times \boldsymbol{B})\cdot\nabla_{\boldsymbol{\varpi}_i}\mathsf{Q}_i^1 &= 2\mathsf{Q}_i^2 \\
(\boldsymbol{\varpi}_i \times \boldsymbol{B})\cdot\nabla_{\boldsymbol{\varpi}_i}\mathsf{Q}_i^2 &= -B^2\mathsf{Q}_i^1 + \mathsf{Q}_i^3 + \mathsf{Q}_i^6 \\
(\boldsymbol{\varpi}_i \times \boldsymbol{B})\cdot\nabla_{\boldsymbol{\varpi}_i}\mathsf{Q}_i^3 &= -\tfrac{4}{3}B^2\mathsf{Q}_i^2 + 2\mathsf{Q}_i^7 \\
(\boldsymbol{\varpi}_i \times \boldsymbol{B})\cdot\nabla_{\boldsymbol{\varpi}_i}\mathsf{Q}_i^4 &= 0 \\
(\boldsymbol{\varpi}_i \times \boldsymbol{B})\cdot\nabla_{\boldsymbol{\varpi}_i}\mathsf{Q}_i^5 &= 0 \\
(\boldsymbol{\varpi}_i \times \boldsymbol{B})\cdot\nabla_{\boldsymbol{\varpi}_i}\mathsf{Q}_i^6 &= \mathsf{Q}_i^7 - \tfrac{2}{3}B^2\mathsf{Q}_i^2 \\
(\boldsymbol{\varpi}_i \times \boldsymbol{B})\cdot\nabla_{\boldsymbol{\varpi}_i}\mathsf{Q}_i^7 &= \mathsf{Q}_i^8 - B^2\mathsf{Q}_i^6 + \tfrac{1}{3}B^2\mathsf{Q}_i^4 - \tfrac{1}{3}B^4\mathsf{Q}_i^1 \\
(\boldsymbol{\varpi}_i \times \boldsymbol{B})\cdot\nabla_{\boldsymbol{\varpi}_i}\mathsf{Q}_i^8 &= 0\,. \qquad (9.143)
\end{aligned}$$

Now put

$$\mathsf{B}_i = \sum_{n=0}^{8} D_i^n\mathsf{Q}_i^n\,, \qquad (9.144)$$

where the D_i^n are scalars. Comparing equation (9.144) with equation (9.137) and using equation (9.142), we find the scalars D_i^n are given in terms of the scalars B_i^n by

$$\begin{aligned}
D_i^0 &= B_i^0 \\
D_i^1 &= B_i^1 - \tfrac{1}{3}B^2 B_i^3 + \tfrac{1}{3}B^2 B_i^4 \\
D_i^2 &= B_i^2 \\
D_i^3 &= B_i^3 \\
D_i^4 &= \tfrac{1}{3}(B_i^3 - B_i^4) \\
D_i^5 &= \tfrac{1}{3}(-B_i^3 + B_i^4 + B^2 B_i^6) \\
D_i^6 &= B_i^4 \\
D_i^7 &= B_i^5 \\
D_i^8 &= B_i^6 \,.
\end{aligned} \tag{9.145}$$

Problem 9.6 Use equations (9.137), (9.142), and (9.144) to prove the relationships of equation (9.145).

We now assume that the scalars B_i^n and D_i^n are independent of $(\boldsymbol{\varpi}_i \cdot \boldsymbol{B})^2$, i.e., that they only depend on ϖ_i^2 and B^2. This assumption is equivalent to the one we made in Section 9.3. We can make it very plausible that this assumption is correct as follows. Suppose we tried to solve equation (9.138) by some iteration process starting from the tensors T_i^n. Then in this iteration process the operator $(\boldsymbol{\varpi}_i \times \boldsymbol{B}) \cdot \partial/\partial\boldsymbol{\varpi}_i$ would generate all the tensors T_i^n again, multiplied in general by functions of B^2 but, from equation (9.140), it would not generate any of these tensors multiplied by $(\boldsymbol{\varpi}_i \cdot \boldsymbol{B})^2$. Hence if we started this iteration process with the tensors T_i^n multiplied by scalar functions which were independent of $(\boldsymbol{\varpi}_i \cdot \boldsymbol{B})^2$, we should never create such a dependence. Strictly speaking, because we make this assumption we should, after obtaining our solution, consider if we can prove a uniqueness theorem but we shall not do this.

From the assumption it follows that when equation (9.144) is substituted into equation (9.138) using equation (9.143), the equation which results can be separated into nine simultaneous equations corresponding to the nine tensors Q_i^n. These equations are:

$$\begin{aligned}
0 &= \mathcal{I}(D_i^0) \\
-2f_i \boldsymbol{\varpi}_i^0 \boldsymbol{\varpi}_i &= \mathcal{I}(\boldsymbol{\varpi}_i^0 \boldsymbol{\varpi}_i D_i^1) - \frac{e_i}{m_i} f_i \left(-B^2 D_i^2 - \tfrac{1}{3}B^4 D_i^7\right) \boldsymbol{\varpi}_i^0 \boldsymbol{\varpi}_i \\
0 &= \mathcal{I}(\boldsymbol{\varpi}_i^0 \boldsymbol{\varpi}_i D_i^2) - \frac{e_i}{m_i} f_i \left(2D_i^1 - \tfrac{4}{3}B^2 D_i^3 - \tfrac{2}{3}B^2 D_i^6\right) \boldsymbol{\varpi}_i^0 \boldsymbol{\varpi}_i \\
0 &= \mathcal{I}(\boldsymbol{\varpi}_i^0 \boldsymbol{\varpi}_i D_i^3) - \frac{e_i}{m_i} f_i D_i^2 \boldsymbol{\varpi}_i^0 \boldsymbol{\varpi}_i \\
0 &= \mathcal{I}(\boldsymbol{\varpi}_i^0 \boldsymbol{\varpi}_i D_i^4) - \frac{e_i}{m_i} f_i \tfrac{1}{3} B^2 D_i^7 \boldsymbol{\varpi}_i^0 \boldsymbol{\varpi}_i
\end{aligned}$$

$$
\begin{aligned}
0 &= \mathcal{I}(\varpi_i^2 D_i^5) \\
0 &= \mathcal{I}(\boldsymbol{\varpi}_i^0 \boldsymbol{\varpi}_i D_i^6) - \frac{e_i}{m_i} f_i (D_i^2 - B^2 D_i^7) \boldsymbol{\varpi}_i^0 \boldsymbol{\varpi}_i \\
0 &= \mathcal{I}(\boldsymbol{\varpi}_i^0 \boldsymbol{\varpi}_i D_i^7) - \frac{e_i}{m_i} f_i (D_i^6 + 2D_i^3) \boldsymbol{\varpi}_i^0 \boldsymbol{\varpi}_i \\
0 &= \mathcal{I}(\boldsymbol{\varpi}_i^0 \boldsymbol{\varpi}_i D_i^8) - \frac{e_i}{m_i} f_i D_i^7 \boldsymbol{\varpi}_i^0 \boldsymbol{\varpi}_i \,.
\end{aligned} \tag{9.146}
$$

Problem 9.7 Show that the nine equations of the set (9.146) follow from the steps indicated.

The coefficients D_i^n must also satisfy subsidiary conditions which come from requiring that n_i, $\boldsymbol{u}$, and T really are the number densities, drift velocity, and temperature. The number density is

$$
n_i = \int \mathrm{d}\boldsymbol{v}\, F_i = \int \mathrm{d}\boldsymbol{v}\, f_i (1 + \varphi_i)\,.
$$

Using equation (9.25) this becomes

$$
\int \mathrm{d}\boldsymbol{v}\, f_i \mathsf{B}_i^{\alpha\beta} \nabla_\alpha u_\beta = 0\,,
$$

or using equation (9.144),

$$
\sum_{n=0}^{8} \int \mathrm{d}\boldsymbol{v}\, f_i D_i^n \mathsf{Q}_i^n = 0\,. \tag{9.147}
$$

Now it is clear that for those tensors Q_i^n which can be expressed in terms of $\boldsymbol{\varpi}_i^0 \boldsymbol{\varpi}_i$ alone, that is all except Q_i^0 and Q_i^5, we have

$$
\int \mathrm{d}\boldsymbol{v}\, f_i D_i^n \mathsf{Q}_i^n = 0 \quad [n \neq 0,\; n \neq 5]\,,
$$

whatever the dependence of D_i^n on ϖ_i^2 and B^2. Hence the sum in equation (9.147) reduces to just two terms, $n = 0$ and $n = 5$. These must be zero separately so we have

$$
\int \mathrm{d}\boldsymbol{v}\, f_i D_i^0 = 0 \quad \text{and} \quad \int \mathrm{d}\boldsymbol{v}\, f_i \varpi_i^2 D_i^5 = 0\,. \tag{9.148}
$$

The drift velocity, $\boldsymbol{u}$, is defined by

$$
\boldsymbol{u} \equiv \frac{1}{\rho} \int \mathrm{d}\boldsymbol{v}\, F_i m_i \boldsymbol{v}_i\,,
$$

so that

$$
\int \mathrm{d}\boldsymbol{v}\, F_i m_i \boldsymbol{w}_i \mathsf{B}_i^{\alpha\beta} \nabla_\alpha u_\beta = 0\,.
$$

This equation is automatically satisfied because $\mathsf{B}_i^{\alpha\beta}$ is even in ϖ_i^2 so the integrals vanish.

The temperature is defined by

$$\tfrac{3}{2}nk_BT \equiv \sum_i \int d\boldsymbol{v}\, F_i \tfrac{1}{2}m_i w_i^2 \,.$$

Hence

$$\sum_i \int d\boldsymbol{v}\, f_i \tfrac{1}{2}m_i w_i^2 \mathsf{B}_i^{\alpha\beta} = 0 \,,$$

which along with equation (9.144) leads to two further conditions on D_i^0 and D_i^5, namely,

$$\sum_i \int d\boldsymbol{v}\, f_i \tfrac{1}{2}m_i w_i^2 D_i^0 = 0 \quad \text{and} \quad \sum_i \int d\boldsymbol{v}\, f_i \tfrac{1}{2}m_i w_i^2 D_i^5 = 0 \,. \tag{9.149}$$

From the first equation of the set (9.146) it follows that D_i^0 is a collision invariant. The only scalar collision invariants are unity and energy so

$$D_i^0 = \alpha + \beta \tfrac{1}{2}m_i w_i^2 \,,$$

where α and β are independent of velocity and the subscript i. But the only values of α and β which are consistent with the conditions of equations (9.148) and (9.149) are zero. Hence

$$D_i^0 = 0 \,.$$

Similarly, from the fifth equation of equation (9.146) and the conditions of Equations (9.148) and (9.149), it follows that

$$D_i^5 = 0 \,.$$

We now substitute the expressions of equation (9.145) into equation (9.146) to obtain equations for the fourteen scalars B_i^n. These are:

$$\begin{aligned} B_i^0 &= 0 \\ B_i^6 &= \frac{1}{B^2}(B_i^3 - B_i^4) \end{aligned} \tag{9.150}$$

$$\begin{aligned} -2f_i \boldsymbol{\varpi}_i^0 \boldsymbol{\varpi}_i &= \mathcal{I}(\boldsymbol{\varpi}_i^0 \boldsymbol{\varpi}_i B_i^1) - \frac{e_i}{m_i} f_i(-B^2 B_i^2)\boldsymbol{\varpi}_i^0 \boldsymbol{\varpi}_i \\ 0 &= \mathcal{I}(\boldsymbol{\varpi}_i^0 \boldsymbol{\varpi}_i B_i^2) - \frac{e_i}{m_i} f_i(2B_i^1 - 2B^2 B_i^3)\boldsymbol{\varpi}_i^0 \boldsymbol{\varpi}_i \\ 0 &= \mathcal{I}(\boldsymbol{\varpi}_i^0 \boldsymbol{\varpi}_i B_i^3) - \frac{e_i}{m_i} f_i B_i^2 \boldsymbol{\varpi}_i^0 \boldsymbol{\varpi}_i \\ 0 &= \mathcal{I}(\boldsymbol{\varpi}_i^0 \boldsymbol{\varpi}_i B_i^4) - \frac{e_i}{m_i} f_i(B_i^2 - B^2 B_i^5)\boldsymbol{\varpi}_i^0 \boldsymbol{\varpi}_i \\ 0 &= \mathcal{I}(\boldsymbol{\varpi}_i^0 \boldsymbol{\varpi}_i B_i^5) - \frac{e_i}{m_i} f_i(2B_i^3 + B_i^4)\boldsymbol{\varpi}_i^0 \boldsymbol{\varpi}_i \,. \end{aligned} \tag{9.151}$$

We get fourteen equations, since the real parts and the imaginary parts of Equations (9.150) and (9.151) must be satisfied separately, just sufficient to determine all the complex B_i^n, in place of the original eighteen equations of equation (9.146) because several of the equations expressed in terms of the B_i^n turn out to be identical. This proves that it is valid to look for a solution of the form of equation (9.137). Now define

$$\begin{aligned} L_i &\equiv B_i^1 + B^2 B_i^3 \\ G_i &\equiv B_i^1 + \mathrm{i}BB_i^2 - B^2 B_i^3 \\ 2P_i &\equiv 2B_i^1 + \mathrm{i}BB_i^2 + B^2 B_i^4 + \mathrm{i}B^3 B_i^5 \,, \end{aligned} \tag{9.152}$$

so the set of equations (9.151) can be reduced to the simple complex equations

$$-2f_i \varpi_i^0 \varpi_i = \mathcal{I}(L_i \varpi_i^0 \varpi_i) \tag{9.153}$$

$$-2f_i \varpi_i^0 \varpi_i = \mathcal{I}(G_i \varpi_i^0 \varpi_i) - 2\mathrm{i}B\frac{e_i}{m_i} f_i G_i \varpi_i^0 \varpi_i \tag{9.154}$$

$$-2f_i \varpi_i^0 \varpi_i = \mathcal{I}(P_i \varpi_i^0 \varpi_i) - \mathrm{i}B\frac{e_i}{m_i} f_i P_i \varpi_i^0 \varpi_i \,. \tag{9.155}$$

Problem 9.8 *Reducing the equations.*

1. Show that the set of equations (9.146) may be reduced to the set (9.151).
2. Then show that the relations of equation (9.152) can reduce the set of equations (9.151) to the set of equations (9.153) – (9.155).

These equations are of the same form as those we considered in Section 9.4 and we can solve them by the same variation method. Notice that in these equations there are no terms corresponding to the second term on the right-hand side of equation (9.48). It therefore follows from the discussion given at the beginning of Section 9.4 that these equations completely determine the quantities L_i, G_i, and P_i. This, of course, is just what we want because there are no subsidiary conditions to be applied to these (the subsidiary conditions only served to make D_i^0 and D_i^5 zero) whereas in Section 9.4 there was a subsidiary condition to be applied.

We expand L_i, G_i, and P_i in generalized Laguerre polynomials

$$\begin{aligned} L_i(\varpi_i^2, B) &= \sum_{n=0}^{\infty} \ell_i^n L_n^{\frac{5}{2}}(\varpi_i^2) \\ G_i(\varpi_i^2, B) &= \sum_{n=0}^{\infty} g_i^n L_n^{\frac{5}{2}}(\varpi_i^2) \\ P_i(\varpi_i^2, B) &= \sum_{n=0}^{\infty} p_i^n L_n^{\frac{5}{2}}(\varpi_i^2) \,, \end{aligned} \tag{9.156}$$

and we can anticipate the result to be derived later that only the coefficients ℓ_i^0, g_i^0, and p_i^0 will enter into the expressions for the pressure tensor. Now we note from equation (9.153) that in evaluating L_i, it is independent of B so that if we were to take the value of $G_i(\varpi_i^2, 0)$ we would have the same expression, so we conclude that $L_i = G_i(\varpi_i^2, 0)$. In a similar manner, if we compare equation (9.155) with equation (9.154), the expressions are equivalent if we replace B by $B/2$ in equation (9.154), so we conclude that $P_i(\varpi_i^2, B) = G_i(\varpi_i^2, B/2)$. We therefore only need to solve for G_i.

9.6.1 Solving for G_i

We start by considering equation (9.156) and construct the Davison function,

$$\begin{aligned}\mathcal{D}(G) = &\sum_i \int \mathrm{d}\boldsymbol{v}(\boldsymbol{\varpi}_i^0\boldsymbol{\varpi}_i)G_i : \left[4f_i\boldsymbol{\varpi}_i^0\boldsymbol{\varpi}_i + \mathcal{I}(\boldsymbol{\varpi}_i^0\boldsymbol{\varpi}_i G_i) - 2\mathrm{i}B\frac{e_i}{m_i}f_i\boldsymbol{\varpi}_i^0\boldsymbol{\varpi}_i G_i\right] \\ = &\, 10\sum_i n_i g_i^0 - 5\mathrm{i}B\sum_i \frac{n_i e_i}{m_i}\sum_{n=0}^{N}(g_i^n)^2\frac{(\frac{5}{2}+n)!}{n!\frac{5}{2}!} \\ & -\frac{1}{2}\left\{\boldsymbol{\varpi}_i^0\boldsymbol{\varpi}_i\sum_{n=0}^{N} g_i^n L_n^{\frac{5}{2}}(\varpi_i^2); \boldsymbol{\varpi}_i^0\boldsymbol{\varpi}_i\sum_{n=0}^{N} g_i^n L_n^{\frac{5}{2}}(\varpi_i^2)\right\}. \end{aligned} \tag{9.157}$$

The first approximation is obtained by setting $N = 0$. $\mathcal{D}(G)$ then becomes

$$\begin{aligned}\mathcal{D}(G) = &\, 10(n_1 g_1^0 + n_2 g_2^0) - 5\mathrm{i}B\left[\frac{n_1 e_1}{m_1}(g_1^0)^2 + \frac{n_2 e_2}{m_2}(g_2^0)^2\right] \\ & -(g_1^0)^2\left\{n_1^2[\boldsymbol{\varpi}_1^0\boldsymbol{\varpi}_1, \boldsymbol{\varpi}_1^0\boldsymbol{\varpi}_1]_1 + n_1 n_2[\boldsymbol{\varpi}_1^0\boldsymbol{\varpi}_1, \boldsymbol{\varpi}_1^0\boldsymbol{\varpi}_1]_{12}\right\} \\ & -(g_2^0)^2\left\{n_2^2[\boldsymbol{\varpi}_2^0\boldsymbol{\varpi}_2, \boldsymbol{\varpi}_2^0\boldsymbol{\varpi}_2]_2 + n_1 n_2[\boldsymbol{\varpi}_2^0\boldsymbol{\varpi}_2, \boldsymbol{\varpi}_2^0\boldsymbol{\varpi}_2]_{12}\right\} \\ & -2g_1^0 g_2^0 n_1 n_2[\boldsymbol{\varpi}_1^0\boldsymbol{\varpi}_1, \boldsymbol{\varpi}_2^0\boldsymbol{\varpi}_2]_{12}, \end{aligned}$$

which in terms of the collision time τ_e defined by equation (9.98) can be written, using Appendix B to give values for the collision integrals, as

$$\begin{aligned}\mathcal{D}(G) \simeq \frac{5n}{\tau_e}\Big\{&\tau_e g_1^0 + \tau_e g_2^0 - \tfrac{1}{2}(g_1^0)^2[-\mathrm{i}\omega_{ce}\tau_e + .3(2+\sqrt{2})] \\ & -\tfrac{1}{2}(g_2^0)^2[.3\sqrt{2M_1} + (1+\mathrm{i}\omega_{ce}\tau_e)M_1] + .4M_1 g_1^0 g_2^0\Big\}, \end{aligned} \tag{9.158}$$

where $\simeq$ signifies that the usual plasma approximations of equation (9.101) have been used, and $M_1 = m_1/(m_1+m_2) \simeq m_1/m_2$. Equation (B6.35) is stationary for the values, using $\omega_{ci}\tau_i \simeq 5\sqrt{M_1}\omega_{ce}\tau_e/2\sqrt{2}$ and $\tau_i = 5\tau_e/2\sqrt{2M_1}$,

$$\begin{aligned}[g_1^0]_1 &\simeq \frac{\tau_e}{.3(2+\sqrt{2}) - \mathrm{i}\omega_{ce}\tau_e} \simeq \frac{4M_1\tau_i}{3(1+\sqrt{2})\sqrt{M_1} - 4\mathrm{i}\omega_{ci}\tau_i} \\ [g_2^0]_1 &\simeq \frac{\tau_e}{\sqrt{M_1}[.3\sqrt{2} + (1+\mathrm{i}\omega_{ce}\tau_e)\sqrt{M_1}]} \simeq \frac{4\tau_i}{3+4\mathrm{i}\omega_{ci}\tau_i}. \end{aligned} \tag{9.159}$$

These are the first approximations to the constants and we can be confident that they are correct to within a few percent by what we know of the problem in the absence of a magnetic field.

Problem 9.9 *Evaluating the coefficients.*

1. Show that the Davison function on line 1 of equation (9.157) is equivalent to line 2 (i.e., show where the factors of 10 and 5 come from).

2. Show that setting $N = 0$ leads to equation (9.158).

3. Find g_1^0 and g_2^0 and show that they reduce to equation (9.159) if we neglect terms of order $\sqrt{M_1}$, depending on the strength of the magnetic field.

From our conclusions about the relationships between L_i, G_i, and P_i, we immediately find

$$\begin{aligned} \left[\ell_1^0\right]_1 &\simeq \frac{\tau_e}{.3(2+\sqrt{2})} \simeq \frac{4\sqrt{M_1}\tau_i}{3(1+\sqrt{2})} \\ \left[\ell_2^0\right]_1 &\simeq \frac{\tau_e}{.3\sqrt{2M_1}} \simeq \frac{4\tau_i}{3} \end{aligned} \tag{9.160}$$

and

$$\begin{aligned} \left[p_1^0\right]_1 &\simeq \frac{2\tau_e}{0.6(2+\sqrt{2}) - \mathrm{i}\omega_{ce}\tau_e} \simeq \frac{4M_1\tau_i}{3(1+\sqrt{2})\sqrt{M_1} - 2\mathrm{i}\omega_{ci}\tau_i} \\ \left[p_2^0\right]_1 &\simeq \frac{\tau_e}{\sqrt{M_1}[.3\sqrt{2} + (1+.5\mathrm{i}\omega_{ce}\tau_e)\sqrt{M_1}]} \simeq \frac{4\tau_i}{3+2\mathrm{i}\omega_{ci}\tau_i}. \end{aligned} \tag{9.161}$$

9.6.2 The pressure tensor elements

We now have to consider how the pressure tensor involves these solutions. The pressure tensor is defined by

$$P_{st} \equiv \sum_i n_i m_i \langle w_{is} w_{it} \rangle = \sum_i m_i \int \mathrm{d}\boldsymbol{v}\, F_i w_{is} w_{it}$$

so

$$\begin{aligned} P_{st} - p\delta_{st} &= -\sum_i \int \mathrm{d}\boldsymbol{v}_i w_{is} w_{it} f_i \mathsf{B}_i^{\alpha\beta} \nabla_\alpha u_\beta \\ &= -2k_B T \sum_i n_i \frac{1}{\pi^{3/2}} \int \mathrm{d}\boldsymbol{\varpi}_i\, \mathrm{e}^{-\varpi_i^2} \varpi_{is}\varpi_{it} \mathsf{B}_i^{\alpha\beta} \nabla_\alpha u_\beta \\ &= -2k_B T \sum_i n_i \sum_{m=1}^{6} \frac{1}{\pi^{3/2}} \int \mathrm{d}\boldsymbol{\varpi}_i\, \mathrm{e}^{-\varpi_i^2} \varpi_{is}\varpi_{it} B_i^m \mathsf{T}_{i,\alpha\beta}^m \nabla_\alpha u_\beta \end{aligned} \tag{9.162}$$

where we have used equation (9.137). The integral over the magnitude of ϖ_i in equation (9.162) is easy to do. We expand the scalar B_i^m in a series of generalized Laguerre polynomials with coefficients which can be related to the coefficients in the expansions of equation (9.156), i.e.,

$$B_i^m = \sum_{n=0}^{\infty} b_i^{m,n} L_n^{\frac{5}{2}}(\varpi_i^2)\,.$$

Recalling that all the tensors T_i^m are quadratic in ϖ_i, equation (9.162) becomes

$$P_{st} - p\delta_{st} = -\tfrac{15}{2} k_B T \sum_{m=1}^{6} (n_1 b_1^{m,0} + n_2 b_2^{m,0})\gamma_{st}^m\,, \tag{9.163}$$

where γ_{st} is given by the integrations over the angles of $\boldsymbol{\varpi}$ and is defined by

$$\gamma_{st}^m \equiv \frac{1}{4\pi}\int \mathrm{d}\Omega_\varpi \frac{1}{\varpi^4}\varpi_s\varpi_t \mathsf{T}_{\alpha\beta}^m \nabla_\alpha u_\beta \tag{9.164}$$

where the tensors T^m are defined by equation (9.136) and because they are all quadratic in ϖ, equation (9.164) is independent of the magnitude of ϖ.

To evaluate these γ_{st}^m we first note that

$$\begin{aligned}
\frac{1}{5} &= \frac{1}{4\pi}\int \mathrm{d}\Omega_\varpi \frac{1}{\varpi^4}\varpi_x^4 = \frac{1}{4\pi}\int \mathrm{d}\Omega_\varpi \frac{1}{\varpi^4}\varpi_y^4 = \text{etc.}\\
\frac{1}{15} &= \frac{1}{4\pi}\int \mathrm{d}\Omega_\varpi \frac{1}{\varpi^4}\varpi_x^2\varpi_y^2 = \text{etc.}\\
\frac{1}{3} &= \frac{1}{4\pi}\int \mathrm{d}\Omega_\varpi \frac{1}{\varpi^4}\varpi_x^2\varpi^2 = \text{etc.}
\end{aligned} \tag{9.165}$$

We then suppose the magnetic field to be in the z-direction and write out the tensors T^m explicitly. With the help of equation (9.165), it is then possible to write down the γ_{st}^m by inspection. We shall describe one typical calculation, that of γ_{st}^2, and merely quote the results for the other cases. The tensor T^2 written explicitly is‡

$$\mathsf{T}^2 = \frac{B}{2}\begin{pmatrix} 2\varpi_x\varpi_y & \varpi_y^2 - \varpi_x^2 & \varpi_y\varpi_z \\ \varpi_y^2 - \varpi_x^2 & -2\varpi_x\varpi_y & -\varpi_x\varpi_z \\ \varpi_y\varpi_z & -\varpi_x\varpi_z & 0 \end{pmatrix},$$

and from this we can write down

$$\gamma_{st}^2 = \frac{B}{15}\begin{pmatrix} -2s_{xy} & s_{xx} - s_{yy} & -s_{yz} \\ s_{xx} - s_{yy} & 2s_{xy} & s_{xz} \\ -s_{yz} & s_{xz} & 0 \end{pmatrix}, \tag{9.166}$$

‡This tensor appears different from that in W. Marshall's work because he chose B to be in the x-direction instead of the z-direction.

where

$$s_{\alpha\beta} = \tfrac{1}{2}(\nabla_\alpha u_\beta + \nabla_\beta u_\alpha) - \tfrac{1}{3}\nabla\cdot\boldsymbol{u}\delta_{\alpha\beta}\,. \qquad (9.167)$$

The argument leading to equation (9.166) is as follows. First consider γ^2_{zz}. Putting $s = t = z$ in equation (9.164), we see that in order to get a nonzero result, we need to find a component of T^2 proportional to ϖ_x^2, ϖ_y^2, or ϖ_z^2. But the only components of T^2 involving the square of components of ϖ are T_{xy} and T_{yx} and in these the integral coming from the product $\varpi_z^2\varpi_y^2$ must be exactly cancelled by that from $\varpi_z^2\varpi_x^2$ which occurs with opposite sign. Hence γ^2_{zz} is zero. Next consider γ^2_{xz}. Putting $s = x$, $t = z$ in equation (9.164), we see we need a component of T^2 proportional to $\varpi_x\varpi_z$ in order to get a nonzero result. The only such components are T^2_{yz} and T^2_{zy}. Using the second of equation (9.165), we can therefore write immediately

$$\gamma^2_{xz} = \tfrac{1}{15}\left(-\tfrac{1}{2}B\nabla_y u_z - \tfrac{1}{2}B\nabla_z u_y\right) = -\tfrac{1}{15}Bs_{yz}\,.$$

By similar arguments the other elements of γ^2_{st} can be written down to give equation (9.166).

Problem 9.10 Work out the remaining elements of $\gamma^2_{\alpha\beta}$ and the rest of the $\gamma^m_{\alpha\beta}$ for $m = 1$ and $m = 3$ through $m = 6$.

Problem 9.11 Work out the diagonal elements of the pressure tensor, p_{xx}, p_{yy}, and p_{zz} after finding the diagonal elements of all of the γ^m_{st}.

Having worked out all the γ^m_{st} in problems 9.10 and 9.11, it is easy to evaluate equation (9.163). The results we obtain are

$$\begin{aligned}
p_{xx} &= p - \tfrac{1}{2}k_BT\sum_i n_i\left(2b_i^{1,0}s_{xx} - 2Bb_i^{2,0}s_{xy} + 2B^2b_i^{3,0}s_{yy}\right),\\
p_{yy} &= p - \tfrac{1}{2}k_BT\sum_i n_i\left(2b_i^{1,0}s_{yy} + 2Bb_i^{2,0}s_{xy} + 2B^2b_i^{3,0}s_{xx}\right),\\
p_{zz} &= p - \tfrac{1}{2}k_BT\sum_i n_i\left(2b_i^{1,0} + 2B^2b_i^{3,0}\right)s_{zz}\,, \qquad (9.168)\\
p_{xy} &= p_{yx} = -\tfrac{1}{2}k_BT\sum_i n_i\left[2(b_i^{1,0} - B^2b_i^{3,0})s_{xy} + Bb_i^{2,0}(s_{xx} - s_{yy})\right]\\
p_{xz} &= p_{zx} = -\tfrac{1}{2}k_BT\sum_i n_i\left[(2b_i^{1,0} + B^2b_i^{4,0})s_{xz} - B(b_i^{2,0} + B^2b_i^{5,0})s_{yz}\right]\\
p_{yz} &= p_{zy} = -\tfrac{1}{2}k_BT\sum_i n_i\left[(2b_i^{1,0} + B^2b_i^{4,0})s_{yz} + B(b_i^{2,0} + B^2b_i^{5,0})s_{xz}\right] \qquad (9.169)
\end{aligned}$$

since $B^2b_i^{6,0} = b_i^{3,0} - b_i^{4,0}$ from equation (9.150). We now use this relation and equation (9.152) to express $b_i^{m,0}$ in terms of ℓ_1^0, g_1^0, and p_1^0 as follows:

$$b_i^{1,0} = \tfrac{1}{2}[\ell_i^0 + \mathrm{Re}(g_i^0)]$$

$$\begin{aligned}
Bb_i^{2,0} &= \mathrm{Im}(g_i^0)\\
B^2 b_i^{3,0} &= \tfrac{1}{2}[\ell_i^0 - \mathrm{Re}(g_i^0)]\\
B^2 b_i^{4,0} &= 2\,\mathrm{Re}(p_i^0) - \ell_i^0 - \mathrm{Re}(g_i^0)\\
B^3 b_i^{5,0} &= 2\,\mathrm{Im}(p_i^0) - \mathrm{Im}(g_i^0)\\
B^4 b_i^{6,0} &= \tfrac{3}{2}\ell_i^0 + \tfrac{1}{2}\mathrm{Re}(g_i^0) - 2\,\mathrm{Re}(p_i^0)\,,
\end{aligned}$$

and remembering from equation (9.167) that

$$s_{yy} + s_{zz} = -s_{xx}\,,$$

equations (9.168) through (9.169) become

$$\begin{aligned}
p_{xx} &= p - k_B T \sum_i n_i \left[\tfrac{1}{2}\mathrm{Re}(g_i^0)(s_{xx} - s_{yy}) - \mathrm{Im}(g_i^0)s_{xy} - \ell_i^0 s_{zz}\right]\\
p_{yy} &= p - k_B T \sum_i n_i \left[\tfrac{1}{2}\mathrm{Re}(g_i^0)(s_{yy} - s_{xx}) + \mathrm{Im}(g_i^0)s_{xy} - \ell_i^0 s_{zz}\right]\\
p_{zz} &= p - k_B T \sum_i n_i \ell_i^0 s_{zz}\\
p_{xy} &= p_{yx} = -k_B T \sum_i n_i \left[\mathrm{Re}(g_i^0)s_{xy} + \tfrac{1}{2}\mathrm{Im}(g_i^0)(s_{xx} - s_{yy})\right]\\
p_{xz} &= p_{zx} = -k_B T \sum_i n_i \left[\mathrm{Re}(p_i^0)s_{xz} - \mathrm{Im}(p_i^0)s_{yz}\right]\\
p_{yz} &= p_{zy} = -k_B T \sum_i n_i \left[\mathrm{Re}(p_i^0)s_{yz} + \mathrm{Im}(p_i^0)s_{xz}\right]\,. \qquad (9.170)
\end{aligned}$$

Problem 9.12 Show that the elements of the pressure tensor may be written in terms of the g_i^0, ℓ_i^0, and p_i^0 as given in equation (9.170).

All these expressions involve sums over i of the coefficients ℓ_i, g_i, or p_i weighted with the number densities n_i. But since we can put $n_1 \simeq n_2 \simeq \frac{1}{2}n$, they effectively involve just the simple sums, $\ell_1^0 + \ell_2^0$, $g_1^0 + g_2^0$, and $p_1^0 + p_2^0$. Now from Equations (9.159) and (9.161) we see that g_2^0 and p_2^0 are greater than g_1^0 and p_1^0 respectively by a factor which is of the order of magnitude of $\sqrt{m_2/m_1}$ in weak fields and a factor of (m_2/m_1) in strong fields, and from equation (9.160) we see that ℓ_2^0 is greater than ℓ_1^0 by a factor $\sim \sqrt{m_2/m_1}$. This clearly shows that the viscosity of the plasma is almost entirely due to the ions, the electrons contributing a negligible amount. This of course is exactly as we found in Chapter 8 when we considered the problem in the absence of a magnetic field and is due to the fact that ions can transport momentum much more efficiently than electrons. In view of this it is not surprising that equations (9.159), (9.160), and (9.161), giving g_i^0, ℓ_i^0, and p_i^0 in terms of ω_{ce} and τ_e, are cumbersome because ω_{ce} is essentially a cyclotron frequency for electrons and τ_e is a suitably defined collision time for electrons whereas the coefficients are really determined by ions and collisions between ions. It is therefore appropriate to express g_i^0, ℓ_i^0, and p_i^0, not in terms of ω_{ce} and τ_e, but in terms of some suitably defined quantities relating to the ions. Defining

$$\omega_{ci} \equiv \frac{e_2 B}{m_2} = \frac{eB}{m_2} = \frac{m_1}{m_2}\omega_{ce} \simeq M_1 \omega_{ce}$$

and

$$\tau_i = \frac{5}{2\sqrt{2}}\left(\frac{m_2}{m_1}\right)^{1/2} \tau_e \simeq \frac{5}{2\sqrt{2}}\frac{\tau_e}{\sqrt{M_1}} ,$$

then ω_{ci} is the cyclotron frequency for the ions and τ_i is an appropriately defined collision time for ion collisions. From equations (9.159), (9.160), and (9.161) it is apparent that g_1^0, ℓ_1^0, and p_1^0 are small so they will not enter into the solutions. We then write the real and imaginary parts for g_2^0 and p_2^0 as

$$\begin{aligned}
\mathrm{Re}(g_2^0) &= \frac{12\tau_i}{9 + 16\omega_{ci}^2\tau_i^2} \\
\mathrm{Im}(g_2^0) &= -\frac{16\tau_i(\omega_{ci}\tau_i)}{9 + 16\omega_{ci}^2\tau_i^2} \\
\mathrm{Re}(p_2^0) &= \frac{12\tau_i}{9 + 4\omega_{ci}^2\tau_i^2} \\
\mathrm{Im}(p_2^0) &= -\frac{8\tau_i(\omega_{ci}\tau_i)}{9 + 4\omega_{ci}^2\tau_i^2} .
\end{aligned}$$

Substituting these results into equation (9.170) we obtain[§]

$$p_{xx} = p - \frac{2\mu}{1 + \frac{16}{9}\omega_{ci}^2\tau_i^2}\left(s_{xx} + \tfrac{4}{3}\omega_{ci}\tau_i s_{xy} - \tfrac{8}{9}\omega_{ci}^2\tau_i^2 s_{zz}\right) \tag{9.171}$$

$$p_{yy} = p - \frac{2\mu}{1 + \frac{16}{9}\omega_{ci}^2\tau_i^2}\left(s_{yy} - \tfrac{4}{3}\omega_{ci}\tau_i s_{xy} - \tfrac{8}{9}\omega_{ci}^2\tau_i^2 s_{zz}\right) \tag{9.172}$$

$$\begin{aligned}
p_{zz} &= p - 2\mu s_{zz} \\
p_{xy} = p_{yx} &= -\frac{2\mu}{1 + \frac{16}{9}\omega_{ci}^2\tau_i^2}\left[s_{xy} - \tfrac{2}{3}\omega_{ci}\tau_i(s_{xx} - s_{yy})\right] \\
p_{xz} = p_{zx} &= -\frac{2\mu}{1 + \frac{4}{9}\omega_{ci}^2\tau_i^2}\left(s_{xz} + \tfrac{2}{3}\omega_{ci}\tau_i s_{yz}\right) \\
p_{yz} = p_{zy} &= -\frac{2\mu}{1 + \frac{4}{9}\omega_{ci}^2\tau_i^2}\left(s_{yz} - \tfrac{2}{3}\omega_{ci}\tau_i s_{xz}\right) ,
\end{aligned} \tag{9.173}$$

where

$$\mu = \tfrac{1}{3} n k_B T \tau_i$$

is the coefficient of viscosity and $s_{\alpha\beta}$ is defined by equation (9.167). The signs of some of the terms are given incorrectly by Chapman and Cowling. This

---

[§]These results are different from those of W. Marshall because he had the magnetic field in the x-direction. Otherwise we agree.

can be seen very directly by appealing to the reciprocal relations of Onsager. These state that if we write

$$p_{\alpha\beta} = \sum_{<\gamma\varphi>} L(B;\alpha\beta;\gamma\varphi)s_{\gamma\varphi}$$

where the sum goes over all pairs xx, yy, zz, xy, xz, yz, then

$$L(B;\alpha\beta;\gamma\varphi) = L(-B;\gamma\varphi;\alpha\beta)\,.$$

Hence since the coefficients of s_{xy} in p_{xx} and s_{xy} in p_{yy} are odd in the field B, it follows that they must be equal in magnitude but opposite in sign. This is just as given in equations (9.172) and (9.173) but not as given by Chapman and Cowling. This mistake in Chapman and Cowling was first noticed by G. J. Hymen, P. Masseur, and S. R. De Groot[54].

Problem 9.13 Fill in the steps between equations (9.170) and (9.173).

9.7 Summary of results

It is intended that this section be entirely self-contained. All the symbols we use will therefore be redefined as needed.

Maxwell's equations, in SI units in vacuum, are

$$\begin{aligned}
\nabla \cdot \boldsymbol{B} &= 0 \\
\nabla \cdot \boldsymbol{E} &= Q/\epsilon_0 \\
\nabla \times \boldsymbol{E} &= -\frac{\partial \boldsymbol{B}}{\partial t} \\
\nabla \times \boldsymbol{B} &= \mu_0 \boldsymbol{J} + \frac{1}{c^2}\frac{\partial \boldsymbol{E}}{\partial t}\,,
\end{aligned}$$

where c is the velocity of light. The total current $\boldsymbol{J}$ is the sum of a convection current $Q\boldsymbol{u}$ and a conduction current $\boldsymbol{j}$, i.e.,

$$\boldsymbol{J} = Q\boldsymbol{u} + \boldsymbol{j}\,,$$

where $\boldsymbol{u}$ is the velocity of the center of mass of a fluid element. The equation of continuity for the density of the plasma, ρ, is

$$\frac{\partial}{\partial t}\rho + \nabla \cdot (\rho \boldsymbol{u}) = 0\,.$$

The equation of motion is

$$\rho\left(\frac{\partial}{\partial t} + \boldsymbol{u}\cdot\nabla\right)u_\alpha = -\nabla_\beta p_{\beta\alpha} + QE_\alpha + (\boldsymbol{J}\times\boldsymbol{B})_\alpha + \sum_i \rho_i X_{i\alpha}\,.$$

In the first term on the right-hand side of this equation, $p_{\alpha\beta}$ is the pressure tensor and the repeated suffix implies a summation. $\boldsymbol{X}_i$ is any non-electromagnetic force per unit mass which may act on particles i. ρ_i is the density of particles and $i = 1$ for electrons and $i = 2$ for ions.

The energy equation is

$$\left(\frac{\partial}{\partial t} + \boldsymbol{u} \cdot \nabla\right) \frac{3}{2} n k_B T = -\frac{3}{2} n k_B T \nabla \cdot \boldsymbol{u} - \beta p_{\alpha\beta} \nabla_\alpha u_\beta - \nabla \cdot \boldsymbol{q}$$
$$+ \boldsymbol{j} \cdot \left[\boldsymbol{E} + \boldsymbol{u} \times \boldsymbol{B} - \frac{m_1}{e}(\boldsymbol{X}_1 - \boldsymbol{X}_2)\right].$$

Here n is the total number density $n_1 + n_2$, k_B is Boltzmann's constant, and T is the temperature. In the second term, a summation over both α and β is implied, i.e., this term is really nine terms. $\boldsymbol{q}$ is the heat flux vector.

The purpose of kinetic theory is to provide expressions for $\boldsymbol{j}$, $\boldsymbol{q}$, and $p_{\alpha\beta}$. The conduction current is given by

$$\boldsymbol{j} = \sigma^I \boldsymbol{D}^{\parallel} + \sigma^{II} \boldsymbol{D}^{\perp} + \sigma^{III} \hat{\boldsymbol{b}} \times \boldsymbol{D}^{\perp} + \varphi^I (\nabla T)^{\parallel} + \varphi^{II} (\nabla T)^{\perp} + \varphi^{III} \hat{\boldsymbol{b}} \times (\nabla T)^{\perp} \tag{9.174}$$

where $\boldsymbol{D}$ is a "generalized" electric field given by

$$\boldsymbol{D} = \boldsymbol{E} + \boldsymbol{u} \times \boldsymbol{B} + \frac{1}{ne} \nabla p - \frac{m_1}{e}(\boldsymbol{X}_1 - \boldsymbol{X}_2),$$

and the superscripts $\parallel$ and $\perp$ denote components parallel and perpendicular to the magnetic field. $\hat{\boldsymbol{b}}$ is a unit vector in the direction of $\boldsymbol{B}$. The three terms involving the temperature gradients in equation (9.174) are thermal diffusion terms. $p = n k_B T$ is the pressure, m_1 the electron mass, and e the electronic charge. The electrical conductivities are given by

$$\begin{aligned}
\left[\sigma^I\right]_2 &= \frac{n_1 e^2 \tau_e}{m_1} 1.932 \\
\left[\sigma^{II}\right]_2 &\simeq \frac{n_1 e^2 \tau_e}{m_1} \frac{\omega_{ce}^2 \tau_e^2 + 1.802}{\omega_{ce}^4 \tau_e^4 + 6.281 \omega_{ce}^2 \tau_e^2 + .9325} \\
\left[\sigma^{III}\right]_2 &\simeq \frac{n_1 e^2 \tau_e}{m_1} \frac{\omega_{ce} \tau_e (\omega_{ce}^2 \tau_e^2 + 4.381)}{\omega_{ce}^4 \tau_e^4 + 6.281 \omega_{ce}^2 \tau_e^2 + .9325},
\end{aligned} \tag{9.175}$$

where

$$\omega_{ce} = \frac{eB}{m_1} = 1.759 \times 10^{11} B \text{ sec}^{-1} \tag{9.176}$$

is the electron cyclotron frequency with the magnetic field in Tesla, and

$$\tau_e = 3(2\pi)^{3/2} \frac{\sqrt{m_1} \epsilon_0^2 (k_B T)^{3/2}}{n_1 e^4 \ln \Lambda} = \frac{2.76 \cdot 10^5 T^{3/2}}{n_e \ln \Lambda} \text{ sec} \tag{9.177}$$

is a suitable defined collision time for electrons. This value is equivalent to that given by Balescu[47] and Braginskii[43] if $\Lambda = 9 N_D$. Plots of σ^{II} and σ^{III} are given in Figure 9.1.

Of particular interest is a "perpendicular conductivity," $\sigma^\perp$, defined in the absence of temperature and density gradients as

$$\boldsymbol{j}^\perp \equiv \sigma^\perp \boldsymbol{E}^\perp ,$$

where $\boldsymbol{E}^\perp$ is an applied electric field perpendicular to $\boldsymbol{B}$ and the experimental arrangement is such that additional electric fields are set up inside the plasma so as to prevent any current flowing in a direction perpendicular to both $\boldsymbol{E}^\perp$ and $\boldsymbol{B}$. $\sigma^\perp$ is given by

$$\sigma^\perp = \sigma^{II} \left[1 + \omega_{ce}^2 \tau_e^2 \left(\frac{\omega_{ce}^2 \tau_e^2 + 4.381}{\omega_{ce}^2 \tau_e^2 + 1.802} \right)^2 \right] .$$

Notice that in the limit of very strong magnetic fields such that $\omega_{ce}\tau_e \gg 1$,

$$\sigma^\perp = \frac{n_1 e^2 \tau_e}{m_1} = \frac{\sigma^I}{1.932} ,$$

whereas in weak fields ($\omega_{ce}\tau_e \ll 1$),

$$\sigma^\perp = \frac{n_1 e^2 \tau_e}{m_1} 1.932 = \sigma^I .$$

The ratio $\sigma^\perp / \sigma^I$ is also plotted in Figure 9.1.

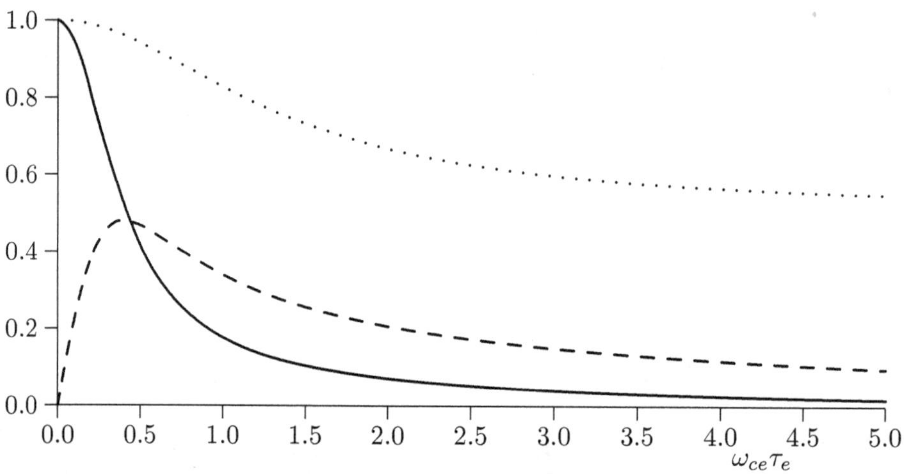

FIGURE 9.1
Electrical Conductivities. σ^{II}/σ^I (solid), σ^{III}/σ^I (dashed), and $\sigma^\perp/\sigma^I$ (dotted).

The thermal diffusion coefficients are given by

$$\left[\varphi^I\right]_2 = 1.553 \frac{k_B n_1 e \tau_e}{m_1}$$

$$\left[\varphi^{II}\right]_2 = -\frac{3k_B n_1 e\tau_e}{2m_1}\frac{\omega_{ce}^2\tau_e^2 - .9657}{\omega_{ce}^4\tau_e^4 + 6.281\omega_{ce}^2\tau_e^2 + .9325}$$

$$\left[\varphi^{III}\right]_2 = \frac{3k_B n_1 e\tau_e}{2m_1}\frac{2.8657\omega_{ce}\tau_e}{\omega_{ce}^4\tau_e^4 + 6.281\omega_{ce}^2\tau_e^2 + .9325}, \tag{9.178}$$

and are plotted in Figure 9.2.

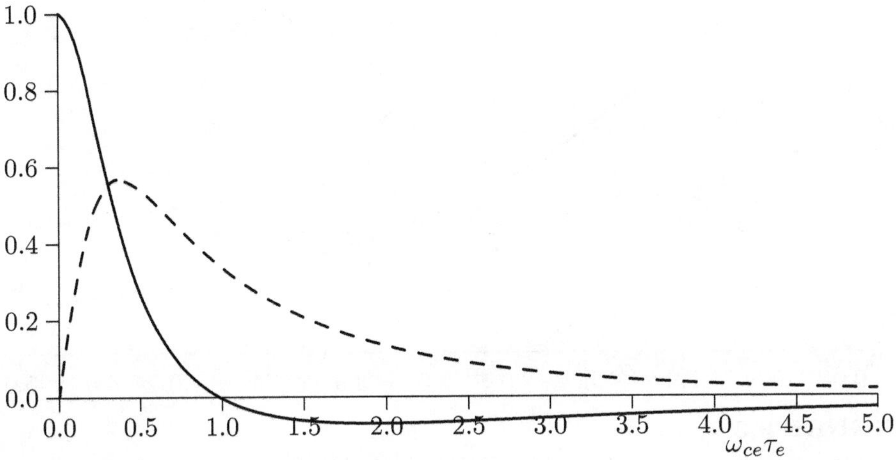

FIGURE 9.2
Thermal diffusion coefficients. φ^{II}/φ^{I} (solid) and $\varphi^{III}/\varphi^{I}$ (dashed).

The heat flux vector can be given in terms of the temperature gradients and the current $\boldsymbol{j}$. These formulas are

$$\boldsymbol{q} = -\lambda^I(\nabla T)^{\parallel} - \lambda^{II}(\nabla T)^{\perp} - \lambda^{III}\hat{\boldsymbol{b}} \times (\nabla T)^{\perp} + \psi^I \boldsymbol{j}^{\parallel} + \psi^{II}\boldsymbol{j}^{\perp} + \psi^{III}\hat{\boldsymbol{b}} \times \boldsymbol{j}^{\perp} \tag{9.179}$$

where

$$\left[\lambda^I\right]_2 = \frac{n_1\tau_e k_B^2 T}{m_1}1.340$$

$$\left[\lambda^{II}\right]_2 = \frac{n_1\tau_e k_B^2 T}{m_1}\left[\frac{4.664}{(\omega_{ce}^2\tau_e^2 + 3.481)}\right]$$

$$\left[\lambda^{III}\right]_2 = \frac{5n_1\tau_e k_B^2 T}{2m_1}\left[\frac{\omega_{ce}\tau_e}{(\omega_{ce}^2\tau_e^2 + 3.481)}\right] \tag{9.180}$$

and

$$\left[\psi^I\right]_2 = -3.304\frac{k_B T}{e}$$

$$\left[\psi^{II}\right]_2 = -\frac{5k_B T}{2e}\frac{\omega_{ce}^2\tau_e^2 + 4.600}{\omega_{ce}^2\tau_e^2 + 3.481}$$

$$\left[\psi^{III}\right]_2 = -\frac{3k_BT}{2e}\frac{\omega_{ce}\tau_e}{\omega_{ce}^2\tau_e^2+3.481}. \tag{9.181}$$

Plots of λ^{II}/λ^I and λ^{III}/λ^I are given in Figure 9.3 while plots of ψ^{II}/ψ^I and $-\psi^{III}/\psi^I$ are given in Figure 9.4.

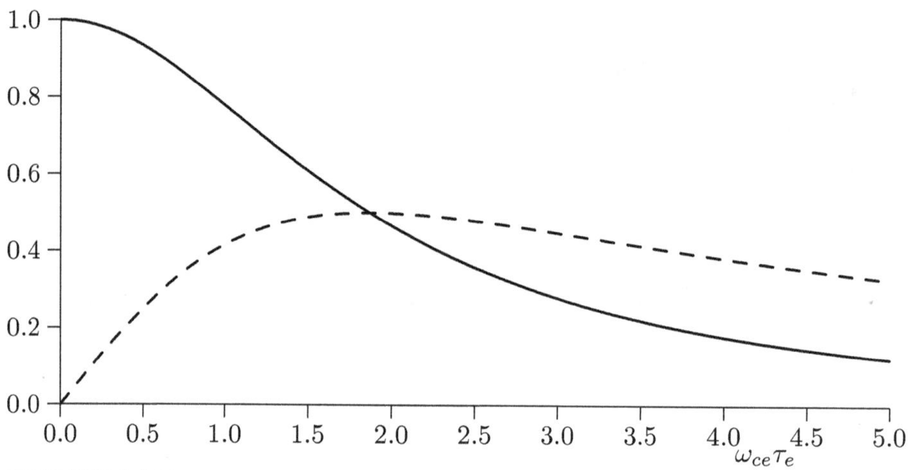

FIGURE 9.3
Heat flux coefficients. λ^{II}/λ^I (solid) and λ^{III}/λ^I (dashed).

The heat flux vector can also be given in terms of the temperature gradient and the "generalized" electric field $\boldsymbol{D}$. These formulas are

$$\boldsymbol{q} = -\theta^I(\nabla T)^{\parallel} - \theta^{II}(\nabla T)^{\perp} - \theta^{III}\hat{\boldsymbol{b}}\times(\nabla T)^{\perp} + \xi^I\boldsymbol{D}^{\parallel} + \xi^{II}\boldsymbol{D}^{\perp} + \xi^{III}\hat{\boldsymbol{b}}\times\boldsymbol{D}^{\perp} \tag{9.182}$$

with

$$\begin{aligned}
\left[\theta^I\right]_2 &\simeq \frac{n_1\tau_ek_B^2T}{m_1}6.472\\
\left[\theta^{II}\right]_2 &\simeq \frac{5n_1\tau_ek_B^2T}{2m_1}\left[\frac{.3657\omega_{ce}^2\tau_e^2+2.415}{\omega_{ce}^4\tau_e^4+6.281\omega_{ce}^2\tau_e^2+.9325}\right]\\
\left[\theta^{III}\right]_2 &\simeq -\frac{5n_1\tau_ek_B^2T}{2m_1}\left[\frac{\omega_{ce}\tau_e(\omega_{ce}^2\tau_e^2+6.199)}{\omega_{ce}^4\tau_e^4+6.281\omega_{ce}^2\tau_e^2+.9325}\right]
\end{aligned} \tag{9.183}$$

and

$$\begin{aligned}
\left[\xi^I\right]_2 &\simeq -\frac{n_1e\tau_ek_BT}{m_1}6.383\\
\left[\xi^{II}\right]_2 &\simeq -\frac{n_1e\tau_ek_BT}{m_1}\frac{\omega_{ce}^2\tau_e^2+5.953}{\omega_{ce}^4\tau_e^4+6.281\omega_{ce}^2\tau_e^2+.9325}\\
\left[\xi^{III}\right]_2 &\simeq -\frac{5n_1e\tau_ek_BT}{2m_1}\frac{\omega_{ce}\tau_e(\omega_{ce}^2\tau_e^2+6.10)}{\omega_{ce}^4\tau_e^4+6.281\omega_{ce}^2\tau_e^2+.9325}.
\end{aligned} \tag{9.184}$$

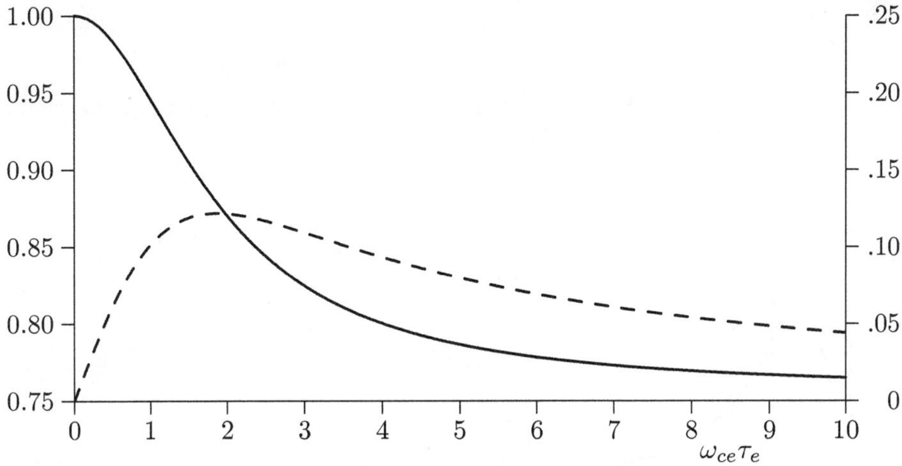

FIGURE 9.4
Heat flux coefficients. ψ^{II}/ψ^{I} (solid with scale on the left) and ψ^{III}/ψ^{I} (dashed with scale on the right).

Plots of θ^{II}/θ^{I} and $-\theta^{III}/\theta^{I}$ are given in Figure 9.5 while plots of ξ^{II}/ξ^{I} and $-\xi^{III}/\xi^{I}$ are given in Figure 9.6.

The pressure tensor is most conveniently written in terms of the cyclotron frequency for ions and an ion-ion collision time

$$\omega_{ci} = \frac{e_2 B}{m_2} = \frac{m_1}{m_2}\omega_{ce} \quad \text{and} \quad \tau_i = \frac{5}{2\sqrt{2}}\left(\frac{m_2}{m_1}\right)^{1/2}\tau_e\,.$$

Defining

$$s_{\alpha\beta} \equiv \tfrac{1}{2}(\nabla_\alpha u_\beta + \nabla_\beta u_\alpha) - \tfrac{1}{3}\nabla\cdot\boldsymbol{u}\delta_{\alpha\beta}\,, \tag{9.185}$$

and supposing the magnetic field to be in the z-direction,

$$\begin{aligned}
p_{xx} &= p - \frac{2\mu}{1+\frac{16}{9}\omega_{ci}^2\tau_i^2}\left(s_{xx} + \tfrac{4}{3}\omega_{ci}\tau_i s_{xy} - \tfrac{8}{9}\omega_{ci}^2\tau_i^2 s_{zz}\right)\\
p_{yy} &= p - \frac{2\mu}{1+\frac{16}{9}\omega_{ci}^2\tau_i^2}\left(s_{yy} - \tfrac{4}{3}\omega_{ci}\tau_i s_{xy} - \tfrac{8}{9}\omega_{ci}^2\tau_i^2 s_{zz}\right)\\
p_{zz} &= p - 2\mu s_{zz}\\
p_{xy} &= p_{yx} = -\frac{2\mu}{1+\frac{16}{9}\omega_{ci}^2\tau_i^2}\left[s_{xy} - \tfrac{2}{3}\omega_{ci}\tau_i(s_{xx} - s_{yy})\right]\\
p_{xz} &= p_{zx} = -\frac{2\mu}{1+\frac{4}{9}\omega_{ci}^2\tau_i^2}\left(s_{xz} + \tfrac{2}{3}\omega_{ci}\tau_i s_{yz}\right)\\
p_{yz} &= p_{zy} = -\frac{2\mu}{1+\frac{4}{9}\omega_{ci}^2\tau_i^2}\left(s_{yz} - \tfrac{2}{3}\omega_{ci}\tau_i s_{xz}\right)
\end{aligned} \tag{9.186}$$

where

$$\mu = \tfrac{1}{3}nk_B T\tau_i\,. \tag{9.187}$$

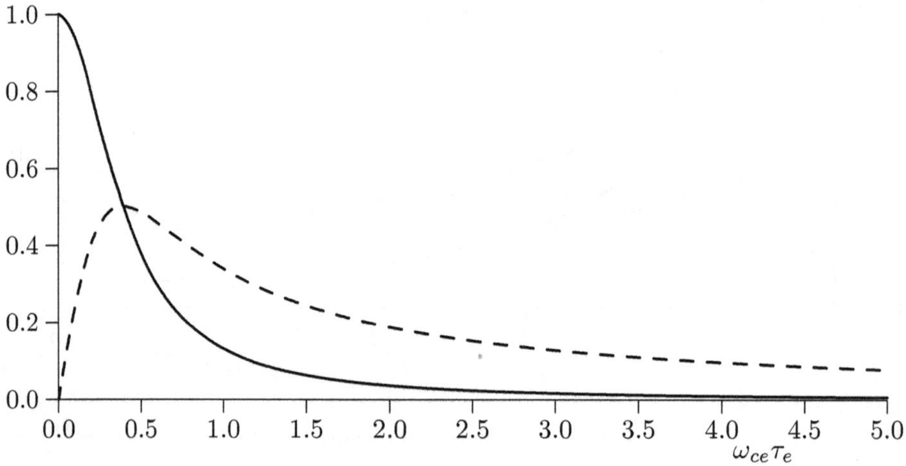

FIGURE 9.5
Alternate heat flux coefficients. θ^{II}/θ^{I} (solid) and $-\theta^{III}/\theta^{I}$ (dashed).

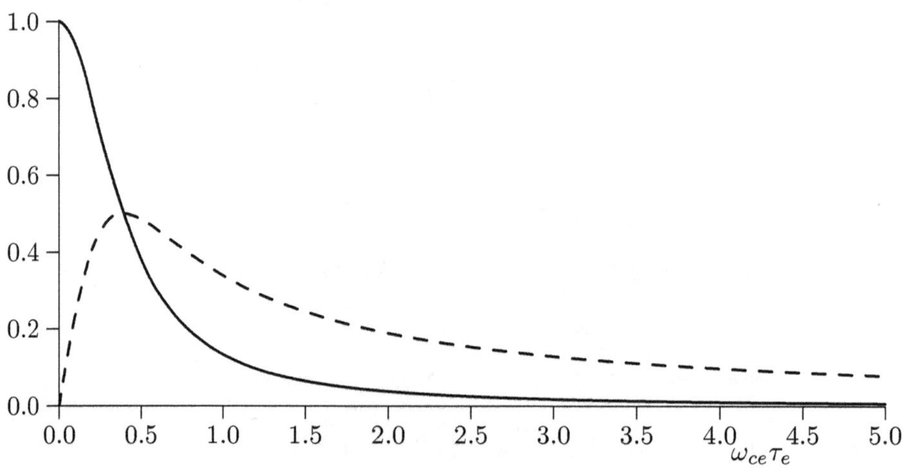

FIGURE 9.6
Alternate heat flux coefficients. ξ^{II}/ξ^{I} (solid) and ξ^{III}/ξ^{I} (dashed).

A

MATHEMATICAL FUNCTIONS

The Plasma Dispersion function is a commonly occurring function in thermal plasmas and is related to the error function for complex arguments and some other representations which occur commonly in the literature. Tabulated by Fried and Conte[20], we list here its properties and show its relationships to other common functions.

A.1 The plasma dispersion function

A.1.1 Definition of the plasma dispersion function

$$Z(\zeta) \equiv \frac{1}{\sqrt{\pi}} \int_{-\infty}^{\infty} \frac{e^{-\xi^2}\, d\xi}{\xi - \zeta}, \qquad \mathrm{Im}(\zeta) > 0\,. \tag{A.1}$$

A.1.1.1 Differential equation

$$Z'(\zeta) = -2[1 + \zeta Z(\zeta)]\,. \tag{A.2}$$

A.1.1.2 Power series

$$\begin{aligned} Z(\zeta) &= \mathrm{i}\sqrt{\pi} \sum_{n=0}^{\infty} \frac{(\mathrm{i}\zeta)^n}{\Gamma\left(\frac{n}{2}+1\right)} \qquad \text{(A.3)} \\ &= \mathrm{i}\sqrt{\pi} e^{-\zeta^2} - 2\zeta \left[1 - \frac{2}{3}\zeta^2 + \frac{4}{15}\zeta^4 \cdots \frac{(-2)^n \zeta^{2n}}{(2n+1)(2n-1)\cdots 3\cdot 1}\right]. \end{aligned}$$

A.1.1.3 Asymptotic series

$$\begin{aligned} Z(\zeta) &= -\frac{1}{\zeta}\left[1 + \sum_{n=1}^{\infty} \frac{1\cdot 3\cdots(2n-1)}{(2\zeta^2)^n}\right] + \mathrm{i}\sigma\sqrt{\pi} e^{-\zeta^2} \\ &= -\frac{1}{\zeta}\left(1 + \frac{1}{2\zeta^2} + \frac{3}{4\zeta^4} + \cdots\right) + \mathrm{i}\sigma\sqrt{\pi} e^{-\zeta^2} \end{aligned} \tag{A.4}$$

where

$$\sigma = \begin{cases} 0\,, & \mathrm{Im}(\zeta) > 0 \\ 1\,, & \mathrm{Im}(\zeta) = 0\,. \\ 2\,, & \mathrm{Im}(\zeta) < 0 \end{cases}$$

The discontinuity in the Stokes' phenomena of the asymptotic series indicates an uncertainty of the same magnitude as the discontinuity in the series expansion, since the asymptotic series must be terminated when the terms no longer decrease, and the smallest term is of the same order as the pole term. The imaginary part of $Z(\zeta)$ is exact only for real ζ.

A.1.1.4 Continued fraction

A continued fraction expansion for the Plasma Dispersion function that is useful in the upper half ζ-plane provided one is not too close to the real axis is given by

$$Z(\zeta) = \frac{z}{\frac{1}{2}z^2-}\;\frac{1\cdot\frac{1}{2}}{\frac{5}{2}z^2-}\;\frac{2\cdot\frac{3}{2}}{\frac{9}{2}z^2-}\;\frac{3\cdot\frac{5}{2}}{\frac{13}{2}z^2-}\cdots\,.$$

A.1.1.5 Symmetry relations

$$Z(-\zeta) = 2\mathrm{i}\sqrt{\pi}\mathrm{e}^{-\zeta^2} - Z(\zeta) \tag{A.5}$$

$$Z(\zeta^*) = [Z(\zeta) - 2\mathrm{i}\sqrt{\pi}\mathrm{e}^{-\zeta^2}]^* \tag{A.6}$$

$$\tilde{Z}(\zeta) = Z(\zeta) - 2\mathrm{i}\sqrt{\pi}\mathrm{e}^{-\zeta^2} \quad \text{for} \quad \mathrm{Im}(\zeta) < 0\,. \tag{A.7}$$

A.1.1.6 Zeroes of $Z(\zeta)$

An infinite number of zeroes are located close to the negative 45° lines as illustrated in Figure A.1 where the first three zeroes are shown. The first five zeroes are located at

$$\zeta_1 = \pm 1.991466835 - 1.354810123\mathrm{i} \tag{A.8}$$

$$\zeta_2 = \pm 2.691149024 - 2.177044906\mathrm{i} \tag{A.9}$$

$$\zeta_3 = \pm 3.235330868 - 2.784387613\mathrm{i} \tag{A.10}$$

$$\zeta_4 = \pm 3.697309702 - 3.287410789\mathrm{i} \tag{A.11}$$

$$\zeta_5 = \pm 4.106107271 - 3.725948729\mathrm{i}\,. \tag{A.12}$$

A.1.2 Relation to the error function for complex argument

We begin the analysis be examining the commonly occurring integral

$$I = \int_{-\infty}^{\infty} \frac{f_0(v)\,\mathrm{d}v}{p + \mathrm{i}kv}\,, \qquad \mathrm{Re}(p) > 0\,, \tag{A.13}$$

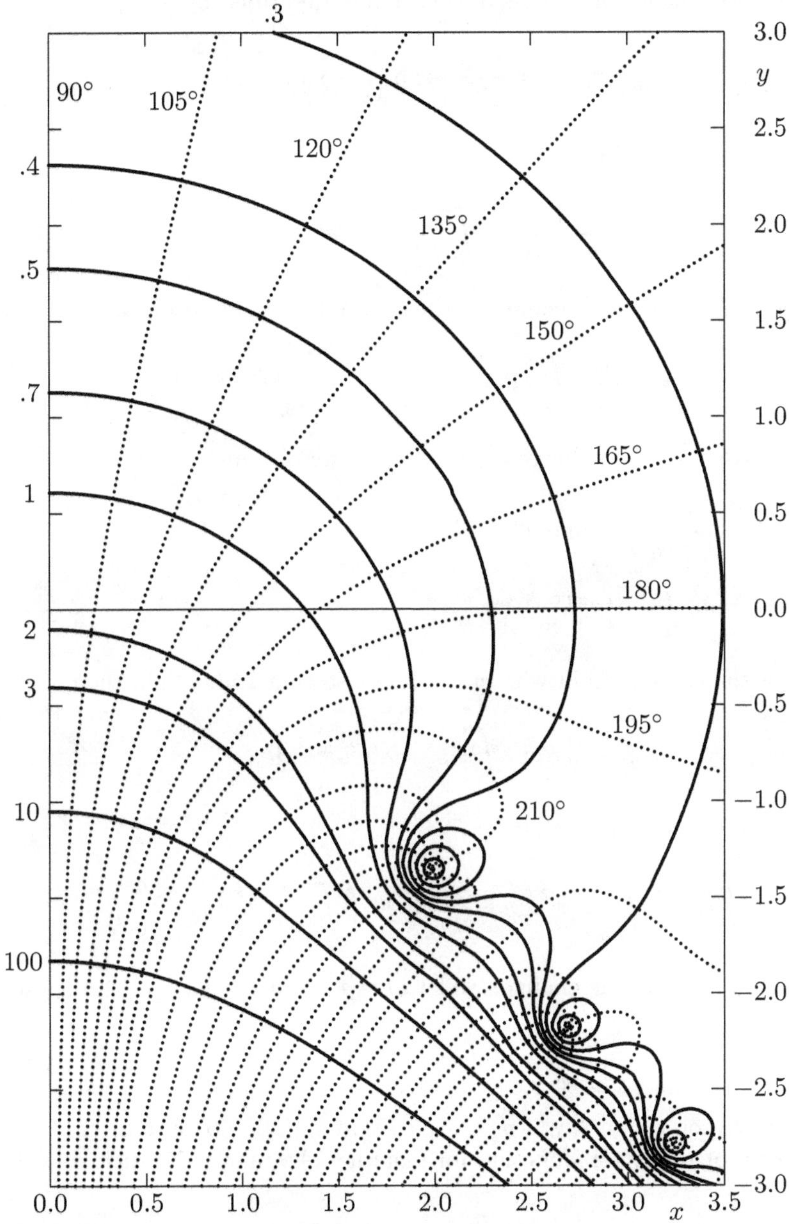

FIGURE A.1
Contours for $|Z(x + \mathrm{i}y)| = 0.1, 0.2, 0.3, 0.4, 0.5, 0.7, 1, 2, 3, 10, 100$ for $x \geq 0$. The altitude contours are symmetric for $x < 0$ but the phase is not, tending toward 0° on the negative x axis. The phase $(\cdots)$ is shown every 15° except in the vicinity of the zeroes where the interval is 30°.

where the distribution function is a Maxwellian described by

$$f_0(v) = \frac{n_0}{\sqrt{\pi} v_0} \exp\left(-\frac{v^2}{v_0^2}\right),$$

and $v_0 = \sqrt{2k_B T/m}$. If we write

$$\frac{1}{p + \mathrm{i}kv} = \int_0^\infty \mathrm{e}^{-(p+\mathrm{i}kv)t}\,\mathrm{d}t$$

with $\mathrm{Re}(p) > 0$ for convergence, then the original integral becomes

$$I = \frac{n_0}{\sqrt{\pi} v_0} \int_0^\infty \mathrm{d}t\,\mathrm{e}^{-pt} \int_{-\infty}^\infty \mathrm{d}v\,\mathrm{e}^{-v^2/v_0^2 - \mathrm{i}kvt}.$$

Completing the square in the velocity integral, we obtain

$$\begin{aligned} I &= \frac{n_0}{\sqrt{\pi} v_0} \int_0^\infty \mathrm{d}t\,\mathrm{e}^{-pt-k^2v_0^2t^2/4} \int_{-\infty}^\infty \mathrm{d}v\,\mathrm{e}^{-\left(v+\mathrm{i}kv_0^2t/2\right)^2/v_0^2} \\ &= n_0 \int_0^\infty \mathrm{d}t\,\mathrm{e}^{-pt-k^2v_0^2t^2/4}. \end{aligned}$$

Completing the square again with $p = -\mathrm{i}kv_0x$ so that $\mathrm{Im}(x) > 0$, then

$$I = n_0 \mathrm{e}^{-x^2} \int_0^\infty \mathrm{e}^{-(kv_0t/2-\mathrm{i}x)^2}\,\mathrm{d}t.$$

Then let $kv_0t/2 - \mathrm{i}x = u$ so that

$$I = \frac{2n_0}{kv_0}\mathrm{e}^{-x^2} \int_{-\mathrm{i}x}^\infty \mathrm{e}^{-u^2}\,\mathrm{d}u = \frac{2n_0}{kv_0}\mathrm{e}^{-x^2}\left(\int_{-\mathrm{i}x}^0 \mathrm{e}^{-u^2}\,\mathrm{d}u + \frac{\sqrt{\pi}}{2}\right),$$

and with the final variable change, we let $\mathrm{i}u = \tau$ so that equation (A.13) may be written as

$$I = \frac{n_0\sqrt{\pi}}{kv_0}\mathrm{e}^{-x^2}\left(1 + \frac{2\mathrm{i}}{\sqrt{\pi}} \int_0^x \mathrm{e}^{\tau^2}\,\mathrm{d}\tau\right) \tag{A.14}$$

with $x = \mathrm{i}p/kv_0$ or $x = \omega/kv_0$ if $p = -\mathrm{i}\omega$ so that $\mathrm{Im}(\omega) > 0$. Equation (A.14) is in the form of the Error Function for Complex Argument[55] such that

$$I = \frac{n_0\sqrt{\pi}}{kv_0} w(x),$$

and $w(x)$ is the tabulated function with $w(0) = 1$. Comparing this with the Plasma Dispersion function, we see that

$$Z(x) = \mathrm{i}\sqrt{\pi} w(x). \tag{A.15}$$

A.1.3 Relation to the Y function

In Soviet literature, the complex error function is tabulated as the Y function where

$$Y(z) \equiv \mathrm{i}\sqrt{\pi}z\mathrm{e}^{-z^2}\left(1+\frac{2\mathrm{i}}{\sqrt{\pi}}\int_0^z \mathrm{e}^{t^2}\,\mathrm{d}t\right), \tag{A.16}$$

so that

$$\begin{aligned} Y &= \mathrm{i}\sqrt{\pi}zw(z) \\ &= zZ(z)\,. \end{aligned} \tag{A.17}$$

Because the Plasma Dispersion function satisfies the differential equation

$$Z'(\zeta) = -2[1+\zeta Z(\zeta)]\,,$$

both the Plasma Dispersion function and its derivative may be simply expressed by means of the Y function as

$$Z(\zeta) = Y(\zeta)/\zeta \tag{A.18}$$

$$Z'(\zeta) = -2[1+Y(\zeta)]\,. \tag{A.19}$$

A.1.4 Relation to the W function

Another equivalent function, similar to the Y function, is the W function, which is defined by[56]

$$W(y) \equiv \frac{1}{\sqrt{2\pi}}\int_{-\infty}^{\infty}\frac{x\mathrm{e}^{-x^2/2}}{x-y}\,\mathrm{d}x\,, \qquad \mathrm{Im}(y) > 0\,. \tag{A.20}$$

If we let $x = \sqrt{2}\xi$, this becomes

$$W(y) = \frac{1}{\sqrt{\pi}}\int_{-\infty}^{\infty}\frac{\xi\mathrm{e}^{-\xi^2}}{\xi - y/\sqrt{2}}\,\mathrm{d}\xi\,, \tag{A.21}$$

so it is related to the Z function either by

$$W(y) = 1 + \frac{y}{\sqrt{2}}Z\left(\frac{y}{\sqrt{2}}\right) = -\frac{1}{2}Z'\left(\frac{y}{\sqrt{2}}\right), \tag{A.22}$$

or letting $y = \sqrt{2}\zeta$, by

$$W(\sqrt{2}\zeta) = 1 + \zeta Z(\zeta) = -\frac{1}{2}Z'(\zeta)\,. \tag{A.23}$$

The W function satisfies the differential equation

$$\frac{\mathrm{d}W}{\mathrm{d}y} = \frac{W-1}{y} - yW\,. \tag{A.24}$$

A.2 Relativistic plasma dispersion functions

A.2.1 Weakly relativistic dispersion function

The weakly relativistic plasma dispersion function is defined by

$$F_q(z) \equiv -\mathrm{i}\int_0^\infty \frac{\mathrm{e}^{\mathrm{i}zt}}{(1-\mathrm{i}t)^q}\,\mathrm{d}t\,. \tag{A.25}$$

A.2.1.1 Relation to other functions

The properties of this function were described by Dnestrovskii et al.[57], and is sometimes called the Dnestrovskii function. Its relation to other functions was noted by Lazzaro et al.[58]. It is equivalent to the confluent hypergeometric function of the second kind by

$$F_q(z) = \psi(1, 2-q, z) = z^{q-1}\psi(q, q, z)\,, \tag{A.26}$$

and this allows us to identify it with the error function so that for $q = \frac{1}{2}$,

$$F_{\frac{1}{2}}(x) = \sqrt{\pi/x}\,\mathrm{e}^x[1-\mathrm{erf}(\sqrt{x})]\,, \tag{A.27}$$

which establishes the connection noted by Shkarofsky[39] between $F_q(z)$ and the Plasma Dispersion function, such that

$$\mathrm{i}F_{\frac{1}{2}}(z) = \int_0^\infty \frac{\mathrm{e}^{\mathrm{i}zt}}{(1-\mathrm{i}t)^{1/2}}\,\mathrm{d}t = \frac{1}{\sqrt{z}}Z(\mathrm{i}\sqrt{z})\,. \tag{A.28}$$

Integrating F_q by parts for $q > \frac{1}{2}$, we find

$$F_q = \sum_{p=0}^{q-\frac{3}{2}}(-z)^p\frac{\Gamma(q-p-1)}{\Gamma(q)} + \frac{\sqrt{\pi}}{\Gamma(q)}(-z)^{q-\frac{3}{2}}[\mathrm{i}\sqrt{z}Z(\mathrm{i}\sqrt{z})]\,. \tag{A.29}$$

A.2.1.2 Properties

A useful recursion relation for finding the properties of the higher order functions is

$$(q-1)F_q(z) = 1 - zF_{q-1}(z)\,, \tag{A.30}$$

so the first few of the functions are

$$F_{\frac{1}{2}}(z) = \begin{cases} \frac{Z(\mathrm{i}\sqrt{z})}{\mathrm{i}\sqrt{z}}, & \mathrm{Re}(z) > 0, \\ \frac{Z^*(\sqrt{-z})}{\sqrt{-z}}, & \mathrm{Re}(z) < 0, \end{cases} \tag{A.31}$$

$$F_{\frac{3}{2}}(z) = 2[1 - zF_{\frac{1}{2}}(z)]\,, \tag{A.32}$$

$$F_{\frac{5}{2}}(z) = \tfrac{2}{3}[1 - zF_{\frac{3}{2}}(z)]\,. \tag{A.33}$$

For large arguments, the relativistic dispersion function with half-integer order varies as

$$F_q(z) \simeq \frac{1}{z}\sum_{n=0}^{N}\frac{(-1)^n\Gamma(n+q)}{\Gamma(q)z^n}, \quad N+q \leq |Z| \tag{A.34}$$

$$= \frac{1}{z}\left[1 - \frac{q}{z} + \frac{q(q+1)}{z^2} - \cdots\right]. \tag{A.35}$$

It is therefore real for real argument.

For small arguments, we first note that for $F_{\frac{1}{2}}$, the function is real for positive real argument, since $Z(\zeta)$ is pure imaginary for pure imaginary $\zeta = \mathrm{i}\sqrt{z}$. For $z < 0$, however, $\mathrm{i}\sqrt{z}$ is real, and $Z(\zeta)$ is complex for real argument. Then using equation (A.29) for $q \geq 3/2$, we may approximate

$$F_q(z) \simeq \frac{1}{q-1} + \frac{\mathrm{i}\pi(-z)^{q-1}\mathrm{e}^z}{\Gamma(q)}. \tag{A.36}$$

A.2.1.3 Zeroes of $F_{\frac{1}{2}}(z)$ and $F_{\frac{3}{2}}(z)$

An infinite number of zeroes are located close to the imaginary axis. The first five zeroes are located at

$$\begin{aligned}
z_{\frac{1}{2}1} &= -2.1304297 \pm 5.3961189\mathrm{i}\\
z_{\frac{1}{2}2} &= -2.5027585 \pm 11.7175045\mathrm{i}\\
z_{\frac{1}{2}3} &= -2.7145514 \pm 18.0168304\mathrm{i}\\
z_{\frac{1}{2}4} &= -2.8630293 \pm 24.3091516\mathrm{i}\\
z_{\frac{1}{2}5} &= -2.9774230 \pm 30.5982903\mathrm{i}\,,
\end{aligned}$$

$$\begin{aligned}
z_{\frac{3}{2}1} &= -1.4326911 \pm 5.4126879\mathrm{i}\\
z_{\frac{3}{2}2} &= -1.8083268 \pm 11.7255352\mathrm{i}\\
z_{\frac{3}{2}3} &= -2.0207916 \pm 18.0221180\mathrm{i}\\
z_{\frac{3}{2}4} &= -2.1695198 \pm 24.3130903\mathrm{i}\\
z_{\frac{3}{2}5} &= -2.2840349 \pm 30.6014277\mathrm{i}\,.
\end{aligned}$$

A.2.2 Generalized relativistic dispersion function

A generalized form of the weakly relativistic Plasma Dispersion function is denoted $\mathcal{F}_q(z,a)$, where $\mathcal{F}_q(z,0) = F_q(z)$. We also have the property that as $a \to \infty$, $-2\sqrt{a}\mathcal{F}_q(z,a) \to Z(\zeta)$ with $z = 2\sqrt{a}\zeta$. This function can also be related to $Z(\zeta)$ through the relations

$$\mathcal{F}_{\frac{1}{2}}(z,a) = \begin{cases} \frac{1}{2\mathrm{i}\sqrt{z-a}}\left[Z(\sqrt{a}+\mathrm{i}\sqrt{z-a}) - Z^*(\sqrt{a}+\mathrm{i}\sqrt{z-a})\right], & z > a, \\ -\frac{1}{2\sqrt{a-z}}\left[Z(\sqrt{a}-\sqrt{a-z}) + Z(-\sqrt{a}-\sqrt{a-z})\right], & z < a, \end{cases} \tag{A.37}$$

$$\mathcal{F}_{\frac{3}{2}}(z,a)=\begin{cases} -\frac{1}{2\sqrt{a}}[Z(\sqrt{a}+\mathrm{i}\sqrt{z-a})+Z^*(\sqrt{a}+\mathrm{i}\sqrt{z-a})]\,,\ z>a, \\ -\frac{1}{2\sqrt{a}}[Z(\sqrt{a}-\sqrt{a-z})-Z(-\sqrt{a}-\sqrt{a-z})]\,,\ z<a, \end{cases} \tag{A.38}$$

and the higher order terms are given by the recursion formula,

$$a\mathcal{F}_{q+2}(z,a)=1+(a-z)\mathcal{F}_q(z,a)-q\mathcal{F}_{q+1}(z,a)\,. \tag{A.39}$$

The various orders are also related by the differential formulas,

$$\frac{\partial\mathcal{F}_q(z,a)}{\partial z}=\mathcal{F}_q(z,a)-\mathcal{F}_{q-1}(z,a)\,, \tag{A.40}$$

$$\frac{\partial\mathcal{F}_q(z,a)}{\partial a}=\mathcal{F}_{q-1}(z,a)+2\mathcal{F}_q(z,a)+\mathcal{F}_{q+1}(z,a)\,. \tag{A.41}$$

The asymptotic expression is given by

$$\begin{aligned}\mathcal{F}_q(z,a)&\simeq\frac{1}{z}\left[1-\frac{q}{z}+\frac{q(q+1)+2a}{z^2}-\frac{q(q+1)(q+2)+6(q+1)a}{z^3}+\cdots\right]\\ &=\frac{1}{z}\left[1-\frac{A_1}{z}+\frac{A_2}{z^2}-\frac{A_3}{z^3}+\frac{A_4}{z^4}-\cdots\right],\end{aligned} \tag{A.42}$$

where the A_n are given by the recursion formulas

$$A_1(q)=q \tag{A.43}$$

$$A_{n+1}(q)=qA_n(q+1)+a[A_n(q+2)-A_n(q)]\,. \tag{A.44}$$

As defined here, these functions are not analytic and defined only for real z.

A.3 Gamma Function, $\Gamma(z)$

A.3.1 Definition

The Gamma function, which for integer values is related to the factorial function, is defined by the integral

$$\Gamma(z)\equiv\int_0^\infty t^{z-1}\mathrm{e}^{-t}\mathrm{d}t\,,\qquad \mathrm{Re}(z)>0\,. \tag{A.45}$$

A.3.1.1 Useful relations

Integrating by parts, it is easy to establish the recursion formula,

$$\Gamma(z+1)=z\Gamma(z)\,, \tag{A.46}$$

which for integer values becomes

$$\Gamma(n+1) = n! \tag{A.47}$$

since $\Gamma(1) = 1$. Some useful special values and relations are

$$\Gamma\left(\frac{1}{2}\right) = \sqrt{\pi} \qquad \Gamma(1+\mathrm{i}y)\Gamma(1-\mathrm{i}y) = \frac{\pi y}{\sinh \pi y}\,.$$

A.3.1.2 Asymptotic expressions

For large arguments, Stirling's formula gives

$$\Gamma(z) = \sqrt{2\pi} z^{z-\frac{1}{2}} \exp\left(-z + \frac{1}{12z} - \frac{1}{360z^3} + \cdots\right). \tag{A.48}$$

For large half-integer values, it is convenient to use the alternate formula

$$\Gamma\left(z+\frac{1}{2}\right) = \sqrt{2\pi} z^{z} \exp\left(-z - \frac{1}{24z} + \frac{7}{2880z^3} - \cdots\right). \tag{A.49}$$

A.3.1.3 Hankel's contour integral

$$\frac{1}{\Gamma(z)} = \frac{\mathrm{i}}{2\pi}\int_C (-t)^{-z}\mathrm{e}^{-t}\mathrm{d}t\,, \qquad (|z| < \infty)\,. \tag{A.50}$$

The path of integration C starts at $+\infty$ on the real axis, circles the origin in the counterclockwise direction, and returns to the starting point.

A.3.2 Incomplete gamma function

The incomplete gamma functions are defined by

$$\gamma(a,x) \equiv \int_0^x \mathrm{e}^{-t}t^{a-1}\,\mathrm{d}t\,, \qquad \mathrm{Re}(a) > 0 \tag{A.51}$$

$$\Gamma(a,x) \equiv \Gamma(a) - \gamma(a,x) = \int_x^\infty \mathrm{e}^{-t}t^{a-1}\,\mathrm{d}t\,. \tag{A.52}$$

The recursion formula is

$$\gamma(a+1,x) = a\gamma(a,x) - x^a\mathrm{e}^{-x}\,. \tag{A.53}$$

The continued fraction expansion for $\Gamma(a,x)$ is

$$\Gamma(a,x) = \mathrm{e}^{-x}x^a\left(\frac{1}{x+}\,\frac{1-a}{1+}\,\frac{1}{x+}\,\frac{2-a}{1+}\,\frac{2}{x+}\right), \qquad x > 0\,. \tag{A.54}$$

The error with this continued fraction does not decrease monotonically until a large number of terms is used, so it must be used with care. Also its accuracy is sensitive to relatively small changes in a. For large $|x|$, the asymptotic expression is

$$\Gamma(a,z) \sim \mathrm{e}^{-z} z^{a-1}\left[1+\frac{a-1}{z}+\frac{(a-1)(a-2)}{z^2}+\cdots\right], \qquad |\arg z| < \frac{3\pi}{2}. \tag{A.55}$$

A.4 Generalized hypergeometric functions

The generalized hypergeometric functions used in Chapter 6 are defined by the power series

$${}_1F_2(a;b_1,b_2;x) \equiv \frac{\Gamma(b_1)\Gamma(b_2)}{\Gamma(a)}\sum_{k=0}^{\infty}\frac{\Gamma(a+k)}{\Gamma(b_1+k)\Gamma(b_2+k)}\frac{x^k}{k!}, \tag{A.56}$$

$${}_2F_3(a_1,a_2;b_1,b_2,b_3;x) \equiv \frac{\Gamma(b_1)\Gamma(b_2)\Gamma(b_3)}{\Gamma(a_1)\Gamma(a_2)} \times \sum_{k=0}^{\infty}\frac{\Gamma(a_1+k)\Gamma(a_2+k)}{\Gamma(b_1+k)\Gamma(b_2+k)\Gamma(b_3+k)}\frac{x^k}{k!}. \tag{A.57}$$

A.4.1 Hypergeometric function integrals — first type

Integrals over the angle that lead to the hypergeometric functions of the first type, ${}_1F_2(a_1;b_1,b_2,x)$, are:

$$\int_0^{\pi}\sin\theta[J_n(b\sin\theta)]^2\,\mathrm{d}\theta = \frac{2b^{2n}{}_1F_2(n+\frac{1}{2};n+\frac{3}{2},2n+1;-b^2)}{(2n+1)!} \tag{A.58}$$

$$\int_0^{\pi}\sin^2\theta J_n(b\sin\theta)J_n'(b\sin\theta)\,\mathrm{d}\theta = \frac{b^{2n-1}{}_1F_2(n+\frac{1}{2};n+\frac{3}{2},2n;-b^2)}{(2n+1)(2n-1)!} - \frac{b^{2n+1}{}_1F_2(n+\frac{3}{2};n+\frac{5}{2},2n+2;-b^2)}{(2n+3)(2n+1)!}$$

$$\int_0^{\pi}\sin^2\theta J_n(b\sin\theta)J_{n-1}(b\sin\theta)\,\mathrm{d}\theta = \frac{2b^{2n-1}{}_1F_2(n+\frac{1}{2};n+\frac{3}{2},2n;-b^2)}{(2n+1)(2n-1)!}$$

$$\int_0^{\pi}\sin^3\theta[J_{n-1}(b\sin\theta]^2\,\mathrm{d}\theta = \frac{4nb^{2n-2}{}_1F_2(n-\frac{1}{2};n+\frac{3}{2},2n-1;-b^2)}{(2n+1)(2n-1)!} - \frac{4b^{2n}{}_1F_2(n+\frac{1}{2};n+\frac{5}{2},2n;-b^2)}{(2n+3)(2n+1)(2n-1)!}$$

$$\int_0^\pi \sin\theta\cos^2\theta[J_n(b\sin\theta)]^2\,d\theta = \frac{2b^{2n}{}_1F_2(n+\frac{1}{2};n+\frac{5}{2},2n+1;-b^2)}{(2n+3)(2n+1)!}.$$

A.4.2 Hypergeometric function integrals — second type

Integrals over the angle that lead to the hypergeometric functions of the second type, ${}_2F_3(a_1,a_2;b_1,b_2,b_3,x)$, are:

$$\int_0^\pi \sin\theta J_a(b\sin\theta)J_{-a}(b\sin\theta)\,d\theta = \frac{2\sin\pi a}{\pi a}\,{}_2F_3(\tfrac{1}{2},1;\tfrac{3}{2},1-a,1+a;-b^2) \tag{A.59}$$

$$\begin{aligned}\int_0^\pi \sin^2\theta J_a(b\sin\theta)J'_{-a}(b\sin\theta)\,d\theta &= \frac{b\sin\pi a}{3\pi a(a-1)}\,{}_2F_3(\tfrac{3}{2},1;\tfrac{5}{2},2-a,1+a;-b^2)\\ &-\frac{\sin\pi a}{\pi b}\,{}_2F_3(\tfrac{1}{2},1;\tfrac{3}{2},-a,1+a;-b^2)\end{aligned} \tag{A.60}$$

$$\int_0^\pi \sin^2\theta J_a(b\sin\theta)J_{1-a}(b\sin\theta)\,d\theta = -\frac{2b\sin\pi a}{3\pi a(a-1)}\,{}_2F_3(\tfrac{3}{2},1;\tfrac{5}{2},2-a,1+a;-b^2) \tag{A.61}$$

$$\int_0^\pi \sin^2\theta J_{a-1}(b\sin\theta)J_{-a}(b\sin\theta)\,d\theta = \frac{2\sin\pi a}{\pi b}\,{}_2F_3(\tfrac{1}{2},1;\tfrac{3}{2},1-a,a;-b^2) \tag{A.62}$$

$$\int_0^\pi \sin^3\theta J_{a-1}(b\sin\theta)J_{1-a}(b\sin\theta)\,d\theta = -\frac{4\sin\pi a}{3\pi(a-1)}\,{}_2F_3(\tfrac{1}{2},2;\tfrac{5}{2},2-a,a;-b^2) \tag{A.63}$$

$$\int_0^\pi \sin\theta\cos^2\theta J_a(b\sin\theta)J_{-a}(b\sin\theta)\,d\theta = \frac{2\sin\pi a}{3a\pi}\,{}_2F_3(\tfrac{1}{2},1;\tfrac{5}{2},1-a,1+a;-b^2). \tag{A.64}$$

A.5 Vector identities

A.5.1 Products of three vectors

$$\begin{aligned}\boldsymbol{A}\cdot(\boldsymbol{B}\times\boldsymbol{C}) &= \boldsymbol{C}\cdot(\boldsymbol{A}\times\boldsymbol{B})\\ &= \boldsymbol{B}\cdot(\boldsymbol{C}\times\boldsymbol{A})\\ &= -\boldsymbol{A}\cdot(\boldsymbol{C}\times\boldsymbol{B}),\quad \text{etc.}\end{aligned}$$

$$\boldsymbol{A}\times(\boldsymbol{B}\times\boldsymbol{C}) = \boldsymbol{B}(\boldsymbol{A}\cdot\boldsymbol{C}) - \boldsymbol{C}(\boldsymbol{A}\cdot\boldsymbol{B}).$$

A.5.2 Vector identities with the ∇ operator

$$
\begin{aligned}
\nabla(\phi\psi) &= \phi\nabla\psi + \psi\nabla\phi \\
\nabla\cdot(\phi\boldsymbol{A}) &= \phi\nabla\cdot\boldsymbol{A} + \boldsymbol{A}\cdot\nabla\phi \\
\nabla\times(\phi\boldsymbol{A}) &= \phi\nabla\times\boldsymbol{A} + \nabla\phi\times\boldsymbol{A} \\
\nabla\times(\nabla\phi) &= 0 \\
\nabla\cdot(\nabla\times\boldsymbol{A}) &= 0 \\
\nabla\cdot(\boldsymbol{A}\times\boldsymbol{B}) &= \boldsymbol{B}\cdot(\nabla\times\boldsymbol{A}) - \boldsymbol{A}\cdot(\nabla\times\boldsymbol{B}) \\
\nabla\times(\boldsymbol{A}\times\boldsymbol{B}) &= \boldsymbol{A}(\nabla\cdot\boldsymbol{B}) - \boldsymbol{B}(\nabla\cdot\boldsymbol{A}) + (\boldsymbol{B}\cdot\nabla)\boldsymbol{A} - (\boldsymbol{A}\cdot\nabla)\boldsymbol{B} \\
\nabla\times(\nabla\times\boldsymbol{A}) &= \nabla(\nabla\cdot\boldsymbol{A}) - \nabla^2\boldsymbol{A} \\
\nabla(\boldsymbol{A}\cdot\boldsymbol{B}) &= (\boldsymbol{A}\cdot\nabla)\boldsymbol{B} + (\boldsymbol{B}\cdot\nabla)\boldsymbol{A} + \boldsymbol{A}\times(\nabla\times\boldsymbol{B}) + \boldsymbol{B}\times(\nabla\times\boldsymbol{A}).
\end{aligned}
$$

B

COLLISION INTEGRALS

In this appendix, we shall consider collisions and the evaluation of the various collision integrals in detail. We shall start from first principles and show how the cross section can be simply expressed for Coulomb interactions and then we shall go on to consider the integrals.

B.1 Rutherford scattering

Suppose we have two particles colliding, the first of mass m_1, charge e_1, position $\boldsymbol{r}_1$, and the second of mass m_2, charge e_2, position $\boldsymbol{r}_2$. Then the velocities of these particles are $\mathrm{d}\boldsymbol{r}_1/\mathrm{d}t$ and $\mathrm{d}\boldsymbol{r}_2/\mathrm{d}t$ and the accelerations $\mathrm{d}^2\boldsymbol{r}_1/\mathrm{d}t^2$ and $\mathrm{d}^2\boldsymbol{r}_2/\mathrm{d}t^2$, respectively, and Newton's law gives

$$m_1\frac{\mathrm{d}^2}{\mathrm{d}t^2}\boldsymbol{r}_1 = \frac{e_1 e_1}{|\boldsymbol{r}_1 - \boldsymbol{r}_2|^3}(\boldsymbol{r}_1 - \boldsymbol{r}_2) = -m_2\frac{\mathrm{d}^2}{\mathrm{d}t^2}\boldsymbol{r}_2 \,. \tag{B.1}$$

The middle term is just the Coulomb force written in vector notation to indicate that it acts along the line joining the particles. In writing equation (B.1) we have assumed that, for the short time the particles are in contact, the Coulomb force between the particles is so much greater than any other force which may also be acting that the latter can be neglected. The validity of this assumption we shall examine later.

One result follows immediately from equation (B.1), namely

$$\frac{\mathrm{d}^2}{\mathrm{d}t^2}(m_1\boldsymbol{r}_1 - m_2\boldsymbol{r}_2) = 0\,, \tag{B.2}$$

so that integrating,

$$\frac{\mathrm{d}}{\mathrm{d}t}(m_1\boldsymbol{r}_1 - m_2\boldsymbol{r}_2) = \text{constant} = m_1\boldsymbol{v}_1 + m_2\boldsymbol{v}_2\,, \tag{B.3}$$

and

$$m_1\boldsymbol{r}_1 - m_2\boldsymbol{r}_2 = (m_1\boldsymbol{v}_1 + m_2\boldsymbol{v}_2)t + \text{constant}\,. \tag{B.4}$$

This tells us that the total momentum is conserved and that the center of mass of the two particles moves with uniform velocity throughout the collision.

The position of particle 1 relative to particle 2 is

$$\boldsymbol{r} = \boldsymbol{r}_1 - \boldsymbol{r}_2 \,. \tag{B.5}$$

Eliminating $\boldsymbol{r}_2$, equation (B.1) becomes

$$m_1 \frac{\mathrm{d}^2}{\mathrm{d}t^2}\boldsymbol{r}_1 = \frac{e_1 e_2}{4\pi\epsilon_0 r^3}\boldsymbol{r} = -\frac{\mathrm{d}^2}{\mathrm{d}t^2}(\boldsymbol{r}_1 - \boldsymbol{r}) \,, \tag{B.6}$$

or that

$$(m_1 + m_2)\frac{\mathrm{d}^2}{\mathrm{d}t^2}\boldsymbol{r}_1 = m_2 \frac{\mathrm{d}^2}{\mathrm{d}t^2}\boldsymbol{r} \,, \tag{B.7}$$

and

$$\mu \frac{\mathrm{d}^2}{\mathrm{d}t^2}\boldsymbol{r} = \frac{e_1 e_2}{4\pi\epsilon_0 r^3}\boldsymbol{r} \,, \tag{B.8}$$

where

$$\mu = \frac{m_1 m_2}{m_1 + m_2} \tag{B.9}$$

is the reduced mass. The problem has thus been reduced to a single equation in a single variable, $\boldsymbol{r}$.

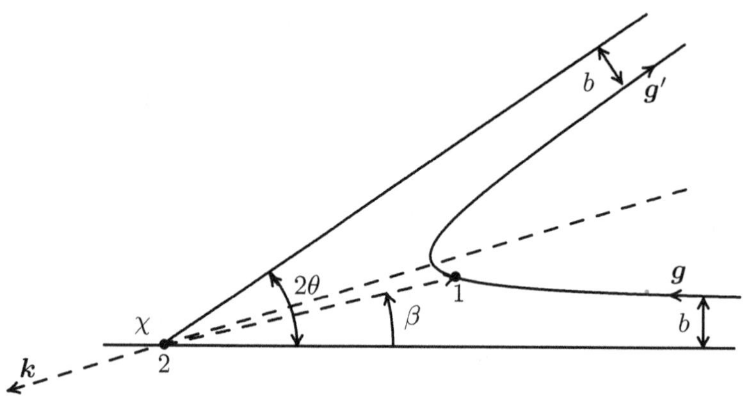

FIGURE B.1
Collision geometry.

The geometry of the collision is shown in Figure B.1. Initially particle 1 has relative velocity

$$\boldsymbol{g} = \boldsymbol{v}_1 - \boldsymbol{v}_2 = \boldsymbol{w}_1 - \boldsymbol{w}_2 \tag{B.10}$$

and an asymptotic distance of approach b. Its relative position vector we suppose makes an angle β with the direction $-\boldsymbol{g}$, so initially $\beta = 0$. Finally, $\beta = 2\theta - \chi$ where χ is the scattering angle we wish to find as a function of g and b. To begin with, we must find an expression for the left-hand side of equation (B.8). To do this, we remember that contributions to the differential of a vector come from the change of magnitude of the vector and from the changes of direction. We write

$$\boldsymbol{r} = r\hat{e}_r$$

where $\hat{e}_r$ denotes a unit vector in the direction $\boldsymbol{r}$. Differentiating,

$$\frac{\mathrm{d}}{\mathrm{d}t}\boldsymbol{r} = \hat{e}_r\frac{\mathrm{d}r}{\mathrm{d}t} + r\frac{\mathrm{d}}{\mathrm{d}t}\hat{e}_r\,,$$

and

$$\frac{\mathrm{d}^2}{\mathrm{d}t^2}\boldsymbol{r} = \hat{e}_r\frac{\mathrm{d}^2 r}{\mathrm{d}t^2} + 2r\frac{\mathrm{d}}{\mathrm{d}t}\boldsymbol{r} + r\frac{\mathrm{d}^2}{\mathrm{d}^2 t}\hat{e}_r\,. \tag{B.11}$$

Now

$$\frac{\mathrm{d}}{\mathrm{d}t}\hat{e}_r = \hat{\boldsymbol{e}}_\beta\frac{\mathrm{d}\beta}{\mathrm{d}t}\,, \qquad \frac{\mathrm{d}}{\mathrm{d}t}\hat{\boldsymbol{e}}_\beta = -\hat{e}_r\frac{\mathrm{d}\beta}{\mathrm{d}t}\,, \tag{B.12}$$

where $\hat{\boldsymbol{e}}_\beta$ is a unit vector in the direction of increasing β. Hence,

$$\frac{\mathrm{d}^2}{\mathrm{d}t^2}\hat{e}_r = \hat{\boldsymbol{e}}_\beta\frac{\mathrm{d}^2\beta}{\mathrm{d}t^2} - \hat{e}_r\left(\frac{\mathrm{d}\beta}{\mathrm{d}t}\right)^2, \tag{B.13}$$

and equation (B.11) becomes

$$\frac{\mathrm{d}^2}{\mathrm{d}t^2}\boldsymbol{r} = \hat{e}_r\left[\frac{\mathrm{d}^2 r}{\mathrm{d}t^2} - r\left(\frac{\mathrm{d}\beta}{\mathrm{d}t}\right)^2\right] + \hat{\boldsymbol{e}}_\beta\frac{1}{r}\frac{\mathrm{d}}{\mathrm{d}t}\left(r^2\frac{\mathrm{d}\beta}{\mathrm{d}t}\right). \tag{B.14}$$

Substituting into equation (B.8) and taking components parallel and perpendicular to $\boldsymbol{r}$ gives

$$\frac{\mathrm{d}^2 r}{\mathrm{d}t^2} - r\left(\frac{\mathrm{d}\beta}{\mathrm{d}t}\right)^2 = \frac{e_1 e_2}{4\pi\epsilon_0 r}, \tag{B.15}$$

and

$$\frac{\mathrm{d}}{\mathrm{d}t}\left(r^2\frac{\mathrm{d}\beta}{\mathrm{d}t}\right) = 0\,. \tag{B.16}$$

Integrating equation (B.16) gives

$$r^2\frac{\mathrm{d}\beta}{\mathrm{d}t} = \text{constant} = \frac{p_\beta}{\mu}\,, \tag{B.17}$$

where p_β is the angular momentum. For the initial conditions,

$$p_\beta = \mu b g\,. \tag{B.18}$$

Using Equations (B.17) and (B.18) to substitute in equation (B.16) gives

$$\frac{\mathrm{d}^2 r}{\mathrm{d}t^2} - \frac{b^2 g^2}{r^2} = \frac{e_1 e_2}{4\pi\epsilon_0\mu r^2}\,, \tag{B.19}$$

which can be solved by the substitution

$$\rho = \frac{1}{r} \,. \tag{B.20}$$

Then,

$$\frac{\mathrm{d}r}{\mathrm{d}t} = \frac{\mathrm{d}\beta}{\mathrm{d}t}\frac{\mathrm{d}r}{\mathrm{d}\beta} = \frac{\mathrm{d}\beta}{\mathrm{d}t}\left(-\frac{1}{\rho^2}\frac{\mathrm{d}\rho}{\mathrm{d}\beta}\right) = -bg\frac{\mathrm{d}\rho}{\mathrm{d}\beta} \tag{B.21}$$

and

$$\frac{\mathrm{d}^2 r}{\mathrm{d}t^2} = \frac{\mathrm{d}\beta}{\mathrm{d}t}\frac{\mathrm{d}}{\mathrm{d}\beta}\left(-bg\frac{\mathrm{d}\rho}{\mathrm{d}\beta}\right) = -b^2 g^2 \rho^2 \frac{\mathrm{d}^2\rho}{\mathrm{d}\beta^2} \,. \tag{B.22}$$

Thus we find equation (B.19) is

$$\frac{\mathrm{d}^2\rho}{\mathrm{d}\beta^2} + \rho = -\frac{e_1 e_2}{4\pi\epsilon_0 \mu b^2 g^2} \,, \tag{B.23}$$

which has the solution

$$\rho = \frac{1}{r} = -\frac{e_1 e_2}{4\pi\epsilon_0 \mu b^2 g^2} + A\cos(\beta - \theta) \,, \tag{B.24}$$

where A and θ are integration constants to be fixed by initial conditions. These are, at $\beta = 0$,

$$\begin{aligned} r \to \infty \quad &\text{or} \quad \rho \to 0 \,, \\ \frac{\mathrm{d}r}{\mathrm{d}t} = -g \quad &\text{or} \quad \frac{\mathrm{d}\rho}{\mathrm{d}t} = g\rho^2 \,. \end{aligned} \tag{B.25}$$

These give

$$A = \frac{e_1 e_2}{4\pi\epsilon_0 \mu b^2 g^2 \cos\theta} \tag{B.26}$$

and

$$\tan\theta = \frac{4\pi\epsilon_0 \mu b g^2}{e_1 e_2} \,. \tag{B.27}$$

Hence the solution is

$$\frac{1}{r} = \frac{e_1 e_2}{4\pi\epsilon_0 \mu b^2 g^2}\left[\frac{\cos(\beta - \theta)}{\cos\theta} - 1\right] . \tag{B.28}$$

From equation (B.28), we see that $r \to \infty$ when

$$\cos(\beta - \theta) = \cos\theta \,,$$

or when $\beta = 0$ or $\beta = 2\theta$. The condition $\beta = 0$, $r \to \infty$ corresponds to the initial condition, and $\beta = 2\theta$, $r \to \infty$ corresponds to the final condition. Hence, θ is the angle marked in Figure B.1. From the figure, we see that

$$\chi = \pi - 2\theta \tag{B.29}$$

so that

$$\cot \tfrac{1}{2}\chi = \frac{4\pi\epsilon_0 \mu b g^2}{e_1 e_2}\,. \tag{B.30}$$

From this we obtain the differential cross section, $\sigma(\chi, \phi)$, for scattering into a unit solid angle about χ and ϕ, using

$$\sigma(\chi,\phi)\mathrm{d}\Omega = \sigma(\chi,\phi)\sin\chi \mathrm{d}\chi \mathrm{d}\phi = b\,\mathrm{d}b\,\mathrm{d}\phi\,, \tag{B.31}$$

so that

$$\sigma(\chi,\phi) = \frac{b}{\sin\chi\left|\dfrac{\mathrm{d}\chi}{\mathrm{d}b}\right|} = \left(\frac{e_1 e_2}{8\pi\epsilon_0 \mu g^2}\right)^2 (\csc \tfrac{1}{2}\chi)^4, \tag{B.32}$$

which is the Rutherford scattering formula.

B.2 Collision integrals

Now let us consider the evaluation of the various collision integrals we require. It is convenient to make use of the unit vector $\hat{\boldsymbol{e}}_k$ drawn as shown in Figure B.1. Clearly,

$$\hat{\boldsymbol{e}}_k \cdot \boldsymbol{g} = g\cos\theta = -\hat{\boldsymbol{e}}_k \cdot \boldsymbol{g}'\,, \tag{B.33}$$

where

$$\boldsymbol{g}' = \boldsymbol{v}_2' - \boldsymbol{v}_1' = \boldsymbol{w}_2' - \boldsymbol{w}_1' \tag{B.34}$$

is the relative velocity after the collision. Also,

$$\begin{aligned}\boldsymbol{g}' &= \boldsymbol{g} - 2(\boldsymbol{g}\cdot\hat{\boldsymbol{e}}_k)\hat{\boldsymbol{e}}_k \\ &= \boldsymbol{g} - 2g\cos\theta\hat{\boldsymbol{e}}_k\,.\end{aligned} \tag{B.35}$$

It is also convenient to introduce the dimensionless numbers,

$$M_1 = \frac{m_1}{m_1+m_2} \quad \text{and} \quad M_2 = \frac{m_2}{m_1+m_2}\,. \tag{B.36}$$

The velocity of the center of mass relative to the drift velocity $\boldsymbol{u}$ is

$$\boldsymbol{G} = M_1\boldsymbol{w}_1 + M_2\boldsymbol{w}_2\,, \tag{B.37}$$

and the following equations can be derived:

$$\begin{aligned}\boldsymbol{w}_1 &= \boldsymbol{G} + M_2\boldsymbol{g}\,,\ \ \boldsymbol{w}_2 = \boldsymbol{G} - M_1\boldsymbol{g}\,, \\ \boldsymbol{w}_1' &= \boldsymbol{G} + M_2\boldsymbol{g}'\,,\ \boldsymbol{w}_2' = \boldsymbol{G} - M_1\boldsymbol{g}'\,,\end{aligned} \tag{B.38}$$

$$\begin{aligned}\boldsymbol{w}_1' &= \boldsymbol{w}_1 - 2gM_2\cos\theta\hat{\boldsymbol{e}}_k\,, \\ \boldsymbol{w}_2' &= \boldsymbol{w}_2 + 2gM_1\cos\theta\hat{\boldsymbol{e}}_k\,.\end{aligned} \tag{B.39}$$

We now define new dimensionless variables $\boldsymbol{x}$ and $\boldsymbol{y}$ by

$$\boldsymbol{x} \equiv \boldsymbol{g}\left(\frac{\mu}{2k_BT}\right)^{1/2} \quad \text{and} \quad \boldsymbol{y} \equiv \boldsymbol{G}\left(\frac{m_1+m_2}{2k_BT}\right)^{1/2}. \tag{B.40}$$

Then the normalized variables become

$$\boldsymbol{\varpi}_1 = \sqrt{M_1}\boldsymbol{y} + \sqrt{M_2}\boldsymbol{x} \tag{B.41}$$

$$\boldsymbol{\varpi}_2 = \sqrt{M_2}\boldsymbol{y} - \sqrt{M_1}\boldsymbol{x} \tag{B.42}$$

$$\boldsymbol{\varpi}_1' = \boldsymbol{\varpi}_1 - 2\sqrt{M_2}x\cos\theta\hat{\mathbf{e}}_k \tag{B.43}$$

$$\boldsymbol{\varpi}_2' = \boldsymbol{\varpi}_2 + 2\sqrt{M_1}x\cos\theta\hat{\mathbf{e}}_k\,, \tag{B.44}$$

and from equation (B.33)

$$\boldsymbol{x}\cdot\hat{\mathbf{e}}_k = x\cos\theta\,, \tag{B.45}$$

so from equation (B.27),

$$\tan\theta = \frac{8\pi\epsilon_0 k_BTb}{e_1e_2}x^2\,. \tag{B.46}$$

In terms of these variables, we find that

$$[G_1, H_2; K_1, L_2]_{12} = -\left(\frac{2k_BT}{\mu}\right)^{1/2}\frac{1}{\pi^{3/2}}\int \mathrm{d}\boldsymbol{x}\mathrm{d}\boldsymbol{y}\mathrm{d}\phi\, b\,\mathrm{d}b\, x\mathrm{e}^{-x^2-y^2}$$
$$\times[G_1(\boldsymbol{\varpi}_1) + H_2(\boldsymbol{\varpi}_2)] : [K_1(\boldsymbol{\varpi}_1') + L_2(\boldsymbol{\varpi}_2') - K_1(\boldsymbol{\varpi}_1) - L_2(\boldsymbol{\varpi}_2)]\,. \tag{B.47}$$

Using this formula, all the integrals for collisions between an electron and an ion can be worked out. Then the integrals for collisions between like particles can be obtained by setting the masses equal. We shall give the details of just one calculation, namely of $[\boldsymbol{\varpi}_1, \boldsymbol{\varpi}_1]_{12}$. From equation (B.47) this is (with $G_1(\boldsymbol{\varpi}_1) = K_1(\boldsymbol{\varpi}_1) = \boldsymbol{\varpi}_1$, and $H_2(\boldsymbol{\varpi}_2) = L_2(\boldsymbol{\varpi}_2) = 0$)

$$[\boldsymbol{\varpi}_1, \boldsymbol{\varpi}_1]_{12} = -\left(\frac{2k_BT}{\mu}\right)^{1/2}\frac{1}{\pi^{3/2}}\int \mathrm{d}\boldsymbol{x}\mathrm{d}\boldsymbol{y}\mathrm{d}\phi\, b\,\mathrm{d}b\, x\mathrm{e}^{-x^2-y^2}\boldsymbol{\varpi}_1\cdot(\boldsymbol{\varpi}_1' - \boldsymbol{\varpi}_1)\,.$$

Now from Equations (B.41) and (B.43),

$$\boldsymbol{\varpi}_1\cdot(\boldsymbol{\varpi}_1' - \boldsymbol{\varpi}_1) = -(\sqrt{M_1}\boldsymbol{y} + \sqrt{M_2}\boldsymbol{x})2\sqrt{M_2}x\cos\theta\cdot\hat{\mathbf{e}}_k\,.$$

Now $\boldsymbol{y}\cdot\hat{\mathbf{e}}_k$ averages to zero on integrating over $\boldsymbol{y}$, so using equation (B.45),

$$[\boldsymbol{\varpi}_1, \boldsymbol{\varpi}_1]_{12} = 2\left(\frac{2k_BT}{\mu}\right)^{1/2}M_2\frac{2\pi}{\pi^{3/2}}\int \mathrm{d}\boldsymbol{x}\, x^3\mathrm{e}^{-x^2}\int b\,\mathrm{d}b\,\cos^2\theta\,. \tag{B.48}$$

Using equation (B.46) this last integral is

$$\int_0^{\lambda_D} b\,\mathrm{d}b\,\cos^2\theta = \int_0^{\lambda_D}\frac{b\,\mathrm{d}b}{1+\left(\frac{8\pi\epsilon_0k_BTx^2}{e_1e_2}\right)^2 b^2}$$
$$= \frac{1}{2}\left(\frac{e_1e_2}{8\pi\epsilon_0k_BT}\right)^2\frac{1}{x^4}\ln\left[1+\lambda_D^2\left(\frac{8\pi\epsilon_0k_BTx^2}{e_1e_2}\right)^2\right]. \tag{B.49}$$

This expression diverges if λ_D tends to infinity but the divergence is only logarithmic and is therefore very slow. It therefore does not matter very much what choice we make for λ_D within reasonable limits.

Because the value we should use for λ_D is not fixed precisely and because the answer is insensitive anyway, we might as well replace x^2 where it appears inside the logarithm by its average value which is

$$\langle x^2 \rangle = \frac{\int \mathrm{d}\boldsymbol{x}\, x^3 \mathrm{e}^{-x^2}}{\int \mathrm{d}\boldsymbol{x}\, x \mathrm{e}^{-x^2}} = \frac{\int \mathrm{d}x\, x^5 \mathrm{e}^{-x^2}}{\int \mathrm{d}x\, x^3 \mathrm{e}^{-x^2}} = 2$$

and hence the right-hand side of equation (B.49) becomes

$$\frac{1}{2}\left(\frac{e_1 e_2}{8\pi\epsilon_0 k_B T}\right)^2 \frac{1}{x^4}\psi$$

where ψ is the logarithmic factor

$$\psi = \ln\left[1 + \left(\frac{16\pi\epsilon_0 k_B T \lambda_D}{e^2}\right)^2\right]. \tag{B.50}$$

The quantity inside the parentheses of equation (B.50) is equivalent to $12N_D$ where

$$N_D = \tfrac{4}{3}\pi n \lambda_D^3 \tag{B.51}$$

is the number of particles in a Debye sphere. Using numerical studies of the integral with large N_D without replacing x^2 by its mean value of 2 indicates a better value is to use 1.526 or replace $12N_D$ by $9.16N_D$, and assuming that N_D is large compared to unity, we may approximate that

$$\psi = 2\ln\Lambda\,, \tag{B.52}$$

where $\Lambda \simeq 9N_D$. This means that equation (B.48) becomes

$$\begin{aligned}[\boldsymbol{\varpi}_1, \boldsymbol{\varpi}_1]_{12} &= \left(\frac{2k_B T}{\mu}\right)^{1/2} M_2 \frac{2\psi}{\pi^{1/2}} \int_0^\infty \mathrm{d}\boldsymbol{x}\, \frac{1}{x}\mathrm{e}^{-x^2} \\ &= \frac{e^4 \ln\Lambda}{\sqrt{2\epsilon_0}(4\pi\epsilon_0 k_B T)^{3/2}} \left[\frac{m_2}{m_1(m_1+m_2)}\right]^{1/2}.\end{aligned} \tag{B.53}$$

Because $m_2 \gg m_1$, this becomes

$$[\boldsymbol{\varpi}_1, \boldsymbol{\varpi}_1]_{12} = \frac{e^4 \ln\Lambda}{4\sqrt{2m_1\epsilon_0}(\pi\epsilon_0 k_B T)^{3/2}}. \tag{B.54}$$

Interchanging the masses in equation (B.53) gives

$$[\boldsymbol{\varpi}_2, \boldsymbol{\varpi}_2]_{12} = \frac{m_1}{m_2}[\boldsymbol{\varpi}_1, \boldsymbol{\varpi}_1]_{12}. \tag{B.55}$$

We can calculate $[\boldsymbol{\varpi}_1, \boldsymbol{\varpi}_2]_{12}$ in a similar fashion. By definition,

$$[\boldsymbol{\varpi}_1, \boldsymbol{\varpi}_2]_{12} \equiv -\left(\frac{2k_B T}{\mu}\right)^{1/2} \frac{1}{\pi^3} \int \mathrm{d}\boldsymbol{x}\mathrm{d}\boldsymbol{y}\mathrm{d}\phi\, b\, \mathrm{d}b\, x\mathrm{e}^{-x^2-y^2} \boldsymbol{\varpi}_1 \cdot (\boldsymbol{\varpi}_2' - \boldsymbol{\varpi}_2) \tag{B.56}$$

and from Equations (B.43) and (B.44)

$$\boldsymbol{\varpi}_2' - \boldsymbol{\varpi}_2 = 2\sqrt{M_1} x \cos\theta \hat{\boldsymbol{e}}_k = -\left(\frac{m_1}{m_2}\right)^{1/2} \boldsymbol{\varpi}_1' - \boldsymbol{\varpi}_1 .$$

Comparing with the formula for $[\boldsymbol{\varpi}_1, \boldsymbol{\varpi}_1]_{12}$ we therefore see that

$$[\boldsymbol{\varpi}_1, \boldsymbol{\varpi}_2]_{12} = -\left(\frac{m_1}{m_2}\right)^{1/2} [\boldsymbol{\varpi}_1, \boldsymbol{\varpi}_1]_{12} . \tag{B.57}$$

From Equations (B.53) and (B.57)

$$[\boldsymbol{\varpi}_1, 0; \boldsymbol{\varpi}_1, \boldsymbol{\varpi}_2]_{12} = \left[1 - \left(\frac{m_1}{m_2}\right)^{\frac{1}{2}}\right] \frac{e^4 \psi}{2\sqrt{2\epsilon_0}(4\pi\epsilon_0 k_B T)^{3/2}} \left[\frac{m_2}{m_1(m_1 + m_2)}\right]^{\frac{1}{2}} .$$

We can now get an expression for $[\boldsymbol{\varpi}_1; \boldsymbol{\varpi}_1]_1$ defined by equation (8.83) or equation (9.93) in this expression. This gives

$$[\boldsymbol{\varpi}_1; \boldsymbol{\varpi}_1]_1 = 0 . \tag{B.58}$$

Similarly,

$$[\boldsymbol{\varpi}_2; \boldsymbol{\varpi}_2]_2 = 0 . \tag{B.59}$$

The integrals for a binary system as given by Hirschfelder et al. [50] for many cases are listed in terms of the following integral forms,

$$c_{i,j}^{(k,\ell)} = \left[L_k^{\frac{3}{2}}(\varpi_i^2)\boldsymbol{\varpi}_i, L_\ell^{\frac{3}{2}}(\varpi_j^2)\boldsymbol{\varpi}_j\right]_{ij} \tag{B.60}$$

$$d_{i,j}^{(k,\ell)} = \left[L_k^{\frac{3}{2}}(\varpi_i^2)\boldsymbol{\varpi}_i^0\boldsymbol{\varpi}_i, L_\ell^{\frac{3}{2}}(\varpi_j^2)\boldsymbol{\varpi}_j^0\boldsymbol{\varpi}_j\right]_{ij} \tag{B.61}$$

as

$$c_{ij}^{(0,0)} = -8\frac{\sqrt{m_i m_j}}{(m_i + m_j)}\Omega_{ij}^{(1,1)} \tag{B.62}$$

$$c_{ii}^{(0,0)} = 8\frac{m_j}{(m_i + m_j)}\Omega_{ij}^{(1,1)} \tag{B.63}$$

$$c_{ij}^{(0,1)} = 8\frac{m_i\sqrt{m_i m_j}}{(m_i + m_j)^2}\left[\Omega_{ij}^{(1,2)} - \tfrac{5}{2}\Omega_{ij}^{(1,1)}\right] \tag{B.64}$$

$$c_{ii}^{(0,1)} = -8\frac{m_j^2}{(m_i + m_j)^2}\left[\Omega_{ij}^{(1,2)} - \tfrac{5}{2}\Omega_{ij}^{(1,1)}\right] \tag{B.65}$$

$$c_{ii}^{(1,1)} = 8\frac{m_j}{(m_i+m_j)^3}\left[\tfrac{5}{4}(6m_i^2+5m_j^2)\Omega_{ij}^{(1,1)} \right.$$
$$\left. -5m_j^2\Omega_{ij}^{(1,2)} + m_j^2\Omega_{ij}^{(1,3)} + 2m_im_j\Omega_{ij}^{(2,2)}\right] \tag{B.66}$$

$$c_{ij}^{(1,1)} = -8\frac{(m_im_j)^{3/2}}{(m_i+m_j)^3}\left[\tfrac{55}{4}\Omega_{ij}^{(1,1)} - 5\Omega_{ij}^{(1,2)} + \Omega_{ij}^{(1,3)} - 2\Omega_{ij}^{(2,2)}\right] \tag{B.67}$$

$$c_{ii}^{(0,2)} = 4\frac{m_j^3}{(m_i+m_j)^3}\left[\tfrac{35}{4}\Omega_{ij}^{(1,1)} - 7\Omega_{ij}^{(1,2)} + \Omega_{ij}^{(1,3)}\right] \tag{B.68}$$

$$c_{ii}^{(1,2)} = 8\frac{m_j^2}{(m_i+m_j)^4}\left[\tfrac{35}{16}(12m_i^2+5m_j^2)\Omega_{ij}^{(1,1)} - \tfrac{21}{8}(4m_i^2+5m_j^2)\Omega_{ij}^{(1,2)}\right.$$
$$\left. +\tfrac{19}{4}m_j^2\Omega_{ij}^{(1,3)} - \tfrac{1}{2}m_j^2\Omega_{ij}^{(1,4)} + 7m_im_j\Omega_{ij}^{(2,2)} - 2m_im_j\Omega_{ij}^{(2,3)}\right] \tag{B.69}$$

$$c_{ij}^{(1,2)} = -8\frac{m_i(m_im_j)^{3/2}}{(m_i+m_j)^4}\left[\tfrac{595}{16}\Omega_{ij}^{(1,1)} - \tfrac{189}{8}\Omega_{ij}^{(1,2)} + \tfrac{19}{4}\Omega_{ij}^{(1,3)} - \tfrac{1}{2}\Omega_{ij}^{(1,4)}\right.$$
$$\left. -7\Omega_{ij}^{(2,2)} + 2\Omega_{ij}^{(2,3)}\right] \tag{B.70}$$

$$c_{ii}^{(2,2)} = 8\frac{m_j}{(m_i+m_j)^5}\left[\tfrac{35}{64}(40m_i^4 + 168m_i^2m_j^2 + 35m_i^4)\Omega_{ij}^{(1,1)}\right.$$
$$-\tfrac{7}{8}m_j^2(84m_i^2+35m_j^2)\Omega_{ij}^{(1,2)} + \tfrac{1}{8}m_j^2(108m_i^2+133m_j^2)\Omega_{ij}^{(1,3)}$$
$$-\tfrac{7}{2}m_j^4\Omega_{ij}^{(1,4)} + \tfrac{1}{4}m_j^4\Omega_{ij}^{(1,5)} + \tfrac{7}{2}m_im_j(4m_i^2+7m_j^2)\Omega_{ij}^{(2,2)}$$
$$\left. -14m_im_j^3\Omega_{ij}^{(2,3)} + 2m_im_j^3\Omega_{ij}^{(2,4)} + 2m_i^2m_j^2\Omega_{ij}^{(3,3)}\right] \tag{B.71}$$

$$c_{ij}^{(2,2)} = -\frac{(m_im_j)^{5/2}}{(m_i+m_j)^5}\left[\tfrac{8505}{64}\Omega_{ij}^{(1,1)} - \tfrac{833}{8}\Omega_{ij}^{(1,2)} + \tfrac{241}{8}\Omega_{ij}^{(1,3)} - \tfrac{7}{2}\Omega_{ij}^{(1,4)}\right.$$
$$\left. +\tfrac{1}{4}\Omega_{ij}^{(1,5)} - \tfrac{77}{2}\Omega_{ij}^{(2,2)} + 14\Omega_{ij}^{(2,3)} - 2\Omega_{ij}^{(2,4)} + 2\Omega_{ij}^{(3,3)}\right] \tag{B.72}$$

$$d_{ii}^{(0,0)} = \frac{16}{3}\frac{m_j}{(m_i+m_j)^2}\left[5m_i\Omega_{ij}^{(1,1)} + \tfrac{3}{2}m_j\Omega_{ij}^{(2,2)}\right] \tag{B.73}$$

$$d_{ij}^{(0,0)} = -\frac{16}{3}\frac{m_im_j}{(m_i+m_j)^2}\left[5\Omega_{ij}^{(1,1)} - \tfrac{3}{2}\Omega_{ij}^{(2,2)}\right] \tag{B.74}$$

where the $\Omega_{ij}^{(\ell,m)}$ are given by Chapman and Cowling[49] for Coulomb interactions as

$$\Omega_{ij}^{(\ell,r)} \equiv \sqrt{\pi}\int_0^\infty \mathrm{e}^{-g^2} g^{2r+2}\phi_{12}^\ell \,\mathrm{d}g \tag{B.75}$$

where

$$\phi_{ij}^\ell \equiv \left(\frac{e^2}{4\pi\epsilon_0}\right)^2 \frac{g^{-3}A_\ell(2)}{2\sqrt{M_1M_2m_0}} \tag{B.76}$$

with

$$A_\ell(2) \equiv \int_0^{v_{01}}\left[1-\left(\frac{v_0^2-1}{v_0^2+1}\right)^\ell\right] v_0\,\mathrm{d}v_0\,. \tag{B.77}$$

Evaluating the first several of the A_ℓ integrals, we find

$$A_1(2) = \ln(1+v_{01}^2) \tag{B.78}$$

$$A_2(2) = 2\left[\ln(1+v_{01}^2) - \frac{v_{01}^2}{1+v_{01}^2}\right] \tag{B.79}$$

$$A_3(2) = 3\ln(1+v_{01}^2) - \frac{6v_{01}^2}{1+v_{01}^2} + \frac{2v_{01}^2(2+v_{01}^2)}{(1+v_{01}^2)^2} \tag{B.80}$$

$$A_4(2) = 4\left[\ln(1+v_{01}^2) - \frac{3v_{01}^2}{1+v_{01}^2} + \frac{2v_{01}^2(2+v_{01}^2)}{(1+v_{01}^2)^2} - \frac{2v_{01}^2(1+v_{01}^2+\frac{2}{3}v_{01}^4)}{(1+v_{01}^2)^3}\right] \tag{B.81}$$

and if we assume both that $v_{01} \gg 1$ and $\ln(1+v_{01}^2) \gg 1$, then

$$A_\ell(2) \simeq \ell\ln(1+v_{01}^2)\,. \tag{B.82}$$

Inserting this result into equation (B.76) and then this into equation (B.75), we find, identifying the logarithmic term with equation (B.50),

$$\Omega_1^{(\ell,r)} \simeq \frac{(r-1)!\ell\varphi}{4\sqrt{2}} \tag{B.83}$$

$$\Omega_{ij}^{(\ell,r)} = \frac{(r-1)!\ell\sqrt{m_1}\varphi}{8\sqrt{M_1 M_2 m_0}} \simeq \frac{(r-1)!\ell\varphi}{8} \tag{B.84}$$

so the first several are, for $m_2 = m_j \gg m_1 = m_i$,

$$\Omega_{12}^{(1,1)} \simeq \frac{\varphi}{8} \tag{B.85}$$

$$\Omega_{12}^{(1,2)} \simeq \frac{\varphi}{8} \tag{B.86}$$

$$\Omega_{12}^{(1,3)} \simeq \frac{\varphi}{4} \tag{B.87}$$

$$\Omega_{12}^{(1,4)} \simeq \frac{3\varphi}{4} \tag{B.88}$$

$$\Omega_{12}^{(1,5)} \simeq 3\varphi \tag{B.89}$$

$$\Omega_{12}^{(2,2)} \simeq \frac{\varphi}{4} \tag{B.90}$$

$$\Omega_{12}^{(2,3)} \simeq \frac{\varphi}{2} \tag{B.91}$$

$$\Omega_{12}^{(2,4)} \simeq \frac{3\varphi}{2} \tag{B.92}$$

$$\Omega_{12}^{(3,3)} \simeq \frac{3\varphi}{4}\,. \tag{B.93}$$

Some of the integrals for a simple gas, where only a single species is considered, are given in terms of the expressions

$$a_k^{(i,j)} \equiv \left[L_i^{\frac{3}{2}}(\varpi^2)\boldsymbol{\varpi}, L_j^{\frac{3}{2}}(\varpi^2)\boldsymbol{\varpi}\right]_k \tag{B.94}$$

and

$$b_k^{(i,j)} \equiv \left[L_i^{\frac{5}{2}}(\varpi^2)\boldsymbol{\varpi}^0\boldsymbol{\varpi}, L_j^{\frac{5}{2}}(\varpi^2)\boldsymbol{\varpi}^0\boldsymbol{\varpi}\right]_k \tag{B.95}$$

where $k = 1, 2$ and

$$a_2^{(i,j)} = \sqrt{M_1}a_1^{(i,j)} \tag{B.96}$$

$$b_2^{(i,j)} = \sqrt{M_1}b_1^{(i,j)} \tag{B.97}$$

which were introduced in Chapter 7 as (from Chapman and Cowling[49] equations 9.6-16 to 9.6-21 for the $b_1^{(i,j)}$ and equations 9.6-8 to 9.6-13 for the $a_1^{(i,j)}$)

$$b_1^{(0,0)} = 4\Omega_1^{(2,2)} \tag{B.98}$$

$$b_1^{(1,0)} = 7\Omega_1^{(2,2)} - 2\Omega_1^{(2,3)} \tag{B.99}$$

$$b_1^{(1,1)} = \tfrac{301}{12}\Omega_1^{(2,2)} - 7\Omega_1^{(2,3)} + \Omega_1^{(2,4)} \tag{B.100}$$

$$b_1^{(2,0)} = \tfrac{63}{8}\Omega_1^{(2,2)} - \tfrac{9}{2}\Omega_1^{(2,3)} + \tfrac{1}{2}\Omega_1^{(2,4)} \tag{B.101}$$

$$b_1^{(2,1)} = \tfrac{1365}{32}\Omega_1^{(2,2)} - \tfrac{321}{16}\Omega_1^{(2,3)} + \tfrac{25}{8}\Omega_1^{(2,4)} - \tfrac{1}{4}\Omega_1^{(2,5)} \tag{B.102}$$

$$b_1^{(2,2)} = \tfrac{25137}{256}\Omega_1^{(2,2)} - \tfrac{1755}{32}\Omega_1^{(2,3)} + \tfrac{381}{32}\Omega_1^{(2,4)} - \tfrac{9}{8}\Omega_1^{(2,5)} + \tfrac{1}{16}\Omega_1^{(2,6)} + \tfrac{1}{2}\Omega_1^{(4,4)} \tag{B.103}$$

$$a_1^{(0,0)} = 0 \tag{B.104}$$

$$a_1^{(0,1)} = 0 \tag{B.105}$$

$$a_1^{(1,1)} = 4\Omega_1^{(2,2)} \tag{B.106}$$

$$a_1^{(1,2)} = 7\Omega_1^{(2,2)} - 2\Omega_1^{(2,3)} \tag{B.107}$$

$$a_1^{(2,2)} = \tfrac{77}{4}\Omega_1^{(2,2)} - 7\Omega_1^{(2,3)} + \Omega_1^{(2,4)} \tag{B.108}$$

$$a_1^{(1,3)} = \tfrac{63}{8}\Omega_1^{(2,2)} - \tfrac{9}{2}\Omega_1^{(2,3)} + \tfrac{1}{2}\Omega_1^{(2,4)} \tag{B.109}$$

$$a_1^{(2,3)} = \tfrac{945}{32}\Omega_1^{(2,2)} - \tfrac{261}{16}\Omega_1^{(2,3)} + \tfrac{25}{8}\Omega_1^{(2,4)} - \tfrac{1}{4}\Omega_1^{(2,5)} \tag{B.110}$$

$$a_1^{(3,3)} = \tfrac{14553}{256}\Omega_1^{(2,2)} - \tfrac{1215}{32}\Omega_1^{(2,3)} + \tfrac{313}{32}\Omega_1^{(2,4)} - \tfrac{9}{8}\Omega_1^{(2,5)} + \tfrac{1}{16}\Omega_1^{(2,6)} + \tfrac{1}{6}\Omega_1^{(4,4)} \tag{B.111}$$

where

$$\Omega_1^{(\ell,r)} = \sqrt{2}\Omega_{12}^{(\ell,r)}\,. \tag{B.112}$$

In a similar way all the other integrals we need can be evaluated and the results are given in the following list. In each case, only the leading terms in the mass ratio are given so each formula is correct only to order m_1/m_2 or one part in two thousand. The uncertainty in the cutoff introduces errors considerably larger than this, so there is no point in including these small terms.

The list is given in terms of the function

$$\varphi = \frac{e^4\psi}{8\sqrt{2m_1\epsilon_0}(\pi\epsilon_0 k_B T)^{3/2}} = \frac{3v_{T1}}{2\sqrt{\pi}n\lambda_D}\frac{\ln\Lambda}{\Lambda} \tag{B.113}$$

as

$$[\boldsymbol{\varpi}_1, \boldsymbol{\varpi}_1]_{12} = \varphi \tag{B.114}$$

$$[\boldsymbol{\varpi}_1, \boldsymbol{\varpi}_2]_{12} = -\sqrt{M_1}\varphi \tag{B.115}$$

$$[\boldsymbol{\varpi}_2, \boldsymbol{\varpi}_2]_{12} = M_1\varphi \tag{B.116}$$

$$\left[\boldsymbol{\varpi}_1^0\boldsymbol{\varpi}_1, \boldsymbol{\varpi}_1^0\boldsymbol{\varpi}_1\right]_{12} = 2\varphi \tag{B.117}$$

$$\left[\boldsymbol{\varpi}_2^0\boldsymbol{\varpi}_2, \boldsymbol{\varpi}_2^0\boldsymbol{\varpi}_2\right]_{12} = \tfrac{10}{3}M_1\varphi \tag{B.118}$$

$$\left[\boldsymbol{\varpi}_1^0\boldsymbol{\varpi}_1, \boldsymbol{\varpi}_2^0\boldsymbol{\varpi}_2\right]_{12} = -\tfrac{4}{3}M_1\varphi \tag{B.119}$$

$$\left[\boldsymbol{\varpi}_1, \boldsymbol{\varpi}_1 L_1^{\frac{3}{2}}(\varpi_1^2)\right]_{12} = \tfrac{3}{2}\varphi \tag{B.120}$$

$$\left[\boldsymbol{\varpi}_2, \boldsymbol{\varpi}_2 L_1^{\frac{3}{2}}(\varpi_2^2)\right]_{12} = \tfrac{3}{2}M_1^2\varphi \tag{B.121}$$

$$\left[\boldsymbol{\varpi}_1, \boldsymbol{\varpi}_2 L_1^{\frac{3}{2}}(\varpi_2^2)\right]_{12} = -\tfrac{3}{2}M_1^{3/2}\varphi \tag{B.122}$$

$$\left[\boldsymbol{\varpi}_1 L_1^{\frac{3}{2}}(\varpi_1^2), \boldsymbol{\varpi}_2\right]_{12} = -\tfrac{3}{2}\sqrt{M_1}\varphi \tag{B.123}$$

$$\left[\boldsymbol{\varpi}_1 L_1^{\frac{3}{2}}(\varpi_1^2), \boldsymbol{\varpi}_1 L_1^{\frac{3}{2}}(\varpi_1^2)\right]_{12} = \tfrac{13}{4}\varphi \tag{B.124}$$

$$\left[\boldsymbol{\varpi}_2 L_1^{\frac{3}{2}}(\varpi_2^2), \boldsymbol{\varpi}_2 L_1^{\frac{3}{2}}(\varpi_2^2)\right]_{12} = \tfrac{15}{2}M_1\varphi \tag{B.125}$$

$$\left[\boldsymbol{\varpi}_1 L_1^{\frac{3}{2}}(\varpi_1^2), \boldsymbol{\varpi}_2 L_1^{\frac{3}{2}}(\varpi_2^2)\right]_{12} = -\tfrac{27}{4}M_1^{3/2}\varphi \tag{B.126}$$

$$\left[\boldsymbol{\varpi}_1, \boldsymbol{\varpi}_1 L_1^{\frac{3}{2}}(\varpi_1^2)\right]_1 = 0 = \left[\boldsymbol{\varpi}_2, \boldsymbol{\varpi}_2 L_1^{\frac{3}{2}}(\varpi_2^2)\right]_2 \tag{B.127}$$

$$\left[\boldsymbol{\varpi}^0\boldsymbol{\varpi}, \boldsymbol{\varpi}^0\boldsymbol{\varpi}\right]_1 = \sqrt{2}\varphi \tag{B.128}$$

$$\left[\boldsymbol{\varpi}^0\boldsymbol{\varpi}, \boldsymbol{\varpi}^0\boldsymbol{\varpi}\right]_2 = \sqrt{2M_1}\varphi \tag{B.129}$$

$$\left[\boldsymbol{\varpi}_1 L_1^{\frac{3}{2}}(\varpi_1^2), \boldsymbol{\varpi}_1 L_1^{\frac{3}{2}}(\varpi_1^2)\right]_1 = \sqrt{2}\varphi \tag{B.130}$$

$$\left[\boldsymbol{\varpi}_2 L_1^{\frac{3}{2}}(\varpi_2^2), \boldsymbol{\varpi}_2 L_1^{\frac{3}{2}}(\varpi_2^2)\right]_2 = \sqrt{2M_1}\varphi\,. \tag{B.131}$$

Problem B.1 Evaluate the $c_{ij}^{(k,\ell)}$ and $d_{ij}^{(k,\ell)}$ collision integrals of equations (B.114) through (B.126).

Problem B.2 Evaluate the $a_k^{(i,j)}$ and $b_k^{(i,j)}$ collision integrals from equations (B.127) through (B.131).

The integral of equation (B.131) is given by Marshall as $\sqrt{2}M_1\varphi$ but this has been corrected by Vaughan-Williams and Haas[59] and the listed result agrees with Rosenbluth and Kaufmann[60]. Garcia-Colin et al. [61] disagree with both equation (B.122) and equation (B.123), listing these as $-\sqrt{2}M_1\varphi$ and $-\frac{3}{2}M_1^{3/2}\varphi$ respectively, but the results of problem B.1 above agree with the Marshall results for these integrals.

It remains to examine the validity of equation (B.1) which assumed that while the particles were interacting, all forces other than their Coulomb interaction could be ignored. This will be valid providing the Debye length, λ_D, is smaller than the electron Larmor radius, or

$$\lambda_D^2 = \frac{\epsilon_0 k_B T}{n_1 e^2} \ll \frac{\langle v^2 \rangle}{\omega_{ce}^2},$$

or when

$$\frac{B^2}{2\mu_0} \ll \frac{3}{2} n_1 m c^2 .$$

C

NOTATION AND LIST OF SYMBOLS

C.1 Mathematical notation

1. Vectors are indicated by bold face symbols. Components are indicated by Greek subscripts, α, β, γ, etc. Thus $\boldsymbol{r}_\alpha$ stands for x, y, or z and $\boldsymbol{v}_\alpha$ stands for v_x, v_y, or v_z.

2. Tensors are indicated by capital sans serif symbols. P stands for the tensor whose components are $p_{\alpha\beta}$.

3. Repeated subscripts are to be summed. Thus the scalar product may be written $\boldsymbol{a} \cdot \boldsymbol{b} = a_\alpha b_\alpha = a_x b_x + a_y b_y + a_z b_z$.

4. The symbol $\mathrm{d}\boldsymbol{r}$ represents a volume element such that $\mathrm{d}\boldsymbol{r} \equiv \mathrm{d}x\mathrm{d}y\mathrm{d}z$, and $\mathrm{d}\boldsymbol{v}$ represents a volume element in velocity space such that $\mathrm{d}\boldsymbol{v} \equiv \mathrm{d}v_x\mathrm{d}v_y\mathrm{d}v_z$. Note the difference between d, representing some quantity, and d, representing the differential of calculus.

5. The nabla symbol $\nabla \equiv \hat{\boldsymbol{x}}\frac{\partial}{\partial x} + \hat{\boldsymbol{y}}\frac{\partial}{\partial y} + \hat{\boldsymbol{z}}\frac{\partial}{\partial z}$ where $\hat{\boldsymbol{x}}$ is a unit vector in the x-direction.

6. The subscripted nabla symbol $\nabla_v \equiv \hat{\boldsymbol{x}}\frac{\partial}{\partial v_x} + \hat{\boldsymbol{y}}\frac{\partial}{\partial v_y} + \hat{\boldsymbol{z}}\frac{\partial}{\partial v_z}$.

7. The symbol $\frac{\mathrm{D}}{\mathrm{D}t} \equiv \frac{\partial}{\partial t} + \boldsymbol{u} \cdot \nabla$.

8. $\delta_{\alpha\beta} \equiv \begin{cases} 1 \text{ if } \alpha = \beta \\ 0 \text{ if } \alpha \neq \beta \end{cases}$ is the Kronecker δ.

C.2 List of symbols

Symbol	Description
c	speed of light.
k_B	Boltzmann constant.
v_{ti}	thermal speed for species i, $v_{ti} \equiv \sqrt{2k_BT_i/m_i}$.
$\boldsymbol{r}$	location of a particle.
$\boldsymbol{v}$	velocity of a particle.
$\boldsymbol{u}$	velocity of a fluid element where $\boldsymbol{u} \equiv \langle \boldsymbol{v} \rangle$.
$\boldsymbol{w}$	random velocity, $\boldsymbol{w} \equiv \boldsymbol{v} - \boldsymbol{u}$.
$\boldsymbol{\varpi} \equiv \boldsymbol{w}_i/v_{ti}$	random velocity normalized to thermal speed.
$F(\boldsymbol{r}, \boldsymbol{v}, t)$	the distribution function.
$\boldsymbol{X}$	the force per unit mass acting on the particles.
Δ	rate of change due to collisions.
Ψ	any property of the particles depending in general. on their position, velocity, and time.
n	the number density of particles.
m	the mass of each particle.
ρ	the mass density where $\rho \equiv nm$.
$\langle \Psi \rangle \equiv \frac{1}{n} \int \mathrm{d}v\, \Psi F$	the mean or average over the velocity of Ψ.
$p_{\alpha\beta}$	the pressure tensor where $p_{\alpha\beta} \equiv \rho \langle w_\alpha w_\beta \rangle$.
p	the static where $p = nkT$.
$\boldsymbol{q}$	the heat flux vector where $\boldsymbol{q} \equiv \frac{1}{2} nm \langle w^2 \boldsymbol{w} \rangle$.
$\boldsymbol{s}$	always used to stand for the "other particle" in a collision.
$\boldsymbol{v}'$, $\boldsymbol{s}'$	represents the velocities of the particles after a collision.
$\boldsymbol{G} \equiv \frac{1}{2}(\boldsymbol{v} + \boldsymbol{s})$	represents the center of gravity velocity in a collision.
$\boldsymbol{g} \equiv \boldsymbol{s} - \boldsymbol{v}$	the relative velocity of particles in a collision.
b	the impact parameter or asymptotic distance of approach in a collision.
χ	the scattering angle in a collision.
ε	the angle the plane of the collision makes with some fixed plane.
λ_D	the Debye length given by $\lambda_D \equiv \sqrt{\epsilon_0 k_B T/ne^2}$.
e	electronic charge.
S	the entropy of the gas.
V	the potential from which $\boldsymbol{X}$ is derived such that $\boldsymbol{X} = \nabla v$.
$\mathrm{d}s$	a unit of area whose orientation is described by the vector $\mathrm{d}\boldsymbol{s}$ drawn normal to it.
τ	a collision time.
$p(t)$	probability that a particle survives making collisions for a time greater than t after making a collision.

Symbol	Description
η	expansion parameter in the formal theory.
f	stands for the Maxwell distribution.
$\boldsymbol{\varpi} = \boldsymbol{w}/v_t$	dimensionless random velocity.
$\boldsymbol{\varpi}^0\boldsymbol{\varpi} \equiv \varpi_\alpha\varpi_\beta - \frac{1}{3}\varpi^2\delta_{\alpha\beta}$	tensor function of random velocities.
a_m	expansion coefficient of $\mathcal{A}$ as in (7.89).
b_m	expansion coefficient of $\mathcal{B}$ as in (7.93).
$L_m^{(n)}$	a Generalized Laguerre polynomial defined by (7.91). $L_m^{(n)} = S_n^m$ where S_n^m is a Sonine polynomial defined by (7.90).
μ	the coefficient of viscosity.
λ	the coefficient of thermal conductivity.
G and H	any properties of the particles.
$[G, H]$	a "collision integral" defined by (7.104).
$[\mu]_1$, $[\mu]_2$, $[\mu]_3 \dots$	denote successive approximations to μ.
ψ	the logarithmic cut-off term defined by (B.50) and (B.52).
$\boldsymbol{A}_i$	a vector making a contribution to φ_i.
$\mathcal{A}$, $\mathcal{A}_i^I$, $\mathcal{A}_i^{II}$, $\mathcal{A}_i^{III}$	scalar functions of ϖ_i^2 and B^2 defined by (9.30) and (9.46).
a_i^m, $a_i^{I,m}$, $a_i^{II,m}$, $a_i^{III,m}$	expansion coefficients of $\mathcal{A}$, $\mathcal{A}_i^I$, $\mathcal{A}_i^{II}$, and $\mathcal{A}_i^{III}$, respectively in Laguerre polynomials.
$\mathcal{E}$, $\mathcal{E}_i^I$, $\mathcal{E}_i^{II}$, $\mathcal{E}_i^{III}$	scalar functions of ϖ_i^2 and B^2 defined by (9.31) and (9.49).
e_i^m, $e_i^{I,m}$, $e_i^{II,m}$, $e_i^{III,m}$	expansion coefficients of $\mathcal{E}$, $\mathcal{E}_i^I$, $\mathcal{E}_i^{II}$, and $\mathcal{E}_i^{III}$, respectively in Laguerre polynomials.
B_i	a tensor contributing $-\mathsf{B}_i^{\alpha\beta}\nabla_\alpha u_\beta$ to φ_i.
B_i^n	scalar functions defined by (9.137).
$b_i^{m,n}$	expansion coefficients of B_i^n in Laguerre polynomials.
$\boldsymbol{D}$	is a "generalized" electric field given by (9.60).
$\mathcal{D}(\mathcal{J})$	the "Davison function" of a variational method with a trial function $\mathcal{J}$.
D_i^n	scalar functions defined by (9.144).
$\boldsymbol{E}$	the electric field.
$\boldsymbol{E}' = \boldsymbol{E} + \boldsymbol{u} \times \boldsymbol{B}$	
e_i	the charge on particle i.
$\boldsymbol{E}_i$	a vector making a contribution $-n\boldsymbol{E}_i \cdot \boldsymbol{d}_1$ to φ_i.
$f_i = \frac{n_i}{\pi^{3/2}v_i^3}\mathrm{e}^{-(\boldsymbol{v}-\boldsymbol{u})^2/v_i^2}$	the Maxwell distribution for species i.
$G(B^2)$, $G^I(B^2)$, $G^{II}(B^2)$	scalar functions of B^2 defined by (9.39), (9.43), and (9.46).
G_i	complex scalar function defined by (9.152).
g_i^n	expansion coefficients of G_i in Laguerre polynomials.

Symbol	Description
$\boldsymbol{B}$	the magnetic field.
$\boldsymbol{b}$	a unit vector in the direction of $\boldsymbol{B}$.
i, j	subscripts labeling electrons and ions, $i, j = 1$ for electrons, $i, j = 2$ for ions.
$\mathcal{I}$	a symbolic integral operator defined by (9.139).
$\boldsymbol{J}$	the total electric current, $\boldsymbol{J} = Q\boldsymbol{u} + \boldsymbol{j}$.
$\boldsymbol{j}$	the conduction current.
$K(B^2)$, $K^I(B^2)$, $K^{II}(B^2)$	scalar functions of B^2 defined by (9.50) and (9.51).
L_i	a scalar function defined by (9.152).
ℓ_i^n	expansion coefficients of L_i in Laguerre polynomials.
m_i	mass of particle i. $m_1 =$ electron mass, $m_2 =$ ion mass.
$M_1 \equiv \frac{m_1}{m_1+m_2}$	$M_1 \simeq m_1/m_2$.
n_i	the number density of particles i.
$n = n_1 + n_2$	the total number density.
$p_{\alpha\beta}$	the pressure tensor.
$p = nk_BT$	the pressure.
P_i	a scalar function defined by (9.152).
p_i^n	expansion coefficients of P_i in Laguerre polynomials.
$Q \equiv (n_2 - n_1)e$	the charge density.
$Q_{i,\alpha\beta}^n$	tensors defined by (9.141).
$s_{\alpha\beta}$	$s_{\alpha\beta} \equiv \frac{1}{2}(\nabla_\alpha u_\beta + \nabla_\beta u_\alpha) - \frac{1}{3}\nabla \cdot \boldsymbol{u}\delta_{\alpha\beta}$.
$T_{i,\alpha\beta}^n$	tensors defined by (9.136).
$\boldsymbol{\varpi}_i = \boldsymbol{w}_i/v_{ti}$	dimensionless random velocity.
$\boldsymbol{\varpi}_i^0\boldsymbol{\varpi}_i \equiv \varpi_{\alpha i}\varpi_{\beta i} - \frac{1}{3}\varpi_i^2\delta_{\alpha\beta}$	function of random velocities.
γ_{st}^m	tensors defined by (9.164).
Δ_j	rate of change with time due to collisions with particles j.
$\Delta = \sum_j \Delta_j$	total rate of change with time due to collisions.
$\epsilon_{\alpha\beta\gamma}$	$+1$ if α, β, γ are cyclic. -1 if α, β, γ are not cyclic. 0 if α, β, γ are not all different.
ξ^I, ξ^{II}, ξ^{III}	coefficients giving the contribution to the heat flux from $\boldsymbol{D}$.
φ_i	the first order correction to F_i defined by $F_i = f_i(1 + \varphi_i)$.
φ^I, φ^{II}, φ^{III}	thermal diffusion coefficients.
ψ^I, ψ^{II}, ψ^{III}	coefficients giving the heat flux due to electric currents.

Symbol	Description
θ^I, θ^{II}, θ^{III}	thermal conduction coefficients (when electric currents are allowed to flow).
λ^I, λ^{II}, λ^{III}	thermal conduction coefficients.
λ'^I, λ'^{II}, λ'^{III}	"true" thermal conduction coefficients. (i.e., if thermal diffusion effects were absent.)
σ	the differential cross section.
σ^I, σ^{II}, σ^{III}	coefficients of electrical conductivity.
ρ_i	mass density of particles i.
$\rho = \rho_1 + \rho_2$	total mass density.
$\mu = \frac{1}{3} n k_B T \tau_i$	the coefficient of viscosity.
$\mathrm{d}\Omega = \mathrm{d}\varepsilon \, \sin\theta \, \mathrm{d}\theta$	element of solid angle in a collision.
$\omega_{ce} = eB/m_1$	cyclotron frequency for electrons.
$\omega_{ci} = eB/m_2$	cyclotron frequency for ions.
$a^{(i,j)}$	collision integral for electrons only. See (7.130).
$b^{(i,j)}$	collision integral for electrons only. See (7.121).
$c^{(i,j)}$	collision integral for a binary system. See (B.60).
$d^{(i,j)}$	collision integral for a binary system. See (B.61).
$\mathcal{I}$	integral operator. See equation (9.139).

Bibliography

[1] Spitzer Lyman Jr. 1962 **Physics of Fully Ionized Gases**, 2nd Edition, (Interscience Publishers, New York)

[2] Bhatnagar P L, Gross E P, and Krook M 1954 *Phys. Rev.* **94** 511

[3] Bernstein I B, Greene J M, and Kruskal M D 1957 *Phys. Rev.* **108** 546

[4] Swanson D G, Gould R W, and Hertel R H 1964 *Phys. Fluids* **7** 269

[5] Swanson D G 2003 **Plasma Waves Second Edition** (Institute of Physics: Bristol and Philadelphia)

[6] Alfvén H 1950 **Cosmical Electrodynamics** (New York: Oxford University Press)

[7] Bogolyubov N N 1946 **Problems of a Dynamical Theory in Statistical Physics** (State University Press, Moscow)

[8] Born M and Green H S 1949 **A General Kinetic Theory of Liquids** (Cambridge University Press, England)

[9] Kirkwood J G 1946 *J. Chem. Phys.* **14** 180; *J. Chem. Phys.* **15**, 72(1947)

[10] Yvon J 1935 **La Theorie des Fluides et l'Equation d'Etat** (Hermann et Cie, Paris)

[11] Dreicer H, 1959 *Phys. Rev.* **115** 238

[12] Rosenbluth M N, MacDonald W M, and Judd D L 1957 *Phys. Rev.* **107** 1

[13] Watson G N 1944 **Theory of Bessel Functions, 2nd Edition** (Cambridge University Press, London)

[14] Stix T H 1975 *Nucl. Fusion* **15** 737

[15] Kennel C F and Engelmann F 1966 *Phys. Fluids* **9** 2377

[16] Cohen R S, Spitzer L Jr., and Routly P M 1950 *Phys Rev.* **80** 230

[17] Swanson D G 1998 **Theory of Mode Conversion and Tunneling in an Inhomogeneous Plasma** (Wiley, New York)

[18] Swanson D G 2002 *Phys. Plasma* **9** 2854

[19] Vlasov A A 1945 *J. Phys.(U.S.S.R.)* **9** 25

[20] Fried B D and Conte S D 1961 **The Plasma Dispersion Function** (Academic Press, Inc., New York)

[21] Landau L D 1946 *J. Phys.(U.S.S.R.)* **10** 25

[22] Ng C S, Bhattachargee A, and Skiff F 1999 *Phys. Rev. Lett.* **83** 1974

[23] Short R W and Simon A 2002 *Phys. Plasmas* **8** 3245

[24] Rostoker N 1960 *Nucl. Fusion* **1** 101

[25] Hagler M and Swanson D G 1967 Private communication

[26] Montgomery G Joyce and Sugihara R 1968 *Plasma Physics* **10** 681

[27] Cooper G 1969 *Phys. Fluids* **12** 2707

[28] Drummond J E 1958 *Phys. Rev.* **110** 293

[29] Sagdeev R Z and Shafranov V D 1958 *Proc. 2nd Int. Conf., (Geneva)* **31** 118

[30] Rosenbluth M N and Rostoker R 1958 *Proc. 2nd Int. Conf., (Geneva)* **31** 144

[31] Stix T H 1962 **The Theory of Plasma Waves** (New York: McGraw-Hill)

[32] Newberger B S 1982 *J. Math. Phys.* **23** 1278; 1983 *J. Math. Phys.* **24** 2250(E)

[33] Bernstein I B 1958 *Phys. Rev.* **109** 10

[34] Stix T H 1958 *Phys. Fluids* **1** 308

[35] Stix T H and Palladino R W 1958 *Proc. 2nd Int. Conf., (Geneva)* **31** 282

[36] Dougherty J P 1964 *Phys. Fluids* **7** 1788

[37] Trubnikov B A 1959 **Plasma Physics and the Problem of Controlled Thermonuclear Reactions** vol III, ed M A Leontovich (Oxford: Pergamon) 122

[38] Brambilla M 1998 **Kinetic Theory of Plasma Waves** (Oxford: Clarendon) 138

[39] Shkarofsky I P 1966 *Phys. Fluids* **9** 561

[40] Swanson D G 2002 *Plasma Phys. Control. Fusion* **44** 1329

[41] Bohm D and Gross E P 1949 *Phys. Rev.* **75** 1851

[42] Stringer T E 1963 *Plasma Physics (J. Nucl Energy)* C **5** 89

[43] Braginskii S I 1965 **Reviews of Plasma Physics** vol 1, ed M A Leontovich (New York: Consultants Bureau) 205

[44] Marshall W 1957 *The Kinetic Theory of an Ionized Gas I* U.K.A.E.A. Technical Report 2277

[45] Marshall W 1957 *The Kinetic Theory of an Ionized Gas II* U.K.A.E.A. Technical Report 2352

[46] Marshall W 1957 *The Kinetic Theory of an Ionized Gas III* U.K.A.E.A. Technical Report 2419

[47] Balescu R 1988 **Transport Processes in Plasmas** Vol 1 (North-Holland, Amsterdam)

[48] Kaneko S J andYamao A 1980 *J. Phys. Soc. Jpn.* **48** 2098

[49] Chapman Sydney and Cowling T G 1952 **The Mathematical Theory of Non-Uniform Gases, 2nd Ed.** (Cambridge University Press, Cambridge)

[50] Hirschfelder J O, Curtiss C F and Bird R B 1954 **The Molecular Theory of Gases and Liquids** Chapters VII and VIII (John Wiley and Sons, Chapman and Hall Ltd., London)

[51] Landshoff R 1951 *Phys. Rev.* **82**, 442

[52] Spitzer L (Jr.) and Harm R 1953 *Phys. Rev.* **89** 977

[53] B. Davison, 1956 **Neutron Transport Theory** (Oxford Univ. Press, London)

[54] Hooyman G J, Mazur P, and de Groot S R 1955 *Physica* XXI 325

[55] Abramowitz M and Stegun I 1964 **Handbook of Mathematical Functions** (National Bureau of Standards, Washington, D. C.)

[56] Ichimaru S 1973 **Basic Principles of Plasma Physics** (Benjamin, Reading, Mass.)

[57] Dnestrovskii Yu N, Kostomarov D P, and Skrydlov N V 1964 *Sov. Phys.-Tech. Phys.* **8**, 691

[58] Lazzaro E, Ramponi G, and Girruzi G 1982 *Phys. Fluids* **25**, 1220

[59] Vaughan-Williams R W and Haas F A 1961 *Phys. Rev. Lett.* **6** 165

[60] Rosenbluth M N and Kaufmann A N 1958 *Phys. Rev.* **109**, 1

[61] Garcia-Colin L S, Garcia-Perciante A L, and Sandoval-Villalbazo A 2007 *Phys. Plasmas* **14** 012305

Index

FSC
www.fsc.org
MIX
Papier | Fördert gute Waldnutzung
FSC® C083411

Kolibri 360 GmbH
Ferdinand-Jühlke-Straße 7
99095 Erfurt, Deutschland
produktsicherheit@kolibri360.de